David Anderson

Final   Tues — June 6   8-10 Am.
Blood e Immunity in heart

THIRD EDITION

FUNCTION OF THE HUMAN BODY

ARTHUR C. GUYTON, M.D.

Professor and Chairman of the Department of
Physiology and Biophysics, University of Mississippi
School of Medicine

W. B. SAUNDERS COMPANY   Philadelphia · London · Toronto

W. B. Saunders Company: West Washington Square
Philadelphia, Pa. 19105

12 Dyott Street
London, W.C.1

1835 Yonge Street
Toronto, Ontario   295

Reprinted June, 1969 and November, 1969

FUNCTION OF THE HUMAN BODY

PREFACE

An author finds his reward in the degree of acceptance of his work. Therefore, it has been most gratifying that the first edition of this textbook, *Function of the Human Body,* received wide usage and more particularly that usage of the second edition multiplied several times more. Such gratification can lead an author only to work much harder to make each succeeding edition still more valuable to the reader. In view of this, very large segments of the text have been recast or extensively revised. In addition, new figures have been added while a large share of the older ones have been modified to improve their quality. And because physiology is in a dynamic stage of discovery, our knowledge in many areas has changed markedly in the last five years. These changes are evident everywhere in the text.

The purpose of this text is to provide a survey of the major facts and theories in the field of human physiology. Yet the field is so great that it has been possible to cover only a small fraction of it. Therefore, it has been necessary to choose carefully which facts of function are most important and, where precise knowledge of function is lacking, to choose as intelligently as possible the most applicable theories. A special attempt has been made to distinguish fact from theory but not to burden the reader with intricate and insignificant details which more properly belong in a reference textbook of physiology.

Still more, it has been my desire to present the basic philosophy of function of the human body, to demonstrate the beauty of organization of the parts of the body and to fit these together into an overall whole — a thinking, sensing, functioning human being capable of living almost automatically and yet capable of immense diversity that characterizes only higher forms of life. There is no machine yet designed (or ever to be designed) that has more intricacy, more excitement, or more majesty than the human body. Therefore, I hope that it will be with pleasure as well as with instructiveness that the reader learns how his body functions.

A text such as this requires work by many different people, not the least among whom are the teachers who send suggestions to the author and especially who point out errors either of fact or of presentation. Feedback of this type helped immensely in making the second edition better than the first and I hope also in making the third edition still much better. I wish also to express my appreciation to Mrs. Billie Howard for her superb secretarial help in preparing this third edition, to Mrs. Carolyn Hull for drawing most of the figures, and to the staff of the W. B. Saunders Company for their continued excellence in production of this book.

ARTHUR C. GUYTON

CONTENTS

v

SECTION ONE

INTRODUCTION: CELL PHYSIOLOGY

CHAPTER 1

INTRODUCTION TO HUMAN PHYSIOLOGY

WHAT IS PHYSIOLOGY? We could spend the remainder of our lives attempting to define this word alone, for physiology is the study of life itself. It is the study of function of all parts of living organisms, as well as of the whole organism. It attempts to discover answers to such questions as: How and why do plants grow? What makes bacteria divide again and again? How do fish obtain oxygen from the sea and what use do they make of it? How is food digested? What is the nature of the thinking process in the brain?

Even small viruses weighing a million times less than bacteria have the characteristics of life, for they feed on their surroundings, they grow and reproduce, and they excrete by-products. These very minute living structures are the subject of the simplest type of physiology, *viral physiology*. Physiology becomes progressively more complicated and vast as it extends through the study of higher and higher forms of life such as bacteria, cells, plants, lower animals, and, finally, human beings. It is obvious, then, that the subject of this book, "human physiology," is but a small part of this vast discipline.

As small children we begin to wonder what enables people to move, how it is possible for them to talk, how they can see the expanse of the world and feel the objects about them, what happens to the food they eat, how they derive from food the energy for locomotion and other types of bodily activity, by what process they reproduce other beings like themselves so that life goes on, generation after generation. All these and other human activities make up *life*. Physiology attempts to explain them and hence to explain life itself.

ROLE OF THE CELL IN THE HUMAN BODY

One hundred trillion cells make up the human body. Each of these is a living unit in itself, capable of existing, performing chemical reactions, contributing its part to the overall function of the body, and in most instances reproducing itself.

The cells are the building blocks of the organs, and each of the organs performs its own specialized function. One will appreciate the importance of the cell when he realizes that many more millions of years went into the evolutionary development of the cell than into the evolution from the cell to the human being. Therefore, before one can understand how any one of the organs functions or how the organs function together to maintain life, it is prerequisite that he understand the inner workings of the cell itself. The next few chapters will be devoted entirely to discussion of basic cellular function, and throughout the remainder of this book we will allude again and again to cellular function as the basis of organ and system operation.

The Internal Environment and Homeostasis

For the cells of the body to continue living, there is one major requirement: the composition of the body fluids that bathe the outside of the cells must be controlled very exactly from moment to moment and day to day, with no single important constituent ever varying more than a few per cent. Indeed, cells can live even after removal from the body if they are placed in a fluid bath that contains the same constituents and has the same physical conditions as those of the body fluids. Claude Bernard, the great nineteenth century physiologist who originated much of our modern physiological thought, called the extracellular fluids that surround the cells the *milieu interne*, the "internal environment," and Walter Cannon, another great physiologist of the first half of this century, referred to the maintenance of constant conditions in these fluids as *homeostasis.*

Thus, at the very outset of our discussion of physiology of the human body, we are beset with a major problem. How does the body maintain the required constancy of the internal environment? The answer to this is that almost every organ plays some role in the control of one or more of the fluid constituents. For instance, the *circulatory system,* composed of the *heart* and *blood vessels,* transports blood throughout the body; water and dissolved substances diffuse back and forth between the blood and the fluids that surround the cells. Thus, the circulatory system keeps the fluids of all parts of the body constantly mixed. This function of the circulatory system is so effective that hardly any portion of fluid in any part of the body remains unmixed with the other fluids of the body more than a few minutes at a time. The *respiratory system* transfers oxygen from the air to the blood,

and the blood in turn transports the oxygen to all the tissue spaces surrounding the cells, thus maintaining the level of oxygen that is required for life by all the cells. The carbon dioxide excreted by the cells enters the tissue fluids, then becomes mixed with the blood and is finally removed through the lungs.

The *digestive system* performs a similar function for other nutrients besides oxygen; it processes nutrients which are then absorbed into the blood and are rapidly transported throughout the body fluids where they can be used by the cells. The *liver,* the *endocrine glands,* and some of the other organs participate in what is collectively known as *intermediary metabolism,* which converts many of the nutrients absorbed from the gastrointestinal tract into substances that can be used directly by the cells. The *kidneys* remove the remains of the nutrients after their energy has been extracted by the cells, and other organs provide for *hearing, feeling, tasting, smelling,* and *seeing,* all of which aid the animal in his search for and selection of food and also help him to protect himself from dangers so that he can perpetuate the almost Utopian internal environment in which his cells continue their life processes.

Thus, we can emphasize once again that organ functions of the body depend on individual functions of cells, and, in turn, sustained life of the cells depends on maintenance of an appropriate fluid environment.

ORGANS AND SYSTEMS OF THE HUMAN BODY

The Skeleton and Its Muscles

Figure 1 illustrates the skeleton with some of its muscles attached. Each joint of the skeleton is enveloped by a loose *capsule,* and the space within the capsule and between the two respective bones is the *joint cavity.* In the joint cavities is a thick, slippery fluid containing hyaluronic acid which lubricates the joints, promoting ease of movement. On the sides of each capsule are strong fibrous *ligaments* that keep the joints from pulling apart. Often the ligaments are only on two sides of the joint, which allows the joint to move freely in one direction but not so freely in another direction. Other are not just a few very large muscles, but are composed of about one hundred different individual muscles each one of which performs a specific function: one rotates an adjacent vertebra while a second one flexes the vertebra forward, a third backward, and so on. This is analogous to the arrangement of the centipede, for each segment can bend independently of all the others. The joint connecting the head to the spinal column is also supplied with many separate muscles arranged on all sides so that the head can be rotated from side to side or bent in any direction.

In summary, then, the skeleton is literally a bag of bones which can be contorted into many different configurations. Each bone has its own function, and the limitations of angulation of each joint are decreed by the ligaments. The knee joint bends in only one direction, the ankle joint in two, the hip joint in two directions plus an additional rotary motion; and, in general, at least two muscles are available for each motion that the ligaments of a joint allow, one for each of the two directions of movement.

The muscles themselves are composed of long *muscle fibers*. Usually many thousands of these fibers are oriented side by side like the threads in a skein of wool. At each end of the muscle, the muscle fibers fuse with strong *tendon fibers* that form a bundle called the *muscle tendon*. The muscle tendons in turn penetrate and fuse with the bones on the two sides of the respective joint so that any pull exerted will effect appropriate movement.

All muscles are not exactly alike in size and appearance; for instance, the smallest skeletal muscle of the body, the stapedius, is a minute muscle in the middle ear only a few millimeters long, while the longest muscle, the sartorius, extends almost two feet down the entire length of the thigh, connecting the pelvic bone with the lower leg. Some muscles, such as those of the abdominal wall, are arranged in thin sheets, while others are round, cigar-shaped structures, for example, the biceps which lifts the lower arm and the gastrocnemius which pulls the foot down when one wishes to stand on tiptoes.

The precise method by which muscle fibers contract is still not completely clear, but we do know that signals arriving in the muscles through nerves cause each fiber

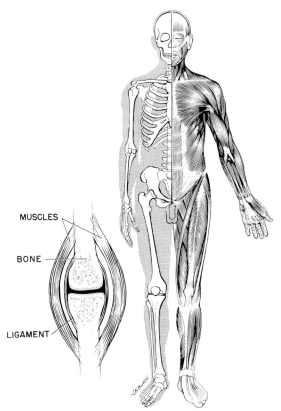

MUSCLES

BONE

LIGAMENT

Figure 1. The skeletal and muscular systems of the human body.

joints, particularly those of the spine, hips, and shoulders, not having very restrictive ligaments, can move in almost any direction; that is, they can bend forward, backward, and to either side, or they can even be rotated. In these instances, loose ligaments merely limit the degree of motion to prevent excessive movement in any one direction.

Muscles move the limbs and other parts of the body in the directions allowed by the ligaments. In the case of movement at the knee joint, for instance, one major muscle is on the front and several muscles are on the back of the joint. Contraction of the anterior muscle pulls the lower leg forward, while contraction of the posterior muscles pulls it backward. There is a similar arrangement of muscles anteriorly and posteriorly about the ankle, except that the ligaments of the ankle allow the ankle joints to move also from side to side, and additional muscles are available to provide the sidewise movements. The muscles of the spine are especially interesting because, contrary to what might be expected, the back muscles to shorten for a brief instant, allowing the entire muscle belly to contract and thereby to perform its function.

The Nervous System

The nervous system, illustrated in Figure 2, is composed of the brain, the spinal cord, and the peripheral nerves that extend throughout the body. A major function of the nervous system is to control many of the bodily activities, especially those of the muscles, but to exert this control intelligently the brain must be apprised continually of the body's surroundings. To perform these varied activities, the nervous system is composed of two separate portions, the *sensory portion* which reports and analyzes the nature of conditions around the body, and the *motor portion* which controls the muscles.

The sensory portion operates through the senses of sight, hearing, smell, taste, and feel. The sense of feel is actually many

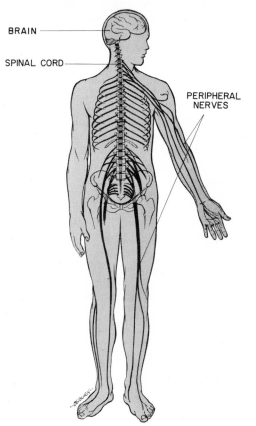

BRAIN

SPINAL CORD

PERIPHERAL NERVES

Figure 2. The nervous system.

different senses, for one can feel light touch, pin pricks, pressure, pain, vibration, position of the joints, tightness of the muscles, and tension on the tendons.

Once information has been relayed to the brain from all the senses, the brain then determines what movement, if any, is most suitable, and the muscles are called into action to implement the decision.

One of the most important functions of the nervous system is to regulate walking. In walking, the body must be supported against gravity, the legs must move rhythmically in a walking motion, equilibrium must be maintained, and the direction of movement of the limbs must be guided. Therefore, the initiation and control of locomotion are very complex functions of the nervous system and require the services of major portions of the brain.

THE AUTONOMIC NERVOUS SYSTEM. The autonomic nervous system, which is really part of the motor portion of the nervous system, regulates many of the internal functions of the body. It operates principally by causing contraction or relaxation of a type of muscle called *smooth muscle*, which constitutes major portions of many of the internal organs. Smooth muscle fibers are much smaller than skeletal muscle fibers, and they usually are arranged in large muscular sheets. For instance, the gastrointestinal tract, the urinary bladder, the uterus, the biliary ducts, and the blood vessels are all composed mainly of smooth muscle sheets rolled into tubular or spheroid structures. Some of the autonomic nerves cause the muscles of these organs to contract while others cause relaxation.

The autonomic nerves also regulate secretion by many of the glands in the gastrointestinal tract and elsewhere in the body, and at times their nerve endings even secrete hormones that can increase or decrease the rates of chemical reactions in the body's tissues.

Finally, the autonomic nervous system helps to control the heart, which is composed of *cardiac muscle,* a type of muscle intermediate between smooth muscle and skeletal muscle. Stimulation of the so-called *sympathetic fibers* causes the rate and force of contraction of the heart to increase, whereas stimulation of the *parasympathetic* fibers causes the opposite effects.

In summary, the autonomic nervous system helps to control most of the body's internal functions.

The Circulatory System

The circulatory system, illustrated in Figure 3, is composed mainly of the heart and blood vessels. The heart consists of two separate pumps arranged side by side. The first pump forces blood into the lungs. From them, the blood returns to the second pump to be forced then into the *systemic arteries* which transport it through the body. From the arteries it flows into the *capillaries*, then into the *veins* and finally back to the heart, thus making a complete circuit. Circulating around and around through the body, the blood acts as a transportation system for conducting various chemical substances from one place to another. It is the circulatory system that carries nutrients to the tissues and then carries excretory products away from the tissues.

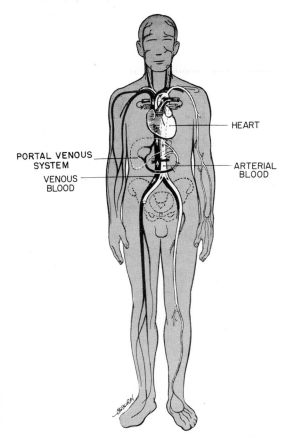

Figure 3. The circulatory system: heart and major vessels.

The capillaries are porous, allowing fluid and nutrients to diffuse into the tissues and excreta from the cells to reenter the blood.

THE LYMPHATICS. Large particles, such as old debris of dead tissues, protein molecules, and dead bacteria cannot pass from the tissues directly into the blood through the small pores of the capillaries. A special accessory circulatory system known as the *lymphatic system* takes care of these materials. Lymphatic vessels originate in small *lymphatic capillaries* which lie beside the blood capillaries, and *lymph*, which is fluid derived from the spaces between the cells, flows along the lymphatic vessels up to the neck where these vessels empty into the neck veins. The lymphatic capillaries are extremely porous so that large particles can enter the lymphatic system and be transported by the lymph. At several points the lymphatic vessels pass through *lymph nodes* where most large particles are filtered out and where bacteria are engulfed and digested by special cells called *reticuloendothelial* cells.

The Respiratory System

Figure 4 is a diagram of the respiratory system, showing the two fundamental portions of this system: (1) the air passages and (2) the blood vessels of the lungs. Air is moved in and out of the lungs by contraction and relaxation of the respiratory muscles, and blood flows continually through the vessels. Only a very thin membrane separates the air from the blood, and since this membrane is porous to gases it allows free passage of oxygen into the blood and of carbon dioxide from the blood into the air.

Oxygen is one of the nutrients needed by the body's tissues. It is carried by the blood and tissue fluids to the cells where it combines chemically with foods to release energy. This in turn is used to promote muscle contraction, secretion of digestive juices, conduction of signals along nerve fibers, and synthesis of many substances needed for growth and function of cells.

When oxygen combines with foods to liberate energy, carbon dioxide is formed. This is carried by tissue fluids to the blood and by the blood to the lungs. Then, the carbon dioxide diffuses from the blood into the lung air to be breathed out into the atmosphere.

Labels on figure: HEART, ARTERIAL BLOOD, PORTAL VENOUS SYSTEM, VENOUS BLOOD

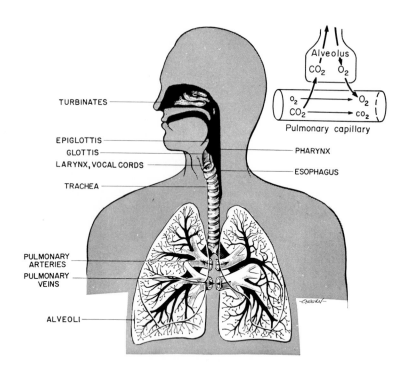

Figure 4. The respiratory system.

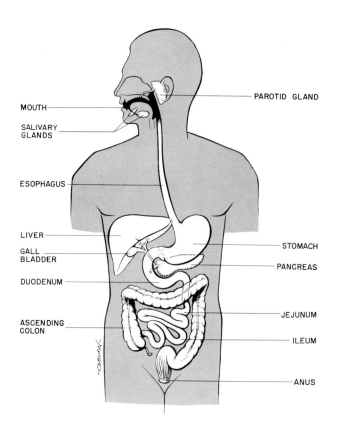

Figure 5. The digestive system.

The Gastrointestinal System

The digestive system is illustrated in Figure 5. Food after being swallowed enters the stomach, then the duodenum, the jejunum, the ileum, and the large intestine, finally to be defecated through the anus. Those portions of the food valuable to the body are chemically and physically extracted by the process of *digestion*, and are transferred into the blood by the process of *absorption*. Along the entire extent of the gastrointestinal tract, special substances are secreted into the gut either continuously or when food is present. These secretions contain *digestive enzymes* which cause the foods to split into chemicals small enough to pass through the pores of the intestinal membrane into underlying capillary and lymphatic vessels. Thence the digestive products enter the circulating blood to be transported and used where they may be needed.

Metabolic Systems

METABOLISM AND GROWTH. The term *metabolism* means simply the chemical reactions that occur in the animal organism. These reactions occur inside the individual cells which make up the tissues, and their functions are to provide energy to perform the bodily activities and to build new structures. It is because of the metabolic processes that the cells grow larger and more numerous. The metabolism of special cells allows them to form structures such as bones and fibrous tissue, enlarging the entire animal. Thus, metabolism is the basis not only for the energy needed by the body but for growth itself.

Intermediary metabolism. Many of the foods entering the blood from the digestive tract can be used by the body's tissues without alteration, but some tissues require special chemicals which are not normally found in the food. To supply these, much of the absorbed food passes to special organs where it is changed into new substances needed by the cells. This process is called *intermediary metabolism.*

THE LIVER. The liver is one of the internal organs especially adapted for intermediary metabolism and storage. It can split fats and proteins into smaller substances so that the tissues can use them for energy, and

it forms products needed for blood coagulation, for transport of fat, for immunity to infection, and for many other purposes. The liver is also capable of storing large quantities of fats, carbohydrates, and even proteins and then releasing these foods when the tissues need them. An animal can live for only a few hours without a liver.

CONTROL OF METABOLISM BY THE HORMONES. Metabolism is an inherent function of every cell of the body. However, the rate of metabolism in each respective cell is very often increased or decreased by the controlling action of *hormones* secreted by *endocrine glands* in different parts of the body. The *thyroid gland*, located in the neck, secretes *thyroxine* which acts on all cells of the body to increase the rate of most metabolic reactions. *Epinephrine* and *norepinephrine*, two hormones secreted by the *adrenal medullas*, also increase the rate of metabolism in all cells. The *ovaries* secrete *estrogens* and *progesterone*, and the *testes* secrete *testosterone*, which helps to control metabolism in the sex organs of the female and male, respectively. *Insulin*, secreted by the *pancreas*, a gland located behind and beneath the stomach, increases the utilization of carbohydrates and decreases the utilization of fats in all the tissues. *Adrenocortical hormones* secreted by the two *adrenal cortices*, located at the upper poles of the kidneys, help to convert proteins to carbohydrates, and they control the passage of proteins, salts, and perhaps other substances through the cell walls. Finally, *parathyroid hormone*, secreted by four minute *parathyroid glands* located behind the thyroid gland in the neck, controls the amount of calcium in the blood by removing calcium from the bones when it is needed and by allowing more deposition in the bones when it is not needed.

The Excretory System

The *kidneys*, illustrated in Figure 6, constitute an excretory system for ridding the blood of unwanted substances. Most of the substances are the end-products of metabolic reactions, including mainly urea, uric acid, creatinine, phenols, sulfates, and phosphates. If they were allowed to collect in the blood in large quantities, the "ashes of the cellular fires" would soon "smother the flames" themselves so that no further metabolic re-

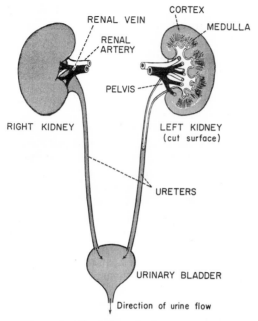

Figure 6. The kidneys and urinary system.

actions could take place. For this reason it is important that the kidneys remove these unwanted substances.

The kidneys have another very valuable function besides that of excretion: they regulate the concentration of most of the electrolytes in the body fluids. A very large proportion of these electrolytes is sodium chloride or common table salt. The kidneys continually adjust the concentrations of both sodium and chloride, and they also regulate very precisely the concentrations of potassium, magnesium, phosphates, and many other substances. The kidneys perform this function by allowing the unwanted substances such as urea to pass easily into the urine while retaining the wanted substances such as glucose. If sodium is already present in the blood in too large a concentration, it becomes an unwanted substance, and much of it is excreted by the kidneys. However, if the concentration of sodium is too low, it is a wanted substance instead, and the kidney's special properties, which will be described in a future chapter, then prevent its loss from the blood.

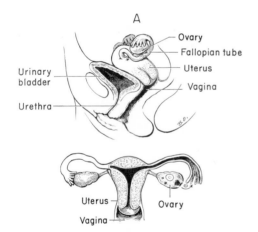

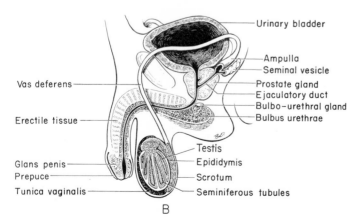

Figure 7. The reproductive systems: (A) female, (B) male.

The Reproductive Systems

All the functions and systems of the body that maintain life would be useless were it not for those that provide for its reproduction. The reproductive systems, shown in Figure 7, are different in the female and male. The female (Fig. 7A) provides the *ovum* (egg) from which a new human being is to develop, but this ovum cannot begin developing until it is fertilized by a *sperm* from the male (Fig. 7B). The fertilized ovum derives half of its developmental characteristics from the mother and half from the fertilizing sperm of the father, so that the offspring owes its characteristics equally to each of the parents.

After the ovum has been fertilized it is at first still a single cell, but soon it divides into two cells, then four cells, and finally into many cells, thus becoming an *embryo*. Gradually, the newly developing cells *differentiate* into the organs of the body.

The mother provides nutrition for the growing embryo by means of the *placenta*, a structure attached to the inner wall of her *uterus*. Nutrients diffuse from the mother's blood into the baby's blood through the *placental membrane* which is very much like the membrane of the respiratory system. In turn, excretory products from the baby pass into the mother's blood. Thus, the embryo is nurtured through a period of nine months in the mother's body until it becomes capable of sustaining life on its own in the outer world. At that time the mother's uterus expels the baby.

COMMUNAL ORGANIZATION OF THE BODY

If this chapter has succeeded in presenting a résumé of the functioning systems of the human body, it should by now be obvious that no single part of the human body can live by itself, but that each of the body's functions is necessary for continuous operation of the others. The human animal is a sensing, thinking, and motile organism which, by virtue of nervous and hormonal systems of control, can adapt itself to most surroundings. Its activities are initiated and controlled in part involuntarily, in part by intuition, and in part by reasoning. In the framework of the organs and other tissues are literally trillions of individual *cells*, each one of which is a living structure. It is the magic of these cells which makes the human body possible. The next few chapters describe function of the cell itself.

GENERAL REFERENCES IN THE FIELD OF PHYSIOLOGY

American Journal of Physiology. This journal, published monthly, contains articles by authors throughout the world on current research projects.

Annual Review of Physiology. Palo Alto, Calif., Annual Reviews, Inc. (One book each year.) Each book contains review articles that cover the literature of the preceding year in almost the entire field of animal physiology.

Best, C. H., and Taylor, N. B.: *The Physiological Basis of Medical Practice.* 8th Ed. Baltimore, The Williams & Wilkins Company, 1966. This text is written principally for the postgraduate student.

Guyton, A. C.: *Textbook of Medical Physiology.* 3rd Ed. Philadelphia, W. B. Saunders Company, 1966. This text presents physiology at the level of the medical student. In general it covers the same material as that in the present text, but in much greater detail and with more emphasis on the medical aspects of human physiology.

Mountcastle, V. B.: *Medical Physiology.* 12th Ed. St. Louis, The C. V. Mosby Company, 1968. This text is most useful for postgraduate students.

Physiological Reviews. This journal, published four times a year, contains reviews of most subjects in the field of physiology every few years.

Ruch, T. C., and Patton, H. D.: *Medical Physiology and Biophysics.* 19th Ed. Philadelphia, W. B. Saunders Company, 1965. This text is written at the level of the postgraduate student.

CHAPTER 2

THE CELL AND ITS COMPOSITION

The human body contains about 100 trillion cells, each of which is a living structure. Since the functions of the body's organs are performed by their constituent cells, the overall function of the human body is actually the combined function of all these 100 trillion cells.

Several hundred basic types of cells exist in the human body, and each type plays a special role in bodily function. Yet, despite this difference between cells, they all have some functions in common, for instance, their abilities to live, grow, and reproduce. The purpose of this chapter is mainly to emphasize these similarities of cells and their functions. Yet, first, let us describe some of the different types of cells and their related structures.

Some Representative Types of Cells

Figure 8 illustrates five types of tissues that perform different functions in the body. Example A depicts *connective tissue* which holds the different structures of the body together. This tissue contains cells called *fibroblasts* that are enmeshed in *collagenous* and *elastic* fibers. The fibroblasts secrete

substances that polymerize to form the fibers, and the fibers in turn provide tensile strength to the tissues, thereby holding them together.

Example B illustrates several *red* and *white blood cells*. The red cells carry oxygen from the lungs to the tissues and carbon dioxide from the tissues back to the lungs, while the white blood cells cleanse the blood and tissues of unwanted materials such as bacteria, debris from degenerating tissues, and so forth.

Example C shows a *nerve cell* and surrounding *supporting cells* in the brain substance. The long projection of the nerve cell is its *axon* that often extends as long as one meter. Electrochemical impulses travel over the surface of the nerve cell and along the membrane of the axon, thereby transmitting information from one part of the body to another.

Example D illustrates *muscle cells* which can also transmit electrochemical impulses over their membranes but which are different from nerve cells in that they contain long myofibrils that extend the entire length of the muscle and contract when an electrochemical impulse travels over the surface of the fibril.

Example E illustrates several different

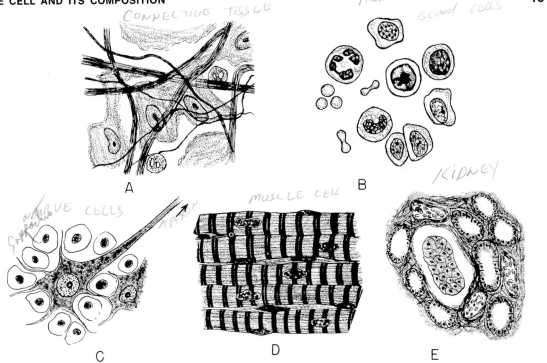

Figure 8. Examples of different types of cells: (A) connective tissue, (B) red and white blood cells, (C) neuronal tissue, (D) muscle cells, (E) kidney tissue.

types of cells and structures in the *kidney.* Connective tissue is present throughout the kidney to hold the different structures together. Several *kidney tubules* are shown, lined by *epithelial cells;* these structures help to form the urine, as is explained in a later chapter. Also, two small *blood vessels* are illustrated in cross-section in this figure; these vessels are filled with red blood cells.

The representative tissues and cells shown in Figure 8 are but a few of the many types found in the body, but they show the wide variability between different cells. These dissimilarities allow cells to perform different functions. The remainder of this chapter, however, presents the *similarities* between cells rather than their dissimilarities. In future chapters, many different types of specialized cells are described in detail, and their functions are presented.

Comparison of the Animal Cell
With Precellular Forms of Life

Many of us think of the cell as the lowest level of animal life. However, the cell is a very complicated organism, which probably required several billion years to develop after the earliest form of life, a form similar to the present-day *virus,* first appeared on Earth. Figure 9 illustrates the relative sizes of the *smallest known virus,* a *large virus,* a *rickettsia,* a *bacterium,* and the *cell,* showing that the cell has a diameter about 1000 times that of the smallest virus, and, therefore, a volume about 1 billion times that of

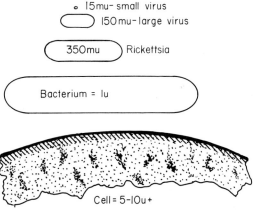

Figure 9. Comparison of sizes of subcellular organisms with that of the average cell in the human body.

the smallest virus. Correspondingly, the functions and anatomical organization of the cell are far more complex than those of the virus.

The principal constituent of the very small virus is a substance called *nucleic acid*. This substance is capable of reproducing itself if appropriate nutrients are available. Thus, the virus is capable of propagating its lineage from generation to generation, and, therefore, is a living structure in the same way that the cell and the human being are living structures.

As life evolved to the large viruses, other chemicals besides nucleic acid became a part of the organism, and specialized functions began to develop in different parts of the virus. A *membrane* formed around the virus, and inside the membrane a fluid *matrix* appeared. Specialized chemicals developed inside the matrix to perform special functions; *protein enzymes* appeared which were capable of catalyzing chemical reactions and, therefore, of controlling the organism's activities.

In the rickettsial and bacterial stages, *organelles* developed, representing aggregates of chemical compounds that perform functions in a more efficient manner than can be achieved by dispersed chemicals throughout the fluid matrix. And, finally, in the cell stage more complex organelles developed, the most important of which is the *nucleus*. The nucleus distinguishes the cell from all lower forms of life; this structure provides a control center for all cellular activities, and it also provides for very exact reproduction of new cells generation after generation, each new cell having essentially the same structure as its progenitor.

ORGANIZATION OF THE CELL— PROTOPLASM

A typical cell, as seen by the light microscope, is illustrated in Figure 10. Its two major parts are the *nucleus*, which is filled with *nucleoplasm*, and the *cytoplasm*. The nucleus is separated from the cytoplasm by a *nuclear membrane*, and the cytoplasm is separated from the surrounding fluids by a *cell membrane*.

The different substances that make up the cell are collectively called *protoplasm*. Protoplasm is composed mainly of five basic

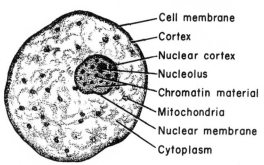

Figure 10. Structure of the cell as seen by the light microscope.

substances: water, electrolytes, proteins, lipids, and carbohydrates.

WATER. The fluid medium of all protoplasm is water, which is present in a concentration between 70 and 85 per cent. Many cellular chemicals are dissolved in the water, while others are suspended in small particulate form. Chemical reactions take place between the dissolved chemicals or at the surface boundaries between the suspended particles and the water. The fluid nature of water allows both the dissolved and suspended substances to diffuse to different parts of the cell, thereby providing transport of the substances from one part of the cell to another.

ELECTROLYTES. The most important electrolytes in the cell are *potassium*, *magnesium*, *phosphate*, *sulfate*, *bicarbonate*, and small quantities of *sodium* and *chloride*. These will be discussed in much greater detail in Chapters 5 and 16, which will consider the interrelationships between the different fluids of the body.

The electrolytes are dissolved in the water of protoplasm, and they provide inorganic chemicals for cellular reactions. Also, they are necessary for operation of some of the cellular control mechanisms. For instance, electrolytes acting at the cell membrane allow transmission of electrochemical impulses in nerve and muscle fibers, and the intracellular electrolytes determine the activity of different enzymatically catalyzed reactions that are necessary for cellular metabolism.

PROTEINS. Next to water, the most abundant substance in most cells is proteins, which normally constitute 10 to 20 per cent of the cell mass. These can be divided into two different types, *structural proteins* and *enzymes*.

To get an idea of what is meant by *structural proteins*, one needs only to note that leather is composed principally of structural proteins, and that hair is almost entirely a structural protein. Proteins of this type are present in the cell membrane, in the nuclear membrane, and in membranes composing special intracellular structures, such as the endoplasmic reticulum and the mitochondria. Thus structural proteins hold the structures of the cell together. Most structural proteins are *fibrillar;* that is, the individual protein molecules are polymerized into long fibrous threads. The threads in turn provide tensile strength for the cellular structures.

Enzymes, on the other hand, are an entirely different type of protein, composed usually of individual protein molecules or at most of aggregates of a few molecules in a *globular* form rather than a fibrillar form. These proteins, in contrast to the fibrillar proteins, are often soluble in the fluid of the cell or are adsorbed to the surface of membranous structures inside the cell. The enzymes come into direct contact with other substances inside the cell and catalyze chemical reactions. For instance, the chemical reactions that split glucose into its component parts and then combine these with oxygen to form carbon dioxide and water are catalyzed by a series of protein enzymes. Thus, enzyme proteins control the metabolic functions of the cell.

Special types of enzyme proteins are present in different parts of the cell. Of particular importance are the *nucleoproteins*, present both in the nucleus and in the cytoplasm. The nucleoproteins of the nucleus are mainly the *genes*, which control both the overall function of the cell and the transmission of hereditary characteristics from cell to cell. These substances are so important that they will be considered in detail in Chapter 4. In addition, the chemical nature of proteins will be considered in Chapter 31, and the different structural and enzymatic functions of proteins will be subjects of discussion at numerous points throughout this text.

LIPIDS. Lipids are several different types of substances that are grouped together because of their common property of being soluble in fat solvents. For instance, the usual fat of animal tissues is a lipid called *neutral fat.* However, in addition to neutral fat two other types of lipids, *phospholipids* and *cholesterol,* are very common throughout the cells.

The average cell contains 2 to 3 per cent lipids which are dispersed throughout the cell but are present in especially high concentrations in the cell membrane, the nuclear membrane, and the membranes lining intracytoplasmic organelles, such as the endoplasmic reticulum and the mitochondria, which will be described later in this chapter.

The special importance of lipids in the cell is that they are either insoluble or only partially soluble in water. They combine with structural proteins to form the membranes that separate the different water compartments of the cell from each other. These membranes contain approximately 40 per cent lipids combined with structural proteins. The lipids of each membrane form a boundary between the solutions on the two sides of the membrane, making it impervious to many dissolved substances.

Some lipids are partially soluble in water as well as in lipid solvents. For instance, the *phospholipid* molecule has a *hydrocarbon portion* that makes it soluble in lipid solvents and a positively charged *polar portion* that makes it soluble in water. Therefore, phospholipid molecules line up at the surfaces of the different membranes with part of their structures protruding into the fat matrix of the membrane. Water-soluble substances then combine with the polar groups of the phospholipids, in this way aggregating at the surfaces of the membranes and in some instances probably being transported through the membranes by the phospholipids.

In summary, the special characteristics of lipids allow the cell to develop many membranous structures that can either limit the transfer of water-soluble substances from one part of the cell to another or can selectively pass substances if they meet appropriate chemical criteria, as will be discussed in Chapter 5. The chemical natures of the different types of fats and their functions in the body will be discussed in Chapter 31.

CARBOHYDRATES. In general, carbohydrates have very little structural function in the cell, but they play a major role in nutrition. Most human cells do not maintain large stores of carbohydrates, usually averaging about 1 per cent of their total mass. However, carbohydrate, in the form of glucose, is always present in the surrounding extracellular fluid so that it is readily available to the cell. The small amount of carbohydrates stored in the cells is almost entirely in the form of *glycogen,* which is an insoluble polymer of glucose.

Glucose, which can be derived by splitting glycogen within the cell or by direct transport through the cell membrane into the cytoplasm, is one of the major nutrients required for cellular function. Enzymes are present throughout the cytoplasm and even in the nucleus for splitting glucose into *smaller carbohydrate compounds,* thereby releasing small amounts of energy that can be used to energize cellular functions. The smaller compounds are then split in the *mitochondria* and combined with oxygen to form *carbon dioxide* and *water.* In this second process, tremendous quantities of energy are released. This energy in turn is utilized by the cell to perform its different functions, such as synthesis of new chemical compounds, muscular contraction, and transmission of nerve impulses.

PHYSICAL STRUCTURE OF THE CELL

The cell is not merely a bag of fluid, enzymes, and chemicals but also contains highly organized physical structures called *organelles,* which are equally as important

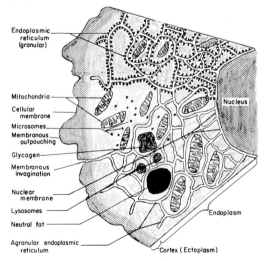

Endoplasmic
reticulum
(granular)

Mitochondria
Cellular
membrane
Microsomes
Membranous
outpouching
Glycogen
Membranous
invagination
Nuclear
membrane
Lysosomes
Neutral fat
Agranular endoplasmic
reticulum

Nucleus

Endoplasm

Cortex (Ectoplasm)

Figure 11. Organization of the cytoplasmic compartment of the cell.

to the function of the cell as the cell's chemical constituents. For instance, without one of the organelles, the *mitochondria,* the energy supply of the cell would cease almost entirely. Some principal organelles of the cell are the *cell membrane, nuclear membrane, endoplasmic reticulum, mitochondria,* and *lysosomes,* as illustrated in Figure 11. Others not shown in the figure are the *Golgi complex, centrioles,* and *cilia.*

Membranes of the Cell

Essentially all physical structures of the cell are lined by a membrane composed primarily of lipids and proteins. The different types of membranes include the *cell membrane,* the *nuclear membrane,* the *membrane of the endoplasmic reticulum,* and the *membranes of the mitochondria, lysosomes, Golgi complexes,* and so forth.

THE CELL MEMBRANE. The cell membrane is thin (approximately 75 to 100 Ångstroms) and elastic. It is composed almost entirely of proteins and lipids, with a percentage composition approximately as follows: proteins, 60 per cent; lipids, 40 per cent. The proteins in the cell are mainly a type of protein called *stromatin,* an insoluble structural protein having elastic properties. The lipids are approximately 65 per cent phospholipids, 25 per cent cholesterol, and 10 per cent other lipids.

The precise molecular organization of the cell membrane is unknown, but many experiments point to the structure illustrated in Figure 12A. This shows a central layer of lipids covered on either face by protein. The presence of protein on the surface supposedly makes the membrane *hydrophilic,* which means that water adheres easily to the membrane. The lipid center of the membrane makes the membrane mainly impervious to lipid-insoluble substances. The small knobbed structures lying at the bases of the protein molecules in Figure 12A are phospholipid molecules; the fat portion of the phospholipid molecule is dissolved in the lipid phase of the cell membrane, and the polar (ionized) portion of the molecule protrudes outward where it is loosely bound electrochemically with the protein lining the outer surface of the membrane.

Pores in the membrane. The membrane has many minute pores that pass from one

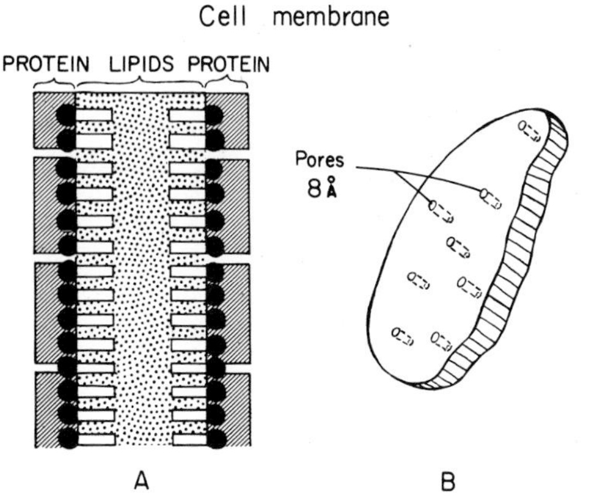

Cell membrane

PROTEIN LIPIDS PROTEIN

Pores
8 Å

A   B

Figure 12. (A) Postulated molecular organization of the cell membrane. (B) Pores in the cell membrane.

side to the other as shown in Figure 12B. These pores have never been demonstrated even with an electron microscope because their sizes are smaller than the resolution of the electron microscope. However, functional experiments to study the movement of molecules of different sizes between the extra- and intracellular fluids have demonstrated free diffusion of molecules up to a size of approximately 8 Ångstroms. It is through these pores that lipid-insoluble substances of very small sizes, such as water and urea molecules, pass with relative ease between the interior and exterior of the cell.

Formation of a new cell membrane. When the membrane of a cell is ruptured, cytoplasm streams out of the opening, but before it can flow far into the surrounding fluids a new membrane is formed having exactly the same properties as those of the original membrane. Exactly how this new membrane is formed is yet unknown. One theory is that lipids and proteins from the cytoplasm precipitate to form the membrane when exposed to the extracellular fluids. However, since the discovery of the endoplasmic reticulum, a large membranous structure inside the cytoplasm which will be described presently, another suggested mechanism has been that the membranes of the endoplasmic reticulum simply fill the rupture and coalesce with the surrounding cell membrane.

Repair of the cell membrane will not occur when the concentration of calcium ions in the extracellular fluids is very low. Therefore, one of the very important functions of calcium ions in the extracellular

fluids is to maintain the integrity of the cell membrane itself. Also the degree of porosity of the membrane is determined to some extent by the concentration of calcium ions—the greater the calcium ion concentration, the less the porosity.

THE NUCLEAR MEMBRANE. The nuclear membrane is actually a double membrane having a wide space between the two membranes. Each half of this membrane is almost identical to the cell membrane, having lipids in its center and protein on its two surfaces. Very little is known about the permeability of the nuclear membrane, but in at least some cells it probably is about the same as that of the cell membrane because ionic concentration differences exist across the nuclear membrane, just as is true of the cell membrane. Yet, one can see in electron micrographs of the nuclear membrane what appear to be large open spaces, as illustrated in Figure 17. These spaces have been called *pores,* even though they are probably covered by a very thin membrane.

THE ENDOPLASMIC RETICULUM. Figure 11 illustrates in the cytoplasm a network of tubular and vesicular structures called the endoplasmic reticulum. The detailed structure of this organelle is illustrated in Figure 13. The space inside the tubules and vesicles is filled with *endoplasmic matrix,* a fluid medium that is different from the fluid outside the endoplasmic reticulum.

Electron micrographs show that the space inside the endoplasmic reticulum is connected with the space between the two membranes of the double nuclear membrane. Also, in rare instances the endoplasmic reticulum probably connects directly through small openings with the exterior of the cell. Some substances formed in the

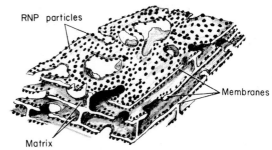

RNP particles

Membranes

Matrix

Figure 13. Structure of the membranes of the endoplasmic reticulum. (Redrawn from DeRobertis, Nowinski, and Saez: *General Cytology.* 4th Ed.)

nucleus are believed to enter the space between the two nuclear membranes and then to be conducted to all parts of the cell through the endoplasmic reticular tubules.

Ribosomes. Attached to the outer surfaces of many parts of the endoplasmic reticulum are large numbers of small granular particles called ribosomes. Where these are attached, the reticulum is frequently called either the *granular endoplasmic reticulum* or the *ergastoplasm.* The ribosomes are composed mainly of ribonucleic acid, which functions in the synthesis of protein in the cells, as is discussed in Chapter 4.

Part of the endoplasmic reticulum has no attached ribosomes. This part is called the *agranular endoplasmic reticulum.* It is believed that the agranular reticulum synthesizes lipid substances and probably also acts as a medium for transporting secretory substances to the exterior of the cell, as is discussed in the following chapter.

GOLGI COMPLEX. The Golgi complex, illustrated in Figure 14, is probably a specialized portion of the endoplasmic reticulum. It has membranes similar to those of the agranular endoplasmic reticulum and is composed of four or more layers of thin vesicles. Electron micrographs show direct connections between the endoplasmic reticulum and parts of the Golgi complex.

The Golgi complex is found most often in

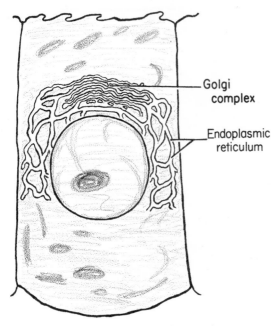

Golgi complex

Endoplasmic reticulum

Figure 14. A typical Golgi complex.

secretory cells. It is ordinarily located on the side of the cell from which substances will be secreted. Its function is believed to be temporary storage of secretory substances and preparation of these for final secretion. The Golgi complex is possibly also the site of synthesis of some complex polysaccharides.

Physical Nature of the Cytoplasm

The cytoplasm is filled with both minute and large dispersed particles ranging in size from a few Ångstroms to 1 micron in diameter. The clear fluid portion of the cytoplasm in which the particles are dispersed is called *hyaloplasm;* this contains mainly dissolved proteins, electrolytes, glucose, and small quantities of phospholipids, cholesterol, and esterified fatty acids.

The portion of the cytoplasm immediately beneath the cell membrane is frequently gelled into a semi-solid called the *cortex,* or *ectoplasm.* The cytoplasm between the cortex and the nuclear membrane is fluid, and is called the *endoplasm.*

Among the large dispersed particles in the cytoplasm are neutral fat globules, glycogen granules, secretory granules, and two organelles that are widely dispersed throughout the cytoplasmic compartment of the cell: the *mitochondria* and *lysosomes,* which are discussed below.

COLLOIDAL NATURE OF THE CYTOPLASM. All the particles dispersed in the cytoplasm, whether the large lysosomes and mitochondria or the small granules, are hydrophilic — that is, attracted to water — because of electrical charges on their surfaces. The lining membranes of the mitochondria and lysosomes have charges on their surfaces, because of the manner in which the protein molecules on the membrane surfaces are oriented. The particles remain dispersed in the cytoplasm because of mutual repulsion of the charges on the different particles. Thus, the cytoplasm is actually a colloidal solution. Likewise, the nucleoplasm is a colloidal solution.

MITOCHONDRIA. Mitochondria are the sites of formation of the energy rich substance adenosine triphosphate, which will be discussed in detail in the following chapter. These organelles are present in

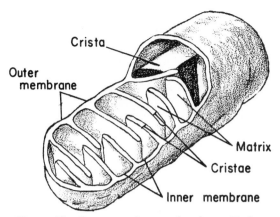

Figure 15. Structure of a mitochondrion. (Redrawn from Robertson: *Biochem. Soc. Symp.* 16:38. Cambridge University Press, 1959.)

the cytoplasm of all cells, as illustrated in Figure 11, but the number per cell varies from a few hundred to a few thousand, depending on the amount of energy required by each cell to perform its functions. Mitochondria are also very variable in size and shape; some are only a few hundred millimicrons in diameter and globular in shape while others are as large as 1 micron in diameter, as long as 7 microns, and filamentous in shape.

The basic structure of a mitochondrion is illustrated in Figure 15, which shows it to be surrounded by a membrane similar in structure to the cell membrane. Many infoldings of the inner surface of the membrane form shelves on which almost all the oxidative enzymes of the cell are believed to be adsorbed. When nutrients and oxygen come in contact with the enzymes in the mitochondrion, they combine to form carbon dioxide and water, and the liberated energy is used to synthesize a substance called *adenosine triphosphate (ATP)*. The ATP then diffuses throughout the cell and releases its stored energy wherever it is needed for performing cellular functions. The function of ATP is so important to the cell that it is discussed in detail later in the chapter.

LYSOSOMES. Another cytoplasmic organelle recently discovered is the lysosome. The lysosome, illustrated in Figure 11, is 250 to 750 millimicrons in diameter, and, like the mitochondrion, is surrounded by a lipoprotein membrane. It is filled with large numbers of small granules 55 to 80 Ångstroms in diameter which are protein aggre-

gates of hydrolytic (digestive) enzymes. A hydrolytic enzyme is capable of splitting an organic compound into two or more parts by combining hydrogen derived from a water molecule with part of the compound and by combining the hydroxyl portion of the water molecule with the other part of the compound. For instance, protein is hydrolyzed to form amino acids, and glycogen is hydrolyzed to form glucose. More than a dozen different *acid hydrolases* have been found in lysosomes, and the principal substances that they digest are proteins, nucleic acids, mucopolysaccharides, and glycogen.

Ordinarily, the membrane surrounding the lysosome prevents the enclosed hydrolytic enzymes from coming in contact with other substances in the cell. However, many different conditions of the cell will break the membranes of some of the lysosomes, allowing release of the enzymes. These enzymes then split the organic substances with which they come in contact into small, highly diffusible substances such as amino acids and glucose. Some of the more specific functions of lysosomes are discussed in the following chapter.

SECRETORY GRANULES. One of the most important functions of many cells is secretion of special substances. The secretory substances are usually formed inside the cell and are held there until an appropriate time for release to the exterior. The storage depots within the cells are called secretory granules; these are illustrated by the dark spots in Figure 16, which shows pancreatic acinar

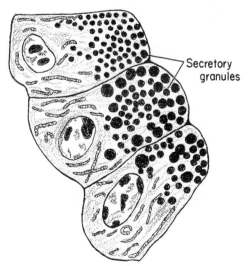

Figure 16. Secretory granules in acinar cells of the pancreas.

cells that have formed and stored enzymes that will be secreted later into the intestinal tract.

Many secretory granules lie inside the tubules and vesicles of the endoplasmic reticulum and Golgi complex, while others are free in the cytoplasm.

OTHER ORGANELLES OF THE CYTOPLASM. The cytoplasm of each cell contains two *centrioles,* which are small cylindrical structures that play a major role in cell division, as will be discussed in Chapter 4. Also, most cells contain small *lipid droplets* and *glycogen granules* that play important roles in energy metabolism of the cell. And certain cells contain highly specialized structures such as the *cilia* of ciliated cells which are actually outgrowths from the cytoplasm, and the *myofibrils* of muscle cells. All of these are discussed in detail at different points in this text.

The Nucleus

The nucleus is the control center of the cell. It controls both the chemical reactions that occur in the cell and reproduction of the cell, though all the precise means by which it effects these controls are still partially obscure. Briefly, the nucleus contains large quantities of *deoxyribonucleic acid,* which we have called *genes* for many years. The genes control the characteristics of the protein enzymes of the cytoplasm, and in this way control cytoplasmic activities. To control reproduction, the genes autocatalytically reproduce themselves, and after this is accomplished the cell splits by a special process called *mitosis* to form two daughter cells, each of which receives one of the two sets of genes. These activities of the nucleus are considered in detail in Chapter 4.

The appearance of the nucleus under the microscope does not give much of a clue to the mechanism by which it performs its control activities. Figure 17 illustrates the interphase nucleus (period between mitoses), showing darkly staining, granular *chromatin material* throughout the *nuclear* sap. During mitosis, the chromatin material becomes readily identifiable as part of the highly structured *chromosomes,* which can be seen easily with a light microscope. Even during the interphase of cellular activity the granular chromatin material is still organized into

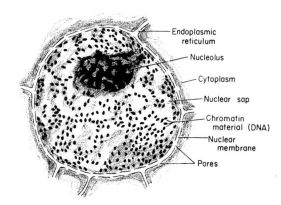

Figure 17. Structure of the nucleus.

chromosomal structures, but this is impossible to see except in a few types of cells.

NUCLEOLI. The nuclei of many cells contain one or more lightly staining structures called nucleoli. The nucleolus, unlike most of the organelles that we have discussed, does not have a limiting membrane. Instead, it is simply an aggregate of loosely bound granules composed mainly of *ribonucleic acid.* It usually becomes considerably enlarged when a cell is actively synthesizing proteins. The genes of the chromosomes are believed to synthesize the ribonucleic acid and then to store it in the nucleolus; this ribonucleic acid later disperses from the nucleolus into the cytoplasm where it controls cytoplasmic function. The details of these mechanisms are discussed in Chapter 4.

REFERENCES

Aronow, L., and Fuhrman, F. A.: Cell structure and function—an introductory course in experimental cell biology. *J. Med. Educ.* 37:737, 1962.

Brachet, J., and Mirsky, A. E.: *The Cell.* 6 volumes. New York, Academic Press, 1961.

Briger, E. M.: *Structure and Ultrastructure of Microorganisms.* New York, Academic Press, 1963.

Ciba Foundation Symposium: *Lysosomes.* Boston, Little, Brown & Company., 1964.

Davson, H.: *A Textbook of General Physiology.* 2nd Ed. London, J. & A. Churchill, 1959.

Fawcett, D. W.: *The Cell.* Philadelphia, W. B. Saunders Company, 1966.

Giese, A. C.: *Cell Physiology.* Philadelphia, W. B. Saunders Company, 1968.

Kinosita, H., and Murakami, A.: Control of ciliary motion. *Physiol. Rev.* 47(1):53, 1967.

Kirschner, L. B.: Comparative physiology. *Ann. Rev. Physiol.* 29:169, 1967.

Porter, K. R., and Bonneville, M. A.: *An Introduction to the Fine Structure of Cells and Tissues.* London, Henry Kimpton, 1963.

Rhodin, J. A. G.: *Atlas of Ultrastructure.* Philadelphia, W. B. Saunders Company, 1963.

Robertson, J. D.: The membrane of the living cell. *Sci. Amer.* 206(4):64, 1962.

Satir, P.: Cilia. *Sci. Amer.* 204(2):108, 1961.

Siekevitz, P.: Protoplasm: endoplasmic reticulum and microsomes and their properties. *Ann. Rev. Physiol.* 25:15, 1963.

Weiss, P.: The cell as a unit. *J. Theor. Biol.* 5:389, 1963.

CHAPTER 3

FUNCTIONAL SYSTEMS OF THE CELL

Each organelle of the cell performs its own specialized functions. For instance, the cell membrane has special abilities to transport selected substances into the cell. The lysosomes digest foreign particles or dead portions of the cell, and the digested end products are then excreted from the cell through the cell membrane or are reused by the cell as nutrients. The mitochondria are called the "powerhouses" of the cell because they burn oxygen with the nutrients and produce the energy required for promoting most chemical reactions in the cell. The endoplasmic reticulum performs such varied functions as synthesis of substances required in the cell and transport of substances from one part of the cell to another. The Golgi complex aids the endoplasmic reticulum in some of these operations. Finally, the nucleus is the control center of the cell, controlling the types and rates of chemical reactions within the cell, the formation of new substances in the cell, and even reproduction of the cell.

In this chapter we will discuss most of the functional systems of the cell, but two of these systems are so important that they deserve special chapters of their own. First, function of the nucleus and its genes in controlling protein synthesis, intracellular chemical reactions, and reproduction will be presented in detail in the following chapter. Second, movement of substances through the cell membrane will be discussed in Chapter 5.

INGESTION BY THE CELL

Pinocytosis

If a cell is to live and grow, it must obtain nutrients and other substances from the surrounding fluids. Substances can pass through a cell membrane in three separate ways: (1) by *diffusion* through the pores in the membrane or through the membrane matrix itself; (2) by *active transport* through the membrane, a mechanism in which enzyme systems and special carrier substances "carry" the substances through the membrane; and (3) by *pinocytosis*, a mechanism by which the membrane actually engulfs some of the extracellular fluid and its contents. Transport of substances through the membrane is such an important subject that it will be considered in detail in Chapter 5, but one of

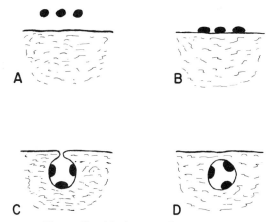

Figure 18. Mechanism of pinocytosis.

these mechanisms of transport, pinocytosis, is such a specialized cellular function that it deserves mention here as well.

Figure 18 illustrates the mechanism of pinocytosis. Figure 18A shows three molecules of protein in the extracellular fluid approaching the surface of the cell. In Figure 18B these molecules have become attached to the surface, presumably by the simple process of adsorption. The presence of these proteins causes the surface tension properties of the cell surface to change in such a way that the membrane invaginates as shown in Figure 18C. Immediately thereafter, the invaginated portion of the membrane breaks away from the surface of the cell, forming a *pinocytic vesicle* which then penetrates deep into the cytoplasm away from the cell membrane.

Pinocytosis occurs only in response to certain types of substances that contact the cell membrane, the two most important of which are *proteins* and *strong solutions of electrolytes.* It is especially significant that proteins cause pinocytosis, because pinocytosis is the only means by which proteins can pass through the cell membrane.

Phagocytosis

Phagocytosis means the ingestion of large particulate matter by a cell, such as the ingestion of (a) a bacterium, (b) some other cell, or (c) particles of degenerating tissue. Pinocytosis was not discovered until the advent of the electron microscope because pinocytic vesicles are smaller than the resolution of the light microscope. However, phagocytosis has been known to occur from the earliest studies using the light microscope.

The mechanism of phagocytosis is almost identical with that of pinocytosis. The particle to be phagocytized must have a surface that can become adsorbed to the cell membrane. Many bacteria have membranes that, on contact with the cell membrane, actually become miscible with the cell membrane so that the cell membrane simply spreads around the bacterium and invaginates to form a *phagocytic vesicle* containing the bacterium, a vesicle that is essentially the same as the pinocytic vesicle shown in Figure 18 but much larger.

Particles whose surfaces will not become adsorbed to the cell membrane are difficult to phagocytize. However, certain proteins in the extracellular fluids often coat these particles with a thin film that then allows the required adsorption to the cell membrane, followed by phagocytosis. This process is called *opsonization.* Phagocytosis and opsonization will be discussed at further length in Chapter 9.

THE DIGESTIVE ORGAN OF THE CELL— THE LYSOSOMES

Almost immediately after a pinocytic or phagocytic vesicle appears inside a cell, one or more lysosomes become attached to the vesicle and empty digestive enzymes called *hydrolases* into the vesicle, as illustrated in Figure 19. Thus, a *digestive vesicle* is formed

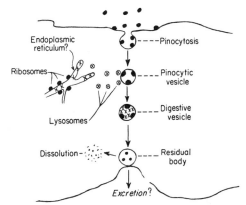

Figure 19. Digestion of substances in pinocytic vesicles by enzymes derived from lysosomes. (Modified from C. De Duve, in *Lysosomes,* Ed. by Reuck and Cameron. Little, Brown and Co., 1963.)

in which the hydrolases begin digesting the proteins, glycogen, nucleic acids, mucopolysaccharides, and other substances in the vesicle. The products of digestion are small molecules of amino acids, glucose, phosphates, and so forth that can then diffuse through the membrane of the vesicle into the cytoplasm. What is left of the digestive vesicle, called the *residual body*, is then either excreted or undergoes dissolution inside the cytoplasm. Thus, the lysosomes may be called the *digestive organ* of the cells.

REGRESSION OF TISSUES AND AUTOLYSIS OF CELLS. Often, tissues of the body regress to a much smaller size than previously. For instance, this occurs in the uterus following pregnancy, in muscles during long periods of inactivity, and in mammary glands at the end of the period of lactation. Lysosomes are responsible for much if not most of this regression, for one can show that the lysosomes become very active during these periods of regression.

Another very special role of the lysosomes is the removal of damaged cells or damaged portions of cells from tissues—cells damaged by heat, cold, trauma, chemicals, disease, or any other factor. Damage to the cell causes lysosomes to rupture, and the released hydrolases begin immediately to digest the surrounding organic substances. If the damage is slight, only a portion of the cell will be removed, followed by repair of the cell. However, if the damage is severe the entire cell will be digested, a process called *autolysis*. In this way, the cell is completely removed and a new cell of the same type ordinarily is formed by mitotic reproduction of an adjacent cell to take the place of the old one.

EXTRACTION OF ENERGY FROM NUTRIENTS

The principal nutrients from which cells extract energy are oxygen and one or more of the foodstuffs—carbohydrates, fats, and proteins. In the human body essentially all carbohydrates are converted in the gut or in the liver into *glucose* before they reach the cell, the proteins are converted into *amino acids*, and the fats are converted into *fatty acids*. Figure 20 shows oxygen and the foodstuffs—glucose, fatty acids, and amino acids—all entering the cell. Inside the cell, the foodstuffs react chemically with the oxygen under the influence of various enzymes that control their rates of reactions and channel the energy that is released in the proper direction.

Formation of Adenosine Triphosphate (ATP)

The energy released from the nutrients is used to form *adenosine triphosphate*, generally called ATP, the formula for which is shown at the top of the following page.

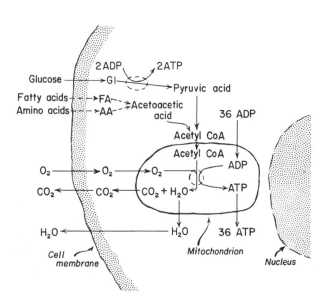

Figure 20.   Formation of adenosine triphosphate in the cell, showing that most of the ATP is formed in the mitochondria.

Note that ATP is a compound composed of three separate parts, the nitrogenous base *adenine*, the pentose sugar *ribose*, and three *phosphate radicals*. The last two phosphate radicals are connected with the remainder of the molecule by so-called *high energy phosphate bonds*, which are represented by the symbol ~. Each of these bonds contains about 8000 calories of energy per mole of ATP, which is many times the energy stored in the average chemical bond, thus giving rise to the term "high energy" bond. Furthermore, the high energy phosphate bond is very labile so that it can be split instantly on demand whenever energy is required to promote other chemical reactions in the cell.

When ATP releases its energy, a phosphoric acid radical is split away at the high energy bond, and *adenosine diphosphate (ADP)* is formed. Then, energy derived from the cellular nutrients causes the ADP and phosphoric acid to recombine to form new ATP, the entire process continuing over and over again. For these reasons, ATP has been called the *energy currency* of the cell, for it can be spent and remade again and again.

Function Mitochondria in the Formation of ATP

On entry into the cells, glucose, fatty acids, and amino acids are subjected to enzymes in the cytoplasm that convert glucose into *pyruvic acid* (a process called *glycolysis*), and convert fatty acids and most amino acids into *acetoacetic acid*. The chemical reactions of these conversions will be presented in Chapter 31. A small amount of ATP is formed by the energy released during these conversions, but this amount is so slight that it plays only a minor role in the overall energy metabolism of the cell.

By far the major portion of the ATP formed in the cell is formed in the mitochondria. The pyruvic and acetoacetic acids are both converted into the compound *acetyl co-A* in the cytoplasm, and this is transported along with oxygen through the mitochondrial membrane into the matrix of the mitochondria. Here this substance is acted upon by a series of enzymes and undergoes dissolution in a sequence of chemical reactions called the *tricarboxylic acid cycle*, or *Krebs cycle*. These chemical reactions will be explained in Chapter 31.

In the tricarboxylic acid cycle, acetyl co-A is split into its component parts, hydrogen atoms and carbon dioxide. The carbon dioxide in turn diffuses out of the mitochondria and eventually out of the cell. The hydrogen atoms combine with carrier substances and are carried to the surfaces of the shelves that protrude into the mitochondria, as shown in Figure 15 of Chapter 2. Attached to these shelves are *oxidative enzymes* which cause the hydrogen atoms to combine with oxygen. The enzymes are arranged on the surfaces of the shelves in such a way that the products of one chemical reaction are immediately relayed to the next enzyme, then to the next, and so on until the complete sequence of reactions has taken place. During the course of these reactions, the energy released from the combination of hydrogen with oxygen is used to manufacture tremendous quantities of ATP from ADP. The ATP then diffuses out of the mitochondria into all parts of the cytoplasm and nucleoplasm where its energy is used to energize the functions of the cell.

ANAEROBIC FORMATION OF ATP DURING GLYCOLYSIS. Note in Figure 20 that no oxygen is required for the splitting of glucose in the cytoplasm to form pyruvic acid, but, even so, a small amount of ATP is formed during this process. Therefore, this mechanism (glycolysis) for providing energy to the cell is called *anaerobic energy metabolism.*

OXIDATIVE ENERGY METABOLISM IN THE MITOCHONDRIA. Note, on the other hand, that tremendous quantities of ATP are formed by the chemical reactions in the mitochondria. Because these reactions require oxygen, this mechanism for providing energy is called *oxidative energy metabolism.*

Approximately 95 *per cent of all ATP* formed in the cell is formed in the mitochondria, in comparison with only 5 per cent by the anaerobic process outside the mitochondria. Because of this tremendous quantity of ATP formed in the mitochondria and because of the need for ATP to supply energy for all cellular functions, the mitochondria are called the *power-houses* of the cell.

Uses of ATP for Cellular Function

ATP is used to promote three major categories of cellular functions: (1) *membrane transport,* (2) *synthesis of chemical compounds* inside the cell, and (3) *mechanical work.* These three different uses of ATP are illustrated in Figure 21: (a) to supply energy for the transport of glucose through the membrane, (b) to promote protein synthesis by the ribosomes, and (c) to supply the energy needed during muscle contraction.

In addition to membrane transport of glucose, energy from ATP is required for transport of sodium ions, potassium ions, and, in certain cells, calcium ions, phosphate ions, chloride ions, uric acid, hydrogen ions, and still many other special substances. Membrane transport is so important to cellular function that some cells utilize as much

as 30 per cent of the ATP formed in the cells for this purpose alone.

In addition to synthesizing proteins, cells also synthesize phospholipids, cholesterol, purines, pyrimidines, and a great host of other substances. Synthesis of almost any chemical compound requires energy. For instance, a single protein molecule might be composed of as many as several thousand amino acids attached to each other by peptide linkages, and the formation of each of these linkages requires the breakdown of an ATP molecule to ADP; thus several thousand ATP molecules must release their energy as each protein molecule is formed. Indeed, cells often utilize as much as 75 per cent of all the ATP formed in the cell simply to synthesize new chemical compounds; this is particularly true during the growth phase of cells.

The final major use of ATP is to supply energy for special cells to perform mechanical work. We shall see in Chapter 7 that each contraction of a muscle fibril requires expenditure of tremendous quantities of ATP. This is true whether the fibril is in skeletal muscle, smooth muscle, or cardiac muscle. Other cells perform mechanical work in two additional ways, by *ciliary* or *ameboid motion,* both of which will be described later in this chapter. The source of energy for all these types of mechanical work is ATP.

In summary, therefore, ATP is always available to release its energy rapidly and almost explosively wherever in the cell it is needed. To replace the ATP used by the cell,

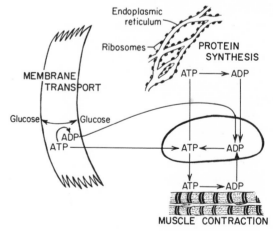

Figure 21. Use of adenosine triphosphate to provide energy for three of the major cellular functions, (1) membrane transport, (2) protein synthesis, and (3) muscle contraction.

other but much slower chemical reactions "burn" carbohydrates, fats, and proteins with oxygen and use the energy derived to form new ATP.

FUNCTIONS OF THE ENDOPLASMIC RETICULUM

The endoplasmic reticulum exists in many different forms in different cells, sometimes highly granular with a large number of ribosomes on its surface, sometimes agranular, sometimes tubular, sometimes vesicular with large shelf-like surfaces, and so forth. Therefore, simply from anatomical considerations alone, it is certain that the endoplasmic reticulum performs a very large share of the cell's functions. Yet the existence of this structure was proved only a dozen years ago, with the advent of the electron microscope, and understanding of its functions has been extremely slow to develop because of the difficulty of studying function with the electron microscope.

On the vast surfaces of the endoplasmic reticulum are adsorbed many of the protein enzymes of the cell, and perhaps still other enzymes are actually integral parts of the reticulum itself. Many of these enzymes synthesize substances in the cells, and others undoubtedly act to transport substances through the membrane of the endoplasmic reticulum from the hyaloplasm of the cytoplasm into the matrix of the reticulum or in the opposite direction. Some of the proved functions of the endoplasmic reticulum are the following:

Secretion of Proteins by Secretory Cells

Many cells, particularly cells in the various glands of the body, form special proteins that are secreted to the outside of the cells. The mechanism for this involves the endoplasmic reticulum and Golgi complex as illustrated in Figure 22. The ribosomes on the surface of the endoplasmic reticulum synthesize the protein that is to be secreted. This protein is either discharged directly into the tubules of the endoplasmic reticulum by the ribosomes or is immediately transported into the tubules to form small protein granules.

These granules then move slowly through the tubules toward the Golgi complex, arriving there a few minutes to an hour or more later. In the Golgi complex the granules are condensed into coalesced granules that then evaginate outward through the membrane of the Golgi complex into the cytoplasm of the cell to form *secretory granules*. Each of these granules carries with it part of the membrane of the Golgi apparatus that provides a membrane around the secretory granule, which prevents it from dispersing in the cytoplasm. Gradually, the secretory granules move toward the surface of the cell where their membranes become miscible with the membrane of the cell itself, and in some way not completely understood they expel their substances to the exterior. It is in this way that protein enzymes, for instance, are secreted by the exocrine glands of the gastrointestinal tract.

Lipid Secretion

Almost exactly the same mechanism applies to lipid secretion. The one major difference is that lipids are synthesized by the agranular portion of the endoplasmic reticulum. The enzymes responsible for lipid synthesis presumably are integral parts of the reticulum and are not simply adsorbed particles as is the case of the ribosomes in the granular portion of the endoplasmic reticulum. The Golgi complex also provides much the same function in lipid secretion as in protein secretion, for here lipids are stored for long periods of time be-

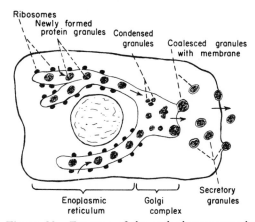

Figure 22. Function of the endoplasmic reticulum and Golgi complex in secreting proteins.

fore finally being extruded into the cytoplasm as lipid droplets and thence to the exterior of the cells.

Release of Glucose from Glycogen Stores of Cells

Electron micrographs show that glycogen is stored in cells as minute granules lying in close apposition to agranular tubules of the endoplasmic reticulum. It is presumed that the endoplasmic reticulum plays a role in the polymerization of glucose to form the glycogen granules and that it again plays a role in later release of glucose from the glycogen. In liver cells, the agranular endoplasmic reticulum contains large quantities of the enzyme *glucose-6-phosphatase*, which is known to be one of the essential enzymes for release of glucose from these cells during periods when the liver is stimulated to release glucose. Following the administration of the hormone glucagon, for instance, the glycogen granules gradually disappear, and glucose is secreted to the outside of the cell, presumably transported by way of the endoplasmic reticular secretory mechanism.

Other Possible Functions of the Endoplasmic Reticulum

The functions presented thus far for the endoplasmic reticulum have involved, first, synthesis of substances and, second, transport of these substances to the exterior of the cell. It is probable that many other substances are synthesized by the endoplasmic reticulum and are then secreted to the interior of the cell rather than to the exterior. It is likewise probable that the Golgi complex plays a role in these internal secretory processes, for large Golgi complexes are found in some cells, such as nerve cells, that do not perform any external secretory function.

Because of the multitude of different anatomical forms of the endoplasmic reticulum, it is certain that we will discover many more functional roles that it plays besides those few that have been discussed here.

CELL MOVEMENT

By far the most important type of cell movement that occurs in the body is that of the specialized muscle cells in skeletal, cardiac, and smooth muscle, which comprise almost 50 per cent of the entire body mass. The specialized functions of these cells will be discussed in Chapter 7. However, two other types of movement occur in other cells, *ameboid movement and ciliary movement.*

Ameboid Motion

Ameboid motion means movement of entire cells by a "streaming" process. That is, the cytoplasm simply streams through its center from one end of the cell toward the other end, and the entire cell moves in that direction. Though this streaming process has been known for many years, it is only in the past few years that the mechanism by which it causes movement has begun to be elucidated as follows:

Figure 23 illustrates an elongated cell with the endoplasmic portion of the cytoplasm streaming through its center from left to right. At the right-hand end of the cell the streaming endoplasm is believed to contract in very much the same way that the fibrils of a muscle cell contract. This contraction pulls still more of the endoplasm toward this end of the cell. The contracted material then moves outward into the ectoplasm and moves backward along the cell wall in the ectoplasm, still in the contracted semisolid state. At the other end of the cell, the contracted ectoplasm undergoes relaxation and streams back through the endoplasmic core once again. One can readily see that this streaming movement inside the cell is analo-

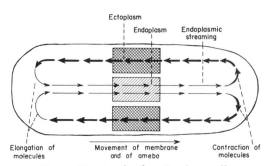

Figure 23. Ameboid motion by a cell.

gous to the revolving movement of the track of a Caterpillar tractor.

TYPES OF CELLS THAT EXHIBIT AMEBOID MOTION. The most common cells to exhibit ameboid motion in the human body are the *white blood cells* moving out of the blood into the tissues in the form of *tissue macrophages* or *microphages*. However, many other types of cells can move by ameboid motion under certain circumstances. For instance, fibroblasts will move into any damaged area to help repair the damage; and even some of the germinal cells of the skin, though ordinarily completely sessile cells, will move by ameboid motion toward a cut area to repair the rent. Finally, ameboid motion is especially important in the development of the fetus, for embryonic cells often migrate long distances from the primordial site of origin to new areas during the development of special structures.

CONTROL OF AMEBOID MOTION – "CHEMOTAXIS." The most important factor controlling the direction and velocity of ameboid motion is the presence of certain chemical substances in the tissues. This phenomenon is called *chemotaxis*, and the chemical substance causing it to occur is called a *chemotaxic substance*. Most ameboid cells move toward the source of the chemotaxic substance – that is, from an area of lower concentration toward an area of higher concentration – which is called *positive chemotaxis*. However, some cells move away from the source, which is called *negative chemotaxis*.

Movement of Cilia

A second type of cellular motion, *ciliary movement,* occurs along the surface of cells in the respiratory tract and in some portions of the reproductive tract. As illustrated in Figure 24, a cilium looks like a minute, sharp-pointed hair that projects 3 to 4 microns from the surface of the cell. Many cilia can project from a single cell.

The cilium is covered by an outcropping of the cell membrane, and it is supported by 11 filaments, 9 located around the periphery of the cilium and 2 down the center, as shown in the cross-section illustrated in Figure 24. Each cilium is an outgrowth of a structure that lies immediately beneath the cell membrane called the *basal body* of the cilium.

In the inset of Figure 24 movement of the cilium is illustrated. The cilium moves forward with a sudden rapid stroke, bending sharply near its base but remaining rigid in its outer portions. Then it moves backward very slowly in whiplike manner without the rigidity of the forward movement. The rapid forward movement pushes the fluid lying adjacent to the cell in the direction that the cilium moves, then the slow whiplike movement in the other direction has almost no effect on the fluid. As a result, fluid is continually propelled in the direction of the forward stroke. Since most ciliated cells have large numbers of cilia on their surfaces, and since ciliated cells on a surface are all oriented in the same direction, this is a very satisfactory means for moving fluids from one part of the surface to another, for instance, for moving mucus out of the lungs or for moving the ovum along the fallopian tube.

MECHANISM OF CILIARY MOVEMENT. The mechanism of ciliary movement is unknown, though we do know some facts about it. First, the movement requires energy, pre-

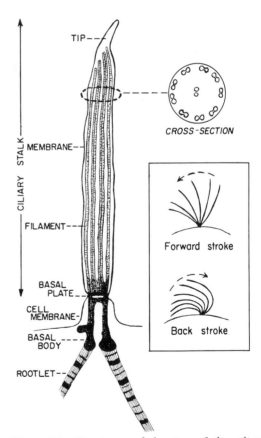

Figure 24. Structure and function of the cilium. (Modified from Satir, *Sci. Amer.*, 204[2]:108, 1961.)

sumably energy supplied by ATP. Second, an electrochemical impulse similar to that transmitted over nerves is transmitted over the membrane of the cilium to produce the movement. Third, the two filaments in the center of the cilium are necessary for movement to occur. One of the theories postulated to explain ciliary movement suggests that the impulse traveling over the membrane of the cilium causes the filaments on one side to contract ahead of the others. Then during the backstroke some reverse process takes place.

REPRODUCTION OF CILIA. Cilia have the peculiar ability to reproduce themselves.

This is achieved by the basal body, which is almost identical to the centriole, an important structure in the reproduction of whole cells, as we shall see in the next chapter. The basal body, as does the centriole, has the ability to reproduce itself by means not yet understood. After it reproduces itself, the new basal body then grows an additional cilium from the surface of the cell.

REFERENCES

See Chapter 2.

CHAPTER 4

GENETIC CONTROL OF PROTEIN SYNTHESIS, CELL FUNCTION, AND CELL REPRODUCTION

Almost everyone knows that the genes control heredity from parents to children, but most persons do not realize that the same genes control day-by-day function of all cells as well. The genes control cell function by determining what substances will be synthesized within the cell—what structures, what enzymes, what chemicals.

Figure 25 illustrates the general schema by which the genes control cellular function. The gene, which is a *nucleic acid* called *deoxyribonucleic acid (DNA)*, automatically controls the formation of another nucleic acid, *ribonucleic acid (RNA)*, which spreads throughout the cell and controls the formation of the different proteins. Some of these proteins are *structural proteins* which, in association with various lipids, form the structures of the various organelles that were discussed in the preceding chapters. But the majority of the proteins are *enzymes* that regulate the different chemical reactions that take place in the cells. For instance, enzymes regulate all the oxidative reactions that supply energy to the cell, and they

regulate the synthesis of various chemicals such as lipids, glycogen, adenosine triphosphate, etc.

The most recent estimates of the number of genes in a nucleus of the human cell is well over a million; however, large numbers

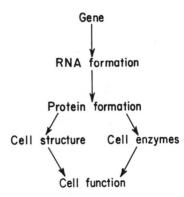

Figure 25. General schema by which the genes control cell function.

of these are exactly alike, causing the formation of precisely the same types of RNA and protein, Yet, even so, the genes probably control the formation of several thousand separate types of protein essential to the function of the different cells.

The Gene

Until the advent of the electron microscope 30 years ago and until the development of proper methods for using this instrument in studying cellular function, which has occurred mainly in the last 10 years, all that was known about the chemical and physical structure of the gene was that many thousands of genes are located in the *chromosomes* of the nucleus and that these are duplicated in each cell prior to cell reproduction. One set of the genes then goes to one daughter cell and the other set goes to the other daughter cell. Fortunately, we now know much more about the gene, both its chemical and physical structure, and even much about the manner in which it functions in the cell.

The gene is a long, double-stranded, helical molecule of *deoxyribonucleic acid (DNA)* having a molecular weight usually measured in the millions. A very short segment of such a molecule is illustrated in Figure 26. This molecule is composed of several simple chemical compounds arranged in a regular pattern, which is explained in the following few paragraphs.

THE BASIC BUILDING BLOCKS OF DNA. Figure 27 illustrates all the basic chemical compounds involved in the formation of DNA. These include *phosphoric acid,* a sugar called *deoxyribose,* and four nitro-

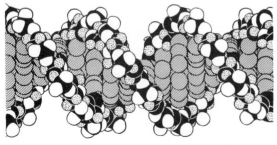

Figure 26. Suggested structure of a deoxyribonucleic acid molecule. The central shaded areas are the purine and pyrimidine bases that constitute the "key" of the gene.

PHOSPHORIC ACID:

DEOXYRIBOSE:

BASES:

Adenine

Thymine

Guanine

Cytosine

PURINES

PYRIMIDINES

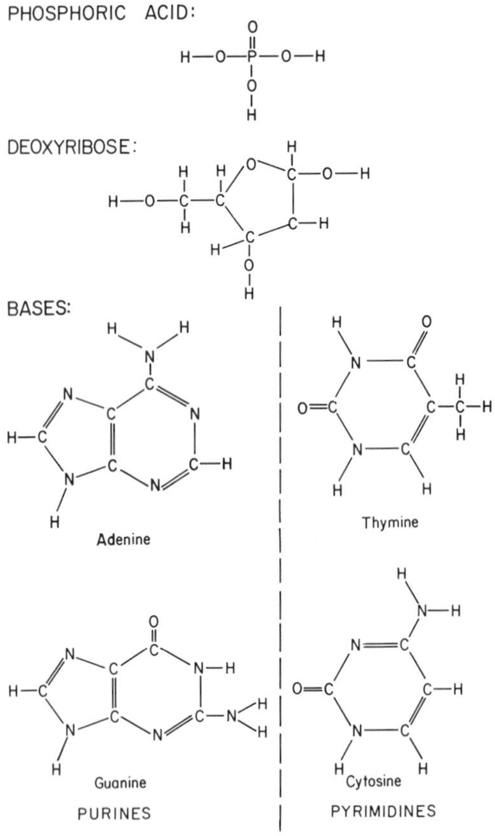

Figure 27. The basic building blocks of DNA.

genous bases (two purines, *adenine* and *guanine,* and two pyrimidines, *thymine* and *cytosine*). The phosphoric acid and deoxyribose form the two helical strands of DNA, and the bases lie between the strands and connect them.

THE NUCLEOTIDES. The first stage in the formation of DNA is the combination of one molecule of phosphoric acid, one molecule of deoxyribose, and one of the four bases to form a nucleotide. Four separate nucleotides are thus formed: *adenylic, thymidylic, guanylic,* and *cytidylic acids.* Figure 28A illustrates the chemical structure of adenylic acid, and Figure 28B illustrates simple symbols for all the four basic nucleotides that form DNA.

Note also in Figure 28B that the nucleotides are separated into two pairs. Adenylic acid, containing the purine adenine, and thymidylic acid, containing the pyrimidine thymine, form one pair. Guanylic acid, containing the purine guanine, and cytidylic acid, containing the pyrimidine cytosine,

A. NUCLEOTIDE

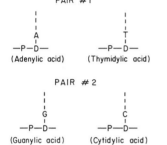

Adenylic acid

B. THE FOUR BASIC NUCLEOTIDES OF DNA

PAIR #1

A   T
—P—D—   —P—D—
(Adenylic acid)   (Thymidylic acid)

PAIR #2

G   C
—P—D—   —P—D—
(Guanylic acid)   (Cytidylic acid)

Figure 28. Combinations of the basic building blocks of DNA to form nucleotides. (*A*, adenine; *C*, cytosine; *D*, deoxyribose; *G*, guanine; *P*, phosphoric acid; *T*, thymine.)

form the other pair. The two nucleotides of each pair are always found together; if one nucleotide of a pair is on one strand of DNA, the other nucleotide of that pair is in a corresponding position on the other strand.

ORGANIZATION OF THE NUCLEOTIDES TO FORM DNA. Figure 29 illustrates the manner in which multiple numbers of nucleotides are bound together to form DNA. Note that these are combined in such a way that phosphoric acid and deoxyribose alternate with each other in the two separate strands, and these strands are held together by the respective pairs of bases. Thus, in Figure 29 the respective pairs of bases are CG, CG,

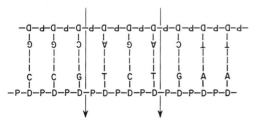

Figure 29. Combination of deoxyribose nucleotides to form DNA.

GC, TA, CG, TA, GC, AT, and AT. However, the pairs are bound together by very loose chemical bonds, represented in the figure by dashed lines. Because of the looseness of these bonds, the two strands can pull apart with ease, and they do so many times during the course of their function in the cell.

Now, to put the DNA of Figure 29 into its proper physical perspective, one needs merely to pick up the two ends and twist them into a helix. Nine pairs of nucleotides are shown in Figure 29; 10 pairs are present in each full turn of the helix in the DNA molecule, as illustrated in Figure 26.

The Genetic Code

The importance of the gene lies in its ability to control the formation of other substances in the cell. It does this by means of the so-called genetic code. When the two strands of a DNA molecule are split apart, this exposes a succession of purine and pyrimidine bases projecting to the side of each strand. It is these projecting bases that form the code.

Research studies in the past few years have demonstrated almost conclusively that the so-called *code words* consist of "triplets" of bases—that is, each three successive bases are a code word. And the successive code words control the sequence of amino acids in a protein molecule to be synthesized in the cell. Note in Figure 29 that each of the two strands of the DNA molecule carries its own genetic code. For instance, the top strand, reading from left to right, has the genetic code GGC, AGA, CTT, the code words being separated from each other by the arrows. As we follow this genetic code through Figures 30 and 31 we shall see that the code words are responsible for placement of the three amino acids, *alanine, serine,* and *glutamic acid,* in a molecule of protein. Furthermore, these three amino acids will be lined up in the protein molecule in exactly the same way that the genetic code is lined up in this strand of DNA.

It is also important that some code words do not cause amino acids to incorporate into proteins but, instead, perform other functions in the synthesis of protein molecules, such as initiating and stopping the formation of a protein molecule. For instance, some DNA molecules of viruses have molecular weights of 40,000,000 or more and cause the formation of many more than a single protein molecule—usually 8 to 20 molecules. It is the code words that do not cause incorporation of amino acids into a protein molecule that signal when to stop the formation of one protein and to start the formation of another. These code words are called *nonsense* words for the simple reason that to early investigators they failed to make sense. But now it is known that a single strand of DNA may represent many different genes, each pair of successive genes separated by a "nonsense" code word.

RIBONUCLEIC ACID (RNA)

Since all DNA is located in the nucleus of the cell and yet most of the functions of the cell are carried out in the cytoplasm, some means must be available for the genes of the nucleus to control the chemical reactions of the cytoplasm. This is done through the intermediary of another type of nucleic acid, ribonucleic acid (RNA), the formation of which is controlled by the DNA of the nucleus. The RNA is then transported into the cytoplasmic cavity where it controls protein synthesis.

At least two separate types of RNA are important to protein synthesis: *messenger RNA* and *soluble RNA*. Several other types of RNA probably also exist, but their functions are yet unknown. Before we describe the functions of messenger and soluble RNA's in the synthesis of proteins, let us see how DNA controls the formation of RNA.

SYNTHESIS OF RNA. One of the strands of the DNA molecule, which constitutes the gene, acts as a template for synthesis of each type of RNA molecule. That is, the code words in DNA cause the formation of *complementary* code words in RNA. The stages in which RNA synthesis takes place are the following.

The basic building blocks of RNA. The basic building blocks of RNA are almost the same as those of DNA except for two differences. First, the sugar deoxyribose is not used in the formation of RNA. In its place is another sugar of similar composition, *ribose.* Referring back to Figure 27, the deoxyribose becomes ribose if the hydrogen radical to the right is replaced by a hydroxyl

radical; this is the only difference between ribose and deoxyribose.

The second difference in the basic building blocks is that thymine is replaced by another pyrimidine, *uracil*. If in Figure 27 the methane radical of the thymine is replaced by a single hydrogen atom, the structure then becomes uracil rather than thymine.

FORMATION OF RNA NUCLEOTIDES. The basic building blocks of RNA first form nucleotides exactly as described above for the synthesis of DNA. Here again, four separate nucleotides are used in the formation of RNA. These four nucleotides contain the bases *adenine, guanine, cytosine,* and *uracil,* respectively, the uracil replacing the thymine found in the four nucleotides that make up DNA. Also, uracil takes the place of thymine in pairing with the purine adenine, as we shall see in the following paragraphs.

ACTIVATION OF THE NUCLEOTIDES. The next step in the synthesis of RNA is activation of the nucleotides. This occurs by addition to each nucleotide of two phosphate radicals to form triphosphates of the nucleotides. These last two phosphates are combined with the nucleotide by *high energy phosphate bonds*. These phosphate radicals and their high energy bonds are derived from the adenosine triphosphate in the cell.

The result of this activation process is that large quantities of energy are made available to each of the nucleotides, and it is this energy that is used in promoting the chemical reactions that eventuate in the formation of the RNA chain.

COMBINATION OF THE ACTIVATED NUCLEOTIDES WITH THE DNA STRAND. The first stage in the formation of RNA is the splitting apart of the two strands of the DNA molecule. Then, activated nucleotides become attached to the bases on one of the DNA strands, as illustrated by the top panel in Figure 30. Note that a ribose nucleotide base always combines with a DNA base in the following combinations: cytosine with guanine, guanine with cytosine, uracil with adenine, and adenine with thymine.

POLYMERIZATION OF THE RNA CHAIN. Once the ribose nucleotides have combined with the DNA strand as shown in the top panel of Figure 30, they are thus lined up in proper sequence to form code words that are complementary to those in the DNA molecule. At this time an enzyme called *RNA polymerase* causes the two extra phosphates

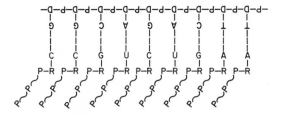

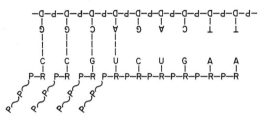

Figure 30. Combination of ribose nucleotides with a strand of DNA to form a molecule of ribonucleic acid (RNA) that carries the DNA code from the gene to the cytoplasm. The wavy lines represent high energy phosphate bonds that energize formation of the RNA strand.

on each nucleotide to split away and at the same time to liberate enough energy to cause bonds to form between the successive ribose and phosphoric acid radicals. As this happens, the RNA strand automatically separates from the DNA strand and becomes a free molecule of RNA. This bonding of the ribose and phosphoric acid radicals and the simultaneous splitting of the RNA from the DNA is illustrated in the lower part of Figure 30.

Once the RNA molecules are formed, they are transported into all parts of the cytoplasm where they perform further functions.

Messenger RNA

The type of RNA that carries the genetic code to the cytoplasm for formation of proteins is called *messenger RNA.* Molecules of messenger RNA are usually composed of several hundred to several thousand nucleotides in a single strand containing code words that are exactly complementary to the code words of the genes. Figure 31 illustrates a small segment of a molecule of messenger RNA. Its code words are CCG, UCU, and GAA. These are the code words for alanine, serine, and glutamic acid. If we now refer back to Figure 30, we see that the events

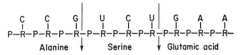

Figure 31. Portion of a ribonucleic acid molecule, showing three "code" words, CCG, UCU, and GAA, which represent the three amino acids *alanine, serine, and glutamic acid.*

recorded in this figure represent transfer of this particular genetic code from the DNA strand to the RNA strand.

Messenger RNA molecules are long straight strands that are suspended in the cytoplasm. These molecules migrate to the ribosomes where protein molecules are manufactured, which is explained below.

RNA CODE WORDS. Table 1 gives the RNA code words for the 20 common amino acids found in protein molecules. Note that several of the amino acids have more than one code word. Also, it is likely that additional code words will be found for some of the other amino acids now listed as having only a single code word. And it was pointed out above that some code words are "nonsense" code words that probably represent such signals as "start manufacturing a protein molecule at this point" or "stop manufacturing a protein molecule at this point."

TABLE 1. *RNA Code Words for the Different Amino Acids*

AMINO ACID	RNA CODE WORDS			
Alanine	CCG	UCG		
Arginine	CGC	AGA	UCG	
Aspargine	ACA	AUA		
Aspartic acid	GUA			
Cysteine	UUG			
Glutamic acid	GAA	AGU		
Glutamine	ACA	AGA	AGU	
Glycine	UGG	AGG		
Histidine	ACC			
Isoleucine	UAU	UAA		
Leucine	UUG	UUC	UUA	UUU
Lysine	AAA	AAG	AAU	
Methionine	UGA			
Phenylalanine	UUU			
Proline	CCC	CCU	CCA	CCG
Serine	UCU	UCC	UCG	
Threonine	CAC	CAA		
Tryptophan	GGU			
Tyrosine	AUU			
Valine	UGU			

Soluble RNA (Transfer RNA)

Another type of RNA that plays a prominent role in protein synthesis is called soluble RNA (or transfer RNA) because it is soluble in the cytoplasm. There is a separate type of soluble RNA for each of the 20 amino acids that are incorporated into proteins. Furthermore, each type of soluble RNA combines with only one type of amino acid—one and no more. Soluble RNA then acts as a *carrier* to transport its specific type of amino acid to the ribosomes where protein molecules are formed. In the ribosomes, each specific type of soluble RNA recognizes a particular code word on the messenger RNA, as is described below, and thereby delivers the appropriate amino acid to the appropriate place in the chain of the newly forming protein molecule.

Very little is known about the precise function of the soluble RNA's except that these nucleic acids are relatively small molecules each containing only about 70 nucleotides. The soluble RNA's are a double-stranded helix like the DNA molecule, except that at one end of the helix the two strands are continuous with each other. At the other end, the two strands are free. It is presumed that one of these ends attaches to a specific amino acid and that the other free end is specific for recognizing a particular code word on the messenger RNA.

Formation of Proteins in the Ribosomes

When a molecule of messenger RNA comes in contact with a ribosome, it penetrates the ribosome at one of its ends, the end probably being determined by one of the "nonsense" code words. Then, as illustrated in Figure 32, the messenger RNA begins to travel through the ribosome; as it travels through, a protein molecule is formed. The exact events in the formation of this protein molecule are still a mystery, though it is presumed that as the messenger RNA travels through the ribosome, its code is "read" in very much the same way that a tape is "read" as it passes through the playback head of a tape recorder. As each code word passes through the ribosome, the specific soluble RNA corresponding to each specific code word supposedly attaches to the code of the messenger RNA, and the

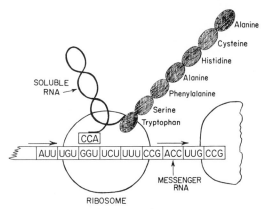

Figure 32. Postulated mechanism by which a protein molecule is formed in ribosomes in association with messenger RNA and soluble RNA.

ribosome in some unexplained way causes the successive amino acids to combine with each other by the process of peptide linkage. At the same time, the soluble RNA is freed from the amino acid, and returns to the cytoplasm to pick up another amino acid of its specific type.

Thus, as the messenger RNA passes through the ribosome, a protein molecule is formed as shown in Figure 32. When an appropriate "nonsense" word passes through the ribosome, the end of a protein molecule is signaled, and the entire molecule is freed into the cytoplasm.

POLYRIBOSOMES. A single messenger RNA molecule can form protein molecules in several different ribosomes at the same time, the RNA molecule entering another ribosome as it leaves the first, as shown in Figure 32. The protein molecules obviously are in different stages of development. As a result, clusters of ribosomes frequently occur, four or five or even more ribosomes being attached together by a single messenger RNA at the same time. These clusters are called polyribosomes.

It is especially important that a messenger RNA can cause the formation of a protein molecule in any ribosome, and that there is no specificity of ribosomes for given types of protein. The ribosome seems to be simply the physical structure in which the chemical reactions take place, and it is the passage of messenger RNA through the ribosome that determines the sequence of amino acids in the protein.

PEPTIDE LINKAGE. The successive amino acids in the protein chain combine with each other by peptide linkages according to the following typical reaction:

$$R-C \overset{\overset{\displaystyle NH_2}{|}}{} \overset{\overset{\displaystyle O}{\parallel}}{C} - \overline{(H + HO)} - \overset{\overset{\displaystyle H}{|}}{N} - \overset{\overset{\displaystyle R}{|}}{C} - COOH \rightarrow$$

$$R-C \overset{\overset{\displaystyle NH_2}{|}}{} \overset{\overset{\displaystyle O}{\parallel}}{C} - \overset{\overset{\displaystyle H}{|}}{N} - \overset{\overset{\displaystyle R}{|}}{C} - COOH + H_2O$$

In this chemical reaction, a hydroxyl radical is removed from the COOH portion of one amino acid while a hydrogen of the NH_2 portion of the other amino acid is removed. These combine to form water, and the two reactive sites left on the two successive amino acids combine, resulting in a single molecule. This process is called *peptide linkage.*

SYNTHESIS OF OTHER SUBSTANCES IN THE CELL

Many hundreds or perhaps a thousand or more protein enzymes formed in the manner just described control essentially all the other chemical reactions that take place in cells. These enzymes promote synthesis of lipids, glycogen, purines, pyrimidines, and hundreds of other substances. We will discuss many of these synthetic processes in relation to carbohydrate, lipid, and protein metabolism in Chapter 31. It is by means of all these different substances that the many functions of the cells are performed. Thus, this description of the hierarchy of cellular control is now complete.

CONTROL OF THE GENES AND THEIR SYNTHETIC ACTIVITIES

Since the genes are considered to be at the apex of the hierarchy that controls cellular function, it will be surprising to many students to learn that function of the genes themselves and their synthetic activities must also be controlled, or otherwise they would kill the cell. One can readily understand this when he remembers that all parts of the cell must be formed in proper proportion to each other for a cell to function normally. For instance, let us assume that

a particular amino acid is missing for the formation of an important enzyme required by the cell. If the cell continues to form all the other types of enzymes besides this particular one, eventually so little of this enzyme will be present in proportion to all the others that the cell will die.

But, fortunately, the functions of the genes are controlled by *feedback* signals from the functional elements of the cell. These feedback signals are generally in the form of chemical products of the different enzyme-controlled reactions, as is explained in the following paragraphs.

ENZYME INHIBITION. One of the most common types of feedback is called enzyme inhibition. Figure 33 illustrates this as follows: It shows first a gene controlling the formation of RNA, RNA controlling the formation of an enzyme, and the enzyme controlling the formation of a certain chemical. If the chemical is formed in excess, this chemical itself often acts directly on the enzyme to inhibit its function. Obviously, the rate of production of the chemical then decreases, thus preventing overabundance of the substance. An example of enzyme inhibition is the following: In some animal cells isoleucine is synthesized from threonine. However, once a small quantity of threonine has been converted to isoleucine, the enzyme system for causing this conversion becomes completely inhibited. Yet if the same enzymes are placed once again in an isoleucine-free medium, the synthetic process begins again immediately. Thus, isoleucine acts as a feedback product to inhibit directly the enzymes for conversion of threonine to isoleucine. Without this inhibition, essentially all the threonine in the cell would be converted to isoleucine, and the cell would have no remaining threo-

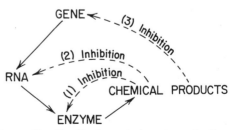

Figure 33. Feedback control systems in the cell that control *rates* of synthesis of the different cellular proteins. Arrow 1 represents enzyme inhibition, while arrows 2 and 3 represent repression of genetic and RNA activity.

nine for other uses, which would be disastrous because threonine is one of the amino acids that is essential for the formation of almost all cell proteins. Therefore, it is important that this feedback mechanism does limit the conversion of threonine to isoleucine and thereby maintain a relatively exact ratio between these two amino acids in the cell.

RNA AND DNA REPRESSION. Figure 33 also shows two other feedback loops besides enzyme inhibition, represented by the dashed arrows 2 and 3, which show *inhibition of gene function* and *of RNA function* by chemical products produced in the cell. For instance, when excess quantities of glucose are present in the cells and this glucose becomes split into its various degradation products, certain of these degradation products act on some of the genes or their corresponding RNA's to decrease their rates of activity. As a result, the genes decrease their rates of formation of RNA's, and the RNA's decrease their rates of formation of enzymes whose physiological role is to aid the energy reactions of the cell. A list of some of the protein enzymes decreased in this way in response to excess glucose is the following: beta galactosidase, inositol dehydrogenase, glycerol dehydrogenase, histidase, tryptophanase, and amino acid deaminases.

Thus far, it has not been possible to tell in the case of most of these repression mechanisms whether the repression occurs at the genetic level or at the RNA level. Nevertheless, it is by such feedback mechanisms that the quantities of enzymes in the cell are controlled to meet the needs of the cell. Thus, if excess glucose is available, the enzymes for supplying energy are reduced. On the other hand, if too small a quantity of glucose is available, the enzymes are increased so that the small amount of glucose that is available can be utilized more easily and so that other substances, such as proteins, might also be utilized for energy in place of glucose. This is an automatic mechanism of the cell, which allows continued normal energy metabolism in the cell despite vast changes in the availability of one of the cellular nutrients.

To summarize, the small molecular chemical products synthesized in the cell control the rates of production of the different protein enzymes. Each chemical product, by enzyme inhibition, by RNA repression, or by DNA repression, limits its own concentration

in the cell, and whenever its concentration falls too low, opposite effects occur, so that increased quantities of the substance are produced. Each feedback mechanism for each chemical substance in the cell has its own critical level of sensitivity so that some chemical products are produced in great quantities in the cell before they become limited, while others are produced only in small quantities.

CELL REPRODUCTION

Most cells of the body are continually growing and reproducing. The new cells take the place of the old ones that die, thus maintaining a complete complement of cells in the body.

Also, one can remove most types of cell from the human body and grow them in tissue culture where they will continue to grow and reproduce so long as appropriate nutrients are supplied and so long as the end-products of the cells' metabolism are not allowed to accumulate in the nutrient medium. Thus, the life lineage of cells is indefinite.

As is true of almost all other events in the cell, reproduction also begins in the nucleus itself. The first step is *replication (duplication) of all genes and of all chromosomes.* The next step is division of the two sets of genes between two separate nuclei. And the final step is splitting of the cell itself to form two new daughter cells.

This process of cell division is called *mitosis.* The complete life cycle of a cell that is not inhibited in some way is about 10 to 30 hours from reproduction to reproduction, and the period of mitosis lasts for approximately one-half hour. The period between mitoses is called *interphase.*

Replication of the Genes

The genes are reproduced several hours before mitosis takes place, and the time required for gene replication is only about one hour. When replication begins, it occurs for all genes at the same time and not for only part of them. Furthermore, the genes are duplicated only once. The net result is two exact duplicates of all genes, which respectively become the genes in the two new daughter cells that will be formed at mitosis. Following replication of the genes, the nucleus continues to function normally for several hours before mitosis begins abruptly.

CHEMICAL AND PHYSICAL EVENTS. The genes are duplicated in almost exactly the same way that RNA is formed from DNA. First, the two strands of the DNA helix of the gene pull apart. Second, each of these strands combines with deoxyribose nucleotides of the four types described early in the chapter as the basic building blocks of DNA. Each of the bases on each strand of DNA in the chain attracts a nucleotide containing the appropriate *complementary* base. In this way, the appropriate nucleotides are lined up side by side. Third, appropriate enzyme mechanisms then provide energy and cause polymerization of the nucleotides to form a new DNA strand. The only difference between this formation of the new strand of DNA and the formation of an RNA strand is that the new strand of DNA remains attached to the old strand that has formed it, thus forming a new double-stranded DNA helix. At the same time the other strand of the original helix forms its complementary DNA strand and thereby also forms a new double-stranded helix.

The Chromosomes and Their Replication

Unfortunately, we know much less about replication of the chromosomes than about replication of the genes themselves. The primary reason for this is that the structure of the chromosome itself is almost unknown.

The chromosomes consist of two major parts: the genes, consisting of DNA, and protein. The DNA is bound loosely with the protein and sometimes during the life cycle of the cell seems to become separated from the protein. The combination of the two is known as a *nucleoprotein.*

Many different theories have been offered for the manner in which the protein is bound with the DNA. One of these is that the protein provides a "backbone" and that the ends of the DNA molecules which constitute the genes are attached to the "backbone."

Another theory is that illustrated in Figure 34 in which the DNA molecules are believed to be polymerized into one long, folded molecule, which during part of its

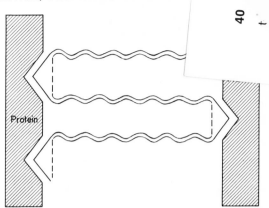

Figure 34. Postulated arrangement of DNA molecules and protein in the chromosome.

life cycle is attached to the protein and yet during other parts of the cycle may become completely detached. It has been calculated that if all the DNA of the human cell nucleus were polymerized into one long strand, it would be about 1 meter long, and the DNA of a single chromosome would be approximately 2 cm. long. The folding theory of Figure 34 represents a means by which all of this DNA of a single chromosome could be compacted into the small length of the chromosome, only $\frac{1}{10,000}$ of 2 cm. It also helps to explain the means by which chromosomes can be replicated, though it must be remembered that this structure is purely hypothetical and has been advanced only as a theory that might lead to the discovery of the precise structure of the chromosome.

To go further with the model of Figure 34, let us assume that the DNA helixes of this model first become replicated. They would then appear as two double helixes lying together side-by-side, as shown in Figure 35A. Then, let us assume that the pro-

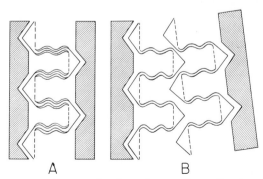

A   B

Figure 35. Postulated mechanism for replication of the chromosome. (A) Replication of DNA strands. (B) Pulling apart of the replicated strands.

ın poles pull the two double helixes apart, as shown in Figure 35B.

This is only a theory, but it does represent a type of hypothesis being considered by research workers who are attempting to unravel the problem of chromosome replication.

NUMBER OF CHROMOSOMES IN THE HUMAN CELL. Each human cell contains 46 chromosomes arranged in 23 pairs. In general, the genes in the two chromosomes of each pair are almost identical with each other, so that it is usually stated that the different genes exist in pairs, though occasionally this is not the case. Also, most genes have several duplicates even on the same chromosome, so that loss of any one usually will not completely remove a particular characteristic from a cell.

Mitosis

The actual process by which the cell splits into two new cells is called mitosis. Once the genes have been duplicated and each chromosome has split to form two new chromosomes, each of which is now called a *chromatid,* mitosis follows automatically, almost without fail, within a few hours.

REPLICATION OF THE CENTRIOLES. Mitosis has generally been considered by biologists to be mainly a nuclear process. However, it is now known that the first event of mitosis actually takes place in the cytoplasm, occurring during the latter part of interphase in the small structures called *centrioles.*

As illustrated in Figure 36A, two pairs of centrioles lie close to each other near one pole of the nucleus. Each centriole is a small cylindrical body about 0.4 micron long and about 0.15 micron in diameter, consisting mainly of nine parallel, tubular-like structures arranged around the inner wall of the cylinder. The two centrioles of each pair lie at right angles to each other.

In so far as is known, the two pairs of centrioles remain dormant during interphase until shortly before mitosis is to take place. At that time, the two pairs begin to move apart from each other. This is probably caused by protein fibers growing between the respective pairs and actually pushing them apart. At the same time, fibers grow radially away from each of the pairs to form the *aster* as illustrated in Figure 36A. Some of these penetrate the nucleus as illustrated

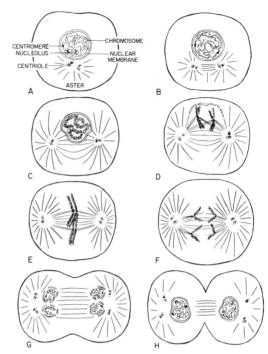

Figure 36. Stages in the reproduction of the cell. A and B, late interphase; C and D, prophase; E, metaphase; F, anaphase; G and H, telophase. (Redrawn from Mazia: *Sci. Amer.* 205[3]:102, 1961.)

in Figures 36B and C. The set of fibers connecting the two centriole pairs is called the *spindle,* and the entire set of fibers plus the two centrioles is called the *mitotic apparatus.*

While the spindle is forming, the *chromatin material* of the nucleus (the DNA) becomes condensed into well-defined but disoriented chromosomes, as shown in Figures 36A and B.

PROPHASE. The first stage of mitosis, called prophase, is shown in Figures 36C and D. The nuclear envelope breaks up, and some of the fibers from the forming mitotic apparatus become attached to the chromosomes. This attachment always occurs at the same point on each chromosome, at a small condensed portion called the *centromere.*

METAPHASE. During metaphase (Fig. 36E) the centriole pairs are pushed far apart by the growing spindle, and the chromosomes are thereby pulled tightly by the attached fibers to the very center of the cell, lining up in the equatorial plane of the mitotic spindle.

ANAPHASE. With still further growth of the spindle, each pair of chromosomes is now broken apart, a stage of mitosis called anaphase (Fig. 36F). A spindle fiber connecting with one pair of centrioles pulls one chromatid, and a fiber connecting with the other centriole pair pulls the opposite chromatid. Thus, all 46 pairs of chromatids are separated, forming 46 *daughter chromosomes* that are pulled toward one mitotic spindle and another 46 identical chromosomes that are pulled toward the outer mitotic spindle.

TELOPHASE. In telophase (Figs. 36G and H) the mitotic spindle grows still longer, pulling the two sets of daughter chromosomes completely apart. Then a new nuclear membrane develops around each set of chromosomes, this membrane perhaps being formed from portions of the endoplasmic reticulum already present in the cytoplasm. Simultaneously, the mitotic apparatus undergoes dissolution, and the cell pinches in two midway between the two nuclei, for reasons totally unexplained at present.

Note, also, that each of the two pairs of centrioles are replicated during telophase, the mechanism of which is not understood. These new pairs of centrioles remain dormant through the next interphase until the mitotic apparatus is required for the next cell division.

Control of Cell Growth and Reproduction

Cell growth and reproduction usually go together, growth occurring simultaneously with replication of the genes and mitosis.

In the normal human body, regulation of cell growth and reproduction is mainly a mystery. We know that certain cells grow and reproduce all the time, such as the blood-forming cells of the bone marrow, the germinal layers of the skin, and the epithelium of the gut. However, many other cells of the body do not reproduce for many years, such as some muscle cells; and a few cells, such as the neurons, do not reproduce during the entire life of the person. In the case of neurons, it is presumed that genetic changes occur within the cells when they are originally formed to prevent any subsequent replication of the genes and, therefore, to prevent further growth and cell division.

On the other hand, most cells of the human body do have the ability to reproduce, though the rate of reproduction usually remains greatly suppressed. Yet if an insufficiency of a given type of cell develops in the body, this type of cell will grow and reproduce very rapidly until appropriate numbers of them are again available. For instance, seven-eighths of the liver can be removed surgically, and the cells of the remaining one-eighth will grow and divide until the liver mass returns almost to normal. The same effect occurs for almost all glandular cells, for cells of the bone marrow, the subcutaneous tissue, the intestinal epithelium, and almost any other tissue except highly differentiated cells, such as nerve cells.

We know almost nothing about the mechanisms that maintain proper numbers of the different types of cells in the body. It is assumed that control substances are secreted by the different cells that cause feedback effects on the same types of cells to stop or slow their growth when too many of them have been formed, though such substances have not yet been found. We know that cells of almost any type removed from the body and grown in tissue culture can grow and reproduce rapidly and indefinitely if the medium in which they grow is continually replenished. Yet they will stop growing when their own secretions are allowed to collect in the medium, which supports the idea that control substances limit cellular growth.

REGULATION OF CELL SIZE. Cell size is determined almost entirely by the amount of DNA in the nucleus. If replication of the genes does not occur, the cell grows to a certain size and thereafter remains at that size. On the other hand, it is possible, by use of the chemical *colchicine*, to prevent mitosis even though replication of the genes continues. In this event, the nucleus then contains far greater quantities of DNA than normally, and the cell grows proportionately larger. It is assumed that this results simply from increased production of RNA and cell proteins, which in turn cause the cell to grow larger.

Cancer

Cancer can occur in any tissue of the body. It results from a change in certain cells that allows them to disrespect normal growth limits, no longer obeying the feedback controls that normally stop cellular growth and reproduction after a given number of such cells have developed. As pointed out above, even normal cells when removed from the body and grown in tissue cultures can grow and proliferate indefinitely if the growth medium is continually changed. Therefore, in tissue culture, normal tissue cells behave exactly as cancer cells, but in the body normal tissue cells behave differently, for they are subject to limits, whereas cancer cells are not.

What is the difference between the cancer cell and the normal tissue cell that allows the cancer cell to grow and reproduce unabated? The answer to this question is not known, but researchers have found the genetic structures of cancer cells to be different from those of normal cells. This has led to the idea that cancer almost certainly results from mutation of part of the genetic system in the nucleus, this mutation eliminating the feedback mechanisms that normally limit growth of the cell. Once even a single such cell is formed, it obviously will grow and proliferate indefinitely, its number increasing exponentially.

Cancerous tissue competes with normal tissues for nutrients, and because cancer cells continue to proliferate indefinitely, their number multiplying day-by-day, one can readily understand that the cancer cells will soon demand essentially all the nutrition available to the body. As a result, the normal tissues gradually suffer nutritive death.

CELL DIFFERENTIATION

A special characteristic of cell growth and cell division is cell differentiation, which means the changes in physical and functional properties of cells as they proliferate in the embryo to form the different bodily struc-

tures. Figure 37 illustrates various stages of cleavage of the fertilized mammalian ovum, the darkened cells representing those that will eventually form the embryo and the lighter cells those that will eventually form the fetal membranes. Note the differences in size and other characteristics of the cells as cell division proceeds during the first few days after fertilization. Obviously, still many more changes in characteristics of the cells will take place before the final postnatal human being has been formed. However, our problem is not to describe the stages of differentiation but simply to discuss the theories of how the cells change their characteristics to form all the different organs and tissues of the body.

The earliest and simplest theory for explaining differentiation was that the genetic composition of the nucleus undergoes changes during successive generations of cells in such a way that one daughter cell develops one set of characteristics while another daughter cell develops entirely different characteristics. This theory probably explains some types of differentiation, because highly differentiated cells cannot be reverted all the way back to the original primordial state. Yet most cells can be reverted at least two or three steps backward in differentiation, which indicates that other factors besides simple changes in genetic potency probably also play a role in differentiation and might even play the dominant role.

The cytoplasm also probably plays a role in differentiation, because cells without nuclei can divide and even differentiate to some extent for a few stages before the cell dies. However, the role of the cytoplasm in differentiation is probably controlled by the nucleus.

Embryological experiments show that certain cells in an embryo control the differentiation of adjacent cells. For instance, the *primordial chordamesoderm* is called the *primary organizer* of the embryo because it forms a focus around which the rest of the embryo develops. It differentiates into a *mesodermal axis* containing segmentally arranged *somites* and, as a result of complex *inductions* in the surrounding tissues, causes formation of essentially all the organs of the body.

Another instance of induction occurs when

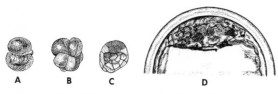

Figure 37. Differentiation of cells in the early embryo during gestation: (A) two blastomeres, (B) six blastomères, (C) a hemisected early blastocyst, (D) a later blastocyst stage showing the early fetus at the top. (Modified from Arey: *Developmental Anatomy.* 6th Ed.)

the developing eye vesicles come in contact with the ectoderm of the head and cause it to thicken into a lens plate which folds inward to form the lens of the eye. It is possible that the entire embryo develops as a result of complex inductions, one part of the body affecting another part, and this part affecting still other parts.

Thus our understanding of cell differentiation is still hazy. We know many different control mechanisms by which differentiation *could* occur. Yet the overall basic controlling factors in cell differentiation are yet to be discovered; when learned, they will make a tremendous difference in our understanding of bodily function.

REFERENCES

Allen, J. M.: *The Molecular Control of Cellular Activity.* New York, McGraw-Hill Book Company, 1962.

Busch, H.: *Biochemistry of the Cancer Cell,* New York, Academic Press, 1962.

Crick, F. H. C.: The genetic code. *Sci. Amer.* 207(4):66, 1962.

Gaze, R. M.: Growth and differentiation. *Ann. Rev. Physiol.* 29:59, 1967.

Harris, R. J. C. (ed.): *Symposia of the International Society for Cell Biology: Cell Growth and Cell Division.* Vol. 2. New York, Academic Press, 1963.

Hurwitz, J., and Furth, J. J.: Messenger RNA. *Sci. Amer.* 206(2):41, 1962.

Mazia, D.: How cells divide. *Sci. Amer.* 205(3):100, 1961.

Nirenberg, M. W.: The genetic code. *Sci. Amer.* 208(3): 80, 1963.

Rich, A.: Polyribosomes. *Sci. Amer.* 209(6):44, 1963.

CHAPTER 5

FLUID ENVIRONMENT OF THE CELL AND TRANSPORT THROUGH THE CELL MEMBRANE

The life processes of the cell all occur in a fluid medium composed of water and dissolved substances. The fluid is mobile so that different substances can be transported from one point to another as needed. Furthermore, dissolved molecules and suspended particles can move within the fluid by the process of diffusion, which will be explained later. Thus, the enzymes within the cell can come into close apposition with chemical substances that react with each other; food substances can be transported from one part of the body to another, and movement of ions through the cell membrane can create electrical currents that are responsible for transmission of nerve impulses or for causing muscle contraction. It will be the purpose of this chapter to characterize anatomically and functionally both the fluid that is present inside the cell, called the *intracellular fluid*, and the fluid outside the cell, called the *extracellular fluid*.

ROLE OF THE BODY FLUIDS IN HOMEOSTASIS. The concept of *homeostasis* was presented in Chapter 1, a concept which states very simply that cells can continue to live and prosper and, in most instances, to reproduce new cells again and again provided that the fluid environment in which they live is maintained with appropriate concentrations of the necessary constituents for cellular life. To state this in another way, so long as the fluid environment of the cell contains the appropriate amount of oxygen, not too little and not too much, so long as it contains the proper amount of glucose, fats, proteins, and other substances necessary to the function of the cell, continued life of the cell is automatic, for the cell has built into its makeup all those necessary ingredients for continuance of its own life and usually for reproducing new cells of its own kind.

Throughout this entire text we will come

again and again to the concept of homeostasis, and a major share of the book will be devoted to explaining how each constituent of the extracellular fluid is maintained at an appropriate and constant concentration. In the present chapter we will characterize both the intracellular and extracellular fluids and explain in particular their interdependence. Especially important will be our consideration of the mechanisms by which substances move through the cell membrane between the extracellular and intracellular fluids.

COMPOSITION OF THE BODY FLUIDS

Constituents of the Intracellular Fluid

Intracellular fluid contains literally hundreds of organic compounds that are responsible for the intracellular chemical reactions discussed in the past few chapters. In addition to these compounds, the intracellular fluid contains, as illustrated in Figure 38, large quantities of the electrolytes potassium, magnesium, and phosphate plus small quantities of *sodium, chloride, bicarbonate,* and *sulphate.* Some of these electrolytes are extremely important as integral parts of the chemical reactions occurring in the cell. The cell membrane has special capabilities for actively transporting many electrolytes between the intracellular and extracellular fluid, as will be discussed later in the chapter. Most important of these are the *sodium pump which transports sodium out of the cell* and the mechanisms for *transport of potassium and phosphate into the cell.*

Intracellular fluid also contains an assortment of nutrients to be used in the cellular metabolic processes. Included in these is a small amount of *glucose* which is mostly converted to intermediate metabolic products for further use in the cell or is stored in the cell in a polymerized form as insoluble *glycogen.* Most cells also contain small quantities of *lipids,* including *neutral fat, phospholipids,* and *cholesterol,* and the cells of fat tissue may contain as much as 95 per cent of their total mass as fat. However, most of the lipids are insoluble in water and, therefore, are present in the cell in a colloidal or deposited form rather than dissolved in the fluids. Another important nutrient in the cell is the *amino acids,*

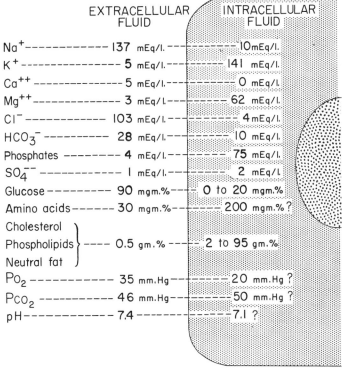

	EXTRACELLULAR FLUID	INTRACELLULAR FLUID
Na+	137 mEq/l.	10 mEq/l.
K+	5 mEq/l.	141 mEq/l.
Ca++	5 mEq/l.	0 mEq/l.
Mg++	3 mEq/l.	62 mEq/l.
Cl-	103 mEq/l.	4 mEq/l.
HCO3-	28 mEq/l.	10 mEq/l.
Phosphates	4 mEq/l.	75 mEq/l.
SO4--	1 mEq/l.	2 mEq/l.
Glucose	90 mgm.%	0 to 20 mgm.%
Amino acids	30 mgm.%	200 mgm.%?
Cholesterol Phospholipids Neutral fat	0.5 gm.%	2 to 95 gm.%
Po2	35 mm.Hg	20 mm.Hg?
Pco2	46 mm.Hg	50 mm.Hg?
pH	7.4	7.1?

Figure 38. Chemical compositions of extracellular and intracellular fluids.

which are probably present in greater concentration than in the extracellular fluid because of active transport through the cell membrane.

Intracellular fluid also contains *oxygen* in a reasonably high concentration. As the oxygen is metabolized with the nutrients to give energy, *carbon dioxide* is liberated into the intracellular fluid, and it then diffuses outward through the cellular membrane into the extracellular fluid.

Constituents of the Extracellular Fluid

Most striking of all the differences between the extracellular and intracellular fluids are the concentrations of the electrolytes. The *sodium, calcium,* and *chloride* concentrations are many times as high in the extracellular fluid as in the intracellular fluid; in contrast, the potassium, magnesium, and phosphate concentrations are many times as high intracellularly as extracellularly. These differences are responsible for electrical potentials that develop across the cell membrane and also for the degree of permeability of the membrane. We shall see in Chapter 6 that it is mainly the concentration gradient of potassium between the two sides of the membrane, 141 milliequivalents per liter intracellularly and 5 milliequivalents per liter extracellularly, that causes electrical potentials to develop across the membranes of nerve fibers, muscle fibers, and probably also of all other cells of the body. Without these membrane potentials, it would be impossible for nerve fibers to conduct impulses or muscle fibers to contract.

The calcium in the extracellular fluid plays a role in determining the degree of permeability of the membrane. A high concentration of calcium decreases the permeability while a low concentration increases the permeability. Furthermore, whenever a cell membrane ruptures, if calcium is present in large enough quantities, a new membrane will be formed immediately, which is called the "surface precipitation reaction"; this was described in Chapter 2.

The nutrients in the extracellular fluid, like those in the intracellular fluid, include glucose, amino acids, and lipids. However, the concentrations of these in the extracellular fluid are quite different from the concentrations inside the cell. For instance,

the extracellular glucose concentration is very high in comparison with that in the intracellular fluid, as shown in Figure 38. However, if we should consider all the *products* of glucose inside the cell—hexose monophosphate, glycogen, and so forth—we would find that the cell actually concentrates glucose in these other forms in very great quantities to be used for energy when needed. Likewise, the concentration of amino acids in the extracellular fluid is low compared to the amount of amino acids inside the cell. Here again few free amino acids are present inside the cell; instead, they are conjugated into simple proteins that can be hydrolyzed rapidly back into amino acids. Also, lipids are usually present in the cell in far greater quantities than in the extracellular fluid, usually deposited in membranes or in inclusions within the cytoplasm.

Finally, oxygen is present in the extracellular fluid in slightly higher concentration than in the intracellular fluid. It is this concentration difference that causes oxygen to diffuse continually from the extracellular fluid into the cell. And *carbon dioxide* is present in intracellular fluid in slightly greater concentration than in extracellular fluid, which allows carbon dioxide to diffuse continually out of the cell as it is formed by the combustion of nutrients with oxygen.

To summarize, the constituents of extracellular fluid have two purposes: first, to maintain membrane potentials and proper membrane permeability so that very important membrane functions can take place, and, second, to provide a system for transport of nutrients and other substances from one part of the body to another. For instance, glucose, amino acids, cholesterol, phospholipids, neutral fat, carbon dioxide, urea, lactate, creatinine, sulfate, and many other substances are all transported by movement of the extracellular fluid.

Movement and Mixing of the Extracellular Fluids Throughout the Body

Fortunately, the extracellular fluids are continually mixed. Were it not for this, the cells would remove all the nutrients from the fluids in their immediate vicinity until none would be left, and cellular excreta on the other hand would accumulate until they would kill the cells. Yet, two different mechanisms provide for continual mixing and movement of the extracellular fluid throughout the body; these are (1) circulation of the blood and (2) the phenomenon of diffusion.

Figure 39 illustrates the general plan of the circulation, showing continuous flow of blood around the system, through the lungs and through the different tissues. The blood picks up oxygen in the lungs and picks up various nutrients in the gut and then carries these to all other areas of the body. On passing through the tissues the blood picks up carbon dioxide, urea, and other excreta from the cells and transports these to the lungs and kidneys to be removed from the body. Thus, the circulatory system provides long-distance transport of the extracellular fluids, in this way keeping the fluids in the different parts of the body mixed with each other.

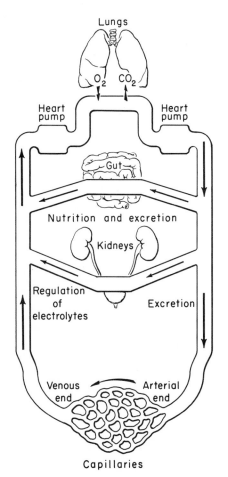

Figure 39. General schema of fluid flow in the circulation.

The phenomenon of diffusion will be described in more detail in a later discussion of transport of substances through the cell membrane. However, the basic principle of diffusion is simply that all molecules in the fluid, including both molecules of water and those of dissolved substances, are continually moving among each other. This motion allows continual mixing of all substances within the blood and also in the fluids of the tissue spaces, the so-called *interstitial fluid*. Furthermore, since there are large numbers of small openings in the capillaries called *pores*, the molecules are continually moving both out of the capillaries into the interstitial fluids and then back again through the pores into the blood. And because the blood is continually moving through all parts of the body, one can readily see that all the extracellular fluids of the body become mixed with each other. Indeed, this process is so effective that almost every portion of the extracellular fluid, even that in the minutest tissue space, becomes mixed with all the other extracellular fluid of the body at least once every 10 to 30 minutes. Because of this rapidity of mixing, the concentrations of substances in the extracellular fluid of one portion of the body are rarely more than a few per cent different from those anywhere else. It is in this way that nutrients from the gut, hormones from the endocrine glands, and oxygen from the lungs are transported to the cells, and it is in this way that the excreta from the cells are carried either to the lungs or kidneys to be removed from the body.

Transport of Substances Through the Cell Membrane

Earlier in the chapter we saw that the dissolved constituents of the intracellular fluid are quite different from those in the extracellular fluid. One of the major reasons for this is selectivity of the cell membrane for transport of substances between the two fluids. A few substances move through the membrane by the simple process of diffusion, while others are "carried" through the membrane by a special mechanism called "active transport." Active transport is unlike the passive process of diffusion in that it can transport substances from a fluid of low concentration on one side of the membrane to one of high concentration on the other side.

In other words, the process of active transport can actually impart energy to cause movement of substances through the membrane, forcing this movement in one particular direction regardless of the concentrations on the two sides of the membrane. This in turn allows development of different concentrations on one side of the membrane with respect to the other, thereby creating appropriate extracellular and intracellular fluid environments for cellular function.

DIFFUSION

All molecules and ions in the body fluids, including both water molecules and dissolved substances, are in constant motion, each particle moving its own separate way. Motion of these particles is what physicists call heat—the greater the motion, the higher is the temperature—and motion never ceases under any conditions except absolute zero temperature. When a moving particle, A, bounces against a stationary particle, B, its electrostatic forces repel particle B, momentarily adding some of its energy of motion to particle B. Consequently, particle B gains kinetic energy of motion while particle A slows down, losing some of its kinetic energy. Thus, as shown in Figure 40, a single molecule in solution bounces among the other molecules first in one direction, then another, then another, and so forth, bouncing hundreds to millions of times each second. At times it travels a far distance before striking the next molecule, but at other times only a short distance.

This continual movement of molecules

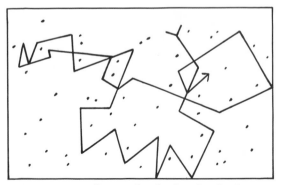

Figure 40. Diffusion of a fluid molecule during a fraction of a second.

among each other in liquids, or in gases, is called *diffusion*. Ions diffuse in exactly the same manner as whole molecules, and even suspended colloid particles diffuse in a similar manner, except that because of their very large sizes they diffuse far less rapidly than molecular substances.

Kinetics of Diffusion—Effect of Concentration Difference

When a large amount of dissolved substance is placed in a solvent at one end of a chamber, it immediately begins to diffuse toward the opposite end of the chamber. If the same amount of substance is placed in the opposite end of the chamber, it begins to diffuse toward the first end, the same amount diffusing in each direction. As a result, the *net rate of diffusion* from one end to the other is zero. If, however, the concentration of the substance is greater at one end of the chamber than at the other end, the net rate of diffusion from the area of high concentration to low concentration is directly proportional to the larger concentration minus the lower concentration. This is called the *concentration difference.*

The rapidity with which a molecule diffuses from one point to another is less the greater the molecular size, because large particles are not so easily impelled by collisions with other molecules. In gases, the rate of diffusion is approximately inversely proportional to the square root of the molecular weight, but in fluids it is more nearly inversely proportional to the molecular weight itself.

If we consider all the different factors that affect the rate of diffusion of a substance from one area to another, they are the following: (1) The greater the concentration difference, the greater is the rate of diffusion. (2) The less the molecular weight, the greater is the rate of diffusion. (3) The shorter the distance, the greater is the rate. (4) The greater the cross-section of the chamber in which diffusion is taking place, the greater is the rate of diffusion. (5) The greater the temperature, the greater is the molecular motion and also the greater is the rate of diffusion. All these can be placed in the following approximate formula for diffusion in solutions:

$$\text{Diffusion rate} \propto \frac{\text{Concentration difference} \times \text{Cross-sectional area} \times \text{Temperature}}{\text{Molecular weight} \times \text{Distance}}$$

Diffusion Through the Cell Membrane

The cell membrane is essentially a sheet of lipid material, called the *lipid matrix*, covered on each surface by a layer of protein, the detailed structure of which was discussed in Chapter 2 and shown in the diagram of Figure 12. The fluids on each side of the membrane penetrate the protein portion of the membrane with ease so that any dissolved substances can diffuse into this portion of the cell membrane without any impedient. However, the lipid portion of the membrane is an entirely different type of medium, acting as a limiting membrane between the extracellular and intracellular fluids.

The two different methods by which the substances can diffuse through the membrane are: (a) by becoming dissolved in the lipid and diffusing through it in the same way that diffusion occurs in water, and (b) by diffusing through minute pores which pass directly through the membrane at wide intervals over its surface.

DIFFUSION IN THE DISSOLVED STATE THROUGH THE LIPID PORTION OF THE MEMBRANE. A few substances are soluble in the lipid of the cell membrane as well as in water. These include oxygen, carbon dioxide, alcohol, and a few other less important ones. When one of these substances comes in contact with the membrane it immediately becomes dissolved in the lipid and continues to diffuse in exactly the same manner that it diffuses within the watery medium on either side of the membrane. The random motion of the molecule may take it on through the membrane or it may take it back out of the membrane on the side from which it came. In other words, the molecule continues its random motion within the substance of the membrane in exactly the same way that it undergoes random motion in the surrounding fluids.

Effect of lipid solubility on diffusion through the lipid matrix. The primary factor that determines how rapidly a substance can diffuse through the lipid matrix of a cell membrane is its solubility in lipids. If it is very soluble, it becomes dissolved in the membrane very easily and therefore passes on through. On the other hand, a substance that dissolves very poorly in lipids will be greatly retarded. For instance, oxygen, which is very soluble, passes through the lipid matrix with ease, as shown in Figure 41, whereas water, which is almost completely insoluble in lipids, passes through the lipid matrix almost not at all.

Facilitated diffusion. Some substances are very insoluble in lipids and yet can still pass through the lipid matrix by a process called facilitated diffusion. This is the means by which the different sugars in particular cross the membrane. The most important of these sugars is glucose, the membrane transport of which is illustrated in the lower part of Figure 41. This shows that glucose Gl combines with a carrier substance C at point 1 to form the compound CGl. This combination is soluble in the lipid so that it can diffuse to the other side of the membrane, where the glucose breaks away from the carrier (point 2) and passes to the inside of the cell. The carrier then diffuses back to the outside surface of the membrane to pick up still more glucose and transport it also to the inside. Thus, the effect of the carrier is to make the glucose soluble in the membrane; without it glucose cannot pass through the membrane.

Specific enzymes are believed to catalyze these chemical reactions; and, in some in-

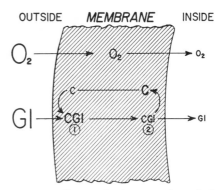

Figure 41. Diffusion of substances through the lipid matrix of the membrane. The upper part of the figure shows *free diffusion* of oxygen through the membrane and the lower part shows *facilitated diffusion* of glucose.

stances of facilitated diffusion, extra energy is required to cause the reactions to take place, though in the case of glucose transport the process takes place without any requirement for extra energy.

The rate at which a substance passes through a membrane by facilitated diffusion depends on the difference in concentration of the substance on the two sides of the membrane, the amount of carrier available, and the rapidity with which the chemical reactions can take place. In the case of glucose transport, the overall rate is greatly increased by *insulin,* which is the primary hormone secreted by the pancreas. Large quantities of insulin can increase the rate of glucose transport about seven- to ten-fold, though it is not known whether this is caused by an effect of insulin to increase the quantity of carrier in the membrane or to increase the rate at which the chemical reactions take place between glucose and the carrier.

DIFFUSION THROUGH THE MEMBRANE PORES. Spread very far apart over the surface of the cell are minute pores approximately 8 Ångstroms in diameter. These are too small to be seen even by the electron microscope but nevertheless large enough to allow free passage of water molecules and very small water-soluble ions and molecules. The total area of all the pores of an average cell is only 1/1600 the total surface area of the cell. Yet, despite this, molecules and ions diffuse through the pores so rapidly that the entire volume of fluid in a cell can easily pass through the pores within a few hundredths of a second, which emphasizes the extreme rapidity of motion of the molecules and ions of the body fluids.

Figure 42 illustrates a postulated structure of a pore, indicating that its surface is probably lined with positively charged proteins. This figure shows several small particles passing through the pore and also shows that the maximum size of the particle that can pass through is approximately equal to the size of the pore itself.

Effect of pore size on diffusion through the pore. Table 2 gives the effective diameters of various substances in comparison with the diameter of the pore. Note that some substances, such as the water molecule, urea molecule, and chloride ion, are considerably smaller in size than the pore. All these pass through the pore with great ease. For instance, the rate per second of diffusion of water in each direction through the pores

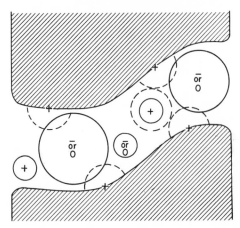

Figure 42. Postulated nature of the cellular pore, showing sphere of influence exerted by positive charges along the surface of the pore. (Modified from Solomon: *Sci. Amer.* Dec. 1960, p. 146.)

of a red blood cell is about one hundred times as great as the volume of the cell itself. It is fortunate that an identical amount of water normally diffuses in each direction, which keeps the cell from either swelling or shrinking despite the rapid rate of diffusion. The rates of diffusion of urea and chloride ions through the membrane are somewhat less than that of water, which is in keeping with the fact that their effective diameters are slightly greater than that of water.

Table 2 also shows that most of the sugars,

TABLE 2. *Relationship of Effective Diameters of Different Substances to Pore Diameter*

SUBSTANCE	DIAMETER Å	RATIO TO PORE DIAMETER	APPROXIMATE RELATIVE DIFFUSION RATE
Water molecule	3	0.38	50,000,000
Urea molecule	3.6	0.45	40,000,000
Hydrated chloride ion	3.86	0.48	36,000,000
Hydrated potassium ion	3.96	0.49	100
Hydrated sodium ion	5.12	0.64	1
Lactate ion	5.2	0.65	?
Glycerol molecule	6.2	0.77	?
Ribose molecule	7.4	0.93	?
Pore size	8 (Ave.)	1.00	—
Galactose	8.4	1.03	?
Glucose	8.6	1.04	0.4
Mannitol	8.6	1.04	?
Sucrose	10.4	1.30	?
Lactose	10.8	1.35	?

including glucose, have effective diameters that are slightly greater than that of the pores. Obviously, not even a single molecule as large as these could go through a pore that is smaller than its size. For this reason essentially none of the sugars can pass through the pores; instead most of these pass through the lipid matrix only by the process of facilitated diffusion.

Effects of electrical charge on transport of ions through the membrane. Negatively charged ions, such as chloride and lactate ions, pass through the pores with the same ease as noncharged water molecules, urea molecules, and others of equivalent size. Therefore, passage of negatively charged ions through the pores is limited only by their diameters in the same way that noncharged molecules are also limited. However, almost all ions have one or more molecules of water attached to them, held in close apposition because of the ion's electrical charge. Therefore, it is this hydrated form of the ion that determines its *effective diameter* and therefore the ease with which it will pass through the pores.

In contrast to negative ions, positively charged ions pass through the membrane with extreme difficulty. For instance, the chloride ion, which is negatively charged, passes through the membrane approximately one million times as easily as the potassium ion, which is positively charged, despite the fact that their hydrated ionic diameters are almost exactly the same, as shown in Table 2. The reason for this difference is believed to be the positive charge of the proteins or adsorbed positive ions, such as calcium ions, lining the pores. Figure 42 illustrates that each positive charge causes a sphere of electrostatic space charge to protrude into the lumen of the pore. Likewise, a positive ion attempting to pass through a pore also exerts a sphere of positive electrostatic charge, and the two positive charges repel each other. This repulsion, therefore, blocks or greatly impedes movement of the positive ion through the pore.

In addition to electrical charge, the effective diameter of a positive ion also determines how easily it can pass through the membrane. For instance, the hydrated diameter of the sodium ion is approximately 30 per cent greater than that of the potassium ion and because of this slight difference the sodium ion diffuses through the pores only $\frac{1}{100}$ as easily as the potassium ion.

Rate of Diffusion Through the Cell Membrane and Factors That Affect it

From the preceding discussion, it is evident that many different substances can diffuse either through the lipid matrix of the cell membrane or through the pores. It should be noted, however, that substances that diffuse in one direction can also diffuse in the opposite direction. It is not the total quantity of substances diffusing in both directions through the cell membrane that is important to the cell but instead the *net quantity* diffusing either into or out of the cell.

Let us assume, for instance, that 1000 cubic microns of water diffuses into a cell each second but exactly the same quantity diffuses to the outside. In this case there is zero net diffusion. On the other hand, if 1000 cubic microns diffuses inward in the same time that 999 cubic microns diffuses outward, there is net diffusion of 1 cubic micron of water to the interior of the cell, and the cell will swell by this amount. The same principle applies to the transport of other substances, such as oxygen and sodium and potassium ions. Therefore, in addition to the permeability of the membrane, which has already been discussed, three other factors determine the rate of net diffusion of a substance: the concentration difference of the substance across the membrane, the electrical difference across the membrane, and the pressure difference across the membrane.

EFFECT OF THE CONCENTRATION DIFFERENCE. Figure 43 illustrates a membrane with a substance in high concentration on the outside and low concentration on the inside. The rate at which the substance diffuses *inward* is proportional to the concentration of molecules on the outside, for this concentration determines how many of the molecules strike the outside of the

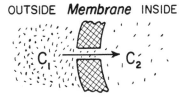

OUTSIDE *Membrane* INSIDE

Figure 43. Effect of concentration gradient on diffusion of molecules and ions through a cell membrane pore.

pore each second. Likewise, the rate at which the molecules diffuse *outward* is proportional to their concentration inside the membrane. Obviously, therefore, the rate of net diffusion is proportional to the concentration on the outside *minus* the concentration on the inside, or *proportional to the concentration difference.*

Net Diffusion of Water Across Cell Membranes – Osmosis

By far the most abundant substance to diffuse through the cell membrane is water. It should be recalled again that enough water ordinarily diffuses in each direction through the membrane of the red blood cell per second to equal about *100 times the volume of the cell itself.* Yet, *normally,* the amount that diffuses in the two directions is so precisely balanced that not even the slightest *net* diffusion of water occurs. Therefore, the volume of the cell remains constant. However, under certain conditions, a *concentration difference for water* can develop across a membrane, just as concentration differences for other substances can also occur. When this happens, net diffusion of water does occur across the cell membrane, causing the cell either to swell or to shrink, depending on the direction of the net diffusion. This process of net diffusion of water caused by a water concentration difference is called *osmosis.*

To give an example of osmosis, let us assume that we have the conditions shown in Figure 44, with pure water on one side of the cell membrane and a solution of sodium chloride on the other side. Referring back to Table 2, we see that water molecules pass through the cell membrane with extreme ease while sodium ions pass through only with extreme difficulty. And chloride ions cannot pass through the membrane because the positive charge of the sodium ions holds the negatively charged chloride ions back. Therefore, sodium chloride solution is actually a mixture of diffusible water molecules and nondiffusible sodium and chloride ions. Yet, the presence of the sodium chloride has displaced many of the water molecules so that the concentration of the water molecules in the sodium chloride solution is considerably less than that in pure water. As a result, in the example of Figure 44, more water molecules strike the

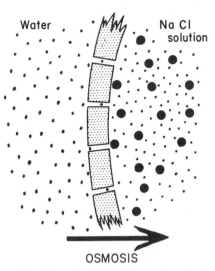

Figure 44. Osmosis occurring at a hypothetical cellular membrane when a sodium chloride solution is placed on one side of the membrane and water on the other side.

pores on the left side where there is pure water than on the right side where the water concentration has been reduced. Thus, osmosis occurs from left to right, from the pure water into the sodium chloride solution.

OSMOTIC PRESSURE. If in Figure 44 pressure were applied to the sodium chloride solution, osmosis of water into this solution could be slowed or even stopped. The amount of pressure required to stop osmosis completely is called the osmotic pressure of the sodium chloride solution.

The principle of a pressure difference opposing an osmotic effect is illustrated in Figure 45, which shows a semipermeable membrane separating two separate columns of fluid, one containing water and the other containing a solution of water and some solute that will not penetrate the membrane. Osmosis of water from chamber B into chamber A causes the levels of the fluid columns to become farther and farther apart, until eventually a pressure gradient is developed that is great enough to oppose the osmotic effect. The pressure gradient across the membrane at this time is the osmotic pressure of the solution containing the nondiffusible solute.

Osmotic pressure of the extracellular and intracellular fluids. The electrolytes dissolved in the extracellular and intracellular fluids, as well as other substances such as glucose, amino acids, free fatty acids, and so forth, can cause osmotic pressure at the

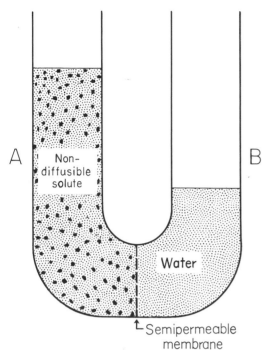

Figure 45. Demonstration of osmotic pressure on the two sides of the semipermeable membrane.

cell membrane. The overall concentration of substances in the extracellular and intracellular fluids is enough to create a total osmotic pressure of approximately 5400 mm. Hg; that is, if clear water should be placed on one side of the cell membrane while either extracellular or intracellular fluid should be placed on the other side, it would require a pressure of 5400 mm. Hg to stop osmosis of water into the body fluid. This amount of pressure is equal to about 7 atmospheres of pressure or approximately equal to the pressure exerted by a column of water 230 feet high. Therefore, one can understand the extreme force developed for movement of water through the pores of cell membranes by the phenomenon of osmosis.

Lack of effect of molecular and ionic mass on osmotic pressure. The osmotic pressure exerted by nondiffusible particles in a solution, whether they be molecules or ions, is determined by the *numbers* of particles per unit volume of fluid and not the mass of the particles. The reason for this is that each particle in a solution, regardless of its mass, exerts, on the average, the same amount of pressure against the membrane.

ISOTONICITY, HYPERTONICITY, AND HYPO-TONICITY OF SOLUTIONS. A solution that,

when placed on the outside of cells, causes no osmosis through the cell membrane in either direction is said to be *isotonic* with the body fluids. For instance, a 0.9 per cent solution of sodium chloride is isotonic. On the other hand, a solution that causes osmosis of fluid out of the cell into the solution is said to be *hypertonic*. Thus, a sodium chloride solution having a concentration greater than 0.9 per cent is hypertonic. Finally, a solution that causes osmosis into cells is *hypotonic*, for instance, a sodium chloride solution of less than 0.9 per cent concentration. In other words, cells placed in an isotonic solution will maintain their volumes at a constant level, cells placed in a hypertonic solution will shrink, and cells placed in a hypotonic solution will swell.

Osmotic Equilibrium of Extracellular and Intracellular Fluids

Except for very short periods of time, usually measured in seconds, the intracellular and extracellular fluids remain in constant osmotic equilibrium; that is, the concentrations of nondiffusible substances on the two sides of the membranes remain almost exactly equal. The reason for this is that, if ever the concentrations become unequal, osmosis of water through the cell membranes occurs so rapidly that equilibrium will be reestablished within a few seconds. This effect is illustrated in Figure 46. Figure 46A shows a cell suddenly placed in a very dilute solution (*hypotonic solution*). Within a few seconds, water passes by osmosis through the cell membrane to the inside of the cell, causing: (1) increase in the intracellular fluid volume, and thus swelling of the cell, (2) decrease in the extracellular fluid volume, (3) dilution of the dissolved substances in the intracellular fluid, and (4) concentration of the dissolved substances in the extracellular fluid. Once these two fluids have reached the same concentrations of osmotically active substances, osmosis through the membrane ceases. Thus, within a few seconds, a new state of osmotic equilibrium has been reestablished, as shown in Figure 46B.

Figure 46C shows exactly the opposite situation, in which a cell having a dilute intracellular fluid is placed in a concentrated extracellular fluid (*hypertonic solu-*

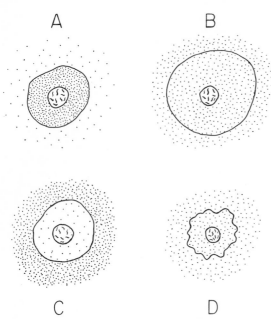

A   B

C   D

Figure 46. (A) A cell suddenly placed in a hypotonic solution. (B) The same cell after osmotic equilibrium has been established, showing swelling of the cell and equilibration of the concentration of extra- and intracellular fluids. (C) A cell placed in a hypertonic solution. (D) The same cell after osmotic equilibrium has been established, showing shrinkage of the cell and equilibration of fluid concentrations.

tion). Water passes by osmosis out of the cell, decreasing the intracellular fluid volume and increasing the extracellular fluid volume. This also concentrates the intracellular fluids while diluting the extracellular fluids. Therefore, within a few seconds, the concentrations of the two fluids reach equilibrium, but in the meantime the cell has decreased in size as shown in Figure 46D. To illustrate how rapidly osmosis can take place at the cell membrane, a red blood cell placed in pure water gains an amount of water equal approximately to its own volume in 2 to 3 seconds. Therefore, one would expect almost complete osmotic equilibrium to take place within 15 to 20 seconds.

ACTIVE TRANSPORT

Often only a minute concentration of a substance is present in the extracellular fluid, and yet a large concentration of the substance is required in the intracellular fluid. For instance, this is true of potassium ions. Conversely, other substances fre-

quently enter cells and must be removed even though the concentration inside is far less than that outside. This is true of sodium ions.

From the discussion thus far it is evident that no substances can diffuse *against a concentration difference,* or, as is often said, "uphill." To cause movement of substances uphill, energy must be imparted to the substance. This is analogous to the compression of air by a pump. After compression, the concentration of the air molecules is far greater than before compression, but to create this greater concentration large amounts of energy must be expended by the piston of the pump as it compresses the air molecules. Likewise, as molecules are transported through a cell membrane from a dilute solution to a concentrated solution, large amounts of energy must be expended. The process of moving molecules uphill against a concentration difference (or uphill against an electrical or pressure difference) is called active transport. Among the different substances that are actively transported through the cell membrane in at least some parts of the body are sodium ions, potassium ions, calcium ions, iron ions, hydrogen ions, iodide ions, ureate ions, several different sugars, and the amino acids.

Basic Mechanism of Active Transport

The mechanism of active transport is believed to be similar for all substances and to be dependent on transport by *carriers.* Figure 47 illustrates the basic mechanism, showing a substance S entering the outside surface of the membrane where it combines

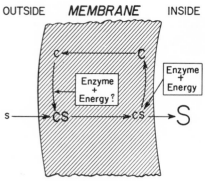

Figure 47. Basic mechanism of active transport.

with carrier C. At the inside surface of the membrane, S separates from the carrier and is released to the inside of the cell. C then moves back to the outside to pick up more S.

One will immediately recognize the similarity between this mechanism of active transport and that of facilitated diffusion discussed earlier in the chapter. The difference, however, is that energy is imparted to the system in the course of transport so that transport can occur *against a concentration difference.*

Unfortunately, little is known about the mechanisms by which energy is utilized in the transport mechanism. The energy could cause combination of the carrier with the substance, or it could cause splitting of the substance away from the carrier. Regardless of which of these is true, the energy is probably supplied to the transport system at the inside surface of the cell membrane, because it is inside the cell that large quantities of the energy-giving substance ATP are available for promoting chemical reactions.

It is also evident from Figure 47 that the chemical reactions are believed to be promoted by specific enzymes which catalyze the chemical reactions.

CHEMICAL NATURE OF THE CARRIER. Unfortunately, the chemical nature of the carrier is not known for any of the transport mechanisms. However, it is believed that the carrier is usually a small molecular weight derivative of a protein, such as a small polypeptide or some similar compound. The reason for believing this is that tremendous quantities of substances are often transported actively through cell membranes in very short periods of time, and only a small molecular weight carrier would have the ability to diffuse back and forth through the membrane rapidly enough to account for such rates of transport.

It is presumed that the carrier has reversibly reactive sites capable of combining easily with the substance to be transported and capable also of releasing the substance at the opposite surface of the membrane.

SPECIFICITY OF CARRIER SYSTEMS. Several different carrier systems exist in cell membranes, each of which transports only certain specific substances. One carrier system, for instance, transports sodium to the outside of the membrane and probably transports potassium to the inside at the same time. Another system actively transports sugars through the membranes of certain cells (intestinal and tubular epithelial cells), whereas still other specific carrier systems transport different ones of the amino acids.

The specificity of active transport systems for substances is determined either by the chemical nature of the carrier, which allows it to combine only with certain substances, or by the nature of the enzymes that catalyze the specific chemical reactions. Unfortunately, almost nothing is known about the precise function of either of these two components of the carrier systems.

ENERGETICS OF ACTIVE TRANSPORT. The amount of energy required to transport a substance actively through a membrane is determined by the degree that the substance is concentrated during transport. For instance, the amount of energy required to concentrate only one gram of sodium 10-fold is 61 calories. One can see, therefore, that the energy expenditure for concentrating substances in cells or for removing substances from cells against a concentration difference can be tremendous. Some cells, such as those lining the renal tubules, expend as much as 30 per cent of their energy for this purpose alone.

Active Transport of Sodium, Potassium, and Other Electrolytes

Referring back to Figure 38, one sees that the sodium concentration outside the cell is very high in comparison with its concentration inside, and the converse is true of potassium. And Table 2 shows that minute quantities of sodium and potassium can diffuse through the pores of the cell. If such diffusion should take place over a long period of time, the concentrations of the two ions would eventually become equal inside and outside the cell unless there were some means to remove the sodium from the inside and to transport potassium back in.

Fortunately, a system for active transport of sodium and potassium is present in all cells of the body. The mechanism is believed to be that illustrated in Figure 48, which shows sodium (Na) inside the cell combining with carrier Y at the membrane surface to form large quantities of the combination NaY. This then diffuses to the outer surface where sodium is released and the carrier Y changes its chemical composition slightly

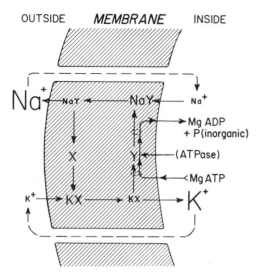

Figure 48. Postulated mechanism for active transport of sodium and potassium through the cell membrane, showing coupling of the two transport mechanisms and delivery of energy to the system at the inner surface of the membrane.

to become carrier X. This carrier then combines with potassium K to form KX, which diffuses to the inner surface of the membrane where energy is provided to split K from X under the influence of the enzyme ATPase, the energy being derived from MgATP.

The transport mechanism in Figure 48 is believed to be more effective in transporting sodium than in transporting potassium, usually transporting about three sodium ions for every two potassium ions.

The sodium transport mechanism is so important to many different functioning systems of the body — such as to nerve and muscle fibers for transmission of impulses, various glands for the secretion of different substances, and all cells of the body to prevent cellular swelling — that it is frequently called the *sodium pump*. We will discuss the sodium pump at many places in this text.

OTHER ELECTROLYTES. Calcium and magnesium are probably transported by all cell membranes in much the same manner that sodium and potassium are transported, and certain cells of the body have the ability to transport still other ions. For instance, the glandular cell membranes of the thyroid gland can transport large quantities of iodide ion; the epithelial cells of the intestine can transport sodium, calcium, iron, bicarbonate, and probably many other ions into the interstitial fluid; and the epithelial cells of the renal tubules can transport

hydrogen, calcium, sodium, potassium, and a number of other ions.

Active Transport of Sugars

Glucose and certain other sugars are transported through the cell membrane by the process of facilitated diffusion in essentially all cells of the body, but *active transport* of sugars occurs in only a few places. For instance, glucose and several other sugars are continually transported into the blood from the intestines and renal tubules, and this occurs even when their concentrations are much smaller in the intestines and renal tubules than in the blood. Therefore essentially none of these sugars is lost either in the intestinal excreta or the urine. Almost all monosaccharides that are important to the body are actively transported, including *glucose, galactose, fructose, mannose, xylose, arabinose,* and *sorbose.* On the other hand, the disaccharides such as sucrose, lactose, and maltose are not actively transported at all.

As is true of essentially all other active transport mechanisms, the precise carrier system and chemical reactions responsible for transport of the monosaccharides are yet unknown. The one common denominator in transport of sugars seems to be the necessity for an intact —OH group on the C_2 carbon of the monosaccharide molecule. It is presumed that the monosaccharide attaches to the carrier at this point.

Active Transport of Amino Acids

Amino acids, the basic building blocks of proteins, are actively transported through the membranes of all cells that have been studied. Active transport through the epithelium of the intestine, the renal tubules, and many glands is especially important. Four different carrier systems for transmitting different groups of amino acids and amines are believed to exist. However, these carrier systems are equally as poorly understood as the systems for transporting other substances. One of the few known features of the amino acid carrier systems is that transport of at least some amino acids depends on pyridoxine (vitamin B_6). As a result of this dependence, deficiency of this vitamin causes protein deficiency.

HORMONAL REGULATION OF AMINO ACID TRANSPORT. At least four different hormones are important in controlling amino acid transport: (1) *Growth hormone*, secreted by the adenohypophysis, increases the transport of some amino acids into essentially all cells. (2) *Insulin* and (3) *glucocorticoids* increase amino acid transport at least into liver cells and possibly into other cells as well. (4) *Estradiol*, the most important of the female sex hormones, causes rapid transport of amino acids into the musculature of the uterus, thereby promoting development of this organ.

Thus, several of the hormones exert much, if not most, of their control effects in the body by controlling active transport of amino acids into all or certain cells.

Pinocytosis

It was pointed out in Chapter 3 that the cell membrane has the ability to imbibe small amounts of substances from the extracellular fluid by the process called *pinocytosis*. This occurs particularly when large quantities of protein or excessive amounts of salt are present in the surrounding fluid. The process seems to be initiated by adsorption of protein or salt to the outside of the membrane, which causes the membrane first to invaginate and then to pinch off inside the cell to form a *pinocytic vesicle*. The pinocytic vesicle either dissolutes inside the cell and discharges its contents into the intracellular space, or it combines with one or more lysosomes whose digestive enzymes digest the substances in the pinocytic vesicle before distributing them intracellularly.

A special importance of pinocytosis to the cell is that this is the only means by which substances of very large molecular weight can be transported into the cell. For instance, whole protein molecules, large hormone molecules, and many others can be transported intact into the cell where they can then perform specific functions without having to be synthesized in each cell.

Another important feature of pinocytosis is that it can concentrate substances during transport to the inside of the cell. For instance, the concentration of transported proteins is sometimes much higher inside the cell than outside. This results from preliminary adsorption of the protein to the cell membrane, thus concentrating the protein on the cell membrane before the pinocytic vesicle forms.

In lower forms of unicellular life, pinocytosis provides many if not most of the nutrients for the cell, but in higher animal life most nutrients are provided to the cells in the form of very small molecular weight substances such as glucose, amino acids, and so forth. These can be transported either by diffusion or active transport, for which reason transport by pinocytosis plays a far less important role in nutrition of the human cell than is true for most unicellular animals.

REFERENCES

Adolph, E. F.: *Physiology of Man in the Desert.* New York, Interscience Publishers, 1947.

Gamble, J. L.: *Chemical Anatomy, Physiology and Pathology of Extracellular Fluid: A Lecture Syllabus.* 6th Ed. Cambridge, Harvard University Press, 1954.

Heinz, E.: Transport through biological membranes. *Ann. Rev. Physiol.* 29:21, 1967.

Potts, W. T. W.: Osmotic and ionic regulation. *Ann. Rev. Physiol.* 30:73, 1968.

Rothstein, A.: Membrane phenomena. *Ann. Rev. Physiol.* 30:15, 1968.

Rustad, R. C.: Pinocytosis. *Sci. Amer.* 204(4):120, 1961.

Skou, J. C.: Enzymatic basis for active transport of Na+ and K+ across cell membrane. *Physiol. Rev.* 45(3):596, 1965.

Solomon, A. K.: Pores in the cell membrane. *Sci. Amer.* 203(6):146, 1960.

Solomon, A. K.: Pumps in the living cell. *Sci. Amer.* 203(2):100, 1962.

CHAPTER 6

THE NERVE, AND MEMBRANE POTENTIALS

The function of a nerve fiber is to conduct signals from one part of the body to another. Nerve fibers are long thread-like extensions of nerve cells, called *neurons,* whose cell bodies are located mainly in the brain and spinal cord, though a few are in ganglia outside the central nervous system.

The purpose of the present chapter is to explain the mechanism by which signals are transmitted over nerve fibers, but to do this it is necessary to explain first how electrical potentials are generated at the nerve fiber membrane. We shall see later that the mechanism for transmission of impulses along nerve fibers also applies to transmission of impulses in muscle fibers and over the surfaces of nerve cell bodies in the central nervous system.

Physiologic Anatomy of the Nerve Fiber

Figure 49A illustrates a neuron with its long fiber extension called an *axon,* and Figure 49B shows a cross-section of the axon. The function of the cell body will be discussed in Chapter 22. The present chapter is concerned with the conduction of im-

pulses along the axon. Referring again to Figure 49B, one sees that the axon contains in its center a gelled substance called *axo-*

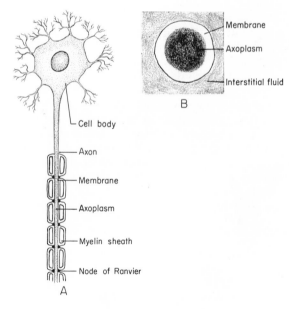

Figure 49. (A) A nerve cell with its thread-like axon, showing the nerve membrane, the axoplasm, and the myelin sheath. (B) Cross-section of an axon.

plasm, and this is surrounded by a *membrane* that separates the axoplasm from the interstitial fluid. This membrane has the same functions as any other cellular membrane, except that it is specifically adapted for transmission of electrochemical impulses.

As illustrated in Figure 49A, some nerve fibers are surrounded on their outsides by an additional layer, the *myelin sheath,* which is discontinuous at the *nodes of Ranvier.* The myelin sheath is an insulator, the function of which is explained later in the chapter.

MEMBRANE POTENTIALS

The first essential for conduction of a nerve impulse is the establishment of an electric potential across the axon membrane. This potential is called a *membrane potential.* It results from concentration differences of certain ions between the two sides of the membrane.

DEVELOPMENT OF CONCENTRATION DIFFERENCES ACROSS THE NERVE MEMBRANE. The membrane of the axon, like other cellular membranes, actively transports some ions from the interstitial fluid to the inside of the fiber, and others from inside back into the interstitial fluid. For nerve conduction, the most important of these transfer processes is the active transport of sodium out of the fiber into the interstitial fluid, which is often referred to as the *sodium pump.* This pump, by making the quantity of sodium inside the fiber very little, causes a membrane potential to develop as follows:

In Figure 50, the concentration of sodium inside the nerve fiber is shown to be approximately 10 milliequivalents, in contrast with its concentration of 137 milliequivalents in the interstitial fluid. On the other hand, the potassium concentration inside the nerve fiber is approximately 35 times its concentration in the interstitial fluid. The high concentration of potassium on the inside of the fiber is mainly a secondary effect of the sodium pump as follows: When sodium ions, which are positive ions, are pumped to the outside a great void of positive ions is left inside the membrane. As a result, the inside becomes strongly electronegative, and, since potassium ions, which are also positive ions, are freely diffusible through the membrane, the electronegativity pulls a large quantity of these ions to the inside, thus mainly filling the deficit left by the expulsed sodium ions.

The electronegative potential created inside the fiber as a result of the sodium pump is the *membrane potential.* In Figure 51A the membrane potential is illustrated by a series of positive and negative charges, positivity outside the membrane and negativity inside. This is the normal resting state of the nerve fiber, and the actual voltage of the membrane potential normally is about 0.085 volt or 85 millivolts.

Depolarization of the Membrane, and the Nerve Impulse

Figure 51B illustrates at its central-most point an area that has suddenly become so permeable that even sodium ions can now diffuse through the membrane with ease. As a result, the large quantity of positively charged sodium ions outside the fiber membrane rushes to the inside, causing the membrane to become suddenly positive inside and negative outside. This is opposite to the usual resting state of the membrane, and it is called *depolarization* because the normal polarized state, with positivity on the outside and negativity on the inside, no longer exists.

THE DEPOLARIZATION WAVE OR NERVE IMPULSE. In Figures 51C and D, the area of depolarization in the middle of the fiber has extended in both directions, and the area of increased permeability has also extended. The cause of this extension is the flow of electrical current from the original depolarized area to adjacent areas. This passage of electrical current, for reasons not yet completely understood, makes the adjacent areas permeable also to sodium. The sodium ions flow through the membrane at

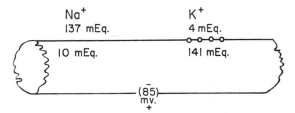

Figure 50. Concentration gradients of sodium and potassium at the axon membrane, showing that the membrane in the resting state is permeable only to potassium.

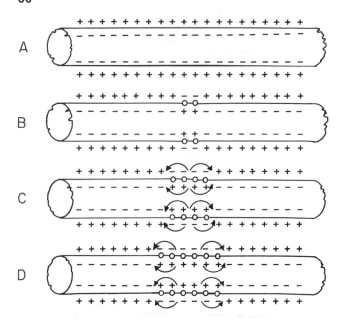

Figure 51. Transmission of the depolarization wave, the initial event in the nerve impulse.

these new areas, depolarizing these areas and causing still more electrical current to spread along the fiber, thereby extending the area of permeability still further. Again more sodium flows through the membrane, more electrical current travels along the fiber, and the cycle repeats itself over and over again. This spread of increased permeability and electrical current along the membrane is called a *depolarization wave* or a *nerve impulse.*

It is obvious from the preceding discussion that once a single point anywhere on the nerve becomes depolarized a nerve impulse travels away from that point in each direction, and each impulse keeps on traveling until it comes to both ends of the fiber. In other words, a nerve fiber can conduct an impulse either toward the nerve cell or away from it.

Immediately after a depolarization wave has traveled along a nerve fiber, the inside of the fiber is positively charged because of the large number of sodium ions that have diffused to the inside. This state of events causes the membrane to become impermeable to sodium ions again, though the exact mechanism of this also is not understood. Yet, potassium can still diffuse through the pores with relative ease, and because of the high concentration of potassium on the inside many potassium ions diffuse outward carrying positive charges with them. This once again creates electronegativity inside the membrane and positivity outside, a process called *repolarization.*

Repolarization usually begins at the same point in the fiber where depolarization had originally begun, and it spreads along the fiber in the manner illustrated in Figure 52. Repolarization occurs a few ten-thousandths of a second after depolarization. That is, the whole cycle of depolarization and repolarization takes place in a very minute fraction of a second, and the fiber is then ready to transmit a new impulse.

Refractory period. When an impulse is traveling along a nerve fiber, the nerve fiber cannot transmit a second impulse until the fiber membrane has been repolarized. Even a strong electrical stimulus causes no effect, for which reason the fiber is said to be in a *refractory* state, and the time that the fiber remains refractory is called the *refractory period.* This varies from about $1/2500$ second for large nerve fibers up to $1/250$ second for small fibers.

Reestablishment of Ionic Concentrations After Nerve Impulses Are Conducted

After the nerve fiber is repolarized the sodium ions that have leaked to the inside of the membrane and the potassium ions that have leaked to the outside must be returned to their original sides of the membrane. This is accomplished by the sodium pump which was discussed previously. That is, this mechanism pumps sodium ions to the outside, creating a deficit of positive charges

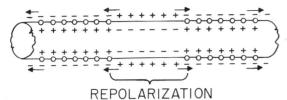

REPOLARIZATION

Figure 52. The repolarization process, the concluding event in the nerve impulse.

inside the membrane. The resulting electronegativity inside the membrane then pulls potassium ions back to the inside. Thus, this process restores the ionic differences to their original levels so that additional nerve impulses can travel along the membrane.

However, it must be emphasized that even when the sodium pump fails to act, a hundred thousand or more impulses can still be transmitted over a single nerve fiber before transmission will cease. The reason for this is that only a few trillionths of a mole of sodium enter the fiber each time a single impulse is transmitted, so that it takes a hundred thousand or more impulses for enough sodium to enter the fiber to cause cessation of impulse transmission. It is evident, then, that the sodium pump is not necessary for repolarization of the membrane after each nerve impulse; this is accomplished by the diffusion of potassium outward through the membrane pores. Instead, the sodium pump simply plods along slowly, reestablishing ionic concentration differences across the membrane whenever a large number of impulses tends to alter these.

Types of Stimuli That Can Excite the Nerve Fiber

Some of the types of stimuli that can initiate a nerve impulse are shown in Figure 53. At point A the fiber membrane is stimulated by a chemical substance such as an acid, a base, or even a strong electrolytic solution. These alter the membrane so that it becomes permeable to sodium, which initiates an impulse in each direction along the fiber. At point B the membrane is excited electrically. A negative charge is applied outside the membrane and a positive charge inside, which is opposite to the normal polarity of the resting membrane potential. When a reverse polarity such as this is applied, the membrane becomes highly permeable, and an impulse begins in each direction. Finally, at point C a needle is shown pricking the membrane. Trauma caused in this manner or by crushing, burning, or otherwise damaging the fiber can increase the permeability of the membrane and result in an impulse.

In the living body, nerve fibers normally are stimulated by both physical and chemical means. For example, pressure on certain of the nerve endings in the skin mechanically stretches these endings, in this way setting off impulses. Heat and cold affect other nerve endings to elicit impulses, and damage to the tissues, such as cutting the skin or stretching the tissues too much, can result in pain impulses.

In the central nervous system, impulses are transmitted from one neuron to another mainly by chemical means. The nerve ending of the first neuron secretes a chemical substance that in turn excites the second neuron. In this way impulses are sometimes passed through many hundred neurons before stopping.

In the laboratory, nerve fibers are usually stimulated electrically. A typical stimulator is illustrated in Figure 54. It emits electrical impulses of any desired voltage either singly or in rapid succession. Usually, a small probe having two wires at its end is placed

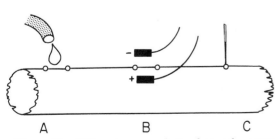

Figure 53. Different means of stimulating the nerve fiber.

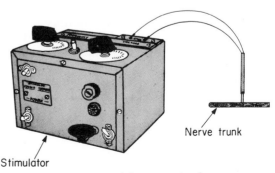

Stimulator

Nerve trunk

Figure 54. A laboratory stimulator.

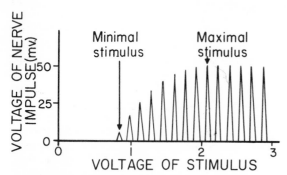

Figure 55.   Minimal and maximal stimuli.

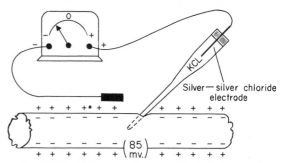

Figure 56.   Measurement of membrane potentials and action potentials of a nerve fiber by means of a microelectrode.

on either side of a nerve trunk, and the electrical stimulus is applied so that electrical current flows through the fibers. As the current passes through the fiber membranes the permeability is altered, eliciting nerve impulses.

THE ALL OR NONE LAW. From the preceding discussion of the nerve impulse it is evident that if a stimulus is strong enough to cause a nerve impulse at all, it causes the entire fiber to fire. In other words, a weak stimulus never causes a weak nerve impulse; either it is strong enough to stimulate the fiber or not. This is called the *all or none law.*

MINIMAL AND MAXIMAL STIMULUS. Some of the fibers in a nerve trunk are more susceptible than others to stimulation. Therefore, as the voltage of an electrical stimulus is progressively increased from zero upward, a voltage will be reached at which a single nerve fiber will be stimulated. This least possible voltage that will cause even a single fiber to fire is called the *minimal stimulus.* For instance, in Figure 55 the minimal stimulus of a particular nerve trunk is shown to be approximately 0.8 volt. As the voltage is increased still further, more and more nerve fibers become stimulated until finally a voltage is reached at which all the fibers in the trunk are stimulated. This is called the *maximal stimulus.* In Figure 55 the maximal stimulus of the particular nerve trunk being studied is 2.1 volts. All stimuli with voltages above the maximal stimulus always cause the same reaction, for all fibers are stimulated in each instance.

ACTION POTENTIALS

The changing electrical voltage at the nerve membrane when an impulse travels

along the fiber is called an *action potential.* The method for recording potentials between the inside and outside of the nerve fiber is illustrated in Figure 56. This figure shows a minute glass pipet having a tip only one micron in diameter and filled with a strong solution of potassium chloride that conducts electricity from the tip. The tip pierces the nerve membrane to make contact with the fluid in the center of the fiber. On the outside of the fiber is another electrode illustrated by the dark rectangle. These two electrodes are connected to a recording meter.

The changing potentials across the membrane are illustrated schematically in Figure 57A. This shows that during the original resting state, a potential of +85 millivolts is recorded outside the fiber with respect to the inside. In Figure 57B a depolarization wave has traveled down the fiber until it is directly at the electrodes. At this instant, the rapid influx of positive sodium ions to the inside of the fiber reverses the potential, causing negativity outside the fiber and posi-

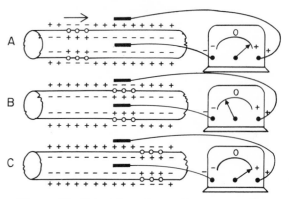

Figure 57. Principles of recording a monophasic action potential.

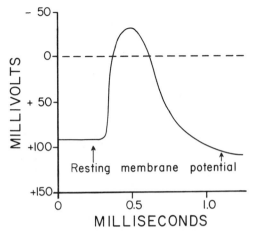

Figure 58. Graphic record of a monophasic action potential recorded from a large nerve fiber.

tivity inside as illustrated by the needle of the meter. Then, soon after the impulse has passed the electrodes, the potential returns again to approximately the original resting potential because of diffusion of positive potassium ions to the outside.

Figure 58 illustrates a continuous recording of the potential changes as a nerve impulse passes the electrodes. This record is called the *monophasic action potential*. It begins with a normal resting membrane potential of +85 millivolts on the outside of the fiber and during the peak of the action potential reverses to become about −35 millivolts. Within approximately one-half millisecond, the normal membrane potential recovers, and the record returns almost to its original level. Actually, it "overshoots"

a little, one or two millivolts, because the sodium pump immediately begins to pump positive sodium ions to the outside, which increases the potential outside slightly above 85 millivolts. This slight overshoot is called the *positive after potential*. It lasts for as long as 50 to 100 milliseconds; during this time all the sodium that has entered the fiber during the action potential is removed from the fiber so that the original resting membrane potential is established again.

USE OF THE OSCILLOSCOPE TO RECORD ACTION POTENTIALS. Mechanical recording apparatus cannot function quickly enough to record the rapidly transient voltages of nerve action potentials. Therefore, a special instrument called the *oscilloscope*, which is similar to a television receiver, is normally used for this purpose. The principal components of the oscilloscope are shown in Figure 59. The basic part of this instrument is a cathode ray tube. An *electron gun* at the base of this tube projects toward the face of the tube a fine beam of electrons having a diameter of about one millimeter. The beam passes between four metal plates, two of which are placed on the two sides of the beam and the other two of which are above and below the beam. On the face of the cathode ray tube is a fluorescent material that glows brightly when the electron beam strikes. The beam in turn can be moved back and forth or up and down across the face of the tube by applying electrical potentials to the horizontal and vertical plates.

An *electronic sweep circuit,* connected to

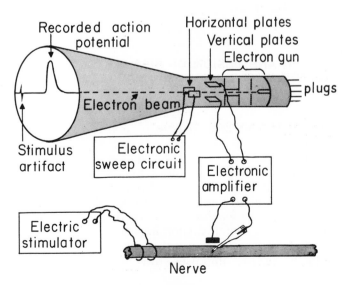

Figure 59. Diagram of an oscilloscope for recording action potentials from nerves.

the horizontal plates, causes the electron beam to move slowly from left to right across the face of the tube. When the beam has moved all the way across it jumps rapidly back to the left side of the face and begins moving across again, retracing the same pathway over and over again. The fluorescence along the path of the moving beam gives the appearance of a horizontal line on the face of the tube.

If, while the beam of electrons is moving across the tube, electrical potentials are applied to the vertical plates, the beam can be made to move up or down. In Figure 59 the beam of electrons deviates from the line slightly at the point called *stimulus artifact,* and then it deviates at the point called *recorded action potential.* The electrical voltages across the nerve membrane are responsible for these deviations. When the nerve is stimulated by an electrical stimulator, some of the electrical current from the stimulator spreads through the fluids of the nerve to the pickup electrodes and causes the stimulus artifact. At the same time, an action potential begins traveling down the nerve fiber toward the two pickup electrodes. Just as soon as the depolarization wave reaches the electrodes, the amplified potential applied to the vertical plates makes the electron beam move upward to record the action potential.

Most action potentials of nerves have a total duration of not more than a few ten-thousandths to a few thousandths of a second. The cathode ray oscilloscope fortunately can record an electrical potential that lasts only a millionth of a second. Therefore, it is quite capable of giving a true record of nerve action potentials.

TRANSMISSION OF SIGNALS OVER NERVE TRUNKS

Types of Fibers in the Nerve Bundle

Figure 60 shows a nerve bundle with fibers of different sizes and of two different types. The large black dots surrounded by white areas are *myelinated fibers;* the white area is a tube of insulator material called *myelin.* This material greatly increases the velocity of the impulses traveling along

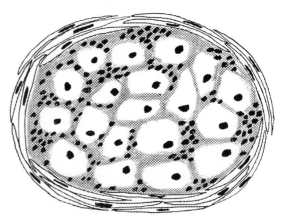

Figure 60. Cross-section of a nerve trunk, showing large, myelinated nerve fibers, small, unmyelinated nerve fibers, and the nerve sheath.

these fibers. In addition to the myelinated fibers are about twice as many small fibers without myelin sheaths. These are called *unmyelinated fibers.*

The large myelinated fibers control the rapid movements of the body and also transmit many of the sensory signals from the body to the brain. The unmyelinated fibers, which conduct impulses quite slowly, cannot cause rapid reactions. These control most of the subconscious activities of the body, such as the excitability of the heart, the contractions of the blood vessels, the gastrointestinal movements, and the emptying of the urinary bladder. Also, the small, unmyelinated fibers transmit those sensory signals that do not require immediate action, such as the aching type of pain sensations, crude touch sensations, and pressure sensations. Further descriptions of the different types of fibers and their functions will be presented in connection with the discussions of sensory functions of the nervous system in Chapter 23 and of motor functions in Chapter 26.

FUNCTION OF THE MYELIN SHEATH. Myelin, which surrounds all large nerve fibers, is a lipid substance that will not conduct electric current. Therefore, it acts as an insulator around the nerve fiber. The myelin sheath is deposited around a nerve fiber by Schwann cells. The Schwann cell wraps itself around and around the fiber, leaving a portion of its own cell membrane each time around; this wrapping of membrane becomes the myelin sheath.

As illustrated schematically in Figure 61, at approximately every millimeter along the

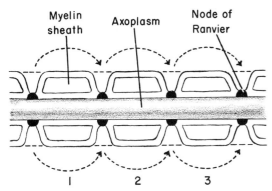

Figure 61. Saltatory conduction in a myelinated axon.

length of the fiber the myelin is broken by a so-called *node of Ranvier*. At these nodes, typical membrane depolarization can occur, but beneath the myelin sheath, membrane depolarization does not take place. Instead, impulses are transmitted along a myelinated nerve by a process called *saltatory conduction* which may be explained as follows: Referring once again to Figure 61, let us assume that the first node of Ranvier becomes depolarized. This causes electrical current to spread as shown by the arrows around the outside of the myelin sheath all the way to the next node of Ranvier, causing it to become depolarized also. Current generated by this node then causes the same effect on the next node; thus, the impulse "jumps" from node to node.

Saltatory conduction is valuable for two reasons: First, by causing the depolarization process to jump long intervals along the nerve fiber, it increases the velocity of conduction along the fiber. Second, and perhaps much more important, saltatory conduction prevents the polarization of large areas of the fiber and thereby prevents leakage of large amounts of sodium to the inside of the fiber. This conserves the energy required by the sodium pump to expel the sodium. Therefore, the myelin sheath greatly decreases the amount of energy required by the nerve for impulse transmission.

VELOCITY OF CONDUCTION IN NERVE FIBERS. The larger the nerve fiber and the thicker the myelin sheath, the more rapidly can the nerve conduct an impulse. The largest nerve fibers are about 20 microns in diameter, and the smallest are about 0.5 micron. The very large fibers conduct impulses at a velocity of 100 meters—about the length of a football field—in 1 second, while the very small fibers conduct impulses at a velocity of only 0.5 meter per second, or approximately the distance from the foot to the knee. All sizes of nerve fibers exist between these smallest and largest sizes, so that a wide spectrum of impulse velocities occurs in the different nerves.

NUMBER OF IMPULSES THAT CAN BE TRANSMITTED PER SECOND. The number of impulses that can be transmitted by any one fiber per second is determined by the refractory period of the fiber, and this depends to a great extent on the size of the fiber. Large fibers (15 to 20 microns in diameter) become repolarized in approximately $1/2500$ second. Therefore, a second nerve impulse can be transmitted $1/2500$ second after the first, or a total of up to 2500 impulses can be conducted each second. At the opposite extreme, the very smallest nerve fibers require as long as $1/250$ second to repolarize, which means that they can transmit no more than 250 impulses per second.

Transmission of Signals of Different Strengths by the Nerve Bundle

The nerve bundle has two means by which it can transmit signals of different strengths. These are (1) to transmit impulses simultaneously over varying numbers of nerve fibers, which is called *spatial summation*, and (2) to transmit impulses in small or large numbers over the same fiber, which is called *temporal summation*. As an example of spatial summation, if 100 nerve fibers are connected between the spinal cord and a foot muscle, stimulation of one of these fibers will cause only a weak response in the muscle, but simultaneous stimulation of all 100 fibers will cause a strong contraction. Obviously, any number of fibers between 1 and 100 can be stimulated at a time, giving any one of 100 different strengths of muscle contraction.

Temporal summation means changing the strength of a signal by sending a large or small number of impulses along the same fiber per second. If one impulse is transmitted each second, only a weak effect usually results, but if 5, 15, 25, 75 or more impulses are transmitted per second, the strength of the effect becomes progressively greater.

Ordinarily the nerve trunk transmits signals of different strengths by a combination of both the spatial and temporal methods. That is, when a strong signal is to be transmitted, large numbers of fibers are utilized and large numbers of impulses are transmitted along each fiber. When a weak signal is to be transmitted, fewer fibers are used and fewer impulses are transmitted.

TRANSMISSION OF IMPULSES BY TISSUES OTHER THAN NERVE FIBERS

TRANSMISSION BY SKELETAL MUSCLE FIBERS. Skeletal muscle fibers transmit impulses exactly as nerve fibers. The normal velocity of transmission in skeletal muscle fibers is about 5 meters per second in contrast to 50 to 100 meters per second in the very large nerve fibers and 0.5 meter per second in the very small fibers. Because the nerve fibers that control the skeletal muscles are a large type, carrying impulses at about 50 meters per second, a signal travels from the brain to the muscle extremely rapidly but then decreases in velocity about 10-fold as it goes into the muscle itself.

TRANSMISSION IN HEART MUSCLE AND SMOOTH MUSCLE. Transmission of impulses in heart muscle and smooth muscle occurs like that in skeletal muscle, but at still lower velocities—about 0.3 meter per second in the heart and only a centimeter or so per second in smooth muscle. In heart muscle and in many smooth muscle masses the fibers interconnect with each other to form latticeworks so that stimulation of any one fiber always causes the impulse to travel over the entire muscle mass, resulting in complete contraction of the whole muscle rather than only part of it. Another difference between these two types of muscle and skeletal muscle is the duration of the refractory period, which in heart muscle is about 0.3 second and in smooth muscle sometimes several seconds. A second impulse cannot stimulate the muscle again for this long period of time. Also, as long as the membranes remain depolarized, the muscle fibers remain contracted. Therefore, the duration of contraction of both heart and smooth muscle is unusually long in comparison with that of skeletal muscle.

TRANSMISSION OF IMPULSES OVER THE MEMBRANES OF GLANDULAR AND OTHER CELLS. Because most glandular secretions of the body are elicited by nerve impulses, it is believed by some physiologists that nerve impulses on reaching the glands cause secondary impulses to travel over the membranes of the secretory cells, this eliciting the secretion. It is also possible that the white blood cells and other cells that exhibit ameboid motion are controlled by electrical potentials that spread over their bodies. Perhaps even the phenomenon of chemotaxis results from membrane potentials initiated by the chemotactic substance. And the migration of fibroblasts into areas that need tissue repair possibly is also controlled by electrical potentials spreading over their surfaces.

THE NEUROMUSCULAR JUNCTION

The *neuromuscular junction* is the connection between the end of a large, myelinated nerve fiber and a skeletal muscle fiber. In general, each skeletal muscle is supplied with at least one neuromuscular junction and occasionally with several. Figure 62A illustrates a neuromuscular junction, showing a nerve fiber passing beneath the mus-

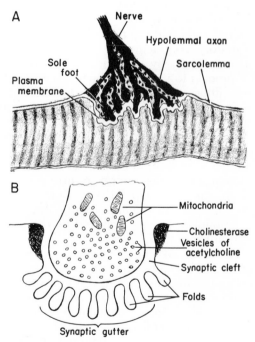

Figure 62. (A) The neuromuscular junction. (B) Invagination of a sole foot into the membrane of the muscle fiber.

cle membrane, called the *sarcolemma*, and then spreading to form many branches, called *hypolemmal axons*. These axons in turn end in club-like feet called *sole feet*. The entire nerve ending, including the hypolemmal axons and the sole feet, is called the *endplate*.

Figure 62B shows the invagination of a sole foot into the muscle fiber, forming a cavity called the *synaptic gutter*. Beneath the sole foot is a small space called the *synaptic cleft*, and along the bottom of the gutter are many large folds of the muscle fiber membrane. It is the plasma membrane that transmits the impulse over the muscle fiber. Stored in the sole feet are many small vesicles containing the chemical *acetylcholine*; this substance is responsible for stimulating the muscle fiber.

Transmission of the Impulse at the Neuromuscular Junction

SECRETION OF ACETYLCHOLINE. When a nerve impulse reaches the neuromuscular junction, passage of the action potential over the sole feet causes many of the small vesicles of acetylcholine stored in the sole feet to rupture into the synaptic cleft between the sole feet and the plasma membrane. The acetylcholine then acts on the folded plasma membrane to increase its permeability. The increased permeability in turn allows instantaneous leakage of sodium to the inside of the fiber, which creates a local potential called the *endplate potential*. When the endplate potential becomes great enough, it stimulates the entire muscle fiber, causing an action potential to travel in both directions along the fiber. In turn, the action potential traveling along the fiber causes muscle contraction, as will be explained in the following chapter.

DESTRUCTION OF ACETYLCHOLINE BY CHOLINESTERASE. If the acetylcholine secreted by the sole feet should remain in contact with the plasma membrane indefinitely, the muscle fiber would transmit a continuous succession of impulses. However, in all the fluids of the body, and located especially in the shoulders around the synaptic gutter of the neuromuscular junction itself, is a substance called *cholinesterase*, which enzymatically splits acetylcholine into acetic acid and choline in about $1/500$ second. Therefore, almost immediately

after the acetylcholine has stimulated the muscle fiber the acetylcholine itself is destroyed. This allows the membrane to repolarize and to become ready for stimulation again as soon as a new nerve impulse approaches.

The acetylcholine mechanism at the neuromuscular junction provides an amplifying system that allows a very weak nerve impulse to stimulate a very large muscle fiber, for the amount of electrical current generated by the nerve fiber is not enough by itself to elicit an impulse in the muscle. Instead, the secreted acetylcholine causes the muscle fiber to generate its own impulse. Thus, each nerve impulse actually comes to a halt at the neuromuscular junction, and in its stead an entirely new impulse begins in the muscle.

Effect of Drugs on the Neuromuscular Junction

EFFECT OF CURARE. *Curare* blocks transmission of the impulse at the neuromuscular junction by making the muscle membrane resistant to the endplate potential caused by acetylcholine. Therefore, the acetylcholine is unable to stimulate the muscle, and the muscle becomes paralyzed.

NEOSTIGMINE AND NEOSTIGMINE-LIKE DRUGS. *Neostigmine* and similar drugs such as *physostigmine* have an opposite effect to that of curare, for they enhance the transmission of impulses at the neuromuscular junction. They do so by inhibiting the action of cholinesterase. As a result, the acetylcholine secreted by the endplate is not destroyed but instead accumulates in progressively larger and larger quantities. Eventually, so much acetylcholine is present that every time the muscle fiber repolarizes it is immediately depolarized again. As a result, the muscle fiber receives a succession of impulses and becomes spastic.

Another drug that has similar effects to those of neostigmine and physostigmine is diisopropyl fluorophosphate, but this drug has a much more prolonged effect. Instead of simply inhibiting the cholinesterase for a few hours, it destroys the cholinesterase, and several weeks are required for resynthesis of new cholinesterase. Therefore, diisopropyl fluorophosphate can cause spastic paralysis for several weeks. This substance is one of the dreaded "nerve" gases that possibly will be used in some future war.

Myasthenia Gravis

Sometimes a person has very poor transmission of impulses at the neuromuscular junction; this condition is called *myasthenia gravis*. Though the exact cause of this condition is unknown, it could result from any one of three abnormalities: (1) deficient acetylcholine secretion, (2) too rapid destruction of acetylcholine by cholinesterase, or (3) depressed responsiveness of the muscle fiber membrane to acetylcholine. Treatment with neostigmine or with some other drug that prevents the destruction of acetylcholine is often very beneficial. These allow acetylcholine to accumulate in the neuromuscular junction and to react with the muscle fiber membrane for a prolonged time. As a result, persons almost totally paralyzed by myasthenia gravis can sometimes be returned to complete normality within less than one minute after a single intravenous injection of neostigmine.

REFERENCES

Eccles, J. C.: *The Physiology of Synapses.* New York, Academic Press, 1963.

Guyton, A. C. and MacDonald, M. A.: Physiology of botulinus toxin. *Arch. Neurol. & Psychiat.* 57:578, 1947.

Hodgkin, A. L.: *The Conduction of the Nervous Impulse.* Springfield, Ill., Charles C Thomas, 1963.

Hodgkin, A. L., and Huxley, A. F.: Quantitative description of membrane current and its application of conduction and excitation in nerve. *J. Physiol. (Lond.)* 117:500, 1952.

Katz, B.: *Nerve, Muscle, and Synapse.* New York, McGraw-Hill Book Company, 1968.

Martin, A. R.: Quantal nature of synaptic transmission. *Physiol. Rev.* 46(1):51, 1966.

McLennan, H. D.: *Synaptic Transmission.* Philadelphia, W. B. Saunders Company, 1963.

Shanes, A. M.: Electrochemical aspects of physiological and pharmacological action in excitable cells. *Pharmacol. Rev.* 10:59, 1958; 10:165, 1958.

von Euler, U. S.: Autonomic neuroeffector transmission. *Handbook of Physiology*, Sec. 1, Vol. 1, p. 215. Baltimore, The Williams & Wilkins Company, 1959.

CHAPTER 7

MUSCLE PHYSIOLOGY

All physical functions of the body involve muscle activity. These functions include skeletal movements, contraction of the heart, contraction of the blood vessels, peristalsis in the gut, and many more. Three different types of muscle are responsible for these activities: skeletal muscle, cardiac muscle, and smooth muscle, all of which have some characteristics in common. For instance, the contractile process is the same or nearly the same in each, but, on the other hand, their reactivities, their durations of contraction, and other features differ greatly and are especially adapted in each type of muscle for the job to be performed.

SKELETAL MUSCLE

The skeletal muscles cause movements of the skeleton and, therefore, are responsible for movement of the different parts of the body. These muscles are controlled by nerve impulses transmitted from the brain.

Physiologic Anatomy of Skeletal Muscle

Figure 63 illustrates a single muscle, showing schematically that the *muscle belly* is

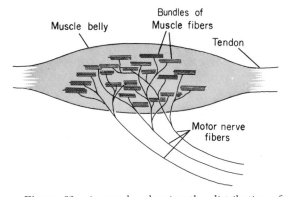

Figure 63. A muscle, showing the distribution of nerve fibers supplying three motor units.

made up of thousands of individual *muscle fibers* attached at each end to *muscle tendons*. When stimulated, the muscle fibers contract, and the force of contraction is transmitted to the bones through the tendons.

Figure 64 illustrates cross-sectional and longitudinal views of a single muscle fiber. Each fiber is between 10 and 100 microns in diameter, and it varies from a few millimeters to 50 centimeters in length. The longitudinal view shows dark and light bands along the fiber, which are characteristic of skeletal and cardiac muscle but not of smooth muscle. A pair of these bands is called a *sarcomere*. Each muscle fiber contains a thousand or more longitudinal *myofibrils* arranged in bundles within the fiber,

69

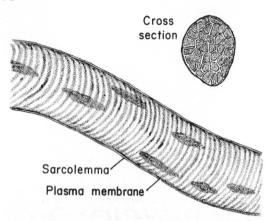

Figure 64. Longitudinal and cross-sectional views of a skeletal muscle fiber.

as shown in the cross-sectional view in Figure 64. The myofibrils in turn are composed of many small filaments composed of *actin* and *myosin* protein molecules which make up the contractile elements of the muscle.

Figure 65 illustrates several sarcomeres of a greatly stretched skeletal muscle fiber, showing by the upper diagram the dark and light bands and by the lower diagram the actual arrangement of actin and myosin filaments in the fiber. The dark band is called the *A band* or *anisotropic band,* and the light band is called the *I band* or *isotropic band.* Note that it is the arrangement of the actin and myosin filaments in the fiber that makes the light and dark bands, the anisotropic band corresponding to the myosin filaments and the isotropic band corresponding to the segment of the fiber between successive myosin filaments. Note also that the myosin and actin filaments interdigitate. During muscle contraction, the actin fila-

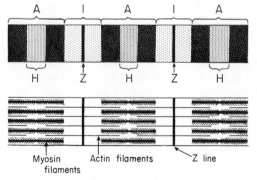

Figure 65. *Above:* Schematic diagram of the light and dark areas in a myofibril. *Below:* Arrangement of the myosin and actin filaments in the sarcomeres.

ments slide inward among the myosin filaments, thus resulting in shortening of the muscle as is explained below.

The muscle fiber is enclosed by a membrane called the *sarcolemma,* and immediately beneath the sarcolemma is a *plasma membrane* which is the true cell membrane of the muscle fiber and which transmits action potentials in the same manner that the nerve membrane transmits action potentials.

SKELETAL MUSCLE CONTRACTION

Excitation of the Skeletal Muscle Fiber

When a nerve impulse reaches the neuromyal junction, acetylcholine is released as explained in the previous chapter, and this acts on the plasma membrane to make it permeable to sodium ions. The rapid influx of sodium ions creates an electrical potential at the neuromyal junction called the *end-plate potential* which, if strong enough, initiates an impulse traveling in both directions along the membrane of the muscle fiber. The impulse travels at a velocity of approximately 5 meters per second. Therefore, after stimulation of a 10 cm. muscle fiber by a neuromyal junction in its middle, the impulse will reach both ends of the fiber in approximately $\frac{1}{100}$ second. It can be seen, therefore, that an impulse excites the entire fiber in a very short period of time.

TRANSMISSION OF THE MUSCLE IMPULSE TO THE INTERIOR OF THE FIBER. When a muscle fiber is excited, an action potential not only travels over the surface of the plasma membrane, but also penetrates deeply into the fiber by way of *T-tubules,* two of which are illustrated in Figure 66. This figure shows diagrammatically a very small segment of the muscle fiber, illustrating two of the many thousands of myofibrils present in a muscle fiber and illustrating in these myofibrils three actin and myosin filaments in contrast to the many thousands that are actually present. The T-tubule passes all the way through the muscle fiber, and it crosses the myofibril where the myosin and actin filaments overlap each other. Parallel with each myofibril but perpendicular to the T-tubule is another tubular structure called the *fine sarcoplasmic reticulum,* which is a

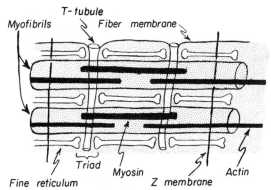

Figure 66. Relation of the T-tubules and of the fine sarcoplasmic reticulum to the myofibrils.

modified type of endoplasmic reticulum. Where the fine sarcoplasmic reticulum comes in contact with the T-tubule, a large bulbous cistern-like structure occurs. This fine reticulum plays the following role in muscle contraction:

FUNCTION OF THE FINE RETICULUM AND OF CALCIUM IONS IN MUSCLE CONTRACTION. When the action potential spreads to the inside of the muscle fiber along the T-tubules, some yet unexplained effect of the action potential on the fine reticulum causes the reticulum to release calcium ions into the fluids surrounding the myofibrils. It is these calcium ions that then elicit the contractile process in the myofibrils, the mechanism of which will be explained in the following section. The contraction lasts only so long as free calcium ions are present.

Immediately after the action potential is over, there is no longer any stimulus to release calcium ions from the reticulum. Instead, the calcium ions that had previously been released now begin to recombine with the reticulum, and in a few thousandths of a second all the calcium ions are bound again to the reticulum. As a result, muscle contraction ceases. If muscle contraction is to be continuous, it is necessary to transmit a succession of action potentials so that sufficient quantities of calcium ions are always available to promote muscle contraction.

The Contractile Process

Figure 67 illustrates the mechanism of skeletal muscle contraction, showing in the upper drawing the relaxed state of the fiber and in the lower drawing the contracted state. This figure illustrates that neither the myosin nor the actin filaments decrease in length but instead that the actin filaments simply slide like pistons inward among the myosin filaments, and the actin filaments overlap each other during normal contraction. The precise manner in which this sliding process is effected at the onset of muscular contraction is still somewhat conjectural, but many of the steps have been discovered. The process is believed to be the following:

When calcium ions suddenly appear in the fluids of the muscle fiber, some of these combine loosely with myosin molecules to form *activated myosin,* which has the property of an enzyme called *ATPase.* This activated myosin ATPase can now react with adenosine triphosphate (ATP) to remove its energy. Attached to the actin filaments is a large amount of ATP, and extending from the myosin filaments toward the actin filaments are numerous *crossbridges,* which can be seen in Figure 67. The activated myosin ATPase releases the energy from the ATP on the actin filaments, probably by interaction between the crossbridges and the actin, which in turn is utilized to pull the actin filaments inward among the myosin filaments. It is presumed that the chemical reactions involved in this process create electrostatic forces between the myosin and actin filaments which cause them to attract each other and thereby promote the sliding process.

The mechanism by which electrostatic forces cause the actin and myosin filaments to slide together is still unknown. One theory assumes that electrostatic forces on the crossbridges of the myosin filaments cause the

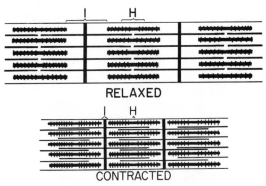

Figure 67. The relaxed and contracted states of a myofibril, showing sliding of the actin filaments into the channels between the myosin filaments.

cross-bridges to attach to reactive sites on the actin filaments. Because of continual kinetic motion of molecules, the bonds persist for only a few microseconds at a time. As each bond breaks, another reactive site a little further down the actin filament then supposedly attaches to the same cross-bridge. Thus, the actin filament slides along the myosin filament by a "ratchet" mechanism, the actin molecule first locking with one cross-bridge, then with the next, then the next, and so forth.

Another theory simply supposes that the myosin filament becomes electrostatically charged to a potential that is different from the potential of the actin filament but that no actual chemical bonds develop between the two filaments. The actin and the myosin filaments attract each other, forming what is called an *electrostatic solenoid* in which the two sets of filaments slide in among each other. It has been calculated that a 70 millivolt difference between the two types of filaments would be sufficient to cause maximum force of skeletal muscle contraction. Since 70 millivolts is less than the action potential of the muscle fiber, this appears to be a reasonable theory.

THE ENERGY FOR CONTRACTION. The immediate source of energy for muscle contraction is the adenosine triphosphate attached to the actin filaments. Adenosine triphosphate (ATP), as was discussed in Chapter 3, is a "high-energy compound" that functions in all cells of the body to provide a ready source of energy for almost any type of cellular function. When one mole of adenosine triphosphate splits to form a mole of phosphoric acid and a mole of adenosine diphosphate (ADP), about 8000 calories of energy are released. It is this energy that in some yet unknown way creates the forces between the myosin and actin filaments that cause muscle contraction.

Once ATP has been split into ADP and phosphoric acid, these substances must be reconverted into ATP for reuse at a later time. This is accomplished by the metabolism of foodstuffs. For instance, the metabolism of glucose with oxygen to form carbon dioxide and water releases 680,000 calories of energy per mole of glucose. This energy is used to synthesize 38 moles of ATP from ADP. The ATP can also be replenished by energy from fat or protein metabolism as well. Thus, increased metabolism in a muscle fiber after contraction is over re-plenishes the chemical stores that are used up during the contractile process. Also, energy must be used to expulse the calcium and sodium ions that enter the fiber during contraction.

Heat liberated by the contractile process. Unfortunately, not all of the energy stored in the form of adenosine triphosphate is converted into muscle work, and, likewise, not all of the energy stored in the glucose molecule is utilized in the formation of ATP. Instead, because of a certain degree of *inefficiency* of all energy transfer processes in the body, much energy is lost in the form of heat. Thus, heat is given off from a muscle both during and after the period of contraction. This heat is divided into four different parts: (1) heat caused by the actual shortening of the muscle, called the *heat of shortening,* (2) heat caused by the chemical processes required to initiate and to maintain contraction, called the *heat of activation and maintenance,* (3) heat caused by viscous changes and other physical changes in the muscle during relaxation, called the *heat of relaxation,* and (4) heat required to replenish the ATP and to remove sodium from the interior of the fiber, called the *heat of recovery.*

Efficiency of muscle contraction. The *efficiency* of an energy exchange process is the percentage of the input energy that becomes output energy. In muscle contraction, the input energy is the potential energy of the nutrients used by the muscle fibers for contraction, and the output energy is the work output of the muscle. Because of inefficiencies of energy transfer in the muscle fibers, the percentage of input energy that eventually becomes work output, even under optimal conditions, is only 20 to 25 per cent; the other 75 to 80 per cent becomes heat. Under some conditions, such as when standing absolutely still, a person has no work output from his muscles, and, yet, a large amount of nutrients is being utilized to maintain the tension in his muscles. Under these conditions the efficiency of muscle contraction is absolute zero.

THE ALL OR NONE PRINCIPLE. The all or none principle applies to contraction of the skeletal muscle fiber in exactly the same way that it applies to transmission of an impulse over a nerve fiber. Once a muscle fiber has become excited and an action potential has begun to spread along its membrane, the entire fiber responds. That is, a muscle fiber cannot be "partially" excited; either it does not become excited at all or it becomes excited in its entirety.

The all or none principle, however, does not imply that the strength of contraction is the same each time the fiber is stimulated. The contractile elements of the muscle fiber are sometimes weaker than at other times because of fatigue, lack of nutrients, or other causes, so that even though the entire fiber responds, the strength may vary from time to time. In essence, then, the all or none principle means that when a muscle fiber is excited the entire fiber contracts to the full extent of its immediate ability.

The Motor Unit

Figure 63 showed three separate nerve fibers entering a muscle. Actually this was merely a figurative representation, because several hundred to several thousand nerve fibers enter each muscle. The figure also showed that each of the nerve fibers divides and spreads throughout the muscle belly to terminate on many different muscle fibers. On the average, a single motor nerve fiber innervates about 150 muscle fibers, which means that stimulation of one nerve fiber will cause contraction of 150 muscle fibers at the same time. All the fibers innervated by the same nerve fiber are called a *motor unit* because they are always excited simultaneously and therefore always contract in unison.

It is especially important that all of the muscle fibers in a motor unit do not necessarily lie side by side, but instead are divided into many bundles of only a few fibers each spread throughout the muscle belly. Because of this, stimulation of the motor unit causes a weak contraction in a broad area of the muscle rather than a strong contraction at one specific point.

Those muscles that must control very fine movements usually have only a few muscle fibers in each motor unit, which means that the ratio of nerve fibers to muscle fibers is very high. For instance, the ocular muscles, which must control the extremely discrete movements of the eyes, have only about 10 muscle fibers to each motor unit. On the other hand, postural muscles, which usually exhibit only very gross movements, may have as many as 200 or more fibers to the motor unit.

The Muscle Twitch

One of the laboratory methods for studying muscle contraction is to elicit a *muscle twitch*. To do this, a single instantaneous stimulus is applied to the nerve supplying the muscle. The duration of the resulting contraction is between $1/10$ and $1/300$ second, depending on the type of muscle, which is so short that it appears to be a "twitch."

ISOMETRIC AND ISOTONIC CONTRACTION. Figure 68 shows the contraction of a muscle under two different conditions. To the left the muscle lifts weights in a pan, becoming shorter in the process. The total amount of weight applied to the muscle is always the same, for which reason the contraction is called *isotonic*, which means "same force." To the right the muscle is attached between a solid bar and a tight torsion spring so that the muscle cannot contract more than about $1/1000$ inch. That is, the muscle belly is suspended between two rigid points. Stimulation of the muscle under these conditions causes it to tighten but not to shorten significantly, and the contraction is called *isometric*, meaning "same length."

The characteristics of isometric and isotonic muscle contraction are somewhat different. The reasons for this are, first, the isometric system has no inertia while the isotonic system does, and, second, during isotonic contraction the shape of the muscle must change so that it can shorten, whereas during isometric contraction the muscle does not need to change shape but only to create force. Therefore, without inertia and without the necessity for changing shape, the isometric muscle twitch usually has a much shorter duration than the isotonic twitch. In expressing relative abilities of

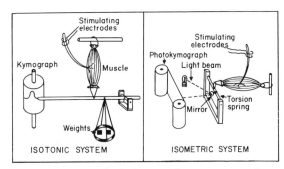

Figure 68. Methods for recording isotonic and isometric muscle twitches.

different muscles to contract, the isometric muscle twitch is the usual criterion employed, because its characteristics are dependent only on intrinsic characteristics of the muscle and not at all on extrinsic factors.

In the human body, muscle contraction is of both the isometric and isotonic types. When one is simply standing he tenses his leg muscles to maintain a fixed position of the joints. This is isometric contraction. On the other hand, when he is walking and moving his legs, or when he is lifting his arms, the contraction is more of the isotonic type.

DURATION OF CONTRACTION OF DIFFERENT SKELETAL MUSCLES. Figure 69 shows recordings of isometric contractions by different muscles. The dashed curve of the figure shows the duration of depolarization of the fiber membrane caused by the action potential traveling over the muscle fiber. This is the period when calcium ions are being released into the fluids of the fiber. Immediately thereafter the contraction begins. The isometric contraction of an ocular muscle lasts for about $1/100$ second while that of a gastrocnemius muscle lasts about $3/100$ second. It is evident, then, that different skeletal muscles have different durations of contraction. The ocular muscles, which must cause extremely rapid movement of the eyes from one position to another, contract more rapidly than almost any other muscle. The gastrocnemius muscle must contract moderately rapidly because it is used in jumping and in performing other rapid downward movements of the foot. The soleus need not contract rapidly at all, be-

cause it is used principally to perform the prolonged movements for support of the body against gravity.

Effect of Initial Muscle Length on the Force of Contraction

The length to which a muscle is stretched before it contracts makes considerable difference in its force of contraction. When the length is much less than normal its force of contraction is greatly weakened, and, also, when it is stretched far beyond its normal limits, it fails to contract with as much force as would otherwise be possible. Figure 70 illustrates these effects, showing that a muscle in its normal stretched state will usually contract with the greatest possible force.

Fortunately, the normal length of a muscle in its most elongated position is almost exactly optimal for maximal strength of contraction. For instance, when the biceps is at its normal full length, it contracts with its greatest force, whereas, as it progressively shortens, its strength of contraction decreases.

THE LEVER SYSTEMS OF THE BODY. Other factors that determine the force of a movement are (a) the manner in which the contracting muscles are attached to the skeletal system and (b) the structure of the joint at which movement will occur. Figure 71, as an example, illustrates movement of the forearm by biceps contraction. The fulcrum of the lever system is at the elbow, and the attachment of the biceps is approximately

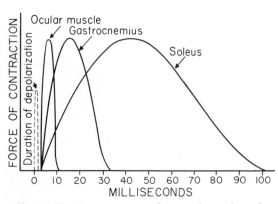

Figure 69. Isometric muscle twitches of ocular, gastrocnemius, and soleus muscles, illustrating the different durations of contraction.

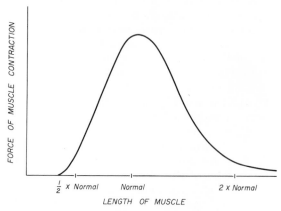

Figure 70. Effect of the initial length of a muscle on the contractile force developed following muscle excitation.

two inches in front of the fulcrum. If we assume that the total length of the forearm lever is about 14 inches, one immediately sees that the force of contraction of the biceps must be at least 7 times as great as the force of movement of the hand. Thus, if the hand is to lift an object that weighs 50 pounds, the total force of contraction of the biceps would have to be about 350 pounds. One can readily see from this tremendous force why muscles sometimes pull their tendons out of the bone substance.

Every muscle of the body has its own peculiar shape and length that suits it to its particular function. For instance, the muscles of the buttocks are extremely broad but do not contract a long distance. They provide tremendous force for movement at the hip joint, and even a very slight distance of movement at this joint can cause tremendous movement of the foot. At the other extreme, some of the muscles of the anterior thigh are very long and can shorten almost a foot, pulling the lower leg upward at the knee joint and flexing the upper leg at the hip joint at the same time.

The study of different types of muscle lever systems and their movements is called *kinesiology;* this is a very important phase of human physioanatomy.

Different Strengths of Muscle Contraction—The Mechanism of "Summation"

In performing the different functions of the body, it is quite important that each muscle be able to contract with varying degrees of strength. This is accomplished by "summing" the contractions of varying numbers of muscle fibers at once. When a weak contraction is desired, only a few muscle fibers are contracted simultaneously. When a strong contraction is desired, a great number of fibers are contracted at the same time. In general, the different "gradations" of muscle contraction are achieved by two different methods of summation called *multiple motor unit summation* and *wave summation.*

MULTIPLE MOTOR UNIT SUMMATION. If only one motor unit contracts at a time, the muscle contraction will be weak, but if all the motor units of the muscle contract simultaneously, the strength of contraction will be maximal. All gradations of muscle contraction between minimal and maximal can be attained by varying the number of motor units contracting. This effect is illustrated in Figure 72, which shows the force of various numbers of motor units contracting simultaneously.

WAVE SUMMATION. Wave summation occurs when each muscle fiber contracts many times in rapid succession, the contractions occurring closely enough together so that a new contraction occurs before the previous one is over. Each succeeding contraction adds to the force of the preceding one, increasing the overall strength of contraction. This effect is illustrated in Figure 73. To the left the contractions occur at a rate of five per second, which is long enough apart so that they do not summate. But, by the time the rate has reached 10 to 15 times per second, successive contractions occur before previous ones are over. As a result, the contractions add to each other, causing an

Figure 71. The lever system activated by the biceps muscle.

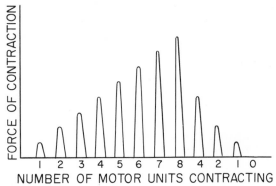

NUMBER OF MOTOR UNITS CONTRACTING

Figure 72. Multiple motor unit summation, illustrating the progressive increase in contractile force caused by increasing numbers of motor units contracting simultaneously.

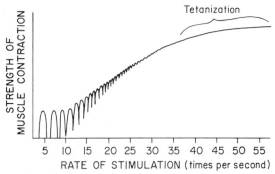

Figure 73. Wave summation, showing progressive summation of successive contractions as the rate of stimulation is increased. Tetanization occurs when the rate of stimulation reaches approximately 35 per second, and maximum force of contraction occurs at approximately 50 per second.

increased force of contraction. In this figure the force of contraction increases progressively until the rate of stimulation reaches about 55 times per second. Some of the rapidly acting muscles, such as the ocular muscles, require as many as 400 to 600 stimulations per second for maximum contraction to take place.

Tetanization. Observing Figure 73 once again, it can be seen that individual contractions are no longer evident after the rate of stimulation has risen above about 35 per second, but instead a smooth, continuous contraction occurs. In other words, the muscle twitches become fused into a single prolonged contraction. This effect is called *tetanization.* The frequencies at which different muscles will tetanize depend on the durations of single contractions. In the soleus muscle tetanization occurs at a frequency of about 20 contractions per second because the soleus is a slowly contracting muscle. Likewise, the gastrocnemius tetanizes at about 50 stimuli per second because it is a moderately rapidly contracting muscle.

ASYNCHRONOUS SUMMATION. In normal muscle the different gradations of contraction are attained by utilizing a mixture of multiple motor unit summation and wave summation. This is effected by contracting the different motor units of a muscle belly in rapid succession one after the other so that the muscle tension is always of a tetanic type rather than of a twitching type. For a weak contraction each motor unit contracts only two to three times per second, but the contractions are spread out one after another

among the different motor units, rather than all units contracting at the same time. In this way some motor units are always contracting, assuring constant though weak tension in the muscle. When a strong contraction is desired each motor unit contracts 50 or more times per second.

Asynchronous summation is caused by neuronal circuits in the spinal cord that automatically distribute the impulses evenly and sequentially among the different nerve fibers to a muscle.

TONE OF SKELETAL MUSCLE. The spinal cord continually transmits a few impulses to all skeletal muscles most of the time, which maintains slight tension in the muscles and keeps them from becoming flabby. This is called *muscle tone,* and it keeps the muscle "taut" so that it will react to a strong stimulus far more rapidly than it could otherwise.

The tone of skeletal muscle may increase or decrease depending on the physiologic activity of the body. When a person is in a state of anxiety, excitement, fear, or certain other emotional conditions, the number of tone impulses from the central nervous system becomes intensified, causing the muscles to become increasingly taut so that they respond to other stimuli with extreme rapidity. This is the basis of the jumpiness of the so-called "nervous" individual. On the other hand, during sleep the number of tone impulses becomes almost nil, allowing the muscles to relax completely.

Effect of Activity on Muscular Development

EXERCISE AND HYPERTROPHY. The more a muscle is used, the greater becomes its size and strength, though the cause is yet unknown. Physical enlargement of the muscles is called *hypertrophy;* examples are (1) the intense muscular development of weight lifters, (2) the great hypertrophy of the leg muscles in skaters, (3) the enlargement of the arm and hand muscles of carpenters, and (4) the enlargement of the thigh muscles of runners.

Associated with muscle hypertrophy is usually an increase in the efficiency of muscular contraction, for the hypertrophied muscle stores increased quantities of glycogen, fatty substances, and other nutrients, and the number of contractile fibrils increases. All these cause the efficiency of the contractile process to increase so that the percentage of energy lost as heat becomes considerably less in the athlete than in the non-athlete.

It should be noted that weak muscular activity, even when sustained over long periods of time, does not result in significant muscle hypertrophy. Instead, hypertrophy is mainly the result of forceful muscle activity even though the activity might occur only a few minutes each day. For this reason, strength can be developed in muscles more rapidly by using "resistive exercises" than by prolonged mild exercise.

MUSCLE DENERVATION AND ATROPHY. When the nerve supply to a muscle is destroyed the muscle begins to *atrophy*— that is, the muscle fibers begin to degenerate. In about six months to two years the muscle will have atrophied to about one-fourth normal size, and the muscle fibers will have been replaced mainly by fibrous tissue.

For some reason nerve stimulation of a muscle keeps the muscle tissue alive. Even when a person does not use his muscles to a great extent, the weak tone impulses are still sufficient to maintain a relatively normal muscle, but without these impulses the muscle fibers soon atrophy entirely. Perhaps this effect is caused by nutritional changes in the denervated muscle, for action potentials traveling down the fiber membrane alter its permeability markedly, which might be necessary for appropriate transfer of nutrients through the membrane. The functional integrity of denervated muscle fibers can be maintained quite satisfactorily by daily electrical stimulation. The action potentials produced in this manner take the place of the nerve induced potentials, and the muscle fibers do not atrophy.

CARDIAC MUSCLE

Figure 74 shows a small microscopic section of cardiac muscle. One can see that cardiac muscle fibers have the same striated appearance as that seen in skeletal muscle. The only significant difference between the appearance of the two types of muscle is that cardiac fibers interconnect with each other in a lattice arrangement called a *functional syncytium*. The heart is composed of two

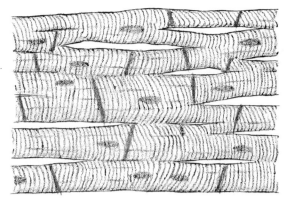

Figure 74. Cardiac muscle, showing the lattice arrangement of the fibers.

separate such syncytiums. One of these forms the walls and septum of the atria, and the other forms the walls and septum of the ventricles. When a single fiber of either of the two functional syncytiums is stimulated, all the connecting muscle fibers in this syncytium become stimulated also. Therefore, in applying the all or none principle to the contraction of cardiac muscle, one does not speak of each muscle fiber separately but instead of the syncytiums. That is, a stimulus applied to the heart either causes the muscle syncytium to contract entirely or not at all. Impulses are transmitted from the atria into the ventricles through the Purkinje system, which will be described in Chapter 11. Because of this connecting link, an impulse beginning anywhere in the atrial syncytium spreads on into the ventricles, causing the entire heart to contract.

RHYTHMICITY OF CARDIAC MUSCLE. One of the major physiologic differences between skeletal muscle and cardiac muscle is that skeletal muscle normally contracts only when stimulated by a nerve, whereas cardiac muscle contracts rhythmically approximately 72 times per minute, never stopping. The contraction is caused by an inherent property of all cardiac muscle to discharge its membrane potential every time the potential builds up. That is, every time the membrane repolarizes after a previous cardiac impulse, the membrane, for yet inexplicable reasons, suddenly becomes permeable again, allowing transmission of a new action potential over the entire heart. This event occurs over and over again approximately once every 8/10 second.

When skeletal muscle is denervated, it also exhibits rhythmicity beginning approximately two to three days after denervation and continuing until the muscle becomes atrophic. The rate of contraction is ten or more times as rapid as that of cardiac muscle, and the resulting movements are called *fibrillations.* This fact simply illustrates that the phenomenon of rhythmicity is probably a fundamental property of all muscle. In skeletal muscle it has been subordinated to nervous control, while in the heart the rhythmic contractions are of primary importance and nervous control is of secondary importance.

PROLONGED DEPOLARIZATION AND PROLONGED CONTRACTION OF CARDIAC MUSCLE. Another difference between cardiac muscle and skeletal muscle is the duration of depolarization and contraction. After stimulation, the membrane of cardiac muscle remains depolarized for about $3/10$ second in contrast to about $1/500$ second in skeletal muscle, and as long as the membrane remains depolarized in the cardiac muscle the contractile process of the actin and myosin molecules is maintained. Therefore, the period of a single cardiac muscle contraction is about $3/10$ second, which is 5 to 50 times as long as a single contraction of a skeletal muscle. This prolonged contraction is a very important characteristic of cardiac muscle, because, for the heart to operate properly, its contraction must last long enough to force blood from its chambers.

SMOOTH MUSCLE

Most of the internal organs of the body contain *smooth muscle,* the third type of muscle. The name is derived from the fact that this muscle does not have striations similar to those in skeletal and cardiac muscle.

In most internal organs the smooth muscle fibers are very tightly bound to each other, as illustrated to the left in Figure 75. This type of muscle is called *visceral smooth muscle* because the internal organs are known as "visceral" organs. Because the smooth muscle fibers are tightly bound, action potentials can pass readily from one fiber to the next. Thus, visceral smooth muscle fibers form a *functional syncytium* similar to that of cardiac muscle.

Occasional types of smooth muscle have discrete fibers without the interconnections commonly found in visceral smooth muscle. This type, called *multiunit smooth muscle,*

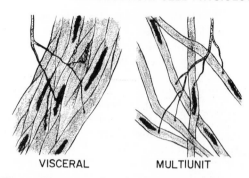

Figure 75. Visceral and multiunit smooth muscle fibers.

is illustrated to the right in Figure 75. The major functional difference between visceral and multiunit smooth muscle is that stimulation of a single smooth muscle fiber in visceral muscle will usually excite the entire muscle mass, while to cause complete contraction of a multiunit smooth muscle mass it is necessary to stimulate each fiber individually in the same way that skeletal muscle fibers are excited. Multiunit smooth muscles are found in blood vessels, in the iris of the eye, and in the nictitating membrane that covers the eyes of some lower animals.

Smooth muscle fibers may be only 20 to 40 microns in length, or they may be several millimeters long, depending on the function to be performed. Quite frequently they are arranged in large sheets rolled into tubes or spheres to form such organs as the gut, the bladder, blood vessels, and many others. Other smooth muscle fibers are arranged like a ring, as around the pupil of the eye, forming a sphincter that can decrease the size of an opening. They occasionally, also, are arranged in tufts similar to the muscle bellies of skeletal muscle. These are seen especially at the bases of hairs where several smooth muscle fibers form each pilo-erector muscle that causes the hairs to stand on end when a person becomes cold or frightened. In summary, except for the movements of the skeleton and the pumping action of the heart, all the physical activities of the body are performed by smooth muscle. This type of muscle is found in almost all internal organs, and its specific characteristics vary with the function to be performed.

THE CONTRACTILE PROCESS IN SMOOTH MUSCLE. Smooth muscle contains actin filaments and probably myosin filaments (though this is not certain), the same as skeletal and cardiac muscle, but the sarcomeric arrangement that causes the striated appearance in these other types of muscle is absent. Smooth muscle also contains the same ionic constituents as the other two types of muscles. Therefore, it is believed that contraction in smooth muscle occurs by some means similar to that in both skeletal and heart muscle, by interaction between myosin and actin filaments initiated by calcium ions. Yet because there is no piston-like arrangement between the two types of filaments in smooth muscle such as one finds in skeletal muscle, some other *mechanical* explanation for the contraction besides the piston movement must be found. Though an exact explanation is unknown, it is assumed that when myosin becomes activated, the actin filaments, however they happen to lie at the moment, are made to slide in among the myosin filaments. Since these filaments are oriented linearly along the muscle fiber, any such sliding process must of necessity shorten the muscle. It is likely that this problem will be solved soon, now that the electron microscope can reveal structural details of the myofibrils.

Control of Smooth Muscle Contraction

MEMBRANE POTENTIAL AND ACTION POTENTIALS IN SMOOTH MUSCLE. The normal resting membrane potential of smooth muscle is about −50 to −65 millivolts in contrast to −85 millivolts usually found in skeletal muscle fibers. At least three different types of transient changes in this membrane potential often occur and are associated with contraction. These are:

(1) *Spike potentials.* Typical spike-like action potentials similar to those that occur in skeletal muscle fibers are seen in certain types of smooth muscle. Immediately following the action potential, a short muscle contraction occurs.

(2) *Action potentials with plateaus.* One of the most common types of action potentials in smooth muscle is that illustrated in Figure 76 having a plateau lasting 0.1 to 1 second. This is also the type of action potential that occurs in cardiac muscle and that will be discussed in more detail in relation to cardiac muscle in Chapter 11. So long as the plateau persists, the smooth muscle remains contracted. This type of action potential is fre-

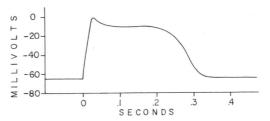

Figure 76. Monophasic action potential from a smooth muscle fiber.

quently seen in visceral smooth muscle, such as that of gut, uterus, and so forth.

(3) *Slow changes in membrane potential.* Under some conditions the membrane becomes either excessively permeable or excessively impermeable to sodium ions, in which case the membrane potential decreases or increases respectively for hours at a time. When the potential becomes very low, the smooth muscle often contracts. In this instance, smooth muscle contraction occurs without a typical action potential.

RHYTHMICITY OF SMOOTH MUSCLE. Most smooth muscle exhibits rhythmicity similar to that in cardiac muscle. This is especially true of the visceral smooth muscle, which comprises perhaps 99 per cent of all smooth muscle. The precise cause of the rhythmicity is not known, but it is presumed that the membrane is very leaky with respect to sodium ions. This causes the membrane potential to fall until it reaches the so-called *threshold voltage,* at which level an action potential is generated. For a few seconds after the action potential, the membrane becomes relatively impermeable to sodium, allowing the membrane potential to redevelop and the entire cycle to begin again.

The rate of rhythmicity of smooth muscle, in general, is quite slow, varying from one contraction every five seconds up to one contraction every few minutes. This is in contrast to the rhythmic contraction of cardiac muscle that occurs more than once every second.

The rhythmic contraction of smooth muscle is responsible for many of the functions of smooth muscle organs. For instance, rhythmic contractions of the uterus at the end of pregnancy are responsible for birth of the baby, and the rhythmic contractions of the gut are responsible for peristalsis that propels food along the intestines.

TONUS OF SMOOTH MUSCLE. Almost all

the smooth muscle in the body remains at least mildly contracted continuously even without stimulation by nerves, unlike skeletal muscle. This continuous contraction is called *tonus* or simply *tone*. Figure 77 illustrates the difference between rhythmic contractions and tonus contraction. Throughout the record the rhythmic contractions cause repetitive increases and decreases in muscle contraction. However, superimposed on these is a single prolonged tonus contraction.

Smooth muscle tonus can vary greatly from time to time because of many factors such as nerve stimulation of the muscle, fatigue of the muscle, availability of nutrients for muscular contraction, hormonal action, and irritation of the muscle.

PLASTICITY OF SMOOTH MUSCLE. Another extremely important characteristic of smooth muscle is its ability to be stretched or shortened and still function equally well in both states. For instance, the urinary bladder when almost entirely empty can yet contract to a still smaller size and empty the remaining few milliliters of urine. On the other hand, after the bladder has been stretched to a volume of as much as one liter, the muscle fibers are still capable of contracting and forcing the urine to the exterior. In other words, smooth muscle is different from skeletal muscle in that its effectiveness of contraction is great at many lengths.

The principle of plasticity applies also to the accumulation of food in the gastrointestinal tract and to the function of the gallbladder, the bile ducts, and almost all hollow organs of the body. All these organs maintain essentially the same internal pressure despite great changes in their volumes.

FACTORS THAT CAN CAUSE SMOOTH MUSCLE CONTRACTION. The most important

stimuli that cause rhythmic and tonic contractions of smooth muscle are (1) the amount of stretch applied to the muscle, (2) local irritation, and (3) the effects of the autonomic nervous system. All of these can elicit one or more types of membrane potential changes as discussed earlier.

Ordinarily, stretch of a smooth muscle organ or local irritation of the organ increases the intensity of both the tonic and rhythmic contractions. Occasionally, though, very strong irritation causes the smooth muscle to go into such an intense state of tonic spasm that the rhythmic contractions are no longer apparent. This is especially characteristic of the gastrointestinal tract when it is irritated by an infectious disease, irritating food or drug, or by trauma.

Control of smooth muscle by the autonomic nervous system. The autonomic nervous system, which will be discussed in Chapter 28, is composed of two separate parts, the *sympathetic* and the *parasympathetic systems*. Nerve fibers pass from each of these to most smooth muscle of the body. When stimulated, the sympathetics secrete a hormone called *norepinephrine* and smaller amounts of a similar hormone, *epinephrine*, and the parasympathetics secrete a hormone called *acetylcholine*. Norepinephrine and epinephrine stimulate the smooth muscle in some organs, but in other organs inhibit it. Acetylcholine, on the other hand, inhibits all smooth muscle that the other two hormones stimulate and stimulates all that the other two inhibit. In other words, the hormones secreted by the two portions of the autonomic nervous system have exactly opposing effects. Some examples of this are the following: (1) The pupil of the eye is constricted by the parasympathetics but dilated by the sympathetics. (2) The coronary arteries are dilated by the sympathetics but constricted by the parasympathetics. (3) The intestines are constricted by the parasympathetics but dilated by the sympathetics. (4) The sphincters of the gut are constricted by the sympathetics but dilated by the parasympathetics. (5) The sphincters that regulate emptying of the urinary bladder are constricted by the sympathetics but dilated by the parasympathetics.

The reason why norepinephrine contracts some but relaxes other smooth muscle, while acetylcholine causes exactly the opposite effects, is not known, but this is believed

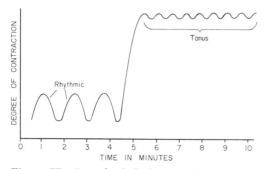

Figure 77. Record of rhythmic and tonic smooth muscle contraction.

to result from a difference in "receptor substances" in different types of smooth muscle cells. If the receptor substance is of an excitatory nature, norepinephrine will cause stimulation; if of an inhibitory nature, it will cause inhibition. It is presumed that the excitatory substance for norepinephrine is an inhibitory substance for acetylcholine, and the inhibitory substance for norepinephrine is an excitatory substance for acetylcholine. This would explain the antagonistic effects of the two hormones on smooth muscle.

REFERENCES

Astrand, P., and Rodahl, K.: Textbook of Work Physiology. New York, McGraw-Hill Book Company, 1968.

Bourne, G. H. (ed.): *The Structure and Function of Muscle*. 3 volumes. New York, Academic Press, 1960.

Brunnstrom, S.: *Clinical Kinesiology*. Philadelphia, F. A. Davis Company, 1962.

Hill, A. L., et al.: Physiology of voluntary muscle. *Brit. Med. Bull.* 12 (Sept.) 1956.

Huxley, A. F.: Muscle. *Ann. Rev. Physiol.* 26:131, 1964.

Huxley, H. E.: The contractile structure of cardiac and skeletal muscle. *Circulation* 24:328, 1961.

McPhedran, A. M., Wuerker, R. B., and Henneman, E.: Properties of motor units in a homogeneous red muscle (soleus) of the cat. *J. Neurophysiol.* 28:71, 1965.

Peachey, L. D.: Muscle. *Ann. Rev. Physiol.* 30:401, 1968.

Rodahl, K., and Horvath, S. M. (eds.): *Muscle as a Tissue*. New York, The Blakiston Company, 1962.

SECTION TWO

BLOOD AND IMMUNITY

CHAPTER 8

THE BLOOD CELLS

Almost all the cells in the blood are *red blood cells,* and their major function is to transport oxygen from the lungs to the tissues. However, approximately one out of every 500 cells is a *white blood cell,* or *leukocyte,* which is called *"white"* because it is not colored by hemoglobin. The leukocytes can be divided into *granulocytes, monocytes,* and *lymphocytes,* and the granulocytes in turn can be divided into *neutrophils, eosinophils,* and *basophils.* White blood cells have several different functions, but the most important of these is to protect the body against invasion by disease organisms.

The blood also contains large numbers of *platelets,* which are often classified as white blood cells. However, the platelets are not really cells but instead very small fragments of a special type of bone marrow cell called the *megakaryocyte.* The platelets are essential for the clotting of blood, which will be explained in detail in Chapter 10.

THE RED BLOOD CELLS

RED BLOOD CELL COUNT AND HEMATOCRIT. The normal number of red blood cells in each cubic millimeter of blood is a little more than 5,000,000 in the male person and a little less than this in the female.

The percentage of the blood made up of red blood cells is called the *hematocrit;* the normal figure is 45 per cent. The hematocrit is determined by centrifuging blood in a graduated tube of the type shown in Figure 78. This figure contrasts the hematocrit of normal blood, 45 per cent, with that of *anemic* blood, 15 per cent, and that of *polycythemic* blood, 65 per cent.

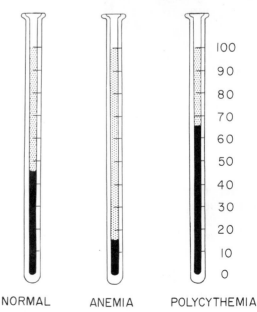

NORMAL   ANEMIA   POLYCYTHEMIA

Figure 78.   Determination of the hematocrit.

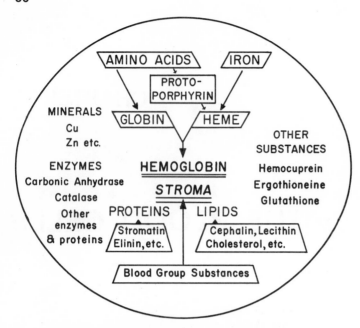

Figure 79. Constituents of the red blood cell. (Redrawn from Wintrobe: *Clinical Hematology.* Lea & Febiger.)

THE RED BLOOD CORPUSCLE

ORGANIZATION OF THE RED BLOOD CELL. The red blood cell is different from all other cells of the body in that it has already lost its nucleus so that it can no longer multiply after entering the blood stream.

The major constituents of the red blood cell are (1) a thin cellular membrane, and (2) a viscid solution of *hemoglobin,* having a concentration of about 33 per cent. In addition to hemoglobin, many other substances, some of which are listed in Figure 79, are present in the red blood cell in small quantities. Some of these are proteins and lipids that make up the physical structures of the cell such as the cellular membrane and the internal framework. Others are enzymes, minerals, and less important compounds that are important for metabolism inside the cell. Metabolism keeps the cell functional even though it is no longer growing or multiplying.

Transport of Oxygen by the Red Blood Cells

The principal function of the red blood cell is to carry oxygen from the lungs to the tissues. Oxygen diffuses through the pulmonary membrane into the blood, then through the cell membrane to combine with the hemoglobin inside the red blood cell. When the cell reaches a tissue capillary, the opposite effects take place: The hemoglobin releases the oxygen which then diffuses out of the capillaries into the tissues. This transport of oxygen from the lungs to the tissues will be discussed in detail in connection with the subject of respiration in Chapter 19.

In some lower animals, hemoglobin circulates freely in solution in the blood plasma rather than being transported inside red blood cells. In mammals (including the human being), however, the hemoglobin molecule is small enough so that once it escapes from the red cell it leaks out of the blood through the capillary pores. Therefore, the function of the red blood cell membrane is to keep the hemoglobin in the circulation, for as long as hemoglobin remains inside the cell, it cannot diffuse through the capillary membranes.

Genesis of the Red Cells

In the fetus red blood cells are formed in (1) the yolk sac of the embryo, (2) the liver, (3) the spleen, and (4) the bone marrow, but by the time the infant is born, red cell formation no longer occurs in any area except the bone marrow. Furthermore, the marrow of the membranous bones—that is, the skull, the vertebrae, the ribs, the ilia, and the sternum—forms more cells than the marrow of the long bones. As illustrated in Figure 80,

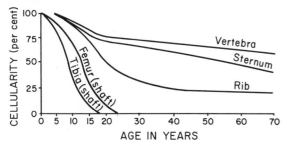

Figure 80. Rate of red blood cell production in the marrow of several different bones. (Redrawn from Wintrobe: *Clinical Hematology.* Lea & Febiger.)

red blood cells continue to be formed in the membranous bones throughout life, whereas they are formed in the shafts of the long bones only during preadolescence. It is evident from this figure, too, that the rate of red blood cell formation in all bones decreases with age. As a result, the number of red blood cells in the circulation also decreases, usually causing mild degrees of anemia in old age.

FORMATION OF THE CELL STRUCTURE. Figure 81 illustrates, to the left, the stages of red blood cell development in the bone marrow, beginning with the *hemocytoblast* and ending with the *erythrocyte*, which is

the mature red blood cell. In the final stage of development—that is, during the change from the *normoblast* to the erythrocyte—the nucleus disintegrates and is absorbed. During this absorption, which continues for a few hours to a few days, the remaining endoplasmic reticulum, and possibly some of the breakdown products of the nucleus, can still be seen as a reticulum, or network, in the cell. The cell at this time is called a *reticulocyte.* Finally, however, this reticulum is absorbed, leaving the cell essentially a bag of hemoglobin.

The erythrocytes, on becoming mature, squeeze through the pores of the capillary walls into the marrow vessels, in this way entering the blood stream. Occasionally, when large numbers of red blood cells are being formed very rapidly, reticulocytes and very rarely normoblasts also enter the circulation before they have turned into erythrocytes. Indeed, the rapidity of red blood cell formation can be estimated simply by counting the number of reticulocytes and normoblasts in the blood.

Special substances needed for red cell formation. Special nutrients are necessary for the formation of the cell structure. These include especially two vitamins, *vitamin B$_{12}$*

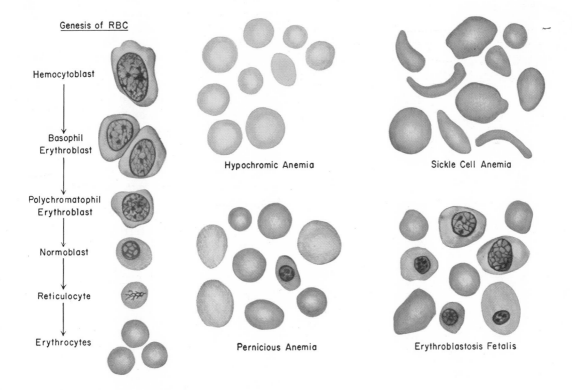

Figure 81. Genesis of red blood cells, and the blood pictures of several types of anemia.

and *folic acid*. Without these the erythro-blasts fail to proliferate rapidly but instead have a tendency to grow larger and larger without dividing as many times as usual. As a result, an oversized erythroblast, called a *megaloblast*, is formed. When the erythrocyte is finally formed it too is greatly oversized, and its membrane is very fragile.

FORMATION OF HEMOGLOBIN. Synthesis of hemoglobin begins in the erythroblast and continues into the erythrocyte stage; even after young erythrocytes leave the bone marrow and pass into the blood stream, they continue to form hemoglobin for many days. The mechanism of synthesis is not completely known, but the five major stages are illustrated in Figure 82. The erythroblast first converts *acetic acid* into *ketoglutaric acid* which then undergoes a series of combinations and changes until the final hemoglobin molecule is formed.

The iron portion (Fe) of the hemoglobin molecule is exceedingly important, for it is with this that oxygen binds, as is discussed below. Experiments in animals have shown

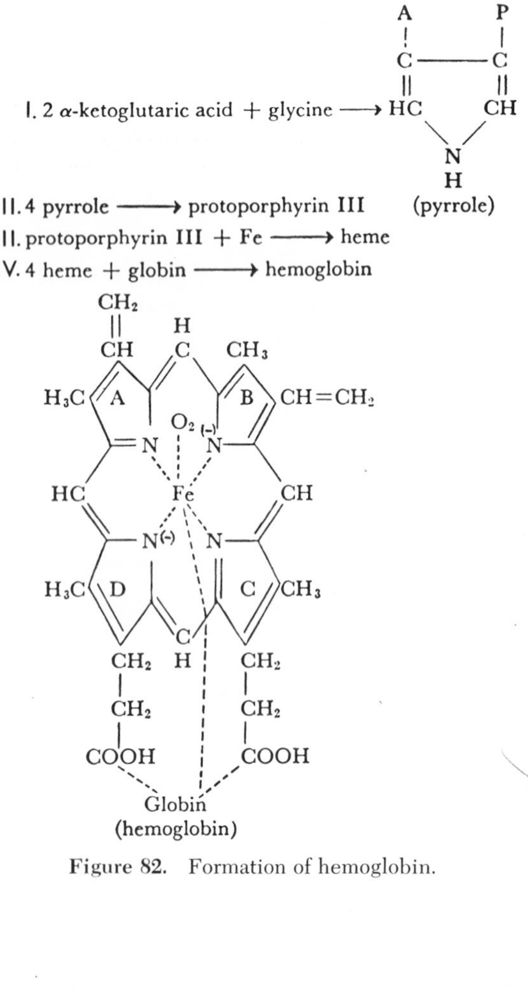

Figure 82.   Formation of hemoglobin.

that *copper, cobalt, nickel,* and the vitamin *pyridoxine* are all necessary for formation of hemoglobin; without any one of these in the diet the quantity of hemoglobin in the red blood cell will be deficient. Yet the precise manner in which each of these helps in the formation of hemoglobin is unknown.

Combination of oxygen with hemoglobin. The most important feature of the hemoglobin molecule is its ability to combine loosely and reversibly with oxygen. The two positive valences of the ferrous iron are firmly bound with other portions of the hemoglobin molecule, but the oxygen molecule (O_2) binds with two of the six so-called "coordination" valences of the iron atom. These valences form an extremely loose bond with the oxygen so that the combination is easily reversible. That is, in the lungs oxygen binds easily with the iron but separates from the iron equally as easily in the tissue capillaries to supply the tissue cells with oxygen.

Iron metabolism. The total quantity of iron in the entire body is about 4.5 grams, and approximately two thirds of this is in the hemoglobin of the blood. When the diet is deficient in iron, the concentration of hemoglobin in the red blood cells falls markedly, sometimes making it difficult to transport adequate amounts of oxygen to the tissues. Therefore, it is important to understand the relationship of iron metabolism to the formation of hemoglobin.

The amount of iron needed each day by the human being is 1 to 3 mg. About 1 mg. is excreted each day into the bowel and urine, and any bleeding can account for loss of another milligram or two of iron from the body. For instance, menstrual bleeding each month in the female accounts for an average loss of about 1 mg. per day when averaged over the entire month.

Only *ferrous* iron can be absorbed from the intestinal tract. Fortunately, however, the acid contents of the stomach can convert most inorganic ferric compounds into ferrous compounds prior to entry into the small intestine.

Iron, like many other nutrients, is absorbed through the intestinal epithelial membrane by an active transport process, though very little is known about the mechanism of this process. As soon as it is absorbed into the blood it combines with a globulin to form the compound, *transferrin*. In this form, iron is carried throughout the body to be used where it is needed, for instance to the bone marrow for use in the formation of hemoglobin.

Regulation of iron absorption from the intestine. Once the body has all the iron that it can use, the plasma globulin which combines with iron to form transferrin becomes saturated. Now, no new iron can be accepted from the gastrointestinal mucosa. Under these circumstances no additional iron is absorbed. Conversely, when the body is deficient in iron, large amounts of plasma globulin again become available to combine with iron from the intestine. Therefore, iron absorption again proceeds rapidly. In this way the absorption of iron from the intestine is automatically regulated so that the proper quantity of iron will be available at all times.

Storage of iron in the liver. A large amount of iron is stored in the liver for use in periods of dietary deficiency. This iron is combined with liver *apoferritin* to form *ferritin.* When great excesses of iron are available to the body all of the apoferritin of the liver becomes ferritin. On the other hand, in times of iron deficiency the iron is released from the ferritin and is carried throughout the body in the form of transferrin to be used where it is needed.

In addition to the major need for iron in the formation of hemoglobin, it is also a key constituent of many other cellular oxidative metabolic substances such as myoglobin and the different cytochromes.

Life of the Red Blood Cell

Once the red blood cell has entered the circulation it circulates an average of 120 days before disintegrating. This long life of the red cell is evident from Figure 83, which

shows the effect on the red blood cell count of cessation of red blood cell formation. The red blood cell count begins to fall rapidly because some of the older cells are already about to disintegrate. By 120 days almost all of the cells are gone, though a few very hardy cells, which had been released into the blood immediately before cessation of red blood cell production, may still continue to circulate for even a few more days.

During the life of the red blood cell some metabolic processes continue to occur. The cytoplasm of the cell contains small amounts of oxidative and other enzymes which liberate the energy needed for maintaining integrity of the cellular structures. Even so, the cell eventually becomes fragile, and when it tries to pass through a tight capillary it finally breaks wide open, emptying its hemoglobin into the circulation. The fragments of the cell structure are engulfed by *reticuloendothelial* cells that line the walls of the capillaries in some organs such as the liver and spleen. These digest the fragments and release the breakdown products into the circulation in a dissolved state.

FATE OF THE HEMOGLOBIN. The hemoglobin released by the rupturing cells at first flows freely in the blood, but within 20 to 60 minutes most of it leaks into the tissue spaces. There it is engulfed by tissue reticuloendothelial cells and is digested by the intracellular enzymes. The breakdown products are *iron* and *bilirubin*. The bilirubin is excreted through the liver into the bowel as will be discussed in more detail in Chapter 30. The iron, on the other hand, is transported in the form of transferrin either to the bone marrow for formation of more hemoglobin or to the liver for storage in the form of ferritin. Thus, the iron is used over and over again and is conserved in the body.

Regulation of Red Blood Cell Production

TISSUE OXYGENATION AS THE BASIC REGULATOR OF RED CELL PRODUCTION. Red blood cell formation is regulated principally by the need of the tissues for oxygen. When a person is exposed to an atmosphere low in oxygen, the rate of red blood cell production increases markedly, or when a person performs very severe and continuous exercise, the relative need for increased quantities of oxygen also increases the rate of red blood

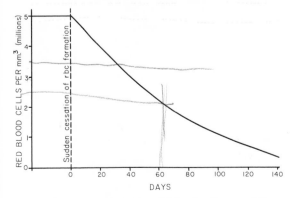

Figure 83. Decrease in red blood cell count following cessation of red blood cell formation.

cell production. Thus, a well trained athlete frequently has a red blood cell count of 6.5 million per cubic mm. in contrast to the normal of 5 million per cubic mm., and persons living at very high altitudes, such as some Peruvian and Chilean natives, often have red cell counts of 7.5 to 8 million. On the other hand, a lazy or bedridden person, whose tissues require little oxygen, may have a red cell count as low as 3 million.

MECHANISM BY WHICH TISSUE OXYGENATION REGULATES RED CELL PRODUCTION— ERYTHROPOIETIN. Even though it is the need of the tissues for oxygen that causes increased red blood cell production, oxygen deficiency does not directly stimulate the bone marrow. Instead, it causes a humoral factor called *erythropoietin* (also known as *erythropoietic stimulating factor* or *hemopoietin*) to be formed. This substance is formed mainly in the kidneys and, to a lesser extent, in the liver and perhaps other tissues. It passes by way of the blood to the bone marrow and mainly stimulates the initial stage in red cell production, the formation of hemocytoblasts from primordial "stem" cells in the bone marrow. Therefore, it is three to four days later before large numbers of red cells are discharged from the bone marrow into the blood stream.

A deficiency of red blood cells in the circulation obviously causes deficient oxygenation of the tissues and therefore greatly enhances the formation of erythropoietin; this in turn accelerates red cell production and returns the total quantity of red cells to normal. Thus, the number of red blood cells in the circulation is automatically regulated in response to the need of the tissues for oxygen, the greater the need, the greater the number of cells, and the less the need, the less the number.

Anemias

Anemia means a diminished concentration of red cells in the blood. Perhaps the most common cause of mild anemia is simple lack of exercise, for, as explained above, exercise is a stimulus to the production of red blood cells. A person with anemia from this cause can be treated with all the substances known to be necessary for red blood cell formation, including iron, vitamin B_{12}, folic acid, and proteins, and still he will continue to have the mild anemia.

In addition to this functional type of anemia, several disease conditions can also cause very serious anemias. Some of these are the following:

BLOOD LOSS ANEMIA. Either acute blood loss or chronic blood loss can cause anemia. Acute loss of blood will immediately decrease the total blood volume, but within approximately 24 hours the plasma portion of the blood will have been replenished by ingested fluids or by fluids derived from the tissues. The red blood cells, however, require several weeks to be replenished. Consequently, after blood loss the red cells are diluted and their concentration in the blood is decreased.

In chronic blood loss the bone marrow becomes hyperactive, attempting to replace the lost red blood cells as rapidly as they are lost, but frequently it cannot keep up with the rate of loss. When blood is lost, some of the body's store of iron is lost as well for approximately two thirds of all the iron in the body is present in the hemoglobin. Therefore, chronic blood loss is likely to deplete the body of iron, leaving insufficient quantities to form new hemoglobin. Consequently, in chronic blood loss anemia, not only is the concentration of cells reduced, but the quantity of hemoglobin in the cells that are formed is often also far less than normal.

IRON DEFICIENCY ANEMIA. Even without chronic blood loss, a person who does not ingest a sufficient quantity of iron may develop anemia. The actual number of red blood cells may be almost normal, but their size will be small because of insufficient hemoglobin, and the concentration of hemoglobin in each cell may be greatly decreased. This gives rise to the condition called *hypochromic* anemia, this term meaning cells with less than normal red color from hemoglobin. Cells of this type are illustrated in Figure 81. In this condition, even though a large number of cells may be circulating in the blood, the amount of oxygen that can be transported to the tissues is far less than normal.

ANEMIAS CAUSED BY FRAGILE CELLS. Anemia can also result from *fragile cells*, whose membranes rupture on the slightest provocation. One of the causes of fragile cells is *familial microcytosis*, a hereditary disease that causes the red blood cells to be much smaller than usual and spherical in shape, rather than having the normal flat baglike shape. When these tightly filled cells are compressed even slightly, such as when they attempt to squeeze through tight capillaries, they rupture.

Another type of anemia caused by destruction of the red blood cells occurs in *erythroblastosis fetalis*. In this condition a pregnant mother becomes immunized against her child's blood, and antibodies (a special type of protein) from the mother's body pass through the placenta into the baby to cause destruction of the child's red blood cells. As the anemia becomes severe in the infant, the infant's production of red blood cells becomes extremely rapid in an attempt to compensate for the anemia, causing many very early cells, normoblasts and even erythroblasts, to be extruded from the bone marrow into the blood. Figure 81 shows the resulting blood picture.

ANEMIA CAUSED BY POOR PRODUCTION OF CELLS. A final cause of anemia is poor production of the red blood cells themselves. In the preceding discussion of the formation of red blood cells, it was pointed out that certain nutrients are required for formation of red blood cells. Lack of any of these nutrients will result in anemia. When vitamin B_{12} or folic acid are deficient, the number of cells is greatly reduced, and those that are produced are large, odd shaped, and quite fragile. This type of anemia is known as *pernicious anemia*, the blood cells of which are shown in Figure 81. When this type of anemia is successfully treated with vitamin B_{12} or folic acid (only a few milligrams a day are usually required) the number of red blood cells begins to increase dramatically within a few days. Reticulocytes, an early form of red blood cell as explained above, immediately become a very large percentage of the total number of cells in the circulation, and within a month the person is completely cured.

On rare occasions the bone marrow completely stops producing red blood cells, in which case the person is said to have *aplastic anemia*. This often occurs very suddenly when the bone marrow is poisoned by some drug, or when the bone marrow is exposed to intense ionizing radiations such as x-rays or the gamma rays emitted by the explosion of a nuclear bomb. In aplastic anemia those cells that are already in the circulation are completely normal, but because the aging cells are no longer replaced by new cells, the red cell count diminishes during the next few weeks, giving a very severe anemia after about a month.

PHYSIOLOGIC EFFECTS OF ANEMIA. Anemia affects the functions of the circulation in two ways. First, the decrease in oxygen-carrying capacity of the blood decreases the rate of oxygen transport to the tissues, which results in poor tissue oxygenation and can cause extensive damage throughout the body or even death. The second effect is decreased viscosity of the blood, for the viscosity is determined almost entirely by the concentration of red blood cells. The resistance to blood flow decreases when the viscosity decreases. Therefore, the low viscosity allows very rapid flow of blood through the peripheral circulation, promoting excessive cardiac output, dilatation of the heart, and overworking of the heart. Consequently, one of the results of anemia is often heart failure because of the extra load imposed.

Polycythemia

Polycythemia means an excess of red blood cells. Mild polycythemia is normal in persons who exercise excessively and also in persons who live at high altitudes, but still other persons develop a condition called *polycythemia vera*, in which the rate of red blood cell production becomes far greater than normal even though there is no apparent physiologic basis for the increased production. It is believed that this disease results from a tumorous condition of the bone marrow. Presumably the hemocytoblasts simply begin producing far too many red blood cells, and the normal regulatory mechanisms fail to stop this production.

PHYSIOLOGIC EFFECTS. All of the important physiologic effects of polycythemia are caused by the increased viscosity of the blood that results. Sometimes the viscosity becomes as much as five times normal, which obviously greatly impedes the flow of blood through the circulatory system. Often the cells conglomerate in the small vessels, stopping flow entirely and resulting in many small blocks in the circulation. In approximately one third of polycythemic persons, blood flows through the systemic circulation so poorly that the arterial pressure rises far above normal, thereby placing an increased load on the heart. In this way, polycythemia like anemia—but for different reasons—can overload the heart and can also result in heart failure.

THE WHITE BLOOD CELLS

The total number of white blood cells in the circulation averages only $1/500$ the number of red blood cells. Yet these cells are so important in protecting the body against disease that each type of white blood cell deserves special consideration.

Genesis of White Blood Cells

Most white blood cells are formed in the bone marrow along with the red blood cells. These include the *neutrophils*, the *eosinophils*, the *basophils*, and the *platelets*. All of these originate from the *myeloblast*, which, throughout the life of the person, continues to divide and to evolve the different types of white cells. The stages of white cell development are shown in Figure 84. When each cell reaches maturity it is emptied from the bone marrow into the circulation.

Platelets, which are also often considered to be white blood cells, are produced in the bone marrow in the following way: The marrow produces very large fragile cells called *megakaryocytes*, illustrated by cell number 3 in Figure 84. When these are mature they suddenly fragment into many minute parts which become the platelets. The platelets have a diameter approximately one-seventh that of a white blood cell and a volume about $1/300$. Though the platelets are extremely important for the coagulation of blood, they have no specific relationship to the other white blood cells. They will be discussed in Chapter 10 in relation to coagulation, instead of in the present chapter.

Lymphocytes and *monocytes* are produced in *lymphoid tissue* throughout the body instead of in the bone marrow. Most of the lymphoid tissue is in the *lymph nodes*, which filter the lymph as it flows through the lymphatics, but large amounts of lymphoid tissue are also present in the spleen, thymus gland, and submucosa of the gastrointestinal and respiratory tracts. After lymphocytes are formed in the lymphoid tissue, they are emptied into the lymphatic channels and eventually flow along with the lymph into the circulation.

Life of the White Blood Cells

On studying the bone marrow with a microscope, it appears that the bone marrow produces approximately as many white blood cells as red blood cells. This makes it difficult at first to understand why the number of white blood cells in the circulation is very small in comparison with the number of red blood cells. However, the white blood cells, as opposed to the red cells, are capable of leaving the circulation by squeezing through the small pores of the capillaries and then passing through the tissue spaces. Furthermore, white blood cells rarely exist in the circulation or in the tissues for longer than a few hours to a few days, for as they perform their functions to protect the body against disease, they themselves are usually destroyed. This explains why the number of white blood cells in the circulation is very slight.

Neutrophils

The neutrophils are the most important white blood cells for protecting the body against *acute* invasion by bacteria. These cells are approximately 12 microns in diameter and can pass rapidly through the pores of the capillaries by the process of *diapedesis* (squeezing through). On entering the tissue spaces they attack almost any agent that may be causing tissue damage.

AMEBOID MOTION OF THE NEUTROPHILS. The neutrophils exhibit more ameboid motion than perhaps any other cell of the body. That is, they move through the tissues by projecting a *pseudopodium*—a fingerlike extension—ahead of the main mass of the cell and then allowing the remainder of the cell contents to "stream" into the pseudopodium as was described in Chapter 3. This process is repeated over and over again until the cell reaches its destination.

The direction in which the pseudopodium projects and moves the neutrophil is determined by chemical substances in the tissue spaces, a phenomenon called *chemotaxis* which may be explained as follows: When a tissue area is damaged it releases several

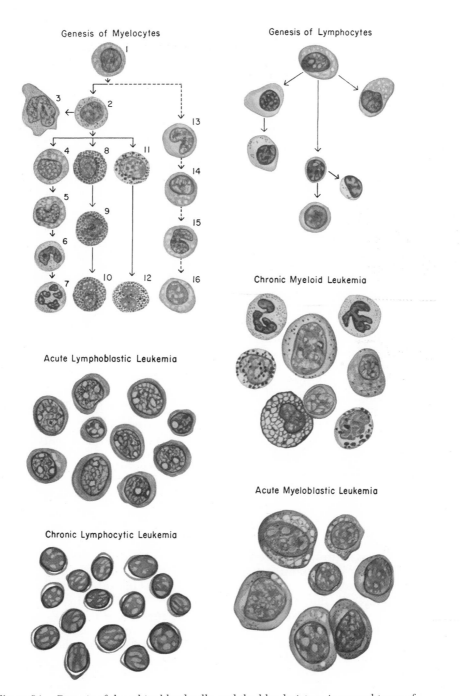

Figure 84.   Genesis of the white blood cells, and the blood picture in several types of leukemia. Cell number 1 is a *myeloblast*, and number 3 is a *megakaryocyte*. Cells 4, 5, 6, and 7 illustrate the formation of *neutrophils;* 8, 9, and 10, the formation of *eosinophils;* 11 and 12, the formation of *basophils;* and 13, 14, 15, and 16, the formation of *monocytes*. (Redrawn in part from Piney: *A Clinical Atlas of Blood Diseases.* The Blakiston Co.)

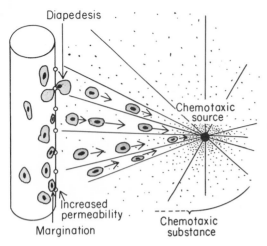

Figure 85. Movement of neutrophils by the process of *chemotaxis* toward an area of tissue damage.

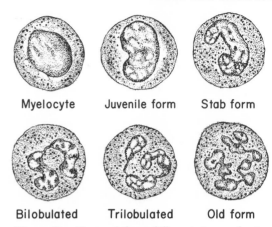

Figure 86. Neutrophils at different stages of aging, showing above the early forms, which appear in the circulation during rapid production of neutrophils, and below the older forms normally present.

chemotaxic substances, one of which is called *leukotaxine*. These substances diffuse in all directions. Obviously, the greatest concentration of the substances will be in the damaged area, and their concentrations become progressively less farther and farther away from the area of damage. The neutrophil, for reasons that are not completely understood, always moves toward an area of higher concentrations of the chemotaxic substances, until it finally reaches the damaged tissue.

Figure 85 exhibits the complete mechanism of the movement of neutrophils to an area of tissue damage, showing (1) increased porosity of the capillaries caused by toxins from damaged tissues, (2) adhesion of neutrophils to the wall of the capillary in the damaged area, a process called *margination*, (3) diapedesis of the neutrophils through the enlarged pores, and (4) ameboid movement of the neutrophils to the damaged area, guided by the process of chemotaxis.

PHAGOCYTOSIS BY THE NEUTROPHILS. Neutrophils have the ability to ingest, or *phagocytize*, particles that are foreign to the tissues, including especially bacteria and to a lesser extent broken down tissue debris. When the neutrophil comes in contact with the particle, its membrane first attaches to the particle and then rapidly spreads in all directions around the particle so that the particle becomes engulfed in a deep well inside the cell.

DIGESTION OF FOREIGN PARTICLES BY NEUTROPHILS. Neutrophils contain in their lysosomes digestive enzymes that are capable of digesting most ingested particles. These enzymes are especially adapted to digest the proteins of bacterial bodies. After digestion, all that remains of a bacterium is many small chemical compounds which are readily dissolved in the intracellular fluids of the neutrophils.

Ordinarily, the neutrophil can phagocytize and digest 5 to 25 bacteria before the neutrophil itself becomes exhausted and dies, but on occasion the bacteria are so toxic to the neutrophils that only one or two can kill a neutrophil.

NEUTROPHILIA. Neutrophilia means an increase in the number of neutrophils in the circulation. It almost always occurs whenever tissues are damaged severely anywhere in the body. The damaged tissues, in addition to releasing leukotaxine and other chemotaxic substances, also release another substance called *leukocytosis promoting factor*. This substance diffuses into the capillaries and flows in the blood to the bone marrow. Here it causes rapid release of leukocytes from the bone marrow into the blood. Because by far the largest number of leukocytes produced by the bone marrow is neutrophils, this substance increases especially the number of circulating neutrophils. Sometimes neutrophils are emptied into the circulation so rapidly that their number increases to two to three times normal in less than an hour.

The more rapid the release of leukocytes from the bone marrow, the earlier the form of the neutrophil. Figure 86 illustrates the early and old types of neutrophils, beginning with the myelocyte and progressing to the very old multilobulated form. Ordinarily neutrophils are not released into the circulation until they have progressed to the bilobulated or trilobulated stages, but under the stimulus of leukocytosis promoting factor, large numbers of stab, juvenile, and even myelocyte forms are also released.

When a much larger number of neutrophils than normal appears in the circulation, and especially when these include many early cells, one can be quite certain that somewhere in the body severe tissue damage is taking place. Most frequently this damage is caused by a serious bacterial infection, but other types of damage, such as that resulting from occlusion of a blood vessel, damage to the tissues by trauma, or nuclear bomb radiation, can also lead to *neutrophilia*.

Monocytes (Macrophages)

Monocytes function much like the neutrophils, though they have a considerably different appearance and have an entirely different origin, from the lymph nodes rather than from the bone marrow. They are often as large as 20 microns in diameter, they do not contain the granular substance characteristic of the neutrophils, and they do not have a multilobulated nucleus. Monocytes move through tissues by ameboid motion, but at first somewhat more slowly than do neutrophils. However, after monocytes have been in the tissues for several hours, they swell to become so-called *tissue histiocytes* or *macrophages*. These cells now develop rapid ameboid motion and proceed quickly to the site of tissue damage. When the monocytes (now converted to macrophages) do finally reach the area of tissue damage, they are capable of engulfing several times as many bacteria and far more tissue debris than can neutrophils, often engulfing as many as 100 bacteria per cell. In other words, the phagocytic properties of macrophages are far superior to those of neutrophils.

Macrophages contain the same digestive enzymes as neutrophils plus others not present in the neutrophils, including especially lipases which can destroy the protective fatty shell of such bacteria as the tubercle bacillus.

When a bacterial disease lasts more than a few days, the proportion of monocytes in the circulating blood becomes greater and greater until finally their number may be as great as that of the neutrophils. Therefore, even though in the earlier stages of tissue damage almost all of the phagocytic cells reaching the damaged area are neutrophils, in the later stages the monocytes often outnumber the neutrophils. Thus, neutrophils are the most important white blood cells for resisting very acute infection, while monocytes are more important for resisting long-term, chronic infections. Also, because monocytes (macrophages) are large enough to phagocytize very large particles of tissue debris, they are much more important than neutrophils in cleaning up the tissues after an infection has been successfully combated.

Eosinophils

Eosinophils are closely related to neutrophils, though they have larger granules that stain strongly with acidic dyes. These cells exhibit chemotaxis, phagocytosis, and ameboid motion much the same as the neutrophils, but to so much less extent that they are not especially effective as scavengers in areas of tissue damage.

Unfortunately, the precise function of eosinophils is not yet known, but it has been suggested that they might detoxify foreign proteins, for they enter the blood in large numbers after foreign proteins are injected. Furthermore, many eosinophils are present in the mucosa of the intestinal tract and in the tissues of the lungs where foreign proteins normally enter the body.

The total number of eosinophils in the circulating blood increases greatly during allergic reactions. Here again, there is no reasonable theory to explain this effect, though it is possible that the allergic reaction releases toxic products from the tissues that might be detoxified by the eosinophils.

Another common cause of extremely large numbers of eosinophils in the blood is infection with parasites. For instance, in the condition known as *trichinosis*, which results from invasion of muscle by the trichinella parasite ("pork worm") after eating inadequately cooked pork, the percentage of eosinophils in the circulating blood occasionally rises to as high as 25 to 50 per cent of all the leukocytes. The cause of this is obscure, though it might again be a reaction in which the eosinophils are detoxifying foreign products produced by the parasites.

Basophils

The basophils are also similar to the neutrophils, except that the granules in their cytoplasm are even larger than those of the eosinophils and the granules stain darkly with basic dyes. The basophils exhibit very mild ameboid motion, phagocytic activity, and chemotaxis, but it is doubtful that these properties make the basophils of any significance in protecting the tissues against damage.

Though the precise function of the basophils is yet unknown, these cells are similar to or identical with the large *mast* cells which are found adjacent to the outside of the capillaries throughout the body. Mast cells are especially abundant in the pericapillary areas of the liver and the lungs, and they secrete a substance called *heparin*. Heparin diffuses into the blood to prevent blood coagulation, which is one of the reasons why the blood normally does not coagulate in the circulation. It is postulated that the basophils circulating in the blood also release small quantities of heparin and thereby help to prevent intravascular blood coagulation.

Lymphocytes

Until the last few years it was believed that the blood lymphocytes had already served their usefulness in the lymph nodes and had been released into the blood to be destroyed, that is, that they were "spent" cells. However, we now know this to be the farthest from the truth, because lymphocytes turn out to be closely akin to the original germ cells responsible in the embryo for forming all the tissues of the body. The lymph nodes release lymphocytes into the blood to be dispersed through the body to perform many functions as follows:

First, the lymphocyte can swell and become a *monocyte*, which can then enter the tissues to become a tissue macrophage (also called tissue histiocyte) where it performs the immensely important protective function against bacterial invasion and other types of tissue damage, as was explained earlier in the chapter.

Second, lymphocytes can enter the bone marrow and become either *hemocytoblasts* or *myeloblasts*, which can then divide to form either red blood cells or granulocytic types of white blood cells.

Third, lymphocytes can enter any tissue of the body and be converted into *fibroblasts*. The fibroblasts in turn secrete substances that later become collagen fibers, elastic fibers, hyaluronic acid gel, and other components of connective tissue. In other words, the lymphocyte is a direct ancestor of our connective and supportive tissues of the body as well as of many other tissues.

Fourth, lymphocytes can become *plasma cells* which in turn have the ability to form and secrete antibodies responsible for immunity against toxins. These cells will be discussed in Chapter 9 in relation to the immune process.

In summary, the lymphocyte is a *multipotential cell* capable of being converted into many other types of cells throughout the body, thereby maintaining the productive capabilities of such proliferative tissues as bone marrow, connective tissue, and immune body producing tissues. At one point in the history of medicine, because of ignorance about the lymphocyte, it was fashionable to irradiate the lymph nodes with x-rays to suppress their activity, on the assumption that this was beneficial in many disease conditions. Now, it appears that even the longevity of the human being might depend on the supply of lymphoid tissue, which points up the folly of drastic actions conceived in ignorance.

REGULATION OF LYMPHOCYTE PRODUCTION. Lymphoid tissue increases when toxic and infectious diseases occur in the body, much the same way that neutrophil and monocyte production also increases. Unfortunately, we do not know the cause of this increase, though it does seem to be a direct response of the lymphoid tissue itself to bacterial proteins or toxins. For instance, if a serious infection occurs in an arm, toxic proteins from bacteria are carried in the lymph to lymph nodes at the elbow and shoulder. Within hours these nodes become tremendously hypertrophied, and inside the nodes many of the lymphoid cells become converted into macrophages that destroy the toxins before they can be emptied by the lymph into the circulatory system.

Differential Count of White Blood Cells

A useful procedure in the study of many disease conditions, especially infectious diseases, is to count the number of each type of white cell in a cubic millimeter of blood. This is called a "differential count." This count is made by accurately diluting a small amount of blood, placing this mixture in a calibrated glass counting chamber, and counting the different types of cells, using a microscope. To illustrate the value of such a procedure, when the number of neutrophils is greatly increased, one has reason to believe that serious tissue damage is occurring, and that more likely than not the damage is being caused by bacterial infection. When the number of monocytes in the circulation is greatly increased, one suspects a chronic infection such as tuberculosis, chronic fallopian tube infection in the female, or some other long-term disease. A high eosinophil count can be of aid in diagnosing allergic conditions or parasitic infestation, and, finally, platelet counts are of special importance in determining the cause of excessive bleeding in certain patients.

Figure 87 illustrates the number of the different white blood cells in each cubic millimeter of blood at different ages. It will be noted that in the adult the total number of white blood cells is normally about 7500 per cu. mm. About 65 per cent of these are granulocytes, and of this 65 per cent approximately 62 per cent are neutrophils, 2 per cent eosinophils, and 1 per cent basophils. The normal proportion of lymphocytes in the white blood cells is approximately 30 per cent. The remaining 5 per cent are monocytes. In severe acute infections the percentage of granulocytes can rise from 65 per cent to as high as 95 per cent, and the total number of white blood cells from 7500 to as many as 30,000 to 50,000. In long-term chronic infections the percentage of monocytes can rise from 5 per cent to as high as 40 per cent and the total number of white blood cells to 15,000 to 20,000.

Leukemia

Leukemia means an excess of white blood cells in the circulation caused by cancer of the white cell producing tissue. The leukemias caused by cancer of the bone marrow are called *myelogenous leukemias,* while those caused by cancer of the lymphoid tissue are called *lymphocytic leukemias.* Leukemias have all degrees of severity. When the cancerous condition is spreading extremely rapidly, the cells released into the circulation are usually very early forms—in the case of myelogenous leukemia *myeloblasts,* and in the case of lymphocytic leukemia *lymphoblasts.* A person having one of these rapidly developing leukemias usually dies within six months to a year. The cause of death is the same as that in most rapidly developing cancers, namely, general debility of the body because of loss of nutrition to the cancerous cells.

In the slowly developing leukemias, the cells released into the circulation may be almost entirely normal and may have the appearance of slightly modified neutrophils,

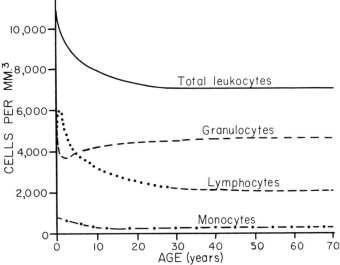

Figure 87. Concentration of the different types of leukocytes in the circulation at different ages.

eosinophils, basophils, monocytes, or small lymphocytes, depending on which particular type of cell in the bone marrow or lymph tissue has become cancerous. Often the total number of white blood cells will rise to as high as 100,000 to 200,000 per cu. mm. instead of the normal number of about 7500. Indeed, the number of circulating white blood cells is usually considerably greater in chronic leukemia than in acute leukemia, because in acute leukemia the cancerous condition frequently limits itself to the bone marrow or lymphoid tissue without releasing large numbers of cells in the circulation. Eventually, even the chronic type of leukemia will lead to inanition and death of the person, though frequently persons with this type of leukemia live as long as 12 to 20 years before this inevitable result.

The white cells found in the circulation in several different types of leukemia are illustrated in Figure 84. These show the large lymphoblasts and myeloblasts characteristic of the acute leukemias, and the small lymphocytes and relatively normal myelocytes characteristic of the more chronic leukemias.

Leukopenia

Leukopenia means decreased number of white blood cells in the circulation. This usually results from damage to the bone marrow by poisons, by toxic reaction to drugs, or by ionizing radiations from x-ray or nuclear bomb exposure. The lack of neutrophils, basophils, and eosinophils is called *agranulocytosis*. When the number of white blood cells falls extremely low the body becomes unprotected against bacterial invasion. Usually within three to four days very serious ulcerative infections begin to appear in the mouth, the colon, and other regions of the gastrointestinal or respiratory tracts, for these tracts always contain many pathogenic bacteria that normally are held in abeyance by continual action of the white blood cells. If the person is not treated with antibiotics or other bacteria-resisting drugs, he usually will die of fulminating infection within another day or two.

REFERENCES

Atamer, M. A.: *Blood Diseases.* New York, Grune & Stratton, 1963.

Athens, J. W.: Blood: leukocytes. *Ann. Rev. Physiol.* 25:195, 1963.

Ciba Foundation Study Group No. 10: *Biological Activity of the Leucocyte.* Boston, Little, Brown and Company, 1961.

Halpern, B. N.: *Physiopathology of the Reticulo-Endothelial System.* Springfield, Ill., Charles C Thomas, 1957.

Harris, J. W.: *The Red Cell—Production, Metabolism, Destruction: Normal and Abnormal.* Cambridge, Harvard University Press, 1963.

Jacobson, L. O., and Doyle, M. (eds.): *Erythropoiesis.* New York, Grune & Stratton, 1962.

Karnovsky, M. L.: Metabolic basis of phagocytic activity. *Physiol. Rev.* 42:143, 1962.

Manis, J. G., and Schachter, D.: Active transport of iron by intestine: features of the two-step mechanism. *Amer. J. Physiol.* 203:73, 1962.

Prankerd, T. A. J.: *The Red Cell: An Account of Its Chemical Physiology and Pathology.* Philadelphia, F. A. Davis Company, 1961.

Rebuck, J. W. (ed.): *The Lymphocyte and Lymphocytic Tissue.* New York, Paul B. Hoeber, 1960.

Riggs, A.: Functional properties of hemoglobins. *Physiol. Rev.* 45(4):619, 1965.

Weissbach, H., and Dickermann, H.: Biochemical role of vitamin B_{12}. *Physiol. Rev.* 45:80, 1965.

Whitby, L. E. H., and Britton, C. J. C.: *Disorders of the Blood.* New York, Grune & Stratton, 1963.

CHAPTER 9

THE RETICULOENDOTHELIAL SYSTEM, IMMUNITY, AND ALLERGY

The role of white blood cells in protecting the body against invading bacteria was discussed in the previous chapter. However, other factors also provide protective roles against either foreign organisms or damaging toxins. These include, especially, the reticuloendothelial cells and the process of immunity, both of which will be described in this chapter. Also, we will present some of the problems of allergy, which is a side effect of immunity.

THE RETICULOENDOTHELIAL CELLS

Reticuloendothelial cells are closely allied to white blood cells. They are primitive cells that have intense ability to phagocytize particles foreign to the body. However, in contrast to white blood cells, reticuloendothelial cells are mainly sessile, usually lining vascular and lymphatic channels. One of the most distinctive types of reticuloendothelial cells is the *Kupffer cells* that line the liver sinusoids, but similar cells line sinusoids of the spleen, the lymph nodes, and the bone marrow.

One will note that the reticuloendothelial cells are propitiously located for removal of invading organisms or toxins before they spread widely. For instance, large numbers of colon bacilli are absorbed from the large intestine into the portal blood, but almost none of these succeeds in passing through the liver sinusoids without being phagocytized by the Kupffer cells, which are illustrated in Figure 88. Indeed, motion pictures have shown that a colon bacillus, on coming in contact with the membrane of a Kupffer cell, enters the cell within the next $1/100$ of a second, illustrating the extreme propensity of these cells for removal of foreign debris from the circulation.

The reticulum cells in the lymph nodes also play a special role in cleansing the body fluids, for all lymph must pass through the sinuses of lymph nodes before entering the blood. Therefore, fluids draining from an infected area, carrying with them live bacteria, are first cleansed by the lymph nodes prior to reaching the general circulation.

In virtually all tissues of the body are cells called *macrophages* or *tissue histiocytes* that have phagocytic properties almost identical with those of the reticulum cells in the lymph nodes and the Kupffer cells in the

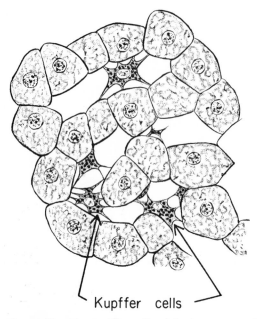

Kupffer cells

Figure 88. The Kupffer cells of the liver sinusoids. These are typical reticuloendothelial cells. (Redrawn from Smith and Copenhaver: *Bailey's Textbook of Histology.* The Williams & Wilkins Company.)

liver sinuses. The tissue histiocytes are also closely related to monocytes and lymphocytes, as was discussed in the previous chapter. Therefore, it is evident that the reticuloendothelial cells and white blood cells are all part of a single large bodily system. This is called the *reticuloendothelial system.*

INFLAMMATION AND WALLING OFF OF INFECTED TISSUE AREAS

Another important process for preventing spread of infection is the process of *inflammation* which occurs in any tissue that becomes damaged. This process may be described as follows:

Damage to the tissues causes several substances to be released from the cells. One of these is *leukotaxine* and another is *necrosin*. Leukotaxine attracts polymorphonuclear neutrophils from the blood stream into the infected area as is described in the previous chapter. Necrosin has several important effects: First, it increases the permeability of the capillaries in the inflamed tissues, which allows fluid, proteins, and white blood cells to leak almost unabated into the area. Second, necrosin activates

the fibrinogen in the leaked fluid, and thereby causes the fluid to clot in the same manner that blood clots, as is described in the following chapter. Clotting of these fluids prevents their flow from the damaged tissue area into the surrounding areas and, therefore, provides so-called "walling off" of the inflamed tissue from the surrounding tissues. This is especially important if the damage to the tissues is caused by bacteria, because it prevents the spread of bacteria.

IMMUNITY

Some bacteria and toxins are not phagocytized by the reticuloendothelial cells but instead must be destroyed in some other way. Fortunately, the body has still another line of defense called *immunity.* The blood and other body fluids contain special proteins called *antibodies* that react chemically with the invading organisms or toxins to destroy them.

NATURAL IMMUNITY. All human beings have natural immunity to certain diseases. For instance, distemper, a disease that kills approximately 50 per cent of all dogs that become infected with it, is totally noninfective to the human being, and the virus disease herpes simplex, which is lethal to rabbits, usually causes nothing more than fever blisters in the human being. This natural immunity results at least partly from the presence of *natural antibodies* in the plasma that have the ability to combine with the invading toxins or bacteria. However, as we shall see later in the chapter, most of the antibodies are not natural but instead are formed only *after a person has been exposed to the disease agent.*

SPECIFIC IMMUNITY. In addition to natural immunity, the human being can develop an extremely high degree of specific immunity against toxins or organisms to which he becomes exposed. For instance, a very minute quantity of tetanus toxin, approximately one millionth of a gram, is sufficient to cause death of a person. However, if a person is exposed several times to less than a lethal dose of this toxin, he can develop so many antibodies to the toxin that as much as 100,000 times the original lethal dose can be injected without causing death.

The Immune Mechanism

ANTIGENS. When bacteria or viruses enter the body, some of them eventually die, then disintegrate, and their proteins are emptied into the circulating fluids. These proteins, as well as the proteins of the various toxins, are called *antigens*, and they initiate the immune process which causes the production of antibodies as is described in the following paragraphs.

FORMATION OF ANTIBODIES. Antibodies are formed to a slight extent in all cells, but certain cells are specialists in this task, the most important of which are the *plasma cells* in lymphoid tissue. In addition, the large lymphocytes and the reticulum cells lining the sinus walls of the bone marrow, spleen, and liver can form small quantities of antibodies.

The mechanism by which antibodies are formed is still conjectural. The first step is phagocytosis of the antigen by the plasma cell. Then the general process is probably that shown in Figure 89. The dumbbell-shaped object is an antigen, which is believed to alter some of the celluar enzymes in such a way that they thereafter produce antibodies, shown in the figure as the small,

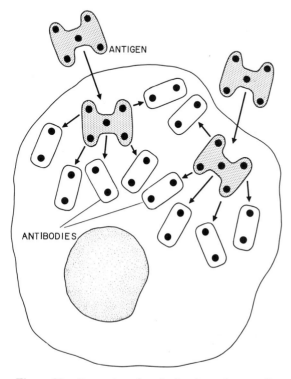

Figure 89.   Formation of antibodies by a plasma cell.

oblong rods. These are proteins that have mirror-image electrical charges on their surfaces, opposite in polarity to those on the antigens. Because of the opposite charges, the antibodies have an intense affinity for the antigen. After the cell has produced a large quantity of antibodies, the cell membrane ruptures and spills most of its cytoplasm into the lymph which then flows into the blood.

Even after the antigen has been destroyed, the plasma cells still continue to form antibodies. The reason for this is that once an antigen has been inside a cell, the cell's enzyme system has become permanently altered to produce the specific antibodies. Therefore, a single exposure to a toxic antigen sometimes makes the person immune to this toxin for long periods of time thereafter. In fact, when the plasma cells or other cells producing antibodies undergo mitosis to form new cells, each of the daughter cells also is still capable of producing antibodies, and this heritage continues for at least several generations of cells. However, the rate of antibody production progressively decreases over a period of months to a few years.

SPECIFICITY OF IMMUNE BODIES. Ordinarily, antibodies formed against a particular antigen will react only with that antigen. For instance, antibodies that protect against tetanus toxin are generally of value only for protection against this single toxin and not against any other type. Occasionally, however, toxins or bacterial proteins are sufficiently alike that antibodies formed against one will react against a second similar toxin or protein. In this instance, it is said that *cross-immunity* has developed between the two different agents.

Function of the Thymus Gland in Immunity

Until the last few years, almost nothing has been known about the function of the thymus gland. It has now been discovered that removal of the thymus gland from a newborn animal or from a fetus shortly before birth causes loss of ability to form significant quantities of antibodies throughout the life of that animal. Therefore, in some mysterious way the thymus plays a key role in the development of the immune mechanism. It is believed that this occurs in the following way:

During late fetal and early neonatal life, the thymus gland is a large organ, composed mainly of a lymphoid type of cell called *thymic cells*. These cells are produced very rapidly and are emptied into the circulating blood. As they pass through the lymph nodes, they become trapped in the sinusoids and become reticulum cells of the type described earlier in the chapter. Throughout life these reticulum cells can be converted immediately into plasma cells whenever a foreign antigen appears. Within the next few days, the plasma cells produce large quantities of antibodies that in turn destroy the antigens.

Therefore, it is obvious that loss of the thymus gland by a fetus or by a newborn infant can be extremely dangerous to the ability of that person to protect himself thenceforth against invading organisms.

IMMUNOLOGIC TOLERANCE TO ONE'S OWN PROTEINS. If the plasma cells should produce antibodies against the proteins of a person's own body, these antibodies would rapidly kill the person. Therefore, it is essential that the immune mechanism be geared to produce antibodies only against *foreign* proteins rather than against one's own proteins. The postulated mechanism by which this occurs is the following: When the thymus gland first begins to produce thymic cells, these cells are strongly exposed to the body's own proteins while they are still in the thymus gland. Furthermore, it is believed that different thymic cells produce different types of antibodies. Those that produce antibodies against the person's own proteins immediately react with these proteins; for reasons not fully understood, this reaction kills these thymic cells. They are subsequently removed by phagocytes also present in the thymus gland. The remaining thymic cells are only those that do not have the potential to react with the person's own proteins, and it is these cells that multiply and are later distributed to the lymph nodes.

As a consequence, the plasma cells produced in the lymph nodes are capable of forming antibodies against almost all types of proteins besides those present in the person's own body. Indeed, even the proteins of another person are sufficiently different that antibodies will be produced against these as well.

Antigen-Antibody Reactions

Antibodies can protect against disease or toxins in a number of different ways. Some-

times the antibodies react directly with the toxins and destroy their toxicity. At other times they cause the noxious agent to precipitate, agglutinate, rupture, or be phagocytized by white blood cells. The mechanisms of these reactions are the following:

NEUTRALIZATION. The method by which most toxins are prevented from damaging the body's tissues is *neutralization*. Figure 90 illustrates a toxin with five separate *reactive areas* indicated by the large black dots. Each of these could cause toxic damage to the body's tissues. An antibody has attached itself to the antigen at each one of these reactive areas, thus preventing the reactive areas from damaging the tissue cells.

PRECIPITATION. A similar method for preventing damage by toxins or other antigens is to cause them to *precipitate*. In this case the antibodies, which usually have two reactive areas, form bridges between the antigen molecules, connecting large numbers of antigens together. When the aggregate becomes large enough, it forms a precipitate and is usually phagocytized and digested by the reticuloendothelial cells.

AGGLUTINATION OF ORGANISMS. The antigens of bacteria are part of the bacterial cell itself, and the antibodies attach themselves to the antigens on the surfaces of the bacteria. Since antibodies normally have two reactive areas, they can form bridges between adjacent bacterial bodies, causing large aggregates of bacteria to clump together. This process is called *agglutination*. Agglutinated bacteria are unable to spread through the body tissues and eventually are phagocytized.

LYSIS OF ORGANISMS. Antibodies can also cause bacteria to become *lysed*, which means rupturing of the cell with spillage of its contents. Lysis begins with many antibodies attaching themselves to protein anti-

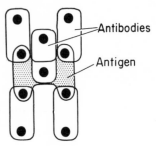

Figure 90. Neutralization of an antigen by antibodies.

gens of the bacterial membrane. Then a substance called *complement,* a fatty material in the extracellular fluid, attaches to the other end of each antibody. Attachment of the antibodies by themselves does not cause lysis of the cells, but once complement attaches, the fat of the complement dissolves the cell membrane and causes the bacterium to break wide open.

OPSONIZATION AND PHAGOCYTOSIS OF BACTERIA. Occasionally, when antibodies attach to the surface of the bacterium, the attachment of complement to the other ends of the antibodies still will not cause rupture of the cell. Yet, the presence of complement causes the bacterium to become especially susceptible to phagocytosis by the reticuloendothelial cells. The probable reason for this is that the complement, being a fatty substance is miscible with the lipid membrane of the phagocytes. This process is called *opsonization,* meaning increased susceptiblity of the bacterial cells to phagocytosis.

Tissue Immunity Versus Humoral Immunity

Humoral immunity means immunity resulting from antibodies flowing freely in the body fluids. However, antibodies can also remain inside cells rather than enter the circulating body fluids. This results in a type of immunity called *tissue immunity.* The body retains tissue immunity many times as long as it retains humoral immunity. For instance, immunity to such infectious agents as smallpox virus may last for as long as ten years to a lifetime. One of the reasons for this seems to be the following: The cells that form the humoral immune bodies are the plasma cells and other reticuloendothelial cells that have life spans of only a few days. With each new generation of these cells, the rate of antibody formation decreases. Therefore, the quantity of humoral antibodies also decreases. On the other hand, many tissue cells, such as neuronal cells, muscle cells, and others, have extremely long life spans, and since these cells, too, can probably produce antibodies that remain in their cytoplasm, they can maintain immunity against noxious agents sometimes for the entire life of the individual. Thus, a person is often immune to tissue damage because of antibodies within his cells even when no humoral immunity can be demonstrated.

Active Immunity Versus Passive Immunity

The development of antibodies in response to a specific antigen is called *active immunity.* However, a person can also be immunized against an antigen by injecting antibodies obtained from another human being or even from an animal that has been actively immunized against the antigen. These injected antibodies usually circulate in the recipient's blood for three to five weeks and then are themselves destroyed by the recipient's reticuloendothelial system. This type of immunity is called *passive immunity.*

Passive immunity has an advantage over active immunity in acute illness, for it can be used clinically to increase the immunity of a person within minutes, whereas the person often will not develop sufficient active immunity to combat a disease for a week or more, in which case the disease frequently will have killed the person before the immunity can be of value. For example, in the case of a snake bite, the person is likely to be dead in only a few hours unless he is treated with immune sera. Immune sera used for instituting passive immunity are usually developed in animals, most frequently in horses.

Vaccination

A method for disease prevention familiar to everyone is *vaccination.* Vaccination causes development of antibodies against specific disease organisms or dangerous toxins without exposing the person to the disease. For instance, injection of dead bacteria that still contain antigenic proteins can cause the development of antibodies that will later attack the live organisms equally as well as the dead organisms. The bacteria used for the vaccination process can be killed by heating them, by destroying them with formalin, destroying them with phenol, grinding them to bits, or even breaking them to pieces with high intensity sound waves.

Vaccines can also be prepared from toxins by exposing them to chemicals that change their structure very slightly, enough to remove their poisonous properties but not enough to destroy their antigenic properties. This can frequently be done by exposing the toxins to formalin, phenol, or some other similar chemical agent. The resulting substance is called a *toxoid*.

A third type of vaccine employs a living agent. For instance, smallpox virus is very similar to cowpox virus. Yet, cowpox virus causes only a single pustule to develop rather than multiple pustules all over the entire body. A person innoculated with cowpox virus develops cross-immunity to smallpox after the cowpox pustule has disappeared. Likewise, non-virulent strains of polio virus, yellow fever virus, rabies virus, measles virus, and many other viruses are used to cause immunity without causing the actual disease, or at least without causing a dangerous form of the actual disease.

TIME REQUIRED TO DEVELOP IMMUNITY. When a person has never been exposed to an antigen previously, either by injection of a vaccine or by exposure to the disease itself, the immune mechanism usually takes one to two weeks after vaccination to build up a significant number of antibodies in the circulating body fluids, as shown by the lower curve of Figure 91. Then, over a period of another four to six weeks, the antibodies disappear. This is called the *"primary" immune response*. When the person is exposed to the same antigen a second time, whether it be a few weeks later or even many years later, the immune response develops in one to two days and is far more lasting; this is called the *secondary immune response*, which is illustrated by the upper curve of Figure 91. This intense secondary effect explains why several injections are often used for vaccination against a disease. It also explains why a very small booster dose of a vaccine is effective year after year in maintaining immunity.

ALLERGY

The Immune Reaction as the Basis of Allergy

If an antigen combines with an antibody that is attached to or adjacent to a cell, the reaction usually damages the cell, causing swelling or even destruction, followed by release of several different substances, including *histamine, heparin, acetylcholine, proteolytic enzymes*, and so forth, as illustrated in Figure 92. These substances then diffuse into the surrounding fluids and throughout the body to cause secondary reactions. The substance that causes the most serious effects is histamine, which is discussed later.

WEAK IMMUNITY AS A CAUSE OF ALLERGY. As pointed out previously in relation to the immune reaction, most antibodies are "bivalent," which means that they have two reactive sites for combining with antigens. However, when antibodies are being formed, one of these valences is formed first and the other later. Sometimes the plasma cells or other antibody producing cells empty partially formed antibodies, called *reagins*, into the body fluids, antibodies that have only one reactive site (univalent antibodies) or antibodies with abnormal molecular structure. This is particularly true when a person has been exposed to very weak antigens that are poor initiators of the immune process.

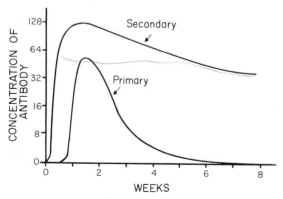

Figure 91. Antibody responses to successive injections of vaccine.

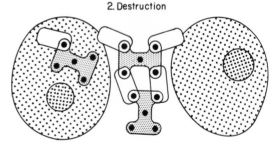

CELLULAR EFFECTS
1. Swelling
2. Destruction

PRODUCTS OF REACTION

1. Histamine	6. Choline
2. Heparin	7. Lysolecithin
3. Bradykinin	8. Serotonin
4. Acetylcholine	9. Proteolytic enzyme
5. Adenosine	

Figure 92. Effects of the antigen-antibody reaction in contact with cells.

The reagins have a special propensity for attaching themselves to tissue cells throughout the body. Therefore, there is far greater chance of immune reactions (and allergic reactions) taking place in contact with tissue cells when such antibodies are present than when normal bivalent antibodies are present.

Allergens. The human body is continually exposed to many substances that cause very weak immunity but not strong immunity. These include such substances as partial proteins, conjugated proteins, or drugs that can combine weakly with proteins in the body. Among the most common of the poorer antigens are *pollens, industrial chemicals, animal dander, feather protein,* certain *food extracts,* and *drugs.* Since these have a tendency to cause allergy, especially in those persons who are normally predisposed to allergy, they are called *allergens.*

Predisposition of certain persons to allergy—the allergic person. The allergic person has a hereditary tendency to produce a higher proportion of reagins (partially formed antibodies) than is true in the normal person. This obviously predisposes such a person to allergy.

Different Types of Allergic Reactions

Many different types of allergic reactions can occur, some of the more important types of which are the following:

ANAPHYLAXIS. Anaphylaxis results from injection of an antigen intravenously into a person who has previously been strongly immunized against the antigen. The direct reaction of the antigen and antibody in the blood stream and in tissues closely adjacent to the blood causes intense cellular damage and release of large quantities of histamine into the circulation. The histamine causes the capillaries to become very permeable to fluid so that plasma leaks into the surrounding tissues, thereby reducing the blood volume, and it causes the blood vessels to become markedly dilated, which causes pooling of large quantities of blood in the peripheral vasculature. The net result is a drastic fall in cardiac output, a condition called *anaphylactic shock,* that can cause death in a few minutes. Also, other tissues closely allied to the circulatory system become seriously damaged. For instance, bronchiolar spasm in the lungs frequently makes it very difficult for the person to breathe.

URTICARIA. Urticaria means the development of "hives" on the skin. It results from much the same type of reaction as that which causes anaphylaxis except that it takes place in many small localized areas throughout the body rather than everywhere, presumably at points where large quantities of partial antibodies are attached to the tissues. At these places the skin first becomes hyperemic (red because of excessive blood flow) because of arteriolar dilation. This is followed by increased leakage of fluid out of the capillaries into the tissues resulting in large welts all over the surface of the body. These effects result mainly from the release of histamine by the damaged cells; antihistaminic drugs are extremely effective in preventing urticaria.

HAY FEVER. Hay fever results from antigen-antibody reactions in the nose that probably cause both direct local tissue damage and effects induced secondarily by histamine. The histamine causes arteriolar dilatation in the local blood vessels, with resultant high capillary pressure and leakage of fluid into the tissues. The tissues become so edematous that the person has difficulty breathing through the nose.

ASTHMA. Asthma results from bronchiolar constriction, usually following respiratory exposure to antigens to which the person is sensitive. Antigens such as ragweed pollen, horse's dander, or even dust in the air can be breathed into the lungs where localized antigen-antibody reactions occur. This in turn causes intense bronchiolar constriction, often making it almost impossible for the person to breathe and almost suffocating him.

Desensitization (Hyperimmunization) as a Physiological Means of Treating Allergy

Most types of allergy, with the exception of anaphylaxis, are caused by weak immunity to an allergen, in which cases the person has many partially formed antibodies (reagins) spread throughout his tissues, a state that predisposes to serious allergy. The person can frequently be made non-allergic by the process called *desensitization.* This is accomplished by injecting a minute quantity of the allergen, either intracutaneously or subcutaneously, every few days, until the person builds up increased immunity to the allergen. Once this has occurred, enough complete bivalent antibodies float around in the body fluids to destroy the allergens before they can ever reach the reagins attached to the tissue cells. Thus, an allergic reaction is prevented.

REFERENCES

Ackroyd, J. F. (ed.): *Immunological Methods.* Philadelphia, F. A. Davis Company, 1964.

Boyd, W. C.: *Introduction to Immunochemical Specificity.* New York, John Wiley & Sons, 1962.

Burnet, F. M.: The mechanism of immunity. *Sci. Amer.* 204(1):58, 1961.

Ciba Foundation, Symposium Volume: *Cellular Aspects of Immunity.* Boston, Little, Brown & Company, 1960.

Haurowitz, F.: Antibody formation. *Physiol. Rev.* 45(1):1, 1965.

Mackay, I. R., and Burnet, F. M.: *Autoimmune Diseases: Pathogenesis, Chemistry, and Therapy.* Springfield, Ill., Charles C Thomas, 1963.

Miller, J. F. A. P., and Osoba, D.: Current concepts of the immunological function of thymus. *Physiol. Rev.* 47(3):437, 1967.

Mongar, J. L., and Schild, H. O.: Cellular mechanisms in anaphylaxis. *Physiol. Rev.* 42:226, 1962.

Najjar, V. A.: Some aspects of antibody-antigen reactions and theoretical considerations of the immunologic response. *Physiol. Rev.* 43:243, 1963.

Uhr, J. W.: Delayed hypersensitivity. *Physiol. Rev.* 46(3):359, 1966.

CHAPTER 10

BLOOD COAGULATION, TRANSFUSION, AND TRANSPLANTATION OF ORGANS

BLOOD COAGULATION

Blood coagulation prevents blood loss when vessels are ruptured. After a vessel begins to bleed, physical and chemical processes are set into operation, leading in approximately two to six minutes to the formation of a clot. The clot plugs the rent in the vessels and prevents further loss of blood.

The Mechanism of Blood Coagulation

The general scheme of blood coagulation is shown by the following physical and chemical reactions:

1. The traumatized tissues at the edge of the broken vessel release *thromboplastin*. Platelets also adhere to the broken edges of the vessel as shown in Figure 93 and then disintegrate to *release platelet factor 3*.

2. *Thromboplastin* or *platelet factor 3* reacts with *protein factors* and *calcium ions* in the blood to form *prothrombin activator*.
3. *Prothrombin activator* + *calcium ions* enzymatically split *prothrombin* into *thrombin*.
4. *Thrombin* + *fibrinogen*→*fibrin*.
5. *Fibrin threads* entrap red blood cells to form the *clot*.

THROMBOPLASTIN, AND ITS ROLE IN THE FORMATION OF PROTHROMBIN ACTIVATOR. Whenever a blood vessel is ruptured, the damaged tissues surrounding the vessel, or the ruptured edges of the vessel itself, release a lipoprotein material called *thromboplastin*. This, in turn, reacts with calcium ions and several protein factors in the blood plasma to produce *prothrombin activator*. Two of the protein factors involved are *factor V* (also called accelerator globulin) and *factor VII* (also called proconvertin). The prothrombin activator that results from this series of reactions initiates development of the clot as will be explained below.

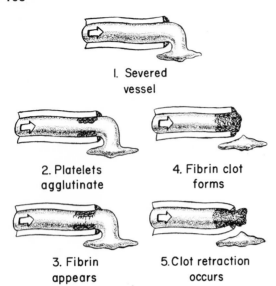

1. Severed
vessel

2. Platelets
agglutinate

4. Fibrin clot
forms

3. Fibrin
appears

5. Clot retraction
occurs

Figure 93.   Stages of clot development and clot retraction following injury to a blood vessel. (Redrawn from Seegers and Sharp: *Hemostatic Agents.* Charles C Thomas.)

PLATELET DISINTEGRATION, AND ITS ROLE IN FORMATION OF PROTHROMBIN ACTIVATOR. *Platelets* are 1 to 2 micron particles formed by disintegration of large *megakaryocytes,* a type of white blood cell produced in the bone marrow. The number of platelets in each cubic millimeter of blood is normally about 300,000. Platelets have a thin membrane that is negatively charged on the outside, and the endothelial walls of the vessels throughout the circulation are also normally negatively charged so that platelets are repelled from the walls. However, when a vessel ruptures, the damaged endothelium loses its negative charge allowing the platelets to stick at the point of rupture as illustrated in Figure 93. Likewise, platelets can stick to foreign bodies in the circulation or even to a clot that is already forming. The process of sticking causes the platelets to disintegrate and to dump their substances into the surrounding fluids.

Platelet factor 3, one of the substances released by disintegrating platelets, initiates prothrombin activator formation in a manner similar to but not identical with its formation under the influence of thromboplastin. The difference is that platelet factor 3 reacts with a different group of protein factors in the blood to produce the prothrombin activator, including (1) *anti-hemophilic factor,* (2) *Hageman factor,* (3) *Stuart factor,* (4) *accelerator globulin,* (5) *plasma thrombo-*

plastin component, and (6) *plasma thromboplastin antecedent.* Also calcium ions are required in this scheme, as well as in the thromboplastin scheme.

ROLE OF PROTHROMBIN ACTIVATOR IN THE FORMATION OF THROMBIN. Prothrombin activator, having been formed in the blood by either the thromboplastin or the platelet factor 3 scheme, acts as an enzyme to initiate the third reaction in the clotting process. That is, it causes *prothrombin* to be split into several fragments, one of which is *thrombin.* For this reaction to occur, calcium ions are again required. The reaction takes place in only a few seconds once an adequate amount of prothrombin activator is available.

ROLE OF THROMBIN IN THE FORMATION OF FIBRIN. Once fibrin has been formed, it causes *fibrinogen* to change into *fibrin.* Fibrinogen is one of the plasma proteins usually present in the plasma in a concentration of about 0.3 per cent. Thrombin acts as an enzyme to activate the fibrinogen molecule, forming what is called *fibrin monomer,* which means a single molecule of activated fibrin. These molecules automatically polymerize with each other into long *fibrin threads* forming a *reticulum* that entraps red blood cells, white blood cells, and platelets to form a *clot.* These events, beginning with the action of prothrombin activator on prothrombin with the eventual formation of the fibrin threads, are shown in Figure 94.

CLOT RETRACTION. Once the clot has formed, the fibrin threads gradually contract, expressing plasma out of the clot but keeping the red blood cells, platelets, and other solid elements. The plasma that is extruded from the clot has no fibrinogen in it because this has been converted into fibrin. Therefore, it is called by another name, *serum.*

Fibrin threads attach themselves to the damaged surfaces of the blood vessels. Therefore, as the blood clot retracts, it pulls

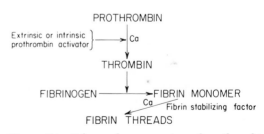

Figure 94.   Schema for conversion of prothrombin to thrombin, and polymerization of fibrinogen to form fibrin threads.

the edges of the opening together, thus contributing to ultimate hemostasis.

For clot retraction to occur to a maximum extent, an adequate number of platelets must be present in the blood. Electron micrographs of blood clots show that platelets form nidi for development of more and more new fibrin threads, which in turn contribute to the contraction of the clot. It is presumed that the platelets that become enmeshed in the clot slowly disintegrate, thereby causing the formation of more and more plasma thromboplastin, which in turn leads to more and more formation of fibrin.

ORGANIZATION OF THE BLOOD CLOT. After a clot is formed, its immediate function of stopping blood loss will have been achieved. To effect permanent repair of the blood vessel, however, the clot becomes *organized*, which means ingrowth of fibrous cells to form connective tissue. This process continues for several days, and endothelial cells simultaneously grow over the vascular surface of the clot to form a new smooth lining on the inside of the vessel.

VASCULAR SPASM AS AN AID TO HEMOSTASIS. The term *hemostasis* means the arrest of blood escape. The blood clotting mechanism is perhaps the most important of all the hemostatic mechanisms, but another one of almost as much importance is spastic contraction of the damaged vessel itself, occurring within seconds after the vessel is ruptured. Trauma to a vessel initiates impulses that travel along the surfaces of the smooth muscle fibers in the vascular wall. These impulses cause contraction of the muscle, with resultant spastic closure of the vessel. This prevents serious bleeding even before the clot can be formed. Indeed, ruptured arteries as large as three-eighths inch in diameter, and veins even larger, sometimes contract so spastically that the person will not bleed to death. Without such contraction enough blood could be lost in less than 20 seconds to cause death.

Prevention of Intravascular
Clotting Under Normal Conditions

Since prothrombin activator and thrombin are both enzymes, it would theoretically be possible for them to cause continual coagulation in the circulation once even a small amount of each of them has been formed.

Fortunately, this does not occur for many different reasons, some of which are the following:

ENDOTHELIAL REPULSION OF PLATELETS. The first factor that prevents intravascular clotting is repulsion of platelets away from the endothelial linings of the blood vessels caused by electro-negative charges on both the endothelial wall and platelets as described above. It is believed that special proteins adhere to the vessel lining with negative electrical charges protrude inward toward the blood. Theoretically, as long as this protein lining remains intact, the platelets cannot adhere to the surface of the vessel wall and, therefore, cannot initiate a clot.

ANTITHROMBIN EFFECTS. A second factor that prevents intravascular clotting is removal of thrombin from the blood in several different ways. First, a large portion of the thrombin is directly adsorbed to the fibrin threads of a clot as they are formed. Second, a special protein in the plasma called *antithrombin* acts enzymatically on thrombin to inactivate it, and, third, the substance *heparin*, which is discussed below, probably also helps to destroy thrombin.

HEPARIN. Another means to prevent intravascular clotting is continual secretion into the blood of the anticoagulant *heparin*. This is a conjugated polysaccharide that is secreted into the circulatory system continually by *mast cells* located in the pericapillary connective tissue throughout the body. These cells appear to be typical basophils, one of the types of white blood cells described in Chapter 8. It is believed that *basophils* circulating in the blood can also secrete heparin, though this has not been proved.

Heparin has several different actions that are important in preventing intravascular clotting. First, it blocks the platelet mechanism for formation of prothrombin activator. Second, it increases the antithrombin effects noted above. Third, it inhibits the action of thrombin on fibrinogen and thereby prevents conversion of fibrinogen into fibrin threads.

The quantity of heparin in the blood stream is very small, barely enough to prevent clotting under normal conditions but not so in the presence of excess prothrombin activator formed either as a consequence of the thromboplastin or the platelet mechanism.

Extravascular Anticoagulants

Some anticoagulants are used artificially to prevent blood coagulation outside the body. The most important of these are chemicals that remove calcium ions from the blood. _Oxalate salts_, for example, when added to blood cause precipitation of calcium oxalate, removing the calcium ions and blocking the clotting process.

Citrate salts also remove calcium ions from the blood by forming calcium citrate, a soluble substance but almost totally non-ionized. Blood that has been rendered incoagulable by citrate can still be reinjected into a person provided that it is given slowly, because almost as rapidly as it enters the circulation the citrate is destroyed by the liver. The blood again becomes capable of coagulating and is relatively normal in all other respects. Therefore, in giving transfusions, blood is usually drawn from a donor and mixed immediately with a mixture of sodium, potassium, and ammonium citrate to prevent coagulation. Upon readministration the citrate is destroyed.

Bleeding Diseases

Lack of any one of the many different blood clotting factors can prevent adequate blood coagulation, so that once blood vessels rupture blood continues to ooze for many hours. A large number of small blood vessels rupture internally each day, and were it not for the blood clotting process, a person would actually bleed to death internally. Fortunately, in most bleeding diseases, clotting does occur to some extent but much more slowly than normally.

HEMOPHILIA. There are several different types of hemophilia, all of which are hereditary but each of which has slightly unique characteristics. These diseases are caused by lack of _antihemophilic factor_, _plasma thromboplastin component_, or _plasma thromboplastin antecedent_, three substances that enter into the platelet mechanism for initiating blood clotting. However, these three factors are not necessary for the tissue thromboplastin mechanism of blood clotting. Therefore, if the tissues of a hemophilic person are torn severely enough to form adequate amounts of tissue thromboplastin, clotting occurs almost normally, and very little blood is lost. On the other hand, simple rupture of a blood vessel does not form much tissue thromboplastin, in which case bleeding often lasts for hours. The gradual oozing of blood frequently allows the hemophiliac to bleed to death.

THROMBOCYTOPENIA. Platelets are also called _thrombocytes_, and a deficiency of thrombocytes is called _thrombocytopenia_. An occasional person develops a deficiency of platelets down to less than one-tenth the normal level. In these persons, the platelet mechanism for blood clotting is again deficient, as occurs in the hemophiliac, so that internal bleeding occurs very frequently. However, an additional effect besides those that take place in hemophilia also occurs. This is oozing of blood from the minute capillaries. It is believed that very small breaks in capillaries are blocked not by clots but instead by direct adherence of platelets to the rents in the vessels. In platelet deficiency, minute rents in the very small vessels cause literally thousands of small bleeding areas throughout the body. Thus, the person's skin becomes studded with many minute hemorrhagic spots only a few millimeters in diameter. Bleeding of this same type also occurs internally in all the organs.

DECREASED PROTHROMBIN OR PROCONVERTIN—VITAMIN K DEFICIENCY. Two other clotting factors frequently absent or deficient in the circulation, prothrombin and proconvertin, are proteins formed by the liver, and both require vitamin K as one of the intermediary substances in their formation. In vitamin K deficiency, the quantities of these substances in the blood become very deficient. Also, diseases of the liver, such as hepatitis, cirrhosis, or acute atrophy of the liver, can depress the rate of prothrombin or proconvertin formation.

When either prothrombin or proconvertin is deficient in the blood, the formation of thrombin is greatly depressed so that a bleeding tendency develops.

Blood Coagulation Tests

BLEEDING TIME. Several tests are used to determine the integrity of the blood coagulation process. The most common of these is a simple bleeding test in which a finger or a lobe of an ear is pricked and the time required for the bleeding to stop is measured.

Following an ordinary prick of the finger, bleeding stops in 2 to 3 minutes, and bleed- of the lobe of the ear normally stops in 4 to 5 minutes.

CLOTTING TIME.   Another test is to deter- mine whether or not blood removed from the person will coagulate as rapidly as normally. To do this, blood is collected in an abso- lutely clean test tube and observed until it clots. Ordinarily, the clotting time is about 6 minutes if the blood has not been agitated too much during the test.

PLATELET COUNT.   If blood fails to coagu- late in the test tube in a reasonable period of time, one searches for the factor that might be missing in the coagulation process. One of the first tests is to count the number of platelets. As long as the number of platelets is above 150,000 per cubic millimeter, the person will have no clotting difficulty as a result of platelet deficiency. However, in persons with thrombocytopenia, the plate- let count frequently falls to as low as 40,000 to 50,000, in which case excessive bleeding does occur.

PROTHROMBIN TIME.   A test to determine the quantity of prothrombin in the blood is performed by first rendering the blood in- coagulable with oxalate which precipitates the ionic calcium. Then large amounts of cal- cium, thromboplastin, accelerator globulin, and proconvertin are all mixed suddenly with the sample of blood. The only addi- tional factor needed for the formation of thrombin that is not added to the blood is prothrombin. Therefore, the length of time required for the blood to clot depends on the amount of prothrombin already in the blood. The longer the time required, the less the amount of prothrombin available. The nor- mal prothrombin time is approximately 12 seconds—that is, this is the time required for the clot to appear after mixing has oc- curred. If the prothrombin time is as long as 20 to 30 seconds, the person can be expected to be a bleeder.

The quantities of proconvertin, accelerator globulin, or any of the other factors that enter into blood coagulation can be estimated by similar procedures.

TRANSFUSION

Often a person loses so much blood that to save his life he must be given a trans- fusion of new blood immediately. At other times transfusions are given to treat anemia or some other blood deficiency. Unfortu- nately, the bloods of different people are not all exactly alike, and failure to inject the appropriate type of blood is likely to cause death of the recipient. For this reason it is very important to understand the physiology of blood grouping and blood matching prior to giving transfusions.

Blood Matching

The reason one person's blood may not be suitable for another person is that the re- cipient may be immune to some of the pro- teins in the blood cells of the donor. Anti- bodies in the recipient's plasma can cause *agglutination* (clumping) and *hemolysis* (rupture) of the injected cells, thus plugging some of the vessels and releasing large quantities of hemoglobin out of the cells into the circulation.

CROSS-MATCHING.   One means by which the suitability of blood for transfusion can be determined is to *cross-match* the bloods of a donor and recipient. To do this, the two bloods are first centrifuged, and the plasma and red cells separated from each other. Then the red blood cells are resuspended in a saline solution. A small quantity of cells from the donor is mixed with a small quantity of plasma from the recipient, and similarly a small quantity of cells from the recipient is mixed with a small quantity of plasma from the donor. These are allowed to remain together for 3 to 15 minutes and then are observed under a microscope. If none of the cells are agglutinated, as shown by the clumps on the microscope slides in Figure 95, it can be assumed that the two bloods *match* each other, or, in other words, that no immune bodies are present in either of the two bloods that can damage the cells of the other blood. However, if on micro- scopic examination clumps of red cells are evident, then it is obvious that an antibody reaction has occurred and that the bloods are mismatched.

THE BLOOD GROUPS—TYPES A, B, AB, AND O.   The different types of blood are classi- fied on the basis of the *cell proteins* that nor- mally cause antibody reactions. The two proteins in the cells that most frequently cause reactions are the type A and type B

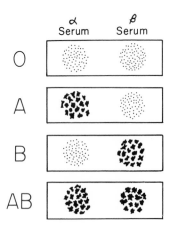

Figure 95. Agglutination of some cells and lack of agglutination of other cells in the process of blood typing.

proteins; these are called *agglutinogens*. All human bloods can be classified into four major groups, which are given in Table 3. These are groups O, A, B, and AB. The red cells of group O blood contain neither A nor B agglutinogen. Group A red blood cells contain type A agglutinogen, and group B red blood cells contain type B agglutinogen. Finally, group AB cells contain both A and B agglutinogens.

In a person's *plasma* are antibodies that agglutinate cells containing agglutinogens different from those in his own cells. These antibodies are called *agglutinins*, and alpha (α) agglutinins agglutinate type A cells, while beta (β) agglutinins agglutinate type B cells. Thus, as shown in Table 3, a person with type O blood has both alpha and beta agglutinins in his plasma. These are capable of agglutinating cells containing either type A or B agglutinogens. Type A blood contains beta agglutinins but not alpha agglutinins, and type B blood contains alpha agglutinins but not beta agglutinins. Finally, type AB blood contains no agglutinins whatsoever because both of the agglutinogens are present in the cells.

TABLE 3. *The Blood Groups*

BLOOD TYPE	AGGLUTINOGENS	AGGLUTININS
O	O	α, β
A	A	β
B	B	α
AB	A, B	none

Blood typing. To determine the type of blood, a procedure called *blood typing* is performed. The red cells from the subject are mixed with two samples of plasma, one containing large amounts of alpha agglutinins and other large amounts of beta agglutinins. Figure 95 illustrates the reactions observed on a microscope slide with each type of blood when this is performed. In the case of type O cells, none of the cells are agglutinated by either the alpha or beta agglutinins because these contain neither A nor B agglutinogens. In the case of type A cells, alpha agglutinins agglutinate the cells but beta agglutinins do not because these cells contain A agglutinogens. In the case of type B cells, beta agglutinins agglutinate the cells but alpha agglutinins do not, and in the case of type AB cells, both alpha and beta agglutinins agglutinate the cells. In other words, by observing which agglutinins agglutinate the cells, one can determine the presence or absence of A or B agglutinogens in the cells and thereby determine the blood type of the subject.

THE RH FACTOR. In addition to the four major blood groups, another protein, the *Rh factor*, present in the red cells of approximately 85 per cent of all persons, sometimes causes mismatching of blood. However, the Rh factor is different from the A and B agglutinogens in the following way: Those persons who do not have Rh factor (the Rh negative persons) do not normally have Rh antibodies. This is different from the A and B blood groups, for those persons who do not have either A or B agglutinogens do have respective agglutinins for agglutinating these substances. Yet, once blood cells containing the Rh factor have been injected into a recipient who does not have the Rh factor, the recipient forms Rh antibodies. On second injection of the Rh cells the Rh antibodies cause agglutination, and a typical transfusion reaction.

To recapitulate, the Rh factor cannot cause a transfusion reaction on first exposure of mismatched blood, but once an *Rh negative person* has been immunized against Rh positive blood, second exposure to the Rh positive blood causes typical agglutination.

Erythroblastosis fetalis caused by the Rh factor. One of the most important consequences of the Rh factor is the abnormality called *erythroblastosis fetalis*, which occurs in about 1 out of every 50 infants born to Rh negative mothers when the father is Rh positive. If a fetus inherits the Rh positive characteristic from the father, its red cells will contain the Rh factor. Many of the fetus' red blood cells continually disintegrate, and some of the Rh protein passes through the placental membrane into the mother. This causes the mother to develop Rh antibodies, and these antibodies in turn pass back through the placental membrane into the baby, resulting in agglutination and destruction of the baby's red blood cells. To offset this rapid destruction, the baby's bone marrow, spleen, and liver produce red blood cells extremely rapidly, so rapidly, in fact, that many of the cells emptied into the circulation are early, nucleated red blood cells. This is the reason why the disease is called "erythroblastosis."

Many infants with erythroblastosis are born dead because of the inability of their anemic blood to provide necessary oxygenation of the tissues. Those that are born alive are likely to die within the first few weeks. The standard treatment is to transfuse the child with Rh negative blood while bleeding him at the same time. In this way his Rh positive blood is gradually removed and is replaced by Rh negative blood that cannot be destroyed by the Rh antibodies in the child's body fluids. Within three to five weeks after birth the Rh antibodies which had passed into the baby from the mother will have been destroyed, and the child's own Rh positive cells will then no longer be destroyed.

OTHER FACTORS THAT LEAD TO MIS-MATCHING OF BLOOD, AND THEIR USE IN FORENSIC MEDICINE. Often present in the red cells are several other proteins that can cause transfusion reactions in the same manner as the Rh factor. Some of these are the Hr, M, N, P, Lewis, and Kell factors. Ordinarily, however, these substances are so poorly antigenic that transfusion reactions that result are very slight. The major importance of these factors is that they are often used in forensic medicine to determine familial relationships between persons. On the average, half of the factors found in a child's blood will also be found in the father's blood and the other half in the mother's blood. Thus, by a process of elimination, it is sometimes possible to disprove fatherhood in cases of disputed parentage.

Techniques of Blood Transfusion

For transfusing blood any method can be used that will transfer the blood from a donor to a recipient without causing coagulation or destruction of the blood in the process. The earliest method was simply direct connection between an artery of the donor, by means of needles and a rubber tube, and a vein of the recipient. Blood was allowed to flow from the donor to the recipient until the donor fainted or until the recipient showed signs either of improvement or of a transfusion reaction.

The most usual technique now used for transfusion is to collect blood from the donor in a container while mixing the blood continually with a small quantity of sodium, potassium, and ammonium citrate and a small quantity of glucose. The citrate compounds prevent blood coagulation, and the glucose provides nutrition for the red cells and keeps them structurally sound. This blood can be given immediately to the recipient, or it can be placed in a refrigerator and kept at 4° C. for as long as three weeks prior to administration. This is the method used by hospital blood banks for storing blood until some patient needs a transfusion.

EFFECT OF ADMINISTERING MISMATCHED BLOOD. When mismatched blood is administered to a recipient, the first effect is usually agglutination of the red blood cells. The cellular clumps then flow into the small vessels, as illustrated in Figure 96, and fail to pass on through the capillaries. This blocks blood flow in the respective vessels temporarily, but after a few minutes to a few hours the wedging of the cells further into the smaller vessels and the destructive effect of antibodies on the cellular membranes cause the red cells to disintegrate. This allows the blood to flow once again, but large quantities of hemoglobin are dumped into the plasma. The final result, therefore, is destruction of many red blood cells, with the release of hemoglobin into the circulation.

Effect of transfusion reactions on the kidneys. Actually, the clumping of the cells and the loss of the cells themselves are rarely of great importance in transfusion reactions. Instead, it is the free hemoglobin released into the plasma that usually causes

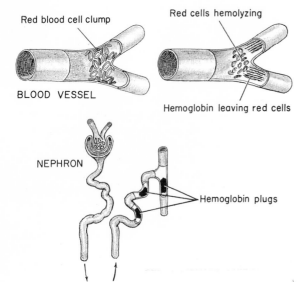

Red blood cell clump

Red cells hemolyzing

BLOOD VESSEL

Hemoglobin leaving red cells

NEPHRON

Hemoglobin plugs

Figure 96. Mechanism of transfusion reactions caused by mismatched blood.

to open up again and the kidneys to become functional.

Blood Substitutes

Many different solutions can be injected into the circulation to increase the blood or plasma volume. Solutions used for this purpose are called *blood substitutes*. Often a blood substitute can be used more advantageously than can blood itself. For instance, on the battlefield it is very difficult to maintain adequate quantities of different types of blood, but blood substitutes can be given regardless of blood type, and, therefore, permit rapid and easy use under stressful conditions. Several blood substitutes are the following:

PLASMA. Plasma is separated from whole blood by centrifuging the red blood cells to the bottom of a container and decanting the plasma. Plasma can be used in place of whole blood to restore normal blood volume in cases of hemorrhage. Because of its protein content, it creates colloid osmotic pressure at the capillary membranes, and this keeps the fluid of the plasma in the circulation. Therefore, except for a resulting anemia, the circulatory system of a person who has hemorrhaged will be essentially normal again. Ordinarily, administration of plasma is completely adequate for treating hemorrhagic shock. However, when extreme quantities of blood have been lost, it is necessary to give at least some whole blood to restore the blood's oxygen-carrying capacity.

SALINE. Saline, prepared in a 0.9 per cent solution by adding common salt to water, is isotonic with the body fluids, and will not cause osmotic changes in the red blood cells or other body cells. On intravenous injection this solution diffuses throughout the entire extracellular fluid compartment, approximately nine-tenths going into the interstitial spaces and approximately one-tenth remaining in the blood. Therefore, saline can be used in treating blood loss, though it causes some edema at the same time.

OTHER SOLUTIONS. A blood substitute that has become very valuable on the battlefield is a solution of *dextran*. This substance is a large polysaccharide compound that has a molecular size equivalent to that of plasma proteins. These molecules, like those of

the damage. The hemoglobin molecule is small enough to pass through capillary pores, and much of it passes through the glomerular capillaries into the kidney tubules. As the tubules reabsorb water and electrolytes, the hemoglobin in the tubular fluid becomes progressively more and more concentrated. Finally, the concentration becomes so great that hemoglobin precipitates and plugs the tubules, as illustrated at the bottom of Figure 96. Obviously, a hemoglobin plug stops further function of the nephron. When this occurs in all the tubules of both kidneys, death due to renal shutdown is likely in 8 to 12 days.

Once a transfusion reaction has occurred, prevention of death is often possible by immediate institution of three physiologic procedures: First, administration of large quantities of water causes water diuresis, making it less likely for hemoglobin to precipitate in the tubules. Second, administration of alkaline substances causes the formation of an alkaline tubular fluid that can hold more hemoglobin in solution than can acidic fluid. Third, administration of diuretics also causes rapid flow of tubular fluids and helps to prevent hemoglobin plugs.

Usually, if death does not occur within two weeks after a transfusion reaction, automatic disintegration of the hemoglobin plugs in the tubules will allow many of the tubules plasma proteins, do not pass through the capillary pores and, therefore, also like plasma proteins, cause colloid osmotic pressure. As a result, dextran solutions have almost exactly the same properties as plasma. They remain in the circulation, increasing the blood volume back to normal while not causing any edema of the interstitial spaces. Further advantages of dextran solutions are that they cause no allergic reactions, they do not require citrate anticoagulant as does plasma, and they can be prepared in large quantities by commercial processes without necessitating removal of blood from donors.

Solutions for intravenous nourishment. Nourishment is often given by intravenous methods. Solutions of glucose, amino acids, and even emulsified fats can be given intravenously for this purpose. Occasionally plasma and whole blood transfusions are given simply because of their nutrient value, and not because of a need for enhanced blood volume.

Solutions for reestablishing electrolyte equilibrium. Occasionally, also, solutions are administered to reestablish electrolyte equilibrium. For instance, a person whose extracellular fluids contain too few potassium ions can be treated by administering a potassium solution. Or a person whose fluids are too acidic can be treated by administering sodium lactate, sodium gluconate, or sodium bicarbonate. The lactate and gluconate radicals are removed by metabolic processes in the liver, and the bicarbonate radical becomes carbon dioxide and is blown off through the lungs. As a result, the sodium ions alkalinize the acidic fluids.

Transplantation of Tissues and Organs

Before discussing transplantation of tissues or organs, we must first note that all cells of the body have antigenic proteins the same as those found in the blood cells. Thus, all cells of a person with type A blood have type A agglutinogens. Likewise, an Rh positive person has Rh positive antigen not only in his blood cells but in all his other cells. Among the many different types of antigenic proteins present in the cells of different persons are the A factor, the B factor, three different Rh factors, three different Hr fac-

tors, and the M, N, S, P, Kell, Lewis, Duffy, Kidd, Diego, and Lutheran factors. Also, besides these well characterized factors still many others exist.

The antigenic factors of a transplanted tissue, if the tissue comes from another person, almost always initiate antibody responses in the host, and within a few weeks enough antibodies are built up to destroy the transplanted tissue.

AUTOLOGOUS, HOMOLOGOUS, AND HETEROLOGOUS TRANSPLANTS. A transplant from one part of a person's body to some other part of his own body is called an *autologous* transplant. Such transplants can be skin, bone, cartilage, and so forth. These transplanted tissues can survive indefinitely, because the body will not build up antibodies against its own tissues.

A *homologous* transplant is one from one member of a species of animal to another member of the same species, such as a transplant from one human being to another. In this case, antibodies usually develop within 3 to 10 weeks in sufficient quantity to destroy the transplant. The one instance in which homologous transplants are successful is from one identical twin to another, for identical twins have precisely the same antigenic substances in their cells. Obviously, therefore, a transplant from one such twin to the second is the same as an autologous transplant.

A *heterologous* transplant is one from a member of one species of animal to a member of another species. In this case, the antigenicity of the transplanted tissue is so great that the tissue or organ is destroyed in two to three weeks.

TRANSPLANTATION OF NON-CELLULAR TISSUES. In a few instances, non-cellular tissues are transplanted from one person to another not for the purpose of providing new cells but for providing a framework into which the recipient's own tissues can grow. For instance, a bone graft from a horse to a human being frequently provides an appropriate latticework into which the human being can grow new bone. All the cells of the transplanted bone die so that only the physical structure of the bone is important. Another transplant that is usually successful is the cornea of the eye from one person to another. The clear part of the cornea has no living cells except for thin linings on the two surfaces, and new linings grow from the host's tissues at the edges of the cornea.

FUNCTION OF THE THYMUS IN RELATION TO TRANSPLANTS. Until very recently, physiologists knew no specific function for the *thymus gland*. Furthermore, removal of the thymus gland from an adult animal causes essentially no changes in growth, development, or function of an animal. However, if a thymus gland is removed during the last few months of fetal life, the animal fails to develop his mechanism for forming antibodies. Therefore, it is believed, as was discussed in the preceding chapter, that the thymus in the fetal animal forms cells that spread by way of the blood to all parts of the body and eventually lodge in the lymphoid tissue to become the precursors of the plasma cells, the cells that in later life form almost all of the antibodies.

Furthermore, the cells of the thymus gland adapt themselves during fetal life so that they will not form antibodies against the animal's own tissues. Also, if the fetus is exposed to foreign proteins continually during fetal life, the immune system will never form antibodies against these same proteins throughout the life of the individual.

These recent studies on the thymus are mentioned here because they provide one of the most likely possibilities for successful transplantation of tissues and organs from one human being to another. Suppose, for instance, that a fetus is injected with a mixture of foreign proteins from many other human beings. The person originating from this fetus would then be unable to form antibodies against antigenic substances from other persons. Therefore, it should be possible to transplant tissues or organs into him at will.

REFERENCES

Biggs, R., and Macfarlane, R. G.: *Human Blood Coagulation and Its Disorders*. 3rd Ed. Philadelphia, F. A. Davis Company, 1962.

Brooks, J. R.: *Endocrine Tissue Transplantation*. Springfield, Ill., Charles C Thomas, 1962.

Calne, R. Y.: *Renal Transplantation*. Baltimore, The Williams & Wilkins Company, 1963.

Hougie, C.: *Fundamentals of Blood Coagulation in Clinical Medicine*. New York, McGraw-Hill Book Company, 1963.

Mollison, P. L.: *Blood Transfusion in Clinical Medicine*. 3rd Ed. Philadephia, F. A. Davis Company, 1962.

Race, R. R., and Sanger, R.: *Blood Groups in Man*. 4th Ed. Philadelphia, F. A. Davis Company, 1962.

Seegers, W. H.: *Prothrombin*. Cambridge, Harvard University Press, 1962.

Wiener, A. S.: *Advances in Blood Grouping*. New York, Grune & Stratton, 1961.

Wiener, A. S., and Wexler, I. B.: *An Rh-Hr Syllabus: The Types and Their Applications*. 2nd Ed. New York, Grune & Stratton, 1963.

SECTION THREE

THE CARDIOVASCULAR SYSTEM

CHAPTER 11

THE PUMPING ACTION OF THE HEART, AND ITS REGULATION

The Heart as a Pump

The function of the heart is to keep blood flowing through the circulatory system. The heart is actually two separate pumps, as shown in Figure 97, one that pumps blood from the systemic circulation into and through the lungs, while the other pumps blood from the lungs through the remainder of the body and back again to the heart. Thus, the blood flows around and around a continuous circuit.

Figure 98 shows the functional details of the heart as a pump. Blood entering the *right atrium* from the large veins is forced by atrial contraction through the *tricuspid valve* into the *right ventricle*. The right ventricle pumps the blood through the *pulmonary valve* into the pulmonary artery, thence through the lungs, and finally through the pulmonary veins into the *left atrium*. Left atrial contraction then forces the blood through the *mitral valve* into the *left ventricle,* from whence it is pumped through the *aortic valve* into the aorta and on through the systemic circulation.

The two atria are *primer pumps* that force

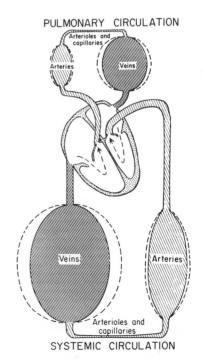

PULMONARY CIRCULATION

Arterioles and capillaries

Arteries

Veins

Veins

Arteries

Arterioles and capillaries

SYSTEMIC CIRCULATION

Figure 97. Schematic representation of the circulation, showing the two sides of the heart and the pulmonary and systemic circulations.

119

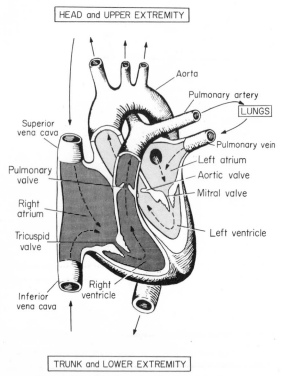

Figure 98. The functional parts of the heart.

the ventricle. No nerves attached to the different portions of the heart make them contract, and no other signals arrive from outside the heart to cause the rhythmicity. In other words, the rhythmicity of the heart is vested in the heart muscle itself, and if any portion of the heart is removed from the body, it will continue to contract as long as it is provided with sufficient nutrition.

THE SINO-ATRIAL NODE AS THE PACE-MAKER OF THE HEART. Referring once again to Figure 99, it will be observed that the sinus of the turtle heart contracts slightly ahead of the atrium, and the atrium slightly ahead of the ventricle. Every time the sinus contracts, an action potential spreads along the muscle fibers from the sinus to the atrium and then to the ventricle, making these parts of the heart contract in succession. Because the sinus initiates the contraction, it is called the *pacemaker* of the heart.

The human heart is different from that of the turtle, for no distinct sinus exists. However, located in the posterior wall of the right atrium immediately beneath the point of entry of the superior vena cava is a small

blood into the respective ventricles. This propulsion of extra blood into the ventricles makes them more efficient as pumps than they would be if they had no special filling mechanisms. However, the ventricles are so powerful that they can still pump large quantities of blood even when the atria fail to function.

CARDIAC RHYTHMICITY AND ITS REGULATION

In order to pump blood the heart must alternately relax and contract, allowing blood to enter its chambers during the relaxation phase and forcing it out during the contraction phase. The alternate contraction and relaxation is provided by an inherent rhythmicity of the cardiac muscle itself, which is seen beautifully by simply observing an exposed heart. Recordings of the contractions of different portions of the turtle heart, for instance, are illustrated in Figure 99, which shows at the top a record from the sinus of the heart, in the middle a record from the atrium, and at the bottom a record from

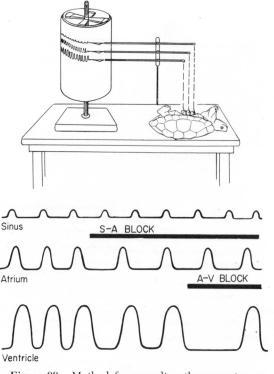

Figure 99. Method for recording the separate contractions of the sinus, atrium, and ventricle of the turtle heart. Below are illustrated the effects of S-A block and A-V block on impulse transmission.

area known as the *sino-atrial node* (S-A node) which is the embryonic remnant of the sinus from lower animals. The rhythmic rate of contraction of muscle fibers excised from the S-A node is approximately 72 times per minute, while muscle excised from the atrium contracts about 60 times per minute and muscle from the ventricle about 20 times per minute. Because the S-A node has a faster rate of rhythm than other portions of the heart, impulses originating in the S-A node spread into the atria and ventricles, stimulating these areas so rapidly that they can never slow down to their natural rates of rhythm. As a result, the rhythm of the S-A node becomes the rhythm of the entire heart, and the S-A node is called the *pacemaker* of the heart.

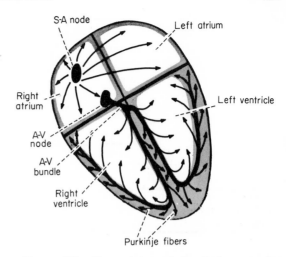

Figure 100. Transmission of the cardiac impulse from the S-A node into the atria, then into the A-V node, and finally through the Purkinje system to all parts of the ventricles.

Conduction of the Impulse Through the Heart

The fibers of heart muscle all interconnect with each other in a latticework as was illustrated in Figure 74 of Chapter 7. Also, the muscle fibers connect with each other so firmly that cardiac muscle is said to be a "functional syncytium." Actually, the heart is composed of two separate muscle masses, that is, two separate functional syncytia. The two atria comprise one of the syncytia and the two ventricles the other. An action potential initiated in any single fiber spreads over the membranes of all the fibers in each respective syncytium. The two syncytia in turn are joined together by the Purkinje system, which conducts impulses from the atrial syncytium into the ventricular syncytium. Thus, once a cardiac impulse begins in the S-A node it travels throughout the entire heart.

THE PURKINJE SYSTEM. Even though the cardiac impulse can travel perfectly well along cardiac muscle fibers, a special conducting medium known as the *Purkinje system* performs two specific functions in conducting the cardiac impulse: First, it conducts the impulse from the atrial muscle syncytium into the ventricular syncytium; second, it increases the velocity of conduction in the ventricles. This system is composed of modified cardiac muscle fibers, called *Purkinje fibers,* that transmit impulses approximately six times as rapidly as normal heart muscle. These fibers, illustrated in Figure 100, begin in the *atrioventricular*

node (A-V node), which is located posteriorly in the lower part of the right atrial wall, and extend through the *A-V bundle* into the ventricular septum where they divide into two major bundles, one spreading along the wall of the right ventricle and the other along the wall of the left ventricle.

One of the major functions of the Purkinje system is to transmit the cardiac impulse throughout the ventricles as rapidly as possible, causing all portions to contract as nearly simultaneously as possible so that they will exert a coordinated pumping effort. Were it not for the Purkinje system, the impulse would travel much more slowly through the ventricular muscle, allowing some of the muscle fibers to contract and relax before others. Obviously, this would result in decreased compression of the blood and, therefore, decreased pumping power.

SEQUENCE OF IMPULSE TRANSMISSION THROUGH THE HEART—IMPULSE DELAY AT THE A-V NODE. After the cardiac impulse originates in the S-A node, it travels first throughout the atria, causing them to contract. Passage of the impulse through the atria is relatively slow because there are no Purkinje fibers to cause rapid conduction. A few hundredths of a second after leaving the S-A node the impulse reaches the A-V node and there enters the Purkinje system. However, the A-V node delays the impulse another few hundredths of a second before allowing it to pass on into the ventricles. This delay allows time for the atria to force blood into the ventricles prior to ventricular contraction. After the delay, the impulse spreads very rapidly through the Purkinje system, causing both ventricles to contract in full force within the next few hundredths of a second.

The A-V node delays the cardiac impulse by the following mechanism: The fibers in this node are so extremely small that they conduct impulses very slowly, at a velocity only one-tenth that of the cardiac muscle and only one-sixtieth that of the Purkinje fibers. Therefore, the cardiac impulse travels so slowly through the A-V node to the Purkinje system that a delay of more than one-tenth second occurs between contraction of the atrium and the ventricle.

BLOCK OF IMPULSE CONDUCTION THROUGH THE HEART. Occasionally the cardiac impulse is blocked at some point in its normal pathway because of damage to the heart. For instance, a portion of heart muscle or of the Purkinje system may be destroyed and replaced by fibrous tissue that cannot transmit the impulse. Figure 99 shows the effect of artificially blocking the impulse at two critical points in the turtle heart. First, a *sino-atrial block* has been effected by tying a string tightly around the heart between the sinus and the atrium. This stops impulse conduction from the sinus into the atrium, and, as is evident from the illustration, the sinus continues to beat at its own natural rate while the atrium and ventricle assume a rate equal to that of the natural rate of the atrium. In other words, the atrium becomes the pacemaker for the ventricle because the ventricle's natural rate of rhythmicity is much slower than that of the atrium. Another ligature is then tied tightly around the heart, this time between the atrium and the ventricle to cause *atrio-ventricular block*. After this, the sinus, the atrium, and the ventricle all beat at their own respective natural rates of rhythm. This experiment shows that the portion of the heart that beats most rapidly controls the rate of rhythm of the remainder of the heart only so long as functioning conductive fibers exist between the different areas.

In the human heart, a block rarely occurs between the S-A node and the atrial muscle, but very frequently conduction from the atria into the ventricles through the A-V bundle is blocked. It is only through this bundle that the normal impulse can pass from the atria into the ventricles, because elsewhere the atria are connected to the ventricles not by conductive fibers, but instead by fibrous tissue that cannot conduct impulses. Therefore, whenever the A-V bundle is blocked, the atria will beat at the rhythm of the S-A node, and the ventricles at their own natural rate. In other words, the rate of the atria will remain at approximately 72 beats per minute, while that of the ventricles will decrease to about 20 beats per minute. Despite this asynchrony of the atria and ventricles, the heart still operates reasonably satisfactorily as a pump, though its pumping ability may be decreased as much as 50 per cent. Nevertheless, it is evident that the atria are not absolutely essential for the heart to pump blood through the circulatory system.

CESSATION OF THE IMPULSE AT THE END OF EACH HEART BEAT—THE REFRACTORY PERIOD. Normally, when an impulse spreads along the membranes of the heart muscle fibers, a second impulse cannot spread along these same membranes until approximately 0.3 second later, during which time the heart is said to be *refractory*. This effect was explained in Chapter 6 in the discussion of action potentials.

After an impulse enters the ventricles from the atria, it spreads through the ventricles in about 0.06 second. Since the ventricular fibers cannot conduct again for 0.30 second, the impulse completely stops. The upper part of Figure 101 illustrates this principle, showing a circular strip of cardiac muscle, in which an impulse starts at the 12 o'clock point, travels around the heart, and finally returns to the 12 o'clock point. When the impulse reaches the starting point the entire heart is still refractory, which causes the impulse to die.

THE CIRCUS MOVEMENT. Occasionally conditions become sufficiently abnormal that a cardiac impulse does continue on and on around the heart, never stopping. For instance, in the lower part of Figure 101, the length of the pathway has been increased so greatly by the larger circle (which represents an enlarged heart) that after traveling all the way around the heart the impulse now returns to the 12 o'clock position more than 0.3 second after it starts. By this time, the originally stimulated portion of the muscle is no longer refractory, which allows the impulse to travel around again. As the

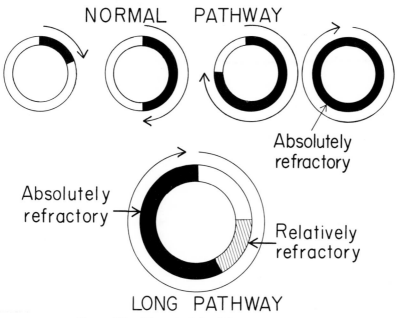

Figure 101.   The principle of the circus movement.

impulse proceeds, the refractory state recedes ahead of it, allowing it to continue indefinitely. This effect is known as a *circus movement*.

Causes of circus movements. Circus movements can result from any of four different abnormalities of the heart. First, as explained in the preceding example, a circus movement is likely to occur when the heart becomes enlarged, thus creating a long pathway. A second cause is slow conduction of the impulse through the heart. For instance, failure of the Purkinje system causes the impulse then to be transmitted by the cardiac muscle itself. This slows impulse transmission about six-fold, often causing the impulse to return to the starting point after the originally excited muscle is no longer refractory. A third cause of a circus movement may be decreased refractory period of the heart muscle. This sometimes results from increased cardiac excitability caused by epinephrine, sympathetic stimulation, or irritation of the heart as a result of disease. Fourth, a very common cause is transmission of impulses in figure 8's, in zigzags, or in any other odd pattern, the impulses sometimes traveling deep in the muscle and then later traveling at shallow levels, recrossing the same area that had already been stimulated at a deeper level. In this way the pathway becomes extremely lengthened, thus

allowing development of an odd-shaped circus movement. Such a pathway, as illustrated in Figure 102B, causes ventricular fibrillation, which is discussed below.

Effect on cardiac pumping. A circus movement is disastrous to the pumping action of the heart because normal pumping requires that the muscle relax as well as contract. A period of total ventricular relaxation is not possible if the impulse continually travels through the muscular mass. Therefore, the entire muscle mass never relaxes or contracts simultaneously, never allowing an alternating filling and squeezing action.

Atrial flutter and fibrillation. Occasionally a circus movement occurs around and around the two atria, as shown in Figure

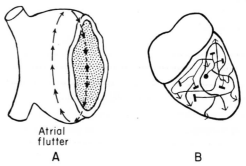

Atrial flutter

A   **B**

Figure 102.   (A) Impulse transmission in *atrial flutter*. (B) Impulse transmission in *ventricular fibrillation*.

102A, at a rate of 200 to 400 times per minute, but not involving the ventricles. This causes *atrial flutter,* during which the atria are "fluttering" rapidly but pumping almost no blood. At other times, an odd-shaped pattern of circus movement occurs in the atria at such a rapid rate that one can see only minute fibrillatory movements of the muscle. This is called *atrial fibrillation,* and when it occurs the atria are of no use whatsoever as primer pumps for the ventricles.

Ventricular fibrillation. A circus movement having the very odd pattern illustrated in Figure 102B frequently develops in the ventricles, causing *ventricular fibrillation,* in which impulses travel in all directions. They go around refractory areas of muscle, dividing into multiple wavefronts in doing so. Some of the impulses die, but for every impulse that does so, another impulse divides to form two impulses, going in separate directions in the heart. This type of circus movement has been called the *chain-reaction movement,* for it is similar to the chain reaction that occurs in nuclear bomb explosions.

Ventricular fibrillation causes the ventricles to contract continuously in very fine, rippling fibrillatory movements. The contracting areas are widespread in the ventricles, interspersed with relaxing areas. As a consequence, the ventricles are incapable of pumping any blood, and the person dies immediately.

Ventricular fibrillation is frequently initiated by electric shock, particularly by 60 cycle alternating current; this causes impulses to go in many different directions at once in the heart, setting up the odd-shaped pattern of impulse transmission illustrated in Figure 102B. Ventricular fibrillation also occurs in diseased ventricles that (a) become overly excitable, (b) develop a damaged Purkinje system, or (c) become greatly enlarged.

The Electrocardiogram

The electrocardiogram is a very important tool for assessing the ability of the heart to transmit the cardiac impulse. When the impulse travels through the heart, electrical current generated by the flowing ions spreads into the fluids surrounding the heart, and a minute portion of the current actually flows as far as the surface of the body. By placing electrodes on the skin over the heart or on any two sides of the heart and connecting these to an appropriate recording instrument, the impulse generated during each heart beat can be recorded.

In the normal electrocardiogram illustrated in Figure 103A, the curve labeled "P" is caused by electrical current generated by passage of the impulse through the atria. The curves marked "Q," "R," and "S" are caused by passage of the impulse through the ventricles, and the curve "T" is caused by the return of the normal membrane potential in the ventricular muscle fibers at the end of their refractory period.

When cardiac abnormalities are caused by various diseases, the electrocardiogram often becomes changed from the normal. Figure 103B shows the effect when some of the ventricular muscle has been damaged. In this record the portion of the electrocardiogram between the S and T waves is depressed. This results from abnormal leakage of electrical current from the heart between heart beats. It indicates damage to the membranes of the ventricular muscle fibers, which often occurs when one has an acute heart attack.

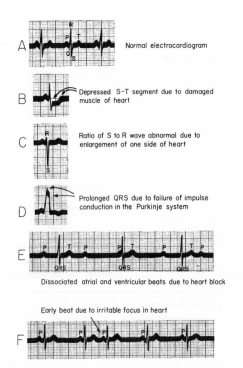

Figure 103.   The normal and several abnormal electrocardiograms.

Figure 103C illustrates the effect seen when one side of the heart is enlarged more than the other. The record shows abnormal enlargement of the S wave and diminishment of the R wave, indicating more current flow from the left side of the heart than from the right. High blood pressure very frequently provides this type of electrocardiogram.

Figure 103D shows what one finds in a person who has a partially blocked Purkinje system. In this instance, the impulse is transmitted through much of the ventricles so slowly that the QRS complex lasts for a prolonged period of time and also develops an abnormal shape.

Figure 103E illustrates the effect of blockage of the impulse at the A-V bundle. The P waves occur regularly, and the QRST waves also occur regularly but in no definite relationship to the P waves. The atria are beating at the natural rate of rhythm of the S-A node, while the ventricles have assumed their own rate of rhythm.

Finally, Figure 103F illustrates a record, indicated by the arrow, of a *premature beat* of the heart. The only abnormality here is that the impulse occurs too soon after the previous heart beat. This is usually caused by an irritable heart resulting from such factors as too much smoking, too much coffee drinking, or lack of sleep.

A study of Figure 103 shows how various abnormalities of heart function can be discovered from electrocardiographic recordings, and why this diagnostic procedure is used in all persons with heart disease.

FUNCTION OF THE HEART VALVES

Referring once again to Figure 98, one can see that the four valves of the heart are all oriented so that blood never flows backward but always forward when the heart contracts. The tricuspid valve prevents backflow from the right ventricle into the right atrium, the mitral valve prevents backflow from the left ventricle into the left atrium, and the pulmonary and aortic valves prevent backflow respectively into the right and left ventricles from the pulmonary and systemic arterial systems. These valves have the same functions as valves in any compression pump, for no pump of this type can possibly operate if fluid is allowed to flow backward during the compression or filling cycle.

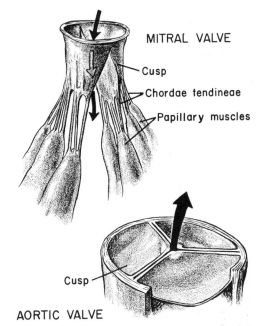

MITRAL VALVE
—Cusp
—Chordae tendineae
—Papillary muscles

Cusp

AORTIC VALVE

Figure 104. Structure of the mitral and aortic valves.

Figure 104 illustrates in more detail the structure of the valves. The tricuspid and mitral valves (the atrioventricular valves) are similar to each other, having rather expansive film-like vanes which are held in place by special ligaments, the *chordae tendineae*, extending from the papillary muscles. The pulmonary and aortic valves (the semilunar valves) are also similar to each other, having small, half-moon-shaped leaflets with very strong structures. The probable reason for these differences in the two types of valves is that blood must flow with great ease from the atria into the ventricles because the atria do not pump with much force. This requires very easily movable valves. The semilunar valves do not have to function with such extreme ease because of the great force of contraction of the ventricles; this allows the valves to be of simpler construction than the A-V valves.

The Heart Sounds

When one listens with a stethoscope to the beating heart he normally hears two sounds that are aptly described as "lub dub, lub dub, lub dub." The "lub" is called the *first heart sound* and the "dub" the *second heart sound.* The first sound is caused by closure of the A-V valves when the ventricles

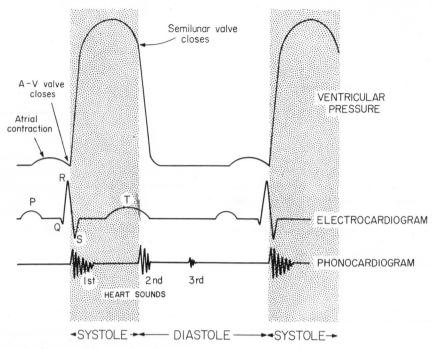

Figure 105. Relationship of ventricular pressure to the electrocardiogram and phono-cardiogram during the cardiac cycle, showing also the periods of systole and diastole.

first contract. This relationship to the cycle of heart beat is illustrated in Figure 105 by a *phonocardiogram,* which is a graphic representation of the heart sounds. When the ventricles contract, the increasing pressures in the two ventricles force the vanes of the A-V valves closed. The sudden stoppage of backflow from the ventricles into the atria creates vibration of the blood and of heart walls, and these vibrations are transmitted through the chest to be heard as the first sound, the "lub" sound. Immediately after the ventricles have discharged their blood into the arterial systems, ventricular relaxation allows blood to begin flowing backward from the arteries into the ventricles, thereby causing the semilunar valves to close suddenly. This also sets up vibrations, this time in the blood of the arteries and ventricles and also in the walls of the vessels and heart. These vibrations are transmitted to the chest causing the "dub" sound.

Valvular Heart Disease

The most frequent cause of valvular heart disease is rheumatic fever, a disease that results from an immune reaction to toxin secreted by streptococcic bacteria as follows: The acute phase of the disease usually occurs two to three weeks after a person has had a streptococcic sore throat, scarlet fever, a streptococcic ear infection, or some other streptococcic infection. Antibodies formed against the toxin attack the valves, causing small, cauliflower-like growths on their edges which, over a period of several months to several years, erode the valves and cause ingrowth of fibrous tissue. The valve sometimes is eaten away completely or becomes so constricted and hardened that it cannot close in a normal manner. Figure 106 illustrates to the left a normal mitral valve and to the right a thickened and calcified valve resulting from several years of the rheumatic fever process. Obviously, such a valve is merely a constricted hole rather than a valve at all.

Valvular stenosis. All degrees of valvular destruction can develop following rheumatic fever. Sometimes the valve openings are so greatly narrowed by scar tissue that blood flows through the opening only with much difficulty. This is called *stenosis.* If the aortic valve becomes stenosed, blood dams up in the left ventricle. If the mitral valve becomes

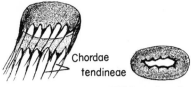

Chordae tendineae

Normal mitral valve   Thickened and calcified mitral valve

Figure 106.   Effect of rheumatic fever on the mitral valve.

stenosed, blood dams up in the left atrium and lungs. Likewise, stenosis of the pulmonary or tricuspid valves causes blood to dam up in the right ventricle or systemic circulation respectively.

Valvular regurgitation.   Equally as often the valves do not become stenotic but instead are eroded so much that they cannot close. As a result, blood leaks backward through the valve that should be stopping this backward flow; this effect is called *regurgitation.* For instance, in aortic regurgitation much of the blood pumped into the aorta by the left ventricle returns to the ventricle during diastole rather than flowing on through the systemic circulation. Likewise, failure of the mitral valve to close when the left ventricle contracts allows blood to flow backward into the left atrium as well as forward into the aorta. Thus, leaking valves are as disastrous to cardiac function as are narrowed valvular openings. Occasionally a valve is both leaky and narrowed, decreasing the effectiveness of cardiac pumping in two ways.

Pulmonary edema.   The valve most frequently affected by rheumatic fever is the mitral valve, though the aortic valve is affected almost as often. On the other hand, the valves of the right heart are rarely damaged severely. Severe damage to the mitral valve leads to excessive damming of blood in the left atrium and lungs, and in aortic valvular disease blood dams up in the left ventricle as well as in the left atrium and lungs. Therefore, following damage of either of these valves, blood engorges the lungs, causing fluid leakage from the pulmonary capillaries into the pulmonary tissues and alveoli, resulting in severe *pulmonary edema* and often drowning the patient in his own fluids. This will be explained in detail in Chapter 15 in the discussion of heart failure.

THE CARDIAC CYCLE

Now that the pumping action of the heart, the rhythmicity of the heart, and the function of the heart valves have been described, it is possible to synthesize this information into a sequence of events called the *cardiac cycle.* The cardiac cycle rightfully begins with initiation of the rhythmic impulse in the S-A node. Then transmission of the impulse through the heart causes the muscle fibers to contract. Thus, as shown in Figure 105, the P wave of the electrocardiogram occurs immediately before the pressure wave caused by atrial contraction. Approximately 0.16 second after the P wave begins, the electrical impulse has completed its passage through the atria and begins to spread rapidly over the ventricles, causing the QRS wave of the electrocardiogram and stimulating the ventricular muscle to contract. The rising ventricular pressure closes the A-V valves, thereby generating the first heart sound, and it opens the semilunar valves. The ventricles remain contracted approximately 0.3 second, and then they relax. During the process of relaxation, ions retransfer through the fiber membranes to reestablish the normal negative electrical charge inside the fibers. This causes the T wave of the electrocardiogram. Immediately after the ventricular muscle relaxes, a small amount of blood flows backward from the arteries into the ventricles, closing the semilunar valves which elicits the second heart sound. Following ventricular relaxation, no further contraction occurs until a new electrical impulse is initiated in the S-A node.

Systole and Diastole

The period during the cardiac cycle when the ventricles are contracting is called *systole,* and the period of relaxation is called *diastole.* A clinician examining the heart can note the periods of systole and diastole either from the electrocardiogram or from the heart sounds; systole begins with the QRS wave and ends with the T wave, or it begins with the first heart sound and ends with the second sound. Diastole on the other hand, begins with the T wave and ends with the QRS wave, or it begins with the second heart sound and ends with the first heart sound.

Sometimes it is quite important to distinguish between systole and diastole. This is particularly true when one is studying valvular disorders or abnormal openings between the two sides of the heart. For instance, leakage of a semilunar valve causes a "swishing" sound called a *murmur* during diastole, because that is the period of the cardiac cycle when abnormal semilunar valves leak instead of closing tightly. On the other hand, a murmur caused by leakage of an A-V valve occurs during systole, because that is the period when these valves leak if they are abnormal.

Pressure Changes During the Cardiac Cycle

Figure 107 shows the changes in pressure in the left atrium, left ventricle, and aorta during a typical cardiac cycle. During diastole, the left atrial pressure is a little higher than that of the left ventricle. This is the reason blood flows continually from the left atrium into the left ventricle. Toward the end of diastole, contraction of the left atrium elevates the left atrial pressure to an even higher value and forces an extra quantity of blood into the ventricle. Then, suddenly, the left ventricle contracts, the A-V valves close, and the ventricular pressure rises rapidly. When the pressure rises higher than that in the aorta, the aortic valve opens and blood flows into the aorta during the entire remainder of systole. When the ventricle relaxes, the ventricular pressure

falls precipitously, allowing a slight backflow of blood that immediately closes the aortic valve. Throughout diastole the aortic pressure remains high because a large quantity of blood has been stored in the very distensible arteries during systole. This blood runs off slowly through the capillaries back to the right atrium, allowing the aortic pressure to fall from a peak during systole of approximately 120 mm. Hg down to a minimum of approximately 80 mm. Hg by the end of diastole. Therefore, the normal arterial blood pressure is said to be 120/80, meaning by this 120 mm. Hg *systolic pressure* and 80 mm. Hg *diastolic pressure.*

THE LAW OF THE HEART

The amount of blood pumped by the heart is normally determined by the amount of blood flowing from the veins into the right atrium. This principle is frequently called the "law of the heart." That is, the heart is simply an automaton that continues to pump all of the time, and whenever blood enters the right atrium it is pumped on through the heart. Of course, there is a maximum rate at which the heart can pump, for which reason, to be completely accurate, the law of the heart is more correctly stated as follows: *Within physiologic limits, the heart pumps all of the blood that flows into it without excessive damming of blood in the veins.* Thus, the heart is analogous to a sump pump, because any time fluid enters the chamber of

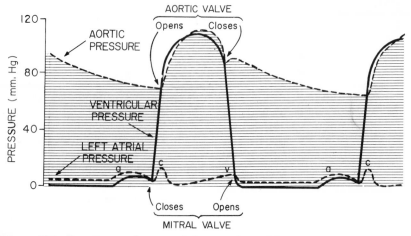

Figure 107. Pressures in the aorta, left ventricle, and left atrium during the cardiac cycle.

a sump pump, it is pumped out of the chamber immediately. In the case of the heart, whenever blood enters one of the atria, it is immediately pumped on into a ventricle and then on into the arteries.

It is a special quality of cardiac muscle that gives the heart this ability to pump varying amounts of blood depending upon the rate of venous inflow. When cardiac muscle is stretched prior to contraction, it contracts with greater force than when it is not preliminarily stretched. When only a small quantity of blood enters the heart, the muscle fibers are not stretched greatly, and the force of contraction is weak. On the other hand, if large quantities of blood enter, the heart chambers dilate greatly, the muscle fibers stretch, and the force of contraction becomes very great. As a result, the increased quantity of blood returning to the heart is pumped on through as a result of this intrinsic adaptation of the heart.

The law of the heart is often inapplicable after the heart has been damaged, for then even normal quantities of blood returning to the heart cannot be pumped with ease. As a consequence, blood begins to dam up in the veins of either the lungs or the systemic circulatory system. In this case the heart is said to be *failing*. The subject of heart failure is so very important that it will be discussed in detail in Chapter 15.

THE HEART-LUNG PREPARATION. One of the methods frequently used in the physiology laboratory to demonstrate the independent ability of the heart to pump blood is the heart-lung preparation illustrated in Figure 108. In this preparation, blood flow from the heart is channeled from the aorta through an external system and then back into the right atrium, rather than through the animal's body. Artificial respiration is supplied to the lungs to keep the blood appropriately oxygenated, and nutrient in the form of glucose is added to the blood. The heart will continue to beat for many hours with the external flow system taking the place of the systemic circulation. One can vary the resistance of the external circuit by tightening or loosening a screw clamp. He can measure the blood flow with a flowmeter, and he can change the amount of blood flowing into the left atrium by raising or lowering the venous reservoir.

Several principles that demonstrate the independent function of the heart can be illustrated beautifully with the heart-lung preparation. First, it can be shown that within physiologic limits raising or lowering the venous reservoir is the main factor that

Figure 108.   The heart-lung preparation.

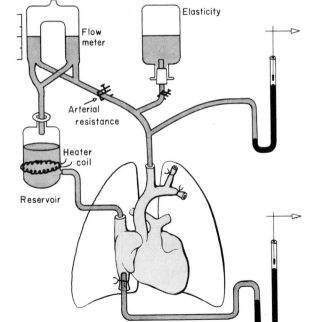

determines how much blood will be pumped. In other words, *the greater the input pressure forcing blood from the veins into the heart, the greater the volume of blood pumped*, which is the "law of the heart."

Second, it can be demonstrated that changing the resistance of the circuit within normal limits *does not* greatly affect the amount of blood pumped. However, the pressure in the aorta increases approximately in proportion to the increase in resistance. That is, *the amount of resistance to blood flow through the circulatory system changes arterial pressure greatly, but hardly affects the amount of blood pumped by the heart* unless the resistance becomes so great that the heart simply cannot pump with enough force to overcome it.

A third interesting effect that can be demonstrated is that of changing the elasticity of the arterial system. This can be done by connecting an air bottle to the arterial system. Blood flows into the bottle and compresses the air when the pressure rises during systole, but during diastole the compressed air forces blood back into the system, thereby helping to maintain a relatively even level of pressure. With the bottle in the system, the arterial pressure pulsates far less than when it is out of the system. This effect illustrates that *elasticity of the arterial walls is very important to smooth out pulsations in the arterial pressure*.

Finally, the heart-lung preparation can be used to demonstrate the effect of temperature, drugs, or abnormal blood constituents on the heart. For instance, *increasing the temperature of the blood 10° F. increases the heart rate approximately 100 per cent*. Also, increasing the amount of *calcium* in the blood makes the heart contract with increased vigor, while increasing the amount of *potassium* decreases the vigor of heart contraction. Finally, when the nutrients of the blood fall to low values, addition of glucose or other nutrients can greatly enhance the function of the heart.

NERVOUS CONTROL OF THE HEART

Though the heart has its own intrinsic control systems and can continue to operate without any nervous influences, there are times when efficacy of heart action can be changed greatly by regulatory impulses from the central nervous system. The nervous system is connected with the heart through two different sets of nerves, the *parasympathetic nerves* and the *sympathetic nerves*. The connections of the parasympathetic (vagi) and sympathetic nerves with the heart are shown in Figure 109.

Stimulation of the parasympathetic nerves causes the following four effects on the heart: (1) decreased rate of rhythmicity of the S-A node, (2) decreased force of contraction of the atrial muscle, (3) decreased rate of conduction of impulses through the A-V node, which lengthens the delay period between atrial and ventricular contraction, and (4) decreased blood flow through the coronary blood vessels which are the vessels that supply nutrition to the heart muscle itself. All of these effects may be summarized by saying that *parasympathetic stimulation decreases all activities of the heart*. Usually, the heart is stimulated by the parasympathetics during periods of rest; this allows the heart to rest at the same time the remainder of the body is resting. This perhaps preserves the resources of the heart, for without periods of rest the heart probably would wear out at a much earlier age than it normally does.

Stimulation of the sympathetic nerves has essentially the opposite effects on the heart:

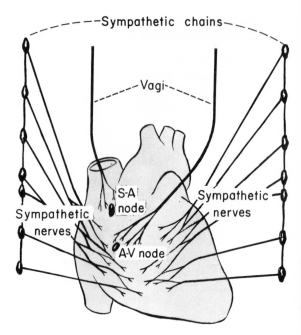

Figure 109.   Innervation of the heart.

(1) increased heart rate, (2) increased vigor of cardiac contraction, and (3) increased blood flow through the coronary blood vessels to supply increased nutrition to the heart muscle. These effects can also be summarized by saying that *sympathetic stimulation increases the activity of the heart as a pump.* This effect is necessary when a person is subjected to stressful situations such as exercise, disease, excessive heat, and other conditions that demand rapid blood flow through the circulatory system. Therefore, the sympathetic effects on the heart are a stand-by mechanism held in readiness to make the heart beat with extreme vigor when necessary.

REFERENCES

Berne, R. M., and Levy, M. N.: Heart. *Ann. Rev. Physiol.* 26:153, 1964.

Brady, A. J.: Excitation and excitation-contraction coupling in cardiac muscle. *Ann. Rev. Physiol.* 26:341, 1964.

Burch, G. E., and Winsor, T.: *A Primer of Electrocardiography.* 4th Ed. Philadelphia, Lea & Febiger, 1960.

Guyton, A. C., and Satterfield, J.: Factors concerned in electrical defibrillation of the heart, particularly through the unopened chest. *Amer. J. Physiol.* 167:81, 1951.

Olson, R. E.: *Physiology of cardiac muscle. Handbook of Physiology,* Sec. II, Vol. I, p. 199. Baltimore, The Williams & Wilkins Company, 1962.

Sarnoff, S. J., and Mitchell, J. H.: The control of the function of the heart. *Handbook of Physiology,* Sec. II, Vol. I, p. 489. Baltimore, The Williams & Wilkins Company, 1962.

Schaper, W.: Heart. *Ann. Rev. Physiol.* 29:259, 1967.

Scher, A. M.: Excitation of the heart. *Handbook of Physiology,* Sec. II, Vol. I, p. 287. Baltimore, The Williams & Wilkins Company, 1962.

Starling, E. H.: *The Linacre Lecture on the Law of the Heart.* London, Longmans Green & Company, 1918.

Woodbury, J. W.: Cellular electrophysiology of the heart. *Handbook of Physiology,* Sec. II, Vol. I, p. 237. Baltimore, The Williams & Wilkins Company, 1962.

CHAPTER 12

BLOOD FLOW THROUGH THE SYSTEMIC CIRCULATION AND ITS REGULATION

All of the circulation besides the heart and the pulmonary circulation is called the *systemic circulation*. The blood flowing through this part of the circulation provides nutrition to the tissues, transport of excreta from the tissues, cleansing of the blood as it passes through the kidneys, absorption of nutrients from the gastrointestinal tract, and mixing of all the fluids of the body, as explained in Chapter 1. The rate of blood flow to each respective tissue is almost exactly the amount required to provide adequate function, no more, no less. The purpose of the present chapter, therefore, is to describe, first, the basic principles of blood flow through the circulation and, second, the mechanisms that control the blood flow to each respective tissue in proportion to its needs.

heart forces blood into the aorta, distending it and creating pressure within it. This pressure then pushes the blood through the arteries, arterioles, capillaries, venules, and veins, finally back to the heart. As long as the animal remains alive, this flow of blood around the continuous circuit never ceases.

The small arteries, arterioles, capillaries, venules, and small veins have such small diameters that blood flows through them with considerable difficulty. In other words, the vessels are said to offer *resistance* to blood flow. Obviously, the smaller the vessel the greater is the resistance, and the larger the vessel the less the resistance.

In essence, the discussion in the present chapter centers around the effects of pressure and resistance on blood flow.

HEMODYNAMICS

The study of the physical principles that govern blood flow through the vessels and the heart is known as *hemodynamics*. The

Blood Flow and Cardiac Output

The amount of blood pumped by the heart when a person is at rest is approximately 5 liters per minute. This is called the *cardiac output,* and it can increase to as much as 20 to 30 liters per minute during the most extreme exercise, or it can decrease to as low as 1.5 liters per minute following severe hemorrhage.

Blood flow to the different parts of the body during rest are given in Table 4. One will note that the brain receives approximately 14 per cent of the total blood flow, the kidneys 22 per cent, the liver 27 per cent, and the muscles, which comprise over half of the body, only 15 per cent. However, during exercise quite a different picture develops, for then essentially all of the increase in blood flow occurs in the muscles, this increasing as much as 15-fold during very intense exercise and then representing as much as 75 per cent of the total blood flow.

MEASUREMENT OF BLOOD FLOW. Several different types of apparatus have been designed to measure blood flow. Some types of measuring apparatus connect with a blood vessel in such a manner that the blood must flow through the apparatus, while others employ indirect methods that do not require opening the vessels.

Figure 110A illustrates the *rotameter* type of flowmeter which must be connected directly into the vascular circuit. Blood enters at the bottom and lifts the float in the conical tube. The greater the flow the higher the float will rise. A steel shaft connected to the float protrudes upward through a coil of wire, and this in turn is connected to an electronic measuring apparatus to record the

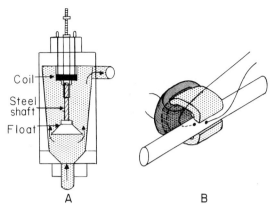

Figure 110. Instruments for recording blood flow: (A) the rotameter, (B) the electromagnetic flowmeter.

rate of blood flow by detecting the change in magnetic properties of the coil.

A second type of flowmeter, the *electromagnetic flowmeter,* is shown in Figure 110B. The electromagnet creates a very strong magnetic field inside the blood vessel. As the blood flows through the field, it "cuts" the magnetic lines of force and develops an electrical potential at right angles to these lines of force. Two electrodes placed at right angles to the magnet pick up this potential and transmit it into an appropriate electronic recording apparatus. The blood vessel does not have to be opened when this type of flowmeter is used, and the response of the instrument is so rapid that it can measure even very transient changes in flow that take place in as little as $1/100$ of a second. Because of these advantages, this type of flowmeter has become very widely used in physiological studies of the circulation.

The plethysmograph. Figure 111A illustrates the *plethysmograph,* which can be used to measure blood flow in a limb or in any other protruding portion of the body. The arm is inserted into the open end of the chamber, and an airtight seal is made between the end of the chamber and the forearm. The cuff around the upper arm is suddenly inflated to create a pressure higher than the pressure in the veins but lower than that in the arteries (about 40 mm. Hg). This allows blood to continue flowing into the arm through the arteries but stops all flow away from the arm through the veins. As a result, the arm swells, forcing air out of the chamber into a recording drum. One can calibrate by injecting known volumes of air into the chamber with a syringe.

TABLE 4. *Blood Flow to Different Organs and Tissues Under Basal Conditions. (Based mainly on data compiled by Dr. L. A. Sapirstein.)*

	PER CENT	ML./MIN.
Brain	14	700
Heart	4	200
Bronchi	2	100
Kidneys	22	1100
Liver	27	1350
Portal	(21)	(1050)
Arterial	(6)	(300)
Muscle	15	750
Bone	5	250
Skin (cool weather)	6	300
Thyroid gland	1	50
Adrenal glands	0.5	25
Other tissues	3.5	175
Total	100.0	5000

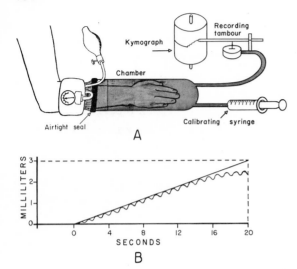

Figure 111. (A) The plethysmograph. (B) A typical recording from the plethysmograph.

Figure 111B shows a typical recording from the plethysmograph, depicting a gradual increase in volume of the arm. A straight line drawn upward along the initial rise in the record illustrates that the initial rate at which the arm swells is approximately 3 ml. in 2 seconds. In other words, the *blood flow* into the arm is 3 ml. per 2 seconds or 90 ml. per minute.

VELOCITY OF BLOOD FLOW IN DIFFERENT PORTIONS OF THE CIRCULATION. The term *blood flow* means the actual quantity of blood flowing through a vessel or group of vessels in a given period of time. In contradistinction to this, the *velocity* of blood flow means the distance that the blood travels along a vessel in a given period of time.

If the quantity of blood flowing through a vessel remains constant, the velocity of blood flow obviously decreases as the size of the vessel increases. The aorta as it leaves the heart has a cross-sectional area of approximately 2 sq. cm. Then it branches into the large arteries, the small arteries, and the capillaries with a portion of the aortic blood flowing into each of these vessels. The total cross-sectional area of the branching vessels is considerably greater than that of the aorta. Therefore, the average velocity of blood flow decreases. For instance, the total cross-sectional area of the arterioles is 15 to 30 times that of the aorta, and the velocity of blood flow decreases this same amount. The cross-sectional area of the capillaries is about 750 times that of the aorta, the venules about

60 times, and the veins about 4 times. It is evident then that the cross-sectional areas of the vessels leaving the heart and entering the heart are rather small, but these spread out like the limbs of a tree into a very broad cross-sectional area in the region of the capillaries. As a consequence, the velocity of blood flow is greatest in the aorta and least in the capillaries. Numerically, the velocities are approximately the following: aorta, 30 cm. per second; arterioles, 1.5 cm. per second; capillaries, 0.4 mm. per second; venules, 5 mm. per second; and venae cavae, 8 cm. per second.

Transit time for blood in the capillaries. The velocity of blood flow in the capillaries is particularly significant because it is here that oxygen, other nutrients, and excreta pass back and forth between the blood and the tissue spaces. The length of the average capillary is about 0.5 to 1 mm. Therefore, at a velocity of blood flow of 0.4 mm. per second, the length of time that blood remains in each capillary averages one to two seconds. Despite this short time, an extremely large proportion of the substances in the blood can transfer through the capillary membrane. This signifies once again the extreme capability of the capillaries as an exchange mechanism between the blood and the tissue fluids. Without this exchange in the capillaries, all the other hemodynamic functions of the circulation would be useless.

Blood Pressure

The pressure in a blood vessel is the force that the blood exerts against the walls of the vessel. This force distends the vessel because all blood vessels are distensible, the veins 6 times as much so as the arteries. Pressure also makes blood attempt to leave a vessel by any available opening, which means that the normally high pressure in the arteries forces blood through the small arteries, then through the capillaries, and finally into the veins. The importance of blood pressure, then, is that it is the force that makes the blood flow through the circulation.

MEASUREMENT OF BLOOD PRESSURE— THE MERCURY MANOMETER. The standard device for measuring blood pressure is the mercury manometer shown in Figure 112. The rubber tube is attached to a blood vessel.

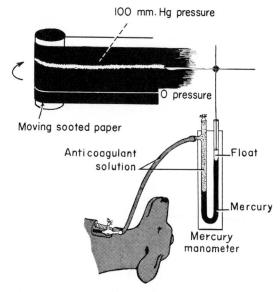

Figure 112.   A mercury manometer.

The blood from the vessel presses downward on the left column of mercury forcing the right column upward. A float and recording arm resting on the mercury rise and fall with the level of the mercury and thereby record the pressure. If the level of mercury in the right column is 100 millimeters above the level in the left column, the blood pressure is said to be 100 millimeters of mercury (mm. Hg). If the level to the right is 200 millimeters above that on the left, then the pressure is 200 mm. Hg.

Because of the convenience of the mercury manometer for measuring pressure, almost all pressures of the circulatory system are expressed in millimeters of mercury. However, pressure can also be expressed as water pressure because one can equally as well connect an artery to a column of water and determine how high the water rises. Mercury weighs 13.5 times water, which means that the level of water in a manometer will rise this many times as high as the level of mercury in a manometer. Therefore, 13.5 mm. or 1.35 cm. of water pressure equals 1 mm. Hg pressure.

High fidelity methods for recording blood pressure.   Unfortunately, the mercury in the mercury manometer has so much *inertia* that it cannot rise and fall rapidly. For this reason, this type of manometer, though excellent for recording mean pressure, cannot respond to pressure *changes* that occur more rapidly than one cycle every two to three seconds. Therefore, when it is desired to record rapidly changing pressures, some other type of pressure recorder such as one of those illustrated in Figure 113 is required. Each of the three transducers illustrated in this figure converts pressure into an electrical signal that is then recorded on a high-speed recorder. A syringe needle is connected to each transducer; this needle is inserted directly into the vessel whose pressure is to be measured, and the pressure is transmitted into a chamber bounded on one side by a thin membrane. In transducer A this membrane forms one plate of a variable capacitor. As the pressure rises, the membrane moves closer to the other plate and changes the electrical *capacitance*, which is detected and recorded by an electronic instrument. In transducer B a small iron slug on the membrane moves upward into a coil and changes the coil's *inductance*, which likewise is detected and recorded electronically. In transducer C the membrane is connected to a thin, stretched wire. As the membrane moves, the length of the wire changes, which changes its *resistance*, once again allowing the pressure to be recorded by an electronic recorder.

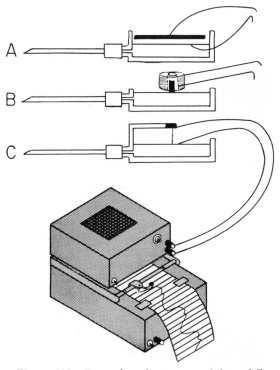

Figure 113.   Principles of operation of three different electronic transducers for recording rapidly changing blood pressure.

RELATIONSHIP OF PRESSURE TO BLOOD FLOW. When the blood pressure is high at one end of a vessel and low at the other end, blood will attempt to flow from the high toward the low pressure area because the two forces are unequal. The rate of blood flow is directly proportional to the difference between the pressures. Figure 114 illustrates this principle, showing that the flow from the spout becomes greater in proportion to the water level in the chamber.

It is not the actual pressure in a vessel that determines how rapidly blood will flow, but the *difference* between the pressures at the two ends of the vessel. For instance, if the pressure at one end of the vessel is 100 mm. Hg and 0 mm. Hg at the other end, the pressure difference is 100 mm. Hg. If the pressures at the two ends of this same vessel are suddenly changed to 5000 mm. Hg at one end and 4950 mm. Hg at the other end, the pressure difference now is only 50 mm. Hg, and the flow will be reduced to one half as much (provided the vessel's diameter does not become changed by the pressure).

Resistance to Blood Flow

Resistance is essentially the same as friction, for it is friction between the blood and the vessel walls that creates impediment to flow. The amount of resistance is dependent on the length of the vessel, the diameter of the vessel, and the viscosity of the blood.

EFFECT OF VESSEL LENGTH ON RESISTANCE TO BLOOD FLOW. The longer a vessel, the greater the vascular surface along which

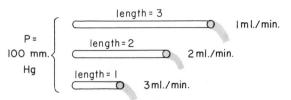

Figure 115.   Effect of vessel length on flow.

the blood must flow, and consequently the greater the friction between the blood and vessel wall. For this reason, the *resistance to blood flow is directly proportional to the length of the vessel.*

This principle is illustrated by the three vessels in Figure 115, each of which has a pressure head of 100 mm. Hg at its inlet. The vessel with a length of 1 has an output of 3 ml. per minute while the one with a length of 3 but of exactly the same diameter has only one-third as much output, illustrating the inverse relationship between the length of the vessel and the amount of fluid that will flow when all other factors remain constant.

EFFECT OF VESSEL DIAMETER ON RESISTANCE TO BLOOD FLOW. Fluid flowing through a vessel is retarded mainly along the walls. For this reason, the velocity of blood flow in the middle of a vessel is very great while the velocity along the surface is very low, and the larger the vessel the more rapidly can the central portion of blood flow. Because of this effect, the blood flow through a vessel—that is, the total quantity of blood that passes through the vessel each minute—increases very markedly as the diameter of the vessel increases. In fact, if all other factors stay constant, the blood flow through the vessel is directly proportional to the *fourth power* of the diameter. Thus, in Figure 116 three vessels of the same length but with diameters of 1, 2, and 4 are shown. Note that even these small changes in diameter increase the flow respectively from 1 to 16 to 256 ml. per minute. Because of this extreme dependence of flow on diameter, very slight changes in vascular diameter can affect

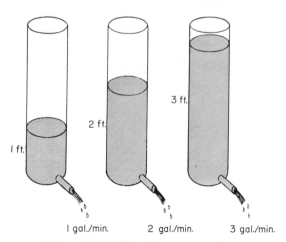

Figure 114.   Effect of pressure on flow.

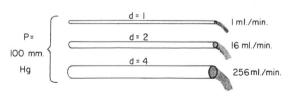

Figure 116.   Effect of vessel diameter on flow.

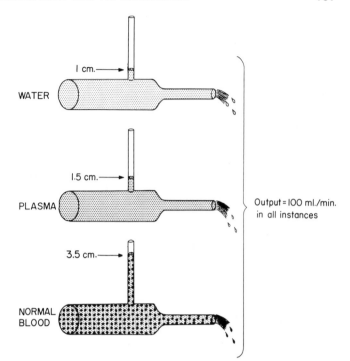

Figure 117. Effect of viscosity on flow.

tremendously blood flow to different vascular regions. The diameter of most vessels can change approximately 4-fold, which can change the blood flow as much as 256-fold.

EFFECT OF VISCOSITY ON RESISTANCE TO BLOOD FLOW. The more viscous the fluid attempting to flow through a vessel the greater the friction with the wall, and, consequently, the greater the resistance. This effect is illustrated in Figure 117, which shows water, plasma, and normal blood attempting to go through spouts of equal dimensions. In each instance the amount of fluid flowing through the spout is 100 ml. per minute, but, because of different viscosities of the three fluids, the pressure required to force each through the spout is also different. The level of fluid in the vertical tube, which is an indicator of the pressure required to cause the flow, is 1 cm. when water is flowing, 1.5 cm. for plasma, and 3.5 cm. for blood. In other words, the *relative* viscosities of water, plasma, and blood are 1, 1.5, and 3.5.

The most important factor governing blood viscosity is the concentration of red blood cells. The viscosity of normal blood is about 3.5 times that of water. However, when the red blood cell concentration falls to one-half normal, the viscosity becomes only 2 times that of water, and when the concentration

rises approximately 70 per cent above normal, the viscosity can increase to as high as 20 times that of water. For these reasons, the flow of blood through the blood vessels of anemic persons—that is, persons who have low concentrations of red blood cells—is extremely rapid, while blood flow is very sluggish in persons who have excess red blood cells (polycythemia).

Interrelationship of Pressure, Flow, and Resistance

It is obvious from the preceding discussions that pressure and resistance oppose each other in affecting blood flow. One can state this relationship mathematically by the following formula:

$$\text{Blood flow} = \frac{\text{Pressure}}{\text{Resistance}}$$

The same formula can be expressed in two other algebraic forms:

$$\text{Pressure} = \text{Blood flow} \times \text{Resistance}$$

$$\text{Resistance} = \frac{\text{Pressure}}{\text{Blood flow}}$$

These formulas are basic in almost all hemodynamic studies of the circulatory system, and therefore must be thoroughly understood prior to any attempt to analyze the operation of the circulation.

POISEUILLE'S LAW. By inserting the various factors that affect resistance into the preceding formulas, one can derive still another formula known as Poiseuille's law:

$$\text{Blood flow} = \frac{\text{Pressure} \times (\text{Diameter})^4}{\text{Length} \times \text{Viscosity}}$$

This formula expresses the ability of blood to flow through any given vessel, showing that the rate of blood flow is directly proportional to the pressure difference between the two ends of the vessel, directly proportional to the fourth power of the vessel diameter, and inversely proportional to the vessel length and blood viscosity.

REGULATION OF BLOOD FLOW THROUGH THE VESSELS

The Arterioles as the Major Regulators of Blood Flow

The rate of flow through each respective tissue is controlled mainly by the arterioles. There are three separate reasons for this:

First, approximately one half of all the resistance to blood flow in the systemic circulation is in the arterioles, which can be seen by studying Figure 118. This figure shows the minuteness of the flow channel through the arteriole, which gives this vessel its high resistance. Therefore, a slight change in arteriolar diameter can change the total resistance to flow through any given tissue far more than can a similar change in diameter in any other vessel.

Second, the arterioles have a very strong muscular wall constructed in such a manner that the diameter of an arteriole can change as much as 3- to 5-fold. Remembering once again that resistance is inversely proportional to the fourth power of a vessel's diameter, it immediately becomes obvious that the resistance to blood flow through the arterioles can be changed as much as several hundred- or even a thousand-fold by simply relaxing or contracting the smooth muscle walls.

Third, the smooth muscle walls of the arterioles respond to two different types of

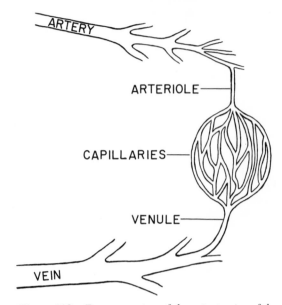

Figure 118. Demonstration of the minute size of the arteriole relative to the remainder of the vascular system, which explains its tremendous resistance.

stimuli that regulate blood flow. First, they respond to the local needs of the tissues, increasing blood flow when the supply of nutrients to the tissues falls too low and decreasing the flow when the nutrient supply becomes too great. This mechanism is called *autoregulation*. Second, autonomic nerve impulses, particularly *sympathetic impulses*, have a profound effect on the degree of contraction of the arterioles.

Autoregulation of Blood Flow — Role of Oxygen

The term autoregulation means automatic adjustment of blood flow in each tissue to the need of that tissue. In most instances the need of the tissue is nutrition, but in a few instances other factors that depend on blood flow are needed even more than nutrition. For instance, in the kidneys, the need to excrete end products of metabolism and electrolytes is prepotent, and the concentrations of these substances in the blood play a major role in controlling renal blood flow. In the brain it is essential that carbon dioxide concentration remain very constant because reactivity of the brain cells increases and decreases with the amount of carbon dioxide; in this tissue it is primarily carbon dioxide that determines the rate of blood flow—the greater the rate of blood flow, the greater the rate of removal of carbon dioxide from the tissue.

However, in most tissues it is the need of the tissue especially for the nutrient *oxygen* that seems to be the most powerful stimulus for autoregulation. For instance, if a tissue becomes more active than usual, the need for oxygen might increase as much as 5- to 10-fold, and the blood flow automatically increases to help supply this extra oxygen.

MECHANISM BY WHICH OXYGEN DEFICIENCY CAUSES ARTERIOLAR DILATATION. Even though we know that blood flow through most tissues is controlled in proportion to their need for oxygen, we still do not know the exact mechanism by which oxygen need can affect the degree of vascular contraction. There are two basic theories for arteriolar vasodilatation. One of these, the "oxygen demand" theory, is illustrated in Figure 119. This shows that the tissue cells lie in close proximity to the arterioles and that they are in competition with the arterioles for the oxygen, which is represented by the small dots. It is assumed that the smooth muscle of the arteriolar wall requires oxygen to contract. Should the oxygen supply diminish, the vessels would dilate simply because of failure to receive adequate oxygen to maintain contraction. Therefore, if the tissue should utilize an increased quantity of oxygen, the amount of oxygen available to the arterioles would decrease and cause increased blood flow. The resulting increase in blood flow then increases the tissue oxygen concentration back toward a normal level.

Another theory is that oxygen lack causes some *vasodilator substance* to be formed in the tissue. That is, decreased availability of oxygen to the metabolic systems of the cells supposedly causes them to secrete a chemical substance that acts directly on the blood vessels to cause active vasodilatation. Some of the suggested vasodilator substances have been histamine, lactic acid, carbon dioxide, and various breakdown products of adenosine triphosphate, such as adenosine. Thus far, however, none of these substances has been proved to be secreted in sufficient quantity to account for the autoregulation response.

LONG-TERM AUTOREGULATION. If the needs of the tissues for oxygen are more than the acute autoregulatory mechanism can supply, a long-term autoregulatory mechanism, requiring several weeks to develop, can cause still more increase in blood flow. This mechanism causes the actual number of blood vessels to increase. For instance, when a blood vessel has become partially occluded such as a coronary vessel of the heart, new vessels grow into the area of poor blood supply, returning the blood supply toward normal after a few weeks. Also, if a person ascends to a high altitude and remains there for many weeks, he gradually develops an increased number of blood vessels in all his tissues, thereby supplying them with increased quantity of oxygen. Conversely, when he comes back to a lower altitude, some of the vessels actually disappear after several more weeks, thus illustrating that this long-term vascular phenomenon helps to autoregulate blood flow in response to the tissues' needs.

Nervous Control of Blood Flow

All the arterioles of the systemic circulation are supplied by nerves from the sympathetic nervous system, as illustrated in Figure 120. Stimulation of the sympathetic nerves causes the arterioles of most tissues to constrict.

In a few tissues, however, sympathetic nerves dilate the arterioles rather than constrict them, for instance in some areas of the

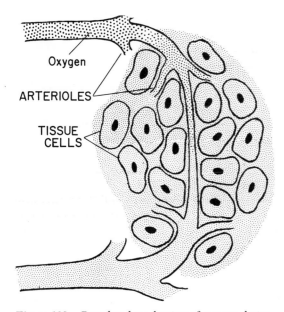

Figure 119.   Postulated mechanism of autoregulation.

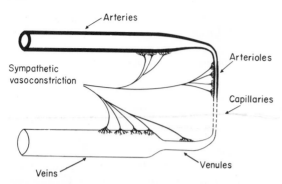

Figure 120. Innervation of the peripheral vascular system, showing especially the abundant innervation of the arterioles.

skin and in the muscles. However, since by far the most important function of sympathetic control of blood vessels is vasoconstriction, the remainder of our discussion at this point concerns the vasoconstrictor sympathetic fibers.

VASOCONSTRICTOR TONE OF THE SYSTEMIC VESSELS. The sympathetic vasoconstrictor nerves normally transmit a continual stream of impulses to the blood vessels, maintaining the vessels in a moderate state of constriction all of the time. This is called *vasomotor tone*. The mechanism by which the sympathetic nervous system constricts vessels more than their normal state of constriction is to increase the number of sympathetic impulses above normal, while the mechanism by which it dilates the vessels is to decrease the impulses. In this manner the sympathetic system can cause both vasoconstriction and vasodilatation, which is the reason special nerves to cause vasodilatation are not essential in most parts of the circulation.

FUNCTION OF NERVOUS CONTROL OF THE BLOOD VESSELS. Nervous control of blood vessels is not generally concerned with regulating local nutrition of the tissues. Instead, it is concerned with the overall distribution of blood flow to major sections of the body. For instance, when one exercises, the muscles require such an extreme increase in blood flow that the heart often cannot pump an adequate quantity of blood both through the dilated muscle vessels and at the same time through such areas as the skin, the kidneys, and the gastrointestinal tract, which normally receive three fourths of the total cardiac output. Therefore, during exercise, sympathetic impulses cause vaso-

constriction in these areas while the muscle vessels become dilated by the autoregulation and sympathetic vasodilator mechanisms. As a result, a major portion of the total blood flow shifts to the muscles, thereby allowing much more muscular activity than could otherwise occur.

The sympathetic nervous system can also shift large amounts of blood flow to help regulate body temperature. When the body temperature rises too high, sympathetic control dilates the arterioles of the skin, increasing the flow of warm blood to the skin; this promotes heat loss until the temperature returns to normal. On the other hand, when the body temperature falls below normal, the opposite effects occur: blood flow decreases, heat loss becomes less, and body temperature rises.

A third instance of important nervous regulation of blood flow occurs when the circulatory system has been damaged so greatly that the heart pumps an insufficient quantity of blood to supply all portions of the body. Under these conditions sympathetic stimulation causes vasoconstriction in the less vital areas, while organs such as the brain and heart continue to receive adequate flow.

And, finally, nervous control of the arterioles plays a very important role in reflex regulation of arterial pressure. This effect is so important that it will be discussed in detail in Chapter 14. Briefly, when the arterial pressure begins to fall, as a result of hemorrhage, cardiac damage, or any other reason, pressure sensitive detectors in the walls of several large arteries, called *baroreceptors*, detect the falling pressure and transmit signals to the brain. The brain in turn reflexly transmits nerve impulses back to the heart to increase heart activity and to the arterioles throughout the body to cause vasoconstriction, thereby elevating the arterial pressure back toward normal.

DISTRIBUTION OF BLOOD IN THE BODY

Figure 121 shows the approximate percentages of the total blood in the different portions of the circulatory system. It is evident from this figure that about three fourths of the blood is in the systemic circulation

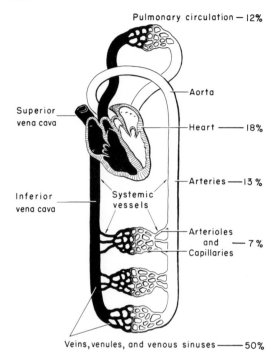

Pulmonary circulation — 12%

Aorta

Superior vena cava

Heart —— 18%

Arteries —— 13%

Inferior vena cava

Systemic vessels

Arterioles and —— 7% Capillaries

Veins, venules, and venous sinuses —— 50%

Figure 121. Distribution of blood in the different portions of the circulatory system.

and one fourth in the pulmonary circulation and heart. Also, a far higher proportion of the blood is in the veins than the arteries. For instance, the arteries of the systemic circulation contain only 13 per cent of the total blood, while the veins, venules, and venous sinuses contain approximately 50 per cent. The arterioles and capillaries, despite their extreme importance to the circulation, contain only a few per cent of the blood.

The Blood Reservoirs

When the total blood volume falls so low that the vessels are no longer adequately filled, blood cannot circulate normally through the tissues. For this reason, it is important to have an extra supply of blood. The entire venous system acts as a *blood reservoir*, for the veins exhibit a plastic quality whereby their walls can distend and contract in response to the amount of blood available in the circulation. Also, the veins are supplied with nerves from the sympathetic nervous system so that any time the tissues begin to suffer for lack of blood flow, nervous reflexes cause a large number of sympathetic impulses to pass to the veins,

constricting them and translocating blood into the heart and other vessels. It is this contractile and expansile quality of the venous system that protects the circulation against the disastrous effects of blood loss.

Certain portions of the venous system are especially important for storing blood: First, the *large veins of the abdominal region* are particularly distensible, and, therefore, normally hold a tremendous amount of blood. Yet they can contract when the blood is needed elsewhere in the circulation.

Second, the *venous sinuses of the liver* can expand and contract many-fold, so that the liver under certain circumstances may hold as much as a liter and a half of blood, but at other times only a few hundred milliliters.

Third, the *spleen* normally contains approximately 200 ml. of blood, but can expand to hold as much as one liter, or can contract to hold as little as 50 ml.

The *venous plexuses of the skin* are a fourth very important blood reservoir. Normally the blood in these plexuses is used to regulate the heat of the body — the more rapidly blood flows through them the greater is the loss of heat. However, when the vital organs need extra blood flow, the sympathetic nervous system can markedly contract the skin's venous plexuses, transferring the stored blood into the main stream of flow.

A fifth blood reservoir is the *pulmonary vessels.* Approximately 12 per cent of the blood is normally in the pulmonary circulation, and much of this can be displaced into other portions of the circulation without impairing the function of the lungs. Therefore, the lungs also act as a major source of blood in times of need.

REFERENCES

Abramson, D. I. (ed.): *Blood Vessels and Lymphatics.* New York, Academic Press, 1962.

Alexander, R. S.: The peripheral venous system. *Handbook of Physiology*, Sec. II, Vol. II, p. 1075. Baltimore, The Williams & Wilkins Company, 1963.

Green, H. D., Rapela, C. E., and Conrad, M. C.: Resistance (conductance) and capacitance phenomena in terminal vascular beds. *Handbook of Physiology*, Sec. II, Vol. II, p. 935. Baltimore, The Williams & Wilkins Company, 1963.

Guyton, A. C.: Peripheral circulation. *Ann. Rev. Physiol.* 21:239, 1959.

Guyton, A. C., Ross, J. M., Carrier, O., Jr., and Walker,

J. R.: Evidence for tissue oxygen demand as the major factor causing autoregulation. *Circ. Res.* 14:60, 1964.

Heymans, C., and Neil, E.: *Reflexogenic Areas of the Cardiovascular System.* Boston, Little, Brown & Company, 1958.

Korner, P. I.: Circulatory adaptations in hypoxia. *Physiol. Rev.* 39:687, 1959.

McDonald, D. A.: Hemodynamics. *Ann. Rev. Physiol.* 30:525, 1968.

Reeve, E. B., and Guyton, A. C.: *Physical Bases of Circulatory Transport: Regulation and Control.* Philadelphia, W. B. Saunders Company, 1967.

Rovick, A. A., and Randall, W. C.: Systemic circulation. *Ann. Rev. Physiol.* 29:225, 1967.

Sagawa, K., Ross, J. M., and Guyton, A. C.: Quantitation of cerebral ischemic pressor response in dogs. *Amer. J. Physiol.* 200:1164, 1961.

Sonnenschein, R. R., and White, F. N.: Systemic circulation. *Ann. Rev. Physiol.* 30:147, 1968.

CHAPTER 13

SPECIAL AREAS OF THE CIRCULATORY SYSTEM

THE PULMONARY CIRCULATION

The pulmonary circulation represents the blood flow through the lungs. Its function is to transport blood through the pulmonary capillaries where oxygen is absorbed into the blood from the alveolar air and where carbon dioxide is excreted from the blood into the alveoli.

The physiologic anatomy of the pulmonary circulation, illustrated in Figure 122, is very simple. The right ventricle pumps blood into the pulmonary artery. From there, the blood flows through the pulmonary capillaries into the pulmonary veins and finally into the left

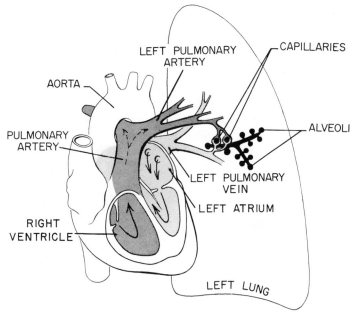

Figure 122. The pulmonary circulation and its relation to the heart.

atrium. Because all portions of the lungs have the same function—that is, to aerate the blood—the arrangement of the vessels is the same in all areas of the pulmonary circulation.

The pulmonary capillaries lie immediately adjacent to the epithelial lining of the alveoli. The combination of the alveolar lining, capillary membrane, and the small amount of tissue that lies between these two is called the *pulmonary membrane,* the total thickness of which is only 0.2 to 0.4 micron. The pores of this membrane are large enough for oxygen and carbon dioxide to diffuse through them with ease. Yet they are small enough that the blood proteins do not leak out of the capillaries.

Flow of Blood Through the Lungs

Because blood flows around a continuous circuit in the body, the same amount of blood must flow through the lungs as through the systemic circulation. The vessels of the lungs are very expansile so that whenever the amount of blood entering the lungs increases, the pulmonary vessels are automatically stretched to allow the more rapid flow demanded of the pulmonary circulation. In other words, the resistance to blood flow through the lungs decreases as the rate of blood flow increases. This allows facile transport of blood through the lungs under widely varying conditions.

REGULATION OF BLOOD FLOW THROUGH THE LUNGS. Since all portions of the lungs perform essentially the same function, there is no need for extensive regulation of the pulmonary blood vessels. However, one feature of blood flow regulation in the lungs is different from that in the remainder of the body. It is quite important for blood to flow only through those portions of the lungs that are adequately aerated and not through portions not aerated. Research studies in the past few years have shown that a low concentration of oxygen in an area of the lungs automatically causes the vessels there to constrict. Therefore, when the bronchi to an area of the lungs become blocked, the alveolar oxygen is rapidly used up, and as a consequence, the vessels constrict, thereby forcing the blood to flow instead through other areas of the lungs that are still aerated.

The resistance to blood flow in the pulmonary circulation is so little that the mean pulmonary arterial pressure is only 13 mm. Hg and the systolic pressure (the highest pressure during contraction of the ventricle) is only 22 mm. Hg, while the diastolic (the lowest pressure between contractions) is 8 mm. Hg. The mean pulmonary arterial pressure is approximately one-seventh the mean systemic arterial pressure which is about 100 mm. Hg, as is discussed in the following chapter.

The pulmonary venous pressure is about 4 mm. Hg. This is also the pressure in the left atrium. The pulmonary capillary pressure has never been measured, but it must be greater than the pulmonary venous pressure and less than the pulmonary arterial pressure —that is, somewhere between 4 and 13 mm. Hg. Indirect measurements of pulmonary capillary pressure indicate that it is probably about 7 mm. Hg.

The mean pressure gradient from the pulmonary artery to the pulmonary vein is 13 mm. Hg minus 4 mm. Hg or 9 mm. Hg. This compares with a pressure gradient in the systemic circulation of approximately 100 mm. Hg. To express this differently, the same amount of blood flows through the lungs as through the systemic circulation with about one-eleventh the force propelling it. This means also that the total resistance of the pulmonary circulation is normally only one-eleventh the total resistance of the systemic circulation.

PRESSURE IN THE PULMONARY ARTERY DURING EXERCISE. When one exercises, the amount of blood flowing through the systemic circulation sometimes increases to as high as five times normal, which means that the amount flowing through the lungs must also increase this much. Still, the pulmonary arterial pressure does not increase more than 25 to 30 per cent because the vessels of the lungs passively enlarge to accommodate whatever amount of blood needs to flow, which keeps the pressure from rising more than a few millimeters of mercury. This obviously keeps the right heart from becoming overloaded during exercise.

PULMONARY CAPILLARY DYNAMICS. Only about one tenth of the blood of the lungs is actually in the capillaries at any one time, and the blood flows through them so rapidly that it remains in the capillaries at most only one to two seconds. During exercise, when blood flow is very rapid, the blood may remain in the pulmonary capillaries only one-quarter to one-half second. Yet, because the pulmonary membrane is extremely permeable to oxygen and carbon dioxide, even during this very minute period of time the blood can become almost completely oxygenated and can excrete the necessary amount of carbon dioxide.

Capillary mechanism for maintaining dry lungs. Another extremely important feature of pulmonary capillary dynamics is the mechanism for keeping the air spaces of the lungs dry. The plasma of the blood has a colloid osmotic pressure of 28 mm. Hg, which causes a continual tendency for fluid to be absorbed into the capillaries. The mean pulmonary capillary pressure, which tends to filter fluid out of the capillaries, is only 7 mm. Hg. Therefore, the tendency for fluid to be absorbed into the capillaries is about 21 mm. Hg greater than for fluid to filter out of the capillaries. Thus, the plasma colloid osmotic pressure continually promotes fluid absorption from the tissues and alveoli of the lungs, and if a small amount of water does enter the alveoli, it is usually absorbed within a few minutes.

Abnormalities of the Pulmonary Circulation

Pulmonary congestion means too much blood and fluid in the lungs. It results from too high a pressure in the pulmonary circulation, which in turn is caused most frequently by failure of the left heart to pump blood adequately from the lungs into the systemic circulation. Once the pulmonary capillary pressure rises above the plasma colloid osmotic pressure, about 28 mm. Hg, fluid leaks out of the capillaries extremely rapidly, and the alveoli no longer remain dry. Thus, severe *pulmonary edema* develops, sometimes so rapidly that it causes death in 1 to 2 hours.

Atelectasis means collapse of a lung or part of a lung. Occasionally some abnormality blocks one or more of the bronchi, cutting off air flow. When this happens, the air in the blocked alveoli is absorbed within a few hours, causing the lung to collapse. Simultaneously, the lack of oxygen in the collapsed alveoli causes vascular constric-

tion, a mechanism that was discussed earlier in the chapter. As a result, the blood flow through the collapsed lung decreases to about one-fifth its previous value, which causes most of the blood to pass through the uncollapsed areas of the lungs where it can be aerated.

Surgical removal of large portions of the lungs causes excessive blood flow through the remaining lung tissue. Ordinarily blood can flow through each lung four times as rapidly as normally before the pulmonary arterial pressure begins to rise. Therefore, a person can lose one whole lung, which increases the blood flow through the opposite lung about two-fold, without causing any serious disability. Yet if this person attempts too heavy exercise, he rapidly reaches the upper limit that the blood can flow before pulmonary hypertension and failure of the right heart develop.

Effect of Congenital Heart Disease on the Pulmonary Circulation

Congenital heart disease means an abnormality of the heart or of closely allied blood vessels that is present at birth. Many types of congenital heart disease affect the pulmonary circulation; two of these, called *patent ductus arteriosus* and *tetralogy of Fallot*, are illustrated in Figures 123 and 124.

During normal fetal life, blood by-passes the lungs by flowing from the pulmonary artery through the *ductus arteriosus* into the aorta. Immediately after birth, expansion of the lungs dilates the pulmonary vessels so that blood can then flow through them with ease. As a result, pulmonary arterial pressure falls, causing the blood to flow backward from the aorta in the pulmonary artery. Also, the blood now flowing through the lungs becomes oxygenated before it enters the aorta. The backward flow of *oxygenated* blood through the ductus arteriosus causes it to become occluded during the first few days or weeks of life in almost all babies. In perhaps 1 out of every 2000 the ductus never closes—that is, the *patent ductus arteriosus*, an open vessel connecting the aorta and the pulmonary artery, is present throughout life (see Figure 123). As the child grows older and his heart becomes stronger,

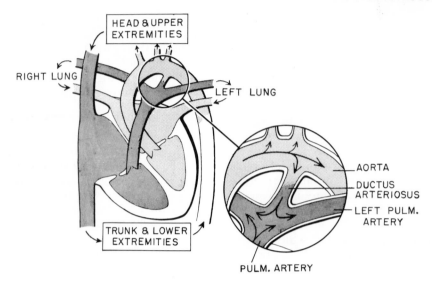

Figure 123.   Patent ductus arteriosus.

the pressure in the aorta becomes much greater than the pressure in the pulmonary artery, forcing a tremendous quantity of blood from the aorta into the pulmonary artery. This blood flows a second time through the lungs, then back through the left heart, through the lungs again, and around and around this circuit several times before finally passing into the systemic circulation. In other words, many times the normal amount of blood is pumped through the lungs. In early life this is a great strain on the heart but does not cause extensive damage to the lungs. However, in early adulthood the very high pressure in the pul-

monary vessels eventually promotes fibrosis of the vessels and congestion of the lungs. Finally, the person has such difficulty aerating his blood that this alone may cause his death.

Figure 124 shows the congenital abnormality called *tetralogy of Fallot*. This disease presents four abnormalities of the heart, including (1) a hole in the septum (partition) between the right and left ventricles, (2) a greatly constricted pulmonary artery, (3) displacement of the aorta to the right, and (4) a hypertrophied right ventricular muscle. Very little blood passes from the right ventricle through the constricted pulmonary

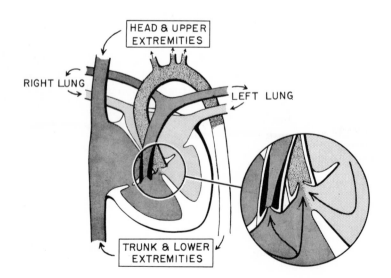

Figure 124.   Tetralogy of Fallot.

artery. Instead, most of it is forced through the hole in the septum and then into the aorta. Thus, most of the blood pumped by the right ventricle by-passes the lungs and is not aerated. Because non-aerated blood is bluish, the whole body assumes this hue. Tetralogy of Fallot is one of the congenital anomalies of the heart that causes the so-called *blue baby.*

Other often encountered congenital abnormalities of the heart are: (1) an abnormal opening through the ventricular septum, (2) an abnormal opening through the atrial septum, (3) constriction or occlusion of the descending aorta, so that blood must flow to the lower part of the body through many small vessels in the body wall rather than through the aorta, (4) abnormal connection of one of the pulmonary veins to the right atrium instead of to the left atrium, and (5) partial constriction of the pulmonary artery, resulting in high right ventricular pressure and overloading of the right ventricle.

In the past few years many advances have been made in the surgical repair of cardiac abnormalities. For instance, any patient with a patent ductus arteriosus can be treated very simply by placing a tight ligature around the ductus. Treatment of tetralogy of Fallot is more complicated but is usually accomplished reasonably satisfactorily by creating an artificial but very small patent ductus; an opening is made between one of the major systemic arteries and the pulmonary artery, allowing much of the blood that by-passes the lungs to flow into the pulmonary system. Surgical procedures have also been devised to transpose blood vessels when necessary and to close abnormal openings in the ventricular or atrial septum.

THE CORONARY CIRCULATION

Contrary to what might be expected, blood does not pass directly from the chambers of the heart into the heart muscle. Instead, the heart has its own special blood supply, which is shown in Figure 125. Two small arteries called the *coronary arteries* originate from the aorta immediately above the aortic valve. Blood flows from these vessels through branches over the *outer surface* of the heart, then into smaller arteries and capillaries in the cardiac muscle, and finally to the right

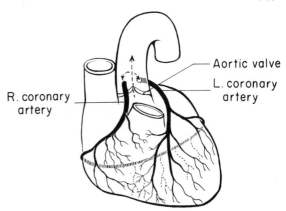

Figure 125.   The coronary circulation.

atrium mainly through a very large vein called the *coronary sinus.*

Flow of Blood Through the Coronary Vessels

The amount of blood flowing through the coronaries each minute is approximately 225 ml., which amounts to approximately 4 to 5 per cent of all the blood pumped by the heart.

In contrast to the flow of blood in other portions of the circulation, flow in the coronaries is greater during diastole than during systole. The reason for this is that systolic contraction of the heart muscle compresses the vessels in the wall of the heart. This momentarily partially or totally occludes these vessels. Therefore, the rate of coronary flow is determined more by the diastolic level of arterial pressure than by either the mean pressure or the systolic pressure. This has importance in those circulatory diseases in which diastolic pressure is especially low.

REGULATION OF CORONARY BLOOD FLOW.   Blood flow through the coronaries is regulated in proportion to the need of the heart for nutrition. The most important method of regulation is the *autoregulation mechanism,* which was discussed in the previous chapter. That is, when the metabolic need for nutrients—especially for oxygen—becomes greater than the supply, the arterioles automatically dilate. As a result, blood flow through the coronaries increases until the level of nutrition matches the degree of cardiac activity.

The sympathetic nervous system also helps slightly to control blood flow through the coronaries. Increased sympathetic stimulation increases coronary flow and decreased sympathetic stimulation decreases flow. Physiologists are not sure, however, how much of this is a direct effect of sympathetic stimulation and how much is an indirect effect as follows: Sympathetic stimulation increases the degree of activity of the cardiac muscle tremendously which in turn would be expected to increase coronary blood flow by the autoregulation mechanism discussed previously. Therefore, some physiologists believe that most of the increased coronary flow following sympathetic stimulation is simply another manifestation of autoregulation.

However, nervous control of coronary blood flow is relatively unimportant in comparison with the autoregulatory control of flow.

Coronary Occlusion

Coronary occlusion means blockage of one of the coronary vessels. When this occurs that portion of the heart formerly supplied by the vessel loses its nutrition, stops contracting, and becomes so flaccid that it actually dilates outward during the period of the cardiac cycle when the pressure inside the heart becomes great. Obviously, when a major portion of the heart is affected, its pumping capability becomes so depressed that it can no longer pump sufficient blood to maintain life.

Coronary occlusion is the cause of approximately one third of all deaths. Occlusion may occur rapidly or slowly, and in almost all instances the cause is *atherosclerosis*. Therefore, to explain why people have heart attacks, it is first necessary to discuss the cause and the effects of atherosclerosis.

ATHEROSCLEROSIS. Atherosclerosis is a disease of fat metabolism in which the body fails to utilize fat in the normal manner. As a result, fatty deposits that contain mainly cholesterol, plus smaller amounts of phospholipids and neutral fat, appear in the walls of the arteries. Gradually, fibrous tissue grows around or into the fatty deposits, and calcium from the body fluids frequently combines with the fat to form solid calcium compounds, eventually evolving into hard plates similar to bone. Thus, in the early stage of atherosclerosis, only fatty deposits occur in the

walls of the vessels, but in the late stage the vessels may become extremely fibrotic and constricted, or even bony hard in consistency, a condition called "hardening of the arteries."

ACUTE CORONARY OCCLUSION. Atherosclerosis can cause acute occlusion of the coronary vessels in three ways: First, the fatty deposits may break through the inside surface of the vessel and cause clotting of the blood, which in turn plugs the vessel. Second, the protruding fat may actually break away from its original deposit and flow into and occlude a smaller vessel. Third, and less common, the fatty deposits may erode into one of the tiny blood vessels called *vasa vasorum*, which supply the walls of the coronaries. When this occurs the minute vessel bleeds into the wall of the coronary pushing the inner lining inward, thereby totally or partially blocking blood flow. In any of these instances, the resulting coronary occlusion can cause sudden diminishment of cardiac function, which is the familiar condition known as a "heart attack." If the occlusion is extremely severe, it will cause immediate death; if it affects only a small vessel, the heart is often only temporarily weakened for one to three months until some of the vascular connections with neighboring blood vessels enlarge to supply new blood. After a heart attack, the heart sometimes returns to full strength, though more often it remains weakened for life.

SLOWLY DEVELOPING OCCLUSION. The slowly developing type of coronary occlusion results from fibrous tissue invading the atherosclerotic fatty deposits. The fibrous tissue contracts slowly and continually, gradually narrowing the vessel. This effect occurs in all people as they become old, though it occurs in some at much earlier ages than in others. Actually, coronary occlusion eventually occurs in everyone who lives long enough. Indeed, almost 100 per cent of hearts examined after death of persons beyond the age of 60, whether the persons had ever had obvious heart attacks or not, show at least one fairly large coronary vessel totally occluded. Ordinarily, accessory blood vessels will have grown into the area supplied by the occluded vessel so that the person may never have known of his coronary disease. The point of this discussion is that all people develop coronary heart disease, which means that it is important for each person as he becomes old to treat himself as if he had heart disease.

RELATIONSHIP OF DIET AND SMOKING TO CORONARY DISEASE. Recent studies have shown that atherosclerosis, and consequently heart attacks as well, can be minimized by two dietary measures: First, the total intake of food should be little enough for the weight to remain either normal or even slightly subnormal. Second, the amount of fat in the diet should be as little as possible. The average American and northwestern European diets contain far more fat than those of other sections of the world, and as a result the incidence of heart attacks is far higher than among such people as the Italians, Japanese, and Chinese. Statistics have also shown that smoking *doubles* the number of deaths from heart disease, but why this occurs is not known.

THE CEREBRAL CIRCULATION

Figure 126 illustrates the arterial system of the brain, showing the two *vertebral arteries* passing upward along the medulla and the two *internal carotid arteries* approaching the brain on its ventral surface. Communicating branches between these four major arteries protect the brain against damage should any one or even two of them become occluded. However, the intermediate-size arteries which spread over the surface of the brain do not have many communications from one to another. Therefore, occlusion of any one of them usually causes destruction of the respective brain tissue.

Cerebral Blood Flow and Its Regulation

The total amount of blood flowing to the brain averages approximately 650 to 700 ml. per minute, which is about 13 per cent of the blood pumped by the heart.

AUTOREGULATION OF CEREBRAL BLOOD FLOW. Cerebral blood flow is autoregulated better than that in almost any other area of the body except in the kidneys. Even though the arterial pressure might fall to as low as 40 mm. Hg or rise to as high as 200 mm. Hg, the blood flow through the brain still varies no more than a few per cent.

Autoregulation of blood flow in the brain results mainly from a carbon dioxide autoregulation mechanism. The cerebral circulation is different from that of other tissues in that it responds very markedly to carbon dioxide concentration as well as to oxygen concentration. When the carbon dioxide concentration rises above normal, cerebral blood flow automatically increases and washes carbon dioxide out of the tissues, bringing the carbon dioxide concentration back toward normal. Conversely, decreased carbon dioxide in the brain decreases the blood flow, which again returns the carbon dioxide concentration toward normal.

The carbon dioxide mechanism is much more powerful in the cerebral circulation than is the oxygen deficiency mechanism of autoregulation (which is far more powerful in most other tissues), and it is mainly because of this mechanism that cerebral blood flow is autoregulated to an almost

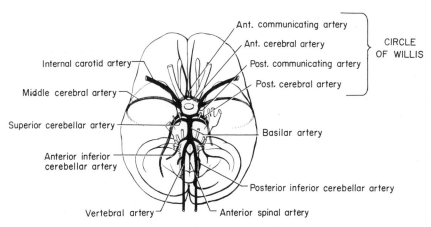

Figure 126. The cerebral circulation.

exactly constant level all the time. The only exception to this constancy of blood flow occurs when the brain itself becomes overly active. Under these conditions cerebral blood flow can on rare occasions rise to as high as 30 to 50 per cent above normal, but even this increase in flow is a manifestation of autoregulation, the flow increasing in proportion to the increase in carbon dioxide formation in the brain.

The unvarying rate of cerebral blood flow is very advantageous to cerebral function, for neurons must never become excessively excitable nor excessively inhibited if the brain is to function properly, and, obviously, changes in the nutritional status of the neurons could quite easily change their degree of excitability. For instance, if the carbon dioxide concentration in the tissues should rise too high, the tissue fluids would become excessively acidic, which is known to depress neuronal function.

Venous Dynamics in the Cerebral Circulation

Function of the veins in the cerebral circulation is somewhat different from that of veins in the remainder of the body because the cerebral veins are non-collapsible. They are formed as *sinuses* in the fibrous lining, the *dura mater,* of the cerebral vault, and the sides of the sinuses are so firmly attached to the skull and other tissues that they cannot collapse. Also, because the head is in an upright position most of the time, the weight of the venous blood makes it run rapidly downward toward the heart, creating a semi-vacuum in the cerebral veins. If one of these veins is inadvertently punctured, air is often sucked into the venous system and occasionally may even flow so rapidly to the heart that it blocks the action of the heart valves and kills the person. Because the veins in all other parts of the body are flimsy and can collapse very easily, any sucking action that might occur simply closes the veins rather than creating a vacuum.

Cerebral Vascular Accidents — the "Stroke"

When a blood vessel supplying blood to an important part of the brain suddenly becomes blocked or when it ruptures, the person is said to have had a *cerebral vascular accident* or a "stroke." This is the cause of death in perhaps 5 to 10 per cent of all people. In about one fourth of all cases the damage is caused by a blood clot developing on an atherosclerotic fatty deposit in a major cerebral artery. In the other three-fourths, an artery ruptures because of excessively high arterial pressure or because the vessel itself has been weakened by some disease process such as atherosclerosis. Such hemorrhages into the tissues of the brain often compress the neuronal cells enough to destroy them or at least to stop their functioning. Also, the blood clot developing in the hemorrhagic area extends backward into the artery, blocking further flow along the inside of the vessel. Therefore, regardless of whether the cerebral vascular accident is initially caused by simple vascular occlusion or by hemorrhage, the results are essentially the same.

The effects of a cerebral vascular accident depend on which vessel is destroyed. Frequently it is the middle cerebral artery supplying the area of the brain that controls muscular function, in which case the opposite side of the body becomes paralyzed. Blockage of the posterior cerebral artery, on the other hand, causes partial blindness. Blockage of arteries in the hindbrain region is likely to destroy nerve tracts connecting the brain and spinal cord, causing any number of abnormalities such as paralysis, loss of sensation, and loss of equilibrium.

THE PORTAL CIRCULATORY SYSTEM

The portal circulatory system, shown in Figure 127, is composed of the veins leading from the intestines and spleen to the liver and thence through the liver into the vena cava. The function of this system is to pass the blood from the intestines through the liver prior to its entering the general circulation. On flowing through the minute sinuses of the liver the blood comes into contact with special phagocytic cells, the *Kupffer cells*, that are capable of removing abnormal debris. These cells, therefore, cleanse the intestinal blood before it reenters the general circulation. This is quite important because

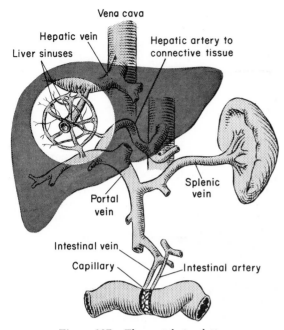

Figure 127.   The portal circulation.

each minute a few of the literally billions of bacteria in the gastrointestinal tract make their way into the portal blood. The Kupffer cells are so effective in removing the bacteria that probably not one in a thousand escapes through the liver into the general circulation.

The portal system also allows the liver to remove various absorbed nutrients from the intestinal blood before they reach the general circulation. For instance, the liver cells normally remove approximately two thirds of the glucose absorbed into the portal blood from the intestines and perhaps as much as one half of the proteins before the blood ever reaches the general circulation. These nutrients are stored in the liver until they are needed later by the remainder of the body and then are released back into the general circulation. The storage of glucose and proteins "buffers" the blood concentrations of these nutrients, for it prevents sudden increases in their concentrations after a meal and also prevents low concentrations during the long intervals between meals.

Vascular Dynamics of the Portal System

The small vessels in the liver offer so much resistance to blood flow that the portal venous pressure is normally about 7 mm. Hg greater than the pressure in the vena cava. Ordinarily this elevated portal pressure is of no major concern, but when some abnormality causes the systemic venous pressure to rise, the portal pressure increases a corresponding amount, always remaining about 7 mm. Hg greater than the systemic venous pressure. As a result, the portal capillary bed is one of the first to suffer from excessively high systemic venous pressures, a condition that occurs most frequently when a failing heart causes blood to dam up in the veins. The high capillary pressure forces fluid to leak through the capillary walls into the intestinal tissues and peritoneal cavity. Therefore, instead of the norm of only a few milliliters of fluid in the abdomen, many liters may accumulate. This is called *ascites.*

PORTAL OBSTRUCTION. Occasionally the portal vein leading to the liver becomes totally occluded by a large blood clot, or more frequently a large proportion of the small vessels in the liver may become blocked because of alcoholic liver disease called liver cirrhosis. In either event, the portal venous pressure rises, which in turn causes high portal capillary pressure, edema of the intestinal walls, enlargement of the spleen, and ascites (free fluid in the abdomen, sometimes as much as 5 gallons!).

Function of the Spleen

RESERVOIR FUNCTION. As was pointed out in Chapter 12, the spleen, which is part of the portal system, is a blood reservoir. It is capable of enlarging up to a total volume of more than 1000 ml. or of contracting down to a minimal volume of less than 50 ml. Its internal structure is illustrated in Figure 128, which shows several small arteries leading into capillaries in the *pulp* of the spleen and finally connecting with large venous sinuses. The blood flows from the venous sinuses into the large circumferential veins and then back into the general circulation. Much of the blood, including the red blood cells, leaks from the capillaries into the pulp and then diffuses through this area, finally reentering the general circulation through the walls of the venous sinuses. Large numbers of cells are often trapped in the pulp while the plasma is returned to the circulation. In this way the spleen stores a much higher per-

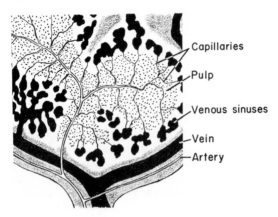

Figure 128. Internal organization of the spleen. (Modified from Bloom and Fawcett: A *Textbook of Histology*. 8th Ed.)

centage of red blood cells than plasma, but when it contracts, the cells are forced from the pulp back into the circulation. Therefore, the spleen is often said to be a *red blood cell reservoir* rather than a general blood reservoir.

In lower animals, sympathetic stimulation causes smooth muscle in the capsule of the spleen to contract, which expels blood both from the splenic pulp and splenic sinuses into the general circulation. In the human being there is very little smooth muscle in the capsule. Therefore, it is believed that sympathetic stimulation causes the spleen of the human being to empty its blood mainly by direct constrictive effects on the vascular walls and perhaps other smooth muscle in the structure of the spleen itself.

Marked splenic contraction occurs during muscular exercise. The extra red blood cells expressed from the spleen into the circulation aid in the transport of oxygen to the active muscles.

PHAGOCYTIC FUNCTION OF THE SPLEEN. Another major function of the spleen is to cleanse the blood. This results from phagocytosis by reticuloendothelial cells that line the splenic sinuses or that are located throughout the splenic pulp.

##

Under resting conditions the blood flow through the muscles is only 15 to 20 per cent of the total flow in the body. During exercise, muscle blood flow increases markedly in response to the increase in muscle metabo-

lism—that is, in response to the increase in use of nutrients by the muscle cells. During extreme exercise, muscle blood flow can increase 15-fold, and sometimes even more than this.

Mechanism of Blood Flow Regulation in the Muscles

AUTOREGULATION. The principal mechanism by which blood flow is regulated in the muscles is the autoregulation mechanism which was described in the previous chapter. That is, increased muscle metabolism increases the utilization of nutrients from the blood, including oxygen, which in turn has a direct vasodilator effect on the blood vessels to increase the rate of blood flow. It is possible that the diminished oxygen causes vasodilator substances to be formed in the muscle, but another likely effect is simply that lack of oxygen directly causes the blood vessels to dilate.

SYMPATHETIC VASODILATOR NERVES TO THE MUSCLES. A second mechanism that dilates the muscle vessels during exercise is stimulation of sympathetic vasodilator nerves. These are controlled by areas of the cerebral cortex in the brain near those that control contraction of the exercising muscles. When the muscles are contracted, impulses are sent simultaneously to the sympathetic vasodilator nerves, thereby increasing the blood flow. However, the importance of the vasodilator fibers has yet to be proved because removal of all sympathetic nerves to muscles hardly affects the degree of vasodilation that occurs during exercise. It is believed, though, that the sympathetic vasodilator nerves perform the following function: The autoregulation mechanism requires 10 to 20 seconds to develop fully after exercise begins. On the other hand, the sympathetic vasodilator mechanism can dilate the blood vessels within the first second or more after exercise begins, and sometimes even *before* exercise begins. Therefore, it is believed that the sympathetic vasodilator mechanism provides an immediate means for increasing blood flow at the onset of muscular exercise, thereby anticipating the increased needs of the muscles for nutrients rather than waiting for a deficiency to occur, as is the case with the autoregulation mechanism.

EFFECT OF INCREASED ARTERIAL PRESSURE. Another factor that increases muscle blood flow during exercise is increased arterial pressure. During very heavy exercise the mean arterial pressure rises from about 100 mm. Hg to 150 or occasionally even to 175 mm. Hg. This increased pressure obviously forces increased amounts of blood through the muscle blood vessels. Also, sympathetic stimulation, which is very powerful during muscle exercise, constricts essentially all the other vessels of the body, such as in the kidneys, the intestines, the skin, and so forth, thereby shunting the blood flow out of other areas and through the muscles.

BLOOD FLOW THROUGH THE SKIN

Blood flow through the skin has two functions: first, to regulate the temperature of the body and, second, to supply nutrition to the skin itself.

RELATIONSHIP OF SKIN BLOOD FLOW TO BODY TEMPERATURE REGULATION. Figure 129 shows an extensive venous plexus lying a few millimeters beneath the surface of the skin. The rate of blood flow through this plexus can probably be altered up or down as much as 100-fold. When the arteries that supply blood to the skin venous plexus are constricted, the blood flow may be as little as 20 to 50 ml. per minute to the skin of the entire body, whereas, when the arteries are completely dilated, the blood flow to the skin may be as great as 2 to 3 liters per minute.

Obviously, if the blood flow to the skin is very rapid, the amount of heat transmitted by the blood from the internal structures to the surface of the body will be great, and large amounts of heat will be radiated from the skin. Conversely, when the quantity of blood flowing to the skin is slight, the amount of heat lost from the body will also be slight. Special temperature regulatory mechanisms, controlled by the hypothalamus in the brain and acting through the sympathetic nerves, can either vasoconstrict or vasodilate the skin vessels, thereby helping to regulate body temperature. This will be discussed in much more detail in Chapter 33.

Certain areas of the body, such as the hands, feet, and ears have special muscular walled vessels, the *arteriovenous anastomoses*, which connect the arteries directly with the venous plexus. When these anastomoses are dilated, blood bypasses the capillary system and flushes extremely rapidly into the plexus. In this way the hands, feet, and ears receive tremendous amounts of blood flow when they are exposed to excessive cold, and the increased blood flow obviously acts to protect the tissues from freezing.

THE NUTRIENT BLOOD FLOW. Ordinarily the blood flow through the skin is some 20 to 30 times that required to supply the required amount of nutrition to the skin tissues. Yet, when the skin becomes extremely cold the body temperature mechanisms constrict the skin blood flow so greatly that nutrition can become impaired. It is only under such conditions as this that one need be concerned with the nutrient vascular supply.

Figure 129 illustrates capillary loops that transmit blood into all the tufts beneath the skin. The nutritive blood flow is subject to the same local autoregulatory mechanism for blood flow as that found elsewhere in the body. For instance, when a person compresses an area of the skin so hard that the blood flow is completely blocked for a few minutes, the skin flow becomes far greater than normal immediately after relief of the pressure, which automatically makes up for the deficiency in nutrients. Fortunately, the skin can be without blood flow for as long as an hour or more before serious damage takes place.

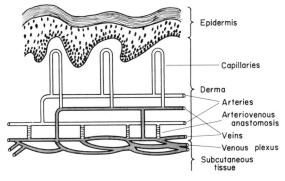

Figure 129. The circulation in the skin.

REFERENCES

Allen, E. V., Barker, N. W., and Hines, E. A., Jr.: *Peripheral Vascular Diseases.* 3rd Ed. Philadelphia, W. B. Saunders Company, 1962.
Barcroft, H.: Circulation in skeletal muscle. *Handbook of Physiology*, Sec. II, Vol. II, p. 1353. Baltimore, The Williams & Wilkins Company, 1963.
Berne, R. M.: Regulation of coronary blood flow. *Physiol. Rev.* 44:1, 1964.

Blaustein, A. (ed.): *The Spleen.* New York, McGraw-Hill Book Company, 1963.

Ciba Foundation Symposium: *Pulmonary Structure and Function.* Boston, Little, Brown & Company, 1962.

Fishman, A. P.: Dynamics of the pulmonary circulation. *Handbook of Physiology,* Sec. II, Vol. II, p. 1667. Baltimore, The Williams & Wilkins Company, 1963.

Greenfield, A. D. M.: The circulation through the skin. *Handbook of Physiology,* Sec. II, Vol. II, p. 1325. Baltimore, The Williams & Wilkins Company, 1963.

Gregg, D. E., and Fisher, L. C.: Blood supply to the heart. *Handbook of Physiology,* Sec. II, Vol. II, p. 1517. Baltimore, The Williams & Wilkins Company, 1963.

Levy, W. N., and Chansky, M.: Collateral circulation after coronary artery constriction. *Amer. J. Physiol.* 208:144, 1965.

Schmidt, C. F.: Central nervous system circulation, fluids and barriers—introduction. *Handbook of Physiology,* Sec. I, Vol. III, p. 1745. Baltimore, The Williams & Wilkins Company, 1960.

Shepherd, J. T.: *Physiology of Circulation in the Limbs,* Philadelphia, W. B. Saunders Company, 1963.

CHAPTER 14

SYSTEMIC ARTERIAL PRESSURE AND HYPERTENSION

When the left ventricle contracts, it forces blood into the systemic arteries, thereby creating pressure that drives the blood through the systemic circulation. The *regulation of arterial pressure* is among the most important subjects of circulatory physiology, and the disease *hypertension*, which means high arterial pressure, is one of the most common of all disorders of the body. These two topics receive special emphasis in the present chapter.

PULSATILE ARTERIAL PRESSURE

Instead of pumping a continuous stream, the heart pumps a small quantity of blood with each beat. As a result, the arterial pressure rises during systole but falls during diastole. Figure 130 shows graphically this pulsatile pressure under both normal and abnormal conditions.

Systolic and Diastolic Pressures

The pressure at its highest point during the pressure cycle is called the *systolic*

pressure, and the pressure at its lowest point is called the *diastolic pressure.* As shown in Figure 130, systolic pressure in a normal young adult is approximately 120 mm. Hg, while the diastolic pressure is approximately 80 mm. Hg. The usual method for writing these pressures is 120/80. As another example, if a person's blood pressure is stated to be 210/125 this means that the

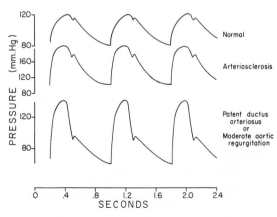

Figure 130. Pressure pulse contours in the normal circulation, in arteriosclerosis, in patent ductus arteriosus, and in moderate aortic regurgitation.

155

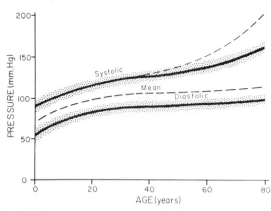

Figure 131. Changes in systolic, diastolic, and mean pressures with age. The shaded areas show the normal range. The upper dashed curve shows the effect of arteriosclerosis on systolic pressure in old age.

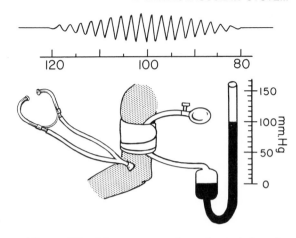

Figure 132. Measurement of systolic and diastolic pressures by the auscultatory method.

systolic pressure is 210 mm. Hg and the diastolic pressure 125 mm. Hg.

The systolic and diastolic pressures change with age, as demonstrated graphically in Figure 131. In a newborn baby the systolic pressure is about 90 mm. Hg and the diastolic about 55 mm. Hg. The pressure usually reaches 120/80 by young adulthood, and in old age an average of about 150/90. The upper dashed line of Figure 131 illustrates what happens to the systolic pressure when a person develops severe arteriosclerosis; this effect is explained in a later section of this chapter.

MEASUREMENT OF SYSTOLIC AND DIASTOLIC PRESSURES. The arterial pressure varies so rapidly during the cardiac cycle that special high-speed recorders designed to follow the rapid changes in pressure must be used to measure it. Several electronic transducers, used with appropriate electronic recorders, were described in Chapter 12 and illustrated in Figure 113. The basic principle of all these transducers is the following: The vessel from which the pressure is to be measured is connected by way of a non-distensible tube with a small chamber bounded on one side by a thin elastic membrane. When the pressure rises, the membrane bulges outward, and when the pressure falls the bulge decreases. The bulge in turn activates a variable capacitor, a variable resistor, or a variable inductor, the electrical output from which is amplified and recorded on a moving strip of paper.

Appropriately designed membrane transducers can record pressure changes that

occur in less than one-hundredth of a second, which is more than adequate for recording any significant pressure change in the circulatory system. The arterial pressure pulse curves illustrated in Figure 130 were recorded using a membrane type electronic transducer.

Indirect measurement of systolic and diastolic pressures by the auscultatory method. Figure 132 illustrates the method usually employed in the doctor's office for measuring systolic and diastolic pressures. An inflatable blood pressure cuff is placed around the upper arm and is connected by a tube to a mercury manometer. When the pressure in the cuff is elevated above that in the brachial artery, the wall of the artery collapses because more pressure is being exerted against the outside of the arterial wall than by the blood on the inside. If the pressure in the cuff is gradually decreased until it falls below systolic pressure, small spurts of blood begin to jet intermittently through the artery each time the arterial pressure rises to its systolic value. However, during diastole no blood flows through the artery. This intermittent flow of blood causes vibrations in the arteries of the lower arm that can be heard with a stethoscope. Therefore, to determine systolic pressure, one simply inflates the cuff until the pressure rises to a high value and then deflates it gradually until the intermittent sounds can be heard. At this instant the pressure recorded by the mercury manometer is a reasonably accurate measure of systolic pressure.

To determine diastolic pressure the cuff pressure is reduced still more. When the pressure falls below the diastolic value, blood then flows through the artery all of the time and no longer in jet-like bursts. As a result, the vibrations disappear, and sounds can no longer be heard with the stethoscope. The pressure recorded at this instant is a close estimate of the diastolic pressure. The intensity of the sounds heard through the stethoscope at different cuff pressure levels is shown graphically at the top of Figure 132.

Factors That Affect Pulse Pressure

The arterial *pulse pressure* is the pressure difference between the systolic and diastolic pressures. The higher the systolic pressure and the lower the diastolic pressure, the greater is the pulse pressure, but the more closely these two pressures approach each other, the less becomes the pulse pressure. The normal pulse pressure is about 40 mm. Hg. Two important factors affect the pulse pressure in the arterial system: These are (1) stroke volume output of the heart, and (2) distensibility of the arterial system.

STROKE VOLUME OUTPUT. The stroke volume output is the amount of blood pumped by the heart with each beat. Normally, it is approximately 70 ml., but it can on occasion fall to as low as 10 to 20 ml. or can rise to as high as 160 ml. Obviously, if a small quantity of blood is pumped into the arterial system with each beat, the pressure will not rise and fall greatly during each cardiac cycle. But if the stroke volume output is great, the pressure will rise and fall tremendously, thereby giving an extremely high pulse pressure. In other words, the pulse pressure is roughly proportional to the stroke volume output.

The slower the rate of the heart, the greater must be the stroke volume output to maintain adequate blood flow through the body. Very slow heart beats are quite often found in well trained athletes, the slowness actually indicating efficient hearts. However, the large stroke volume output is accompanied by a high pulse pressure. On the other hand, persons with fever, toxicity, or weakened hearts very frequently have fast heart beats with accompanying small stroke volume outputs. The total amount of blood pumped per minute may be the same as that pumped by the athlete's heart, but because of the small stroke volume output the pulse pressure is very slight.

DISTENSIBILITY OF THE ARTERIAL SYSTEM. The more distensible the arteries, the greater the quantity of blood that can be compressed into the arterial system by a given amount of pressure. Therefore, each beat of the heart causes far less pressure rise in very distensible arteries than in non-distensible arteries.

CONDITIONS THAT CAUSE ABNORMAL PULSE PRESSURE. In old age, the distensibility of the arterial system decreases tremendously because of arteriosclerotic changes in the walls of the vessels. *Arteriosclerosis* is usually the end result of atherosclerosis, which was described in the previous chapter, and arteriosclerosis actually means "hardening of the arterial walls." In many older persons who have arteriosclerosis, the walls even become calcified, bone-hard tubes. As a result, the arterial system then cannot stretch adequately during systole and cannot relax during diastole. Pulses of blood entering the arterial tree cause tremendous changes in pressure, sometimes causing a pulse pressure of 100 mm. Hg or more. A pulse pressure record from an arteriosclerotic patient is shown in Figure 130.

In *aortic regurgitation* the aortic valve has been partially destroyed. Therefore, after blood is pumped out of the heart into the arterial system during systole, much of it flows backward into the relaxed left ventricle during diastole, thus causing the diastolic pressure to fall very low—often to zero, as shown by the third curve in Figure 130. As a result, the pulse pressure becomes the greatest possible, sometimes as great as 160 mm. Hg.

Transmission of Pressure Pulses to the Smaller Vessels

When the blood pumped by each beat of the heart is suddenly thrust into the aorta, the resulting increase in pressure at first distends only the portion of the aorta adjacent to the heart, as shown in Figure 133. A short period of time is required for some of this blood to push forward along the arterial tree and build up the pressure further peripherally. This movement of the pressure

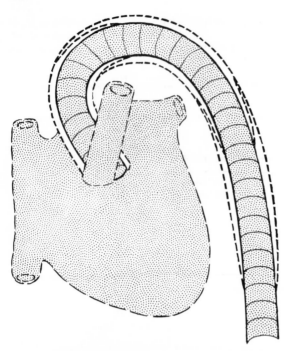

Figure 133. Progressive movement of the pressure pulse wave along the aorta.

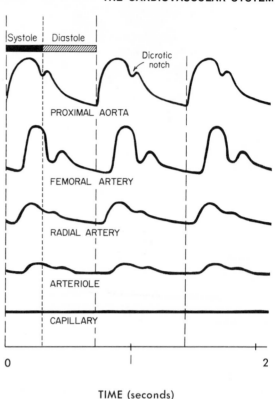

TIME (seconds)

Figure 134. Changes in the pulse pressure contour as the pulse wave travels toward the smaller vessels.

along the arteries is called *transmission of the pressure pulse.*

Transmission of the pressure pulse is analogous to the movement of a wave on the surface of a pond. When a pebble is thrown into the middle of the pond, a wave travels in every direction. When it reaches the shore, water rolls onto the bank and then back into the pond, causing the wave to turn around and go backward. The same happens in the arteries, for the pressure wave travels along the arteries away from the heart and upon reaching the very small arteries is often reflected back into the large arteries. Because of this reflection, the contour of the pressure wave is sometimes quite different in the more peripheral arteries than in the base of the aorta. This is illustrated in Figure 134, which shows the normal pulse pressure contour in the proximal aorta and then a considerably different contour in the femoral artery as the result of a reflected wave. Also, it will be noted that the initial rise in the pressure contour in the femoral artery begins approximately 0.2 second after the beginning of the pressure rise in the aorta. This time lag is caused by the time required for transmission of the pressure wave from the aorta to the femoral artery.

DAMPING OF THE PRESSURE PULSES. Except when the pressure wave is reflected, the pulse pressure becomes less and less as the wave spreads toward the smaller vessels. This is called *damping* of the pulsations, and it results mainly from (1) the resistance to blood flow in the artery and (2) the distensibility of the artery. Figure 134 illustrates the damping of the pulse wave as it spreads peripherally, showing especially the complete absence or almost complete absence of pulsations in the capillary.

PRESSURES AT DIFFERENT POINTS IN THE SYSTEMIC CIRCULATION

Figure 135 shows the pressures in the different vessels from the aorta to the venae cavae. From this chart it is obvious that about one half of the entire fall in pressure is in the arterioles and about one fourth is in the capillaries. The reason for this is that

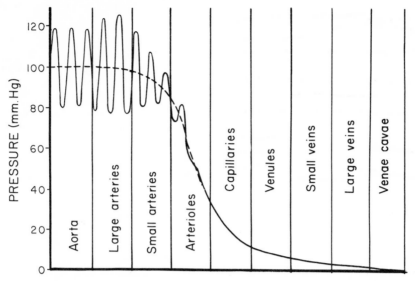

Figure 135. Pressures in different parts of the systemic circulation.

about one half of the resistance to blood flow is in the arterioles and about one fourth is in the capillaries.

In the normal person, the mean pressure in the aorta is approximately 100 mm. Hg, and it falls to about 85 mm. Hg at the end of the small arteries. The pressure then falls another 55 mm. Hg in the arterioles down to 30 mm. Hg at the beginning of the capillaries. At the end of the capillaries the pressure normally is about 10 mm. Hg and it gradually decreases in the venules, small veins, and large veins until it reaches 0 mm. Hg in the right atrium.

Effect of Degree of Arteriolar Constriction on Pressures at Different Points in the Circulation

In the previous chapter it was pointed out that the arterioles are abundantly supplied with nerves that can cause them to contract and relax with ease. Also, the phenomenon of autoregulation can make the arterioles contract and relax in response to local conditions in the tissues. For these reasons, arteriolar resistance can change tremendously from time to time. It is important to understand how these changes affect pressures in the various portions of the circulation.

When the arterioles constrict totally, blood cannot flow into or out of the arterioles. As a result, the pressure in the small arteries rises to equal that in the aorta, and the pressure in the capillaries falls to equal that in the large veins.

When the arterioles relax and allow blood to flow with great ease, the capillary pressure rises to approach the pressure in the small arteries. This effect is especially important because a rise in capillary pressure can cause very rapid transfer of fluid into the tissues, and, conversely, a fall in the capillary pressure can cause rapid flow of fluid from the tissues back into the capillaries, as will be explained in Chapter 16. Therefore, capillary dynamics are dependent to a great extent upon regulation of capillary pressure by the arterioles.

MEAN ARTERIAL PRESSURE

The *mean arterial pressure* is the arterial pressure averaged during a complete pressure pulse cycle. The mean arterial pressure is not always equal to the average of systolic and diastolic pressures, for during each pressure cycle the pressure usually remains at systolic levels for a shorter time than at diastolic levels. Therefore, when the pressures at all stages of the pressure cycle are averaged, the resulting mean pressure is usually slightly nearer diastolic pressure than systolic pressure.

So far as the circulatory system is concerned, *the mean arterial pressure is much more important than is either systolic or diastolic pressure*, because it is the mean pressure that determines the average rate at which blood will flow through the systemic vessels. For this reason, in most physiologic experiments it is not necessary to record the pressure changes throughout the pressure cycle, but instead it is necessary simply to record the mean arterial pressure. This can be accomplished using either an electronic pressure transducer with appropriate electronic damping of the pressure pulse or using the mercury manometer, which was illustrated in Figure 112 of Chapter 12. Mercury has so much inertia that it cannot move rapidly enough to raise the recording arm to the systolic level during systole, and it will not fall enough to register diastolic pressure during diastole. Instead, it maintains a relatively even level with some minor pulsations during the beat of the heart; essentially, it records mean arterial pressure rather than the pulse contour.

Relationship of Mean Arterial Pressure to Total Peripheral Resistance and Cardiac Output

Recalling once again the basic formula given in Chapter 12 for the relationship of blood flow to pressure and resistance, the effect on the arterial pressure of vascular resistance and of the amount of blood pumped per minute is given by the following formula:

Arterial pressure = Cardiac output
× Total peripheral resistance

The *cardiac output* is the rate at which blood is pumped by the heart, and the *total peripheral resistance* is the total resistance of all the vessels in the systemic circulation, from the origin of the aorta back to the veins entering the right atrium. This means, then, that either generalized constriction of all the blood vessels or an increase in cardiac output will raise the arterial pressure. Conversely, dilatation of the vessels or a decrease in cardiac output will lower the pressure.

Regulation of Mean Arterial Pressure

Under resting conditions, the mean arterial pressure is normally regulated at a level of almost 100 mm. Hg. The body has at least four separate types of arterial pressure regulatory systems which are responsible for maintaining the normal pressure. These are: (1) nervous mechanisms that regulate arterial pressure by controlling the degree of arteriolar constriction, (2) a capillary fluid shift mechanism which regulates arterial pressure by altering the blood volume, (3) the kidney excretory mechanism which also regulates arterial pressure by altering the blood volume, and (4) hormonal mechanisms that regulate either the blood volume or the degree of arteriolar constriction.

NERVOUS REGULATION OF ARTERIAL PRESSURE. In Chapter 13, it was pointed out that almost all of the blood vessels of the body are supplied with sympathetic nerve fibers which, when stimulated, cause most of the blood vessels to constrict. Constriction of the vessels obviously impedes blood flow and thereby increases the arterial pressure.

The nervous control of the circulation is illustrated in Figure 136. The insert to the left shows the medullary portion of the brain stem in which is located the so-called *vasomotor center* for control of (a) the degree of vasoconstriction of the blood vessels, (b) the degree of vasodilatation, and (c) the heart rate (cardioacceleration and cardioinhibition).

The vasomotor center controls the circulation mainly through the sympathetic nervous system, which is illustrated to the right in Figure 136. Nerve impulses are transmitted down the spinal cord into the sympathetic chains and then to the heart and blood vessels. These impulses increase the *rate of the heart* and its *strength of contraction*, both of which tend to increase the arterial pressure. They also cause *vasoconstriction* of the blood vessels in most parts of the body, which increases the total peripheral resistance and, therefore, increases the arterial pressure.

The vasomotor system also helps to control the circulation through the vagus nerves that go to the heart. When the sympathetics are stimulated, the parasympathetic fibers in the vagi are inhibited, and, conversely, when the sympathetic fibers are inhibited, the parasympathetic fibers are excited. Parasympathetic stimulation also has almost the opposite effect on the heart to that of sympathetic stimulation; that is, it decreases the heart's activity. Except for this effect on the

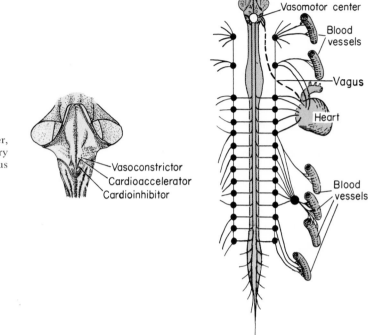

Figure 136. The vasomotor center, and its connections with the circulatory system through the sympathetic nervous system.

heart, the parasympathetics play very little role in control of the circulation.

Vasomotor tone.   Even normally, the sympathetic nervous system transmits impulses at a slow but continuous rate to the heart and vascular system, thus maintaining a moderate degree of vasoconstriction in the blood vessels. The sympathetic system decreases the arterial pressure by simply decreasing the number of impulses transmitted to the blood vessels, thus allowing the arterioles to dilate and reduce the pressure. Conversely, it increases the arterial pressure by increasing the number of im-

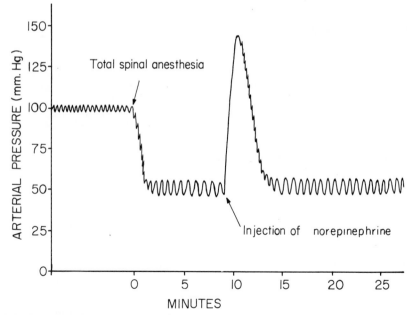

Figure 137.   Effect on arterial pressure of greatly decreased vasomotor tone caused by spinal anesthesia, and of greatly increased vasomotor tone caused by injection of norepinephrine.

pulses above those normally transmitted, thus further constricting the arterioles.

Figure 137 demonstrates the importance of vasomotor tone in the regulation of arterial pressure. At the beginning of this record, the mean arterial pressure was 100 mm. Hg; then total spinal anesthesia was instituted by injecting a local anesthetic all the way up the spinal canal. This blocked all nerves coming from the spinal cord, including the sympathetic nerves. Note that the arterial pressure immediately fell to 50 mm. Hg, which is the pressure in the normal circulatory system when there is no vasomotor tone. After a few minutes, a small amount of norepinephrine (the substance secreted at the sympathetic nerve endings) was injected into the blood. This caused vasoconstriction throughout the body and thereby immediately elevated the arterial pressure up to 150 mm. Hg. Yet, when the norepinephrine wore off after another few minutes, the arterial pressure returned to its basal level. Thus, we see from this experiment that arterial pressure can be reduced far below normal by decreasing the degree of vasomotor tone, and can be increased far above normal by increasing the tone.

Factors that Affect Activity of the Vasomotor Center

EFFECT OF CARBON DIOXIDE. One of the most powerful stimuli affecting the activity of the vasomotor center is carbon dioxide in the blood. When its concentration rises above normal, the vasomotor center becomes greatly excited, sending far more than normal numbers of impulses through the sympathetics to the blood vessels. As a result, the arterial pressure rises. A very large increase in carbon dioxide concentration can make the mean arterial pressure rise to as high as 260 mm. Hg, which is as high a pressure as the normal heart can pump.

This effect of carbon dioxide on arterial pressure helps to prevent malnutrition of the tissues, for, ordinarily, a high concentration of carbon dioxide in the blood is a result of very rapid metabolism in the tissues. The carbon dioxide excites the vasomotor center, the arterial pressure rises, this increases the rate of blood flow through the tissues, and the enhanced flow carries increased quantities of nutrients to the tissues.

EFFECT OF MEDULLARY ISCHEMIA. The term *ischemia* means poor blood supply. When the vasomotor center becomes ischemic it becomes greatly excited, thereby elevating the arterial pressure. This effect is illustrated by the experiment of Figure 138, which shows fluid being injected under pressure into the space around the brain. Compression of the brain by the fluid also compresses the arteries that supply the medulla, and the resultant medullary ischemia excites the vasomotor center, thereby elevating the arterial pressure.

The medullary ischemia mechanism normally protects the brain from being injured because of insufficient blood flow. That is, the resulting rise in pressure increases the blood flow back toward normal. This mechanism is equally as powerful as the effect of carbon dioxide, for it, too, can increase the mean arterial pressure to as high as 260 mm. Hg.

The Baroreceptor System

The pressure regulatory system that has been studied more than any other is the *baroreceptor system* illustrated in Figure 139. In the arch of the aorta and in the internal carotid arteries near their origins in the neck (areas called the *carotid sinuses*) are many small nerve receptors that detect the

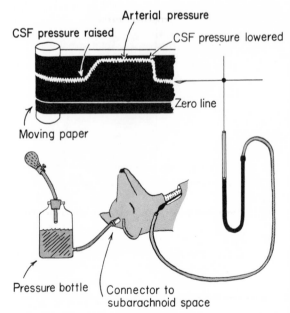

Figure 138. Elevation of arterial pressure caused by increased cerebrospinal fluid pressure.

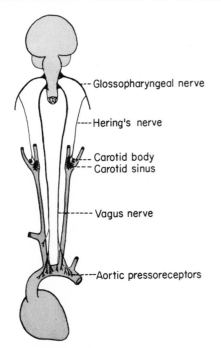

Figure 139. The baroreceptors (pressoreceptors) and their connections with the vasomotor center.

— Glossopharyngeal nerve

—Hering's nerve

— Carotid body
— Carotid sinus

— Vagus nerve

—Aortic pressoreceptors

degree of stretch caused in the walls of the arteries by the pressure. These receptors are called *baroreceptors* or *pressoreceptors.* The number of impulses transmitted from the baroreceptors increases as the arterial pressure rises. On passing to the brain, the impulses inhibit the vasomotor center and thereby decrease the arterial pressure back toward normal. On the other hand, when the pressure in the arteries falls too low, the baroreceptors lose their stimulation so that the vasomotor center now becomes excessively excited, elevating the arterial pressure back to a higher level.

One can readily see that the baroreceptor system opposes either a rise or a fall in pressure. For this reason, it is sometimes called a *moderator system* or a *buffer system.* As an example of this system's effectiveness, if 300 ml. of blood is injected into a person's circulation over a period of about 30 seconds, the arterial pressure will normally rise no more than 20 mm. Hg. However, injection of the same quantity of blood into the circulation of someone whose baroreceptors have been inactivated will cause the pressure to rise about three times this much, or 60 mm. Hg.

FUNCTION OF THE BARORECEPTORS WHEN A PERSON STANDS. The pressure in the

arteries of the upper body tends to fall when one stands because of *pooling* of blood in the lower body, but the falling pressure automatically excites the baroreceptor reflex, returning the pressure in the upper body almost to normal. One of the values of this constancy of pressure in the upper part of the body is that it helps to maintain an adequate blood supply to the brain regardless of the position of the body.

The Capillary Fluid Shift Mechanism for Pressure Regulation

Because of the rapidity with which the various nervous control systems can return arterial pressure toward normal when it becomes abnormal, it is very tempting to explain arterial pressure regulation entirely on the basis of nervous mechanisms alone. However, the nervous mechanisms have a major failing that prevents them from continuing to control arterial pressure over a long period of time. This failing is that the pressure receptors *adapt,* which means that they gradually become insensitive to the change in pressure. Therefore, after a few days, the nervous regulatory mechanisms become ineffective. By that time, nonnervous mechanisms take over the pressure regulation. One of these is the *capillary fluid shift mechanism,* which is particularly important in helping to regulate arterial pressure when the blood volume tends to become either too little or too great.

An increase in blood volume such as might occur following a transfusion increases the pressures in all parts of the systemic circulation, but over a period of a few minutes, the pressures return back toward normal. Much of this decline in pressure is caused by shift of fluid out of the blood through the capillary membranes into the interstitial spaces, thereby decreasing the blood volume back toward normal.

Conversely, when a person hemorrhages severely, fluid shifts out of the interstitial spaces into the circulation, increasing the blood volume back toward normal and returning the pressure also back toward normal.

The mechanism of the capillary fluid shift is simply the following: Too much blood volume increases the pressures in the capillaries as well as in the arteries. Therefore, in accordance with the law of the capillaries, which is explained in Chapter 16, the increased capillary pressure causes fluid to leak out of the circulation into the interstitial spaces. Conversely, when blood is lost from the circulation, the capillary pressure falls, and, here again in accordance with the law of the capillaries, fluid is pulled by the colloid osmotic pressure of the blood from the interstitial spaces into the circulation.

The capillary fluid shift mechanism is much slower to regulate arterial pressure than are the nervous mechanisms, for it requires from 10 minutes to several hours to readjust the arterial pressure back toward normal.

Regulation of Arterial Pressure by the Kidneys

The kidneys, like the capillaries, can regulate arterial pressure by increasing or decreasing the blood volume. After a substantial loss of blood, for instance, the kidneys simply stop forming urine because the glomerular pressure falls too low for glomerular filtration to take place. The fluid and electrolytes taken in by mouth gradually accumulate in the body until the blood volume returns to normal, reestablishing normal arterial pressure and at the same time reestablishing normal glomerular dynamics so that the kidneys begin to form urine again.

Conversely, when the blood volume becomes too great, the urinary output increases. Gradually, over a period of hours, the blood volume decreases back toward normal, causing the arterial pressure also to return toward normal.

FUNCTION OF 'RENIN" IN ARTERIAL PRESSURE REGULATION. Many physiologists believe that the kidneys also have a hormonal effect in regulating arterial pressure. It is postulated that when the arterial pressure falls to a low value, diminished blood flow through the kidneys causes them to secrete into the blood a special hormone called *renin*. Renin in turn combines with one of the plasma proteins of the plasma to form a substance called *angiotensin* which has been postulated to (a) act directly on the blood vessels to cause vasoconstriction, which increases the arterial pressure, or (b) act on the adrenal glands to cause increased aldosterone secretion, which in turn causes the kidneys to retain salt and water; the effect of the salt and water is to increase arterial pressure, as is explained below.

IMPORTANCE OF RENAL REGULATION OF ARTERIAL PRESSURE. Despite the uncertainty of the method by which the kidneys regulate arterial pressure, these organs seem to be the most important of all parts of the body for long-term regulation of arterial pressure. This is shown very dramatically by the experiment of Figure 140. If clamps are placed on the arteries of both kidneys so that the renal blood supply is

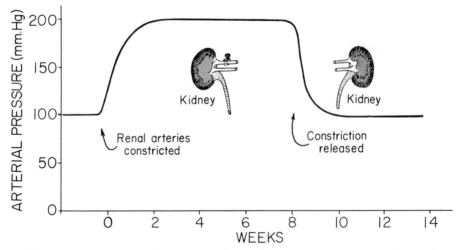

Figure 140. Effect on arterial pressure of constricting the renal arteries and weeks later releasing the constriction.

diminished, the arterial pressure begins to rise. In several weeks, the mean pressure may be twice its original value. The elevated pressure will then force an adequate blood flow through the kidneys despite the continued constriction of the arteries, and the kidneys will function almost normally again. When the constrictions are removed, the arterial pressure falls back to normal within another few days. It is obvious that the kidneys are capable of protecting themselves against diminished blood flow; that is, *diminished flow through the kidneys causes a general arterial pressure rise until the kidneys again receive an adequate blood supply.* On the other hand, if the blood flow through the kidneys becomes excessive, the arterial pressure falls until the renal flow returns to normal.

Hormonal Regulation of Arterial Pressure – The Effect of Aldosterone

At least one major hormonal system is also involved in the regulation of arterial pressure. This is the secretion of aldosterone by the *adrenal cortex.* The adrenal cortex is the outer shell of a small endocrine gland, the adrenal, one located immediately above each of the two kidneys. These cortices secrete adrenocortical hormones, one of which, *aldosterone,* decreases the kidney output of salt and water. The amount of salt and water in the body in turn helps to determine the volumes of blood and interstitial fluids in the body.

The manner in which the aldosterone mechanism enters into blood pressure regulation is the following. When the arterial pressure falls very low, lack of adequate blood flow through the body's tissues, for reasons not yet understood, causes the adrenal cortices to secrete aldosterone. The aldosterone acts to build up the blood volume and return the arterial pressure to normal. Conversely, an elevated arterial pressure reverses this mechanism so that fluid volumes, and consequently the arterial pressure, decreases.

Regulation of Mean Arterial Pressure During Exercise

One of the most severe types of stress to which the circulation can be subjected is strenuous exercise, for a rapidly acting muscle needs a tremendous blood supply. One of the means for increasing blood flow to the muscles during exercise is to elevate the arterial pressure. The mechanisms causing this are: First, increased muscular metabolism during exercise increases the concentration of carbon dioxide in the blood, and this in turn excites the vasomotor center to elevate the arterial pressure. Second, lactic acid and perhaps other metabolites are released from the exercising muscles; these also circulate through the blood to the vasomotor center and excite it, elevating the pressure. Finally, the vasomotor center is stimulated by nerve impulses generated in the motor area of the brain. Therefore, at the same time that the motor area causes the muscles to contract it also excites the vasomotor center and thereby elevates the arterial pressure.

HYPERTENSION

Hypertension means high arterial pressure, and it occurs in approximately one out of five persons. Hypertension can cause rupture of the blood vessels in the brain, in the kidney, or in other vital organs. Also, it can place excessive strain on the heart and cause it to fail. For these reasons one of the most important research problems in physiology is to determine the causes of hypertension. Some of the known causes are the following:

Hypertension Caused by Abnormal Function of the Pressure Regulatory Mechanisms

RENAL HYPERTENSION. Any factor that damages the kidneys can cause *renal hypertension.* For instance, destruction of one whole kidney and two thirds of the second kidney causes the arterial pressure to rise, the greater the degree of destruction the greater the elevation of pressure. Diseases of the kidneys such as kidney infections, sclerosis of the renal blood vessels, kidney inflammation, and so forth can all elevate the pressure.

A particularly interesting type of renal hypertension occurs when the aorta is occluded above the kidneys, which occurs frequently before birth, in which case a person goes through life with an occluded aorta, a condition called *coarctation of the aorta*. Blood finds its way through many smaller arteries from the upper segment of the aorta into the lower segment, but the resistance to flow through these small arteries is so great that the arterial pressure in the lower segment is much lower than that in the upper segment. The insufficient blood flow to the kidneys from the lower aorta causes the pressures throughout the body to rise until the pressure in the lower aorta returns to normal. Obviously, by this time, the pressure in the upper aorta will have risen to a very high level. Thus, the kidneys automatically maintain a normal arterial pressure at the kidney level even at the expense of the arterial pressure in the upper part of the body.

Mechanism of renal hypertension. The precise mechanism of renal hypertension is not known, but there are two prevalent theories. A theory that has been taught for the past 35 years is based on the renin mechanism for arterial pressure regulation. It is claimed that a damaged kidney or a kidney with a poor blood supply secretes very large quantities of renin which, as explained previously, causes either vasoconstriction or increased aldosterone secretion which eventuates in hypertension.

A second theory to explain the mechanism of renal hypertension is based on alterations in fluid volumes of the body when renal function becomes abnormal. Obviously, if renal output becomes reduced as a result of renal disease or poor blood supply, the retention of salt and water would be expected to increase both the interstitial fluid and blood volumes. Experiments have recently demonstrated that removal of large portions of the kidney do indeed increase both of these volumes considerably. It is postulated that these increased volumes increase the arterial pressure. The mechanism of this appears to be the following: (1) The increased blood volume increases the blood flow into the heart, which obviously increases the cardiac output above normal. (2) The excess cardiac output causes too much blood to flow through the tissues. (3) The tissue vessels constrict over a period of several days to reduce the tissue flow back to normal. (4) Constriction of the tissue vessels increases the total peripheral resistance far above normal, thus resulting in hypertension.

HORMONAL HYPERTENSION. Occasionally the adrenal cortices become excessively active, either because of a secreting tumor in one of the adrenal glands or because of excessive stimulation of the adrenals by the anterior pituitary gland. In any event the increased production of aldosterone causes the kidneys to retain excessive quantities of water and salt. The fluid volumes throughout the body increase, and the arterial pressure often rises to as high as two times its normal pressure.

A second type of hormonal hypertension is that caused by a *pheochromocytoma*. This is a tumor of the adrenal *medulla*, which is the central portion of the adrenal gland, entirely distinct from the adrenal cortex. Because the medulla is part of the sympathetic nervous system, the pheochromocytoma secretes large quantities of epinephrine and norepinephrine, the hormones normally secreted by sympathetic nerve endings. These hormones circulate through the blood to all the vessels and promote intense vasoconstriction. As a result, the mean arterial pressure may rise to as high as 200 mm. Hg or higher. The pheochromocytoma usually secretes its hormones only when stimulated by the sympathetic nervous system. This means that when a person who has this type of tumor becomes excessively excited, the secretion of epinephrine and norepinephrine may be tremendous, causing the blood pressure to become very high. Between periods of excitement the blood pressure may be essentially normal.

NEUROGENIC HYPERTENSION. Hypertension often results from abnormalities in the brain. For instance, sudden blockage of one of the arteries to vital portions of the vasomotor center sometimes causes hypertension that lasts for life. Also, experiments in animals have indicated that destruction of certain areas in the hypothalamus can cause prolonged hypertension. These abnormalities cause hypertension by continually exciting the vasomotor center.

Many psychiatrists believe, though it has not yet been proved, that overexcitation of certain of the conscious regions of the brain in neurotic states such as anxiety, worry, and so forth, can excite the vasomotor center and permanently elevate the arterial pressure. Finally, it has been observed many times that excessive pressure in the cranial vault, resulting from vascular hemorrhage, brain tumors, or other causes can elicit the medullary ischemic reflex, thereby causing high arterial pressure.

Essential Hypertension

The types of hypertension described above have known causes. Yet in approximately 90 per cent of all hypertensive persons the cause is unknown, and they are said to have *essential hypertension,* the word essential meaning simply "of unknown cause." Many efforts have been made to prove that essential hypertension is caused by a kidney abnormality, a glandular abnormality, or excessive activity of the vasomotor center. Authentic proof that any of these factors is involved in essential hypertension is yet lacking, which means that the physiologist is still faced with discovering the cause of high blood pressure in most patients. A possible cause of essential hypertension might be a genetic abnormality of the arterioles, causing them always to constrict to excessive degrees. Indeed, it is a known fact that essential hypertension exhibits a very strong hereditary tendency. If one person should inherit excessively active arterioles and another more relaxed arterioles, the first would have hypertension and the second would be normal. Also, it has been suggested that the person with essential hypertension might have a genetic deficiency of the number of

arterioles, which obviously would impede the flow of blood and cause hypertension. Thus far, researchers have not found means of exploring these possibilities, though they must be included among the others as possible causes of essential hypertension.

REFERENCES

Blalock, A.: Experimental hypertension. *Physiol. Rev.* 20:159, 1940.

Brest, A. N., and Moyer, J. H. (eds.): *Hypertension: Recent Advances.* The Second Hahnemann Symposium on Hypertensive Disease. London, Henry Kimpton, 1962.

Dahl, L. K., Smilay, M. G., Silver, L., and Spraragen, S.: Evidence for a prolonged biological half-life of Na^{22} in patients with hypertension. *Circ. Res.* 10:313, 1962.

Douglas, B. H., Guyton, A. C., Langston, J. B., and Bishop, V. S.: Hypertension caused by salt loading. II. Fluid volume and tissue pressure changes. *Amer. J. Physiol.* 207:669, 1964.

Freis, E. D.: Hemodynamics of hypertension. *Physiol. Rev.* 40:27, 1960.

Guyton, A. C.: Acute hypertension in dogs with cerebral ischemia. *Amer. J. Physiol.* 154:45, 1948.

Guyton, A. C.: Physiologic regulation of arterial pressure. *Amer. J. Cardiol.* 8:401, 1961.

Langston, J. B., Guyton, A. C., Douglas, B. H., and Dorsett, P. E.: Effect of changes in salt intake on arterial pressure and renal function in partially nephrectomized dogs. *Circ. Res.* 12:508, 1963.

Page, I. H., and Bumpas, F. M.: Angiotensin. *Physiol. Rev.* 41:331, 1961.

Tobian, L.: Interrelationship of electrolytes, juxtaglomerular cells, and hypertension. *Physiol. Rev.* 40:280, 1960.

Vander, A. J.: Control of renin release. *Physiol. Rev.* 47(3):359, 1967.

CHAPTER 15

CARDIAC OUTPUT, VENOUS PRESSURE, CARDIAC FAILURE, AND SHOCK

CARDIAC OUTPUT

Cardiac output is the rate at which the heart pumps blood. Ordinarily it is expressed in liters or milliliters per minute. Because the function of the circulatory system is to supply adequate nutrition to the tissues, the subject of cardiac output is one of the most important of all phases of physiology.

Normal Values of Cardiac Output

The average cardiac output in a person lying down and in a state of complete rest is about 5 liters per minute. If he walks, it rises to perhaps 7.5 liters per minute. If he performs strenuous exercise, it might rise to as high as 25 liters per minute in the normal person or to as high as 35 liters per minute in the well trained athlete.

One can see, therefore, that the cardiac output varies in proportion to the degree of activity of the person. The ability to

regulate the cardiac output in accord with the needs of the body is one of the most remarkable feats of the circulation. Much of the present chapter is devoted to discussing this regulation.

Regulation of Cardiac Output

The amount of blood that the heart can pump each minute is determined by two major factors: (1) the pumping effectiveness of the heart itself, and (2) the ease with which blood can return to the heart from the systemic circulation after it is pumped by the heart. To understand the regulation of cardiac output it is essential that one be familiar with the way in which each of these factors affects the circulation.

THE PUMPING EFFECTIVENESS OF THE HEART. Normally the heart acts like a "sump" pump, pumping any blood that flows into it from the veins. If the amount flowing from the veins increases, the heart becomes stretched and automatically ad-

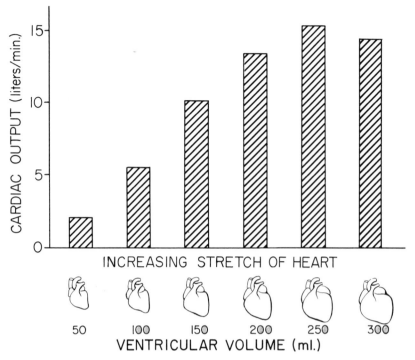

Figure 141.   Effect of increased cardiac filling on the heart's ability to pump blood.

justs to accommodate the extra blood. This ability of the heart to pump any amount of blood, within limits, is called the *law of the heart*, which was explained in detail in Chapter 11. Figure 141 illustrates the principle pictorially and graphically, showing that as the heart becomes progressively stretched by greater and greater inflow of blood, the amount pumped increases up to the point at which the heart muscle finally becomes stretched beyond its physiologic limits.

Effect of nervous stimulation on the effectiveness of the heart as a pump. Stimulation of the sympathetic nerves to the heart greatly enhances the effectiveness of the heart as a pump, while stimulation of the parasympathetic nerves (the vagi) greatly decreases the pumping effectiveness. Under resting conditions parasympathetic stimulation keeps the heart activity depressed, which probably prolongs the life of the heart. During exercise and during other circulatory states when the cardiac output needs to be enhanced very greatly, sympathetic stimulation takes over and makes the heart a very effective pump, about two-thirds more effective than the normal unstimulated heart.

Therefore, the pumping effectiveness of

the heart can be increased in two separate ways: (1) by local adaptation of the heart caused by increased inflow of blood, the mechanism called the law of the heart and (2) by a shift from parasympathetic to sympathetic stimulaion.

FLOW OF BLOOD INTO THE HEART (VENOUS RETURN). Because the heart pumps whatever amount of blood enters its chambers (up to its limit to pump), cardiac output normally is regulated mainly by the amount of blood that returns to the heart from the peripheral vessels. This flow of blood into the heart is called *venous return*. When a person is at rest, venous return averages 5 liters per minute. Therefore, this is the amount of blood that is pumped. However, if 40 liters per minute, a very abnormal amount, should attempt to return to the heart, the normal heart, even one strongly stimulated by the sympathetics, would be able to pump only about 25 liters of this because this is its limit. Therefore, blood would dam up in the veins rather than being pumped into the arteries.

The important factors that determine the amount of venous return per minute are (1) the average pressure of the blood in all parts of the systemic circulation, which is called the *mean systemic pressure*, (2) the right atrial pressure, and (3) the resistance to blood flow through the systemic vessels.

Effect of mean systemic pressure and right atrial pressure on venous return. The *mean systemic pressure* is the average "filling" pressure of the systemic circulation. It is this average pressure that continually pushes blood from the peripheral vessels toward the right atrium.

The *right atrial pressure*, on the other hand, is the input pressure to the heart; this acts as a back pressure to impede blood flow into the heart. When the heart becomes weakened so that it cannot pump blood out of the right atrium with ease, or whenever the heart becomes overloaded for any reason, the right atrial pressure rises and tends to keep blood from entering the heart from the systemic circulation.

Thus, the mean systemic pressure is the average pressure in the periphery tending to push blood toward the heart, and the right atrial pressure is the input pressure against which the blood must flow. Therefore, venous return is determined to a great extent by the difference between these two pressures, *mean systemic pressure* minus *right atrial pressure*. This difference is called the *pressure gradient for venous return*. Figure 142 illustrates this relationship to the control of venous return and cardiac output. In Figure 142A, the mean systemic pressure is 7 mm. Hg and the right atrial pressure 0 mm. Hg. The pressure gradient for venous return in this case is 7–0, or 7 mm. Hg, and the venous return and cardiac output are 5 liters per minute. In Figure 142B, the mean systemic pressure is 18 mm. Hg, and the right atrial pressure, because the heart is now beginning to be overloaded, has risen to 4 mm. Hg. In this case, the pressure gradient for venous return is 18–4 mm. Hg or 14 mm. Hg. This pressure gradient is twice that in Figure 142A. Likewise, the cardiac output and venous return are now twice that in Figure 142A, or 10 liters per minute, showing the importance of the pressure gradient for venous return in determining cardiac output.

Effect of vascular resistance on venous return. If all other factors remain constant, an increase in resistance to blood flow in the systemic circulation, particularly an increase in the resistance in the veins, reduces the amount of blood that returns to the heart, and this correspondingly reduces the cardiac

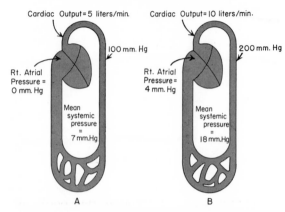

Figure 142. Effect of mean systemic pressure and right atrial pressure on venous return.

output. Figure 143A shows that if the resistance in all vessels of the body is decreased to one-half normal, the cardiac output will become two times normal, or 10 liters per minute. Figure 143B shows that if the resistance increases to two times normal, the cardiac output becomes only one-half normal, or 2.5 liters per minute.

EFFECT OF SPECIAL FACTORS ON CARDIAC OUTPUT. *Transfusion of blood.* Transfusion of blood into a person increases the volume of blood in the peripheral vessels and thereby increases cardiac output because of two very important effects. First, the increased volume *increases the mean*

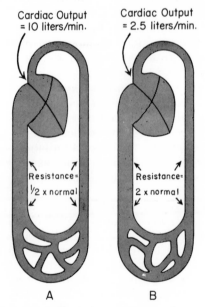

Figure 143. Effect of systemic vascular resistance on venous return.

systemic pressure which forces an increased quantity of blood toward the heart. Second, the increased blood volume dilates the vessels and thereby decreases the resistance to blood flow. This also increases venous return.

Changes in vasomotor tone. Increased tone of the blood vessels normally increases cardiac output. The reason for this is that increased tone contracts the blood reservoirs, thus forcing increased amounts of blood from the veins toward the heart. Therefore, sympathetic stimulation, which increases the vasomotor tone all through the body, increases the cardiac output. Administration of various vasoconstrictor drugs also usually increases the output.

Exercise. Exercise greatly increases the cardiac output for several reasons. First, the sympathetic nervous system becomes strongly stimulated within the first 10 to 20 seconds, causing the blood reservoirs to contract and thereby to increase the mean systemic pressure. As a result, there is a corresponding increase in cardiac output.

Another important factor that increases the output is the effect of local *autoregulation* in the active muscles themselves. The high rate of muscular metabolism during exercise causes a relative deficiency of oxygen and other nutrients in the muscles, which results in muscular vasodilatation, as was discussed in detail in Chapter 12. This in turn allows blood to flow very easily through the muscle vessels back to the heart. Thus, the total resistance to blood flow through the systemic circulation is greatly decreased, thereby promoting markedly increased venous return and correspondingly increased cardiac output. This change in vascular resistance in exercising muscles explains why cardiac output normally increases in proportion to a person's degree of activity.

The most intense degrees of exercise can increase cardiac output to as high as 20 to 25 liters per minute in a normal person and to as high as 30 to 35 liters per minute in the well trained athlete.

SUMMARY OF CARDIAC OUTPUT REGULATION. Cardiac output is regulated, first, by the tendency for blood to return to the heart, and, second, by the effectiveness of the heart as a pump. Ordinarily, the normal heart is capable of pumping all the blood that attempts to return to it from the peripheral circulatory system. Therefore, it is frequently said that cardiac output is "regulated by the venous return." However, the amount of blood attempting to return to the heart often rises above that which the normal heart can pump. As a result, blood dams up in the veins, the venous pressure rises, and the heart is said to be *overloaded,* or said by some physiologists to be *in failure.*

In the preceding sections we have discussed many details of cardiac output regulation but not an overall theory of cardiac output regulation. The basic factor that determines the cardiac output is the need of the tissues for nutrition, and the nutrient that is most limited to the tissues is oxygen. Furthermore, experiments have shown that blood flow through most tissues of the body is directly proportional to the needs of the tissues for oxygen. Thus, each tissue regulates locally its own blood flow. In doing so it increases the venous return when the tissue metabolism increases, which obviously increases the cardiac output. To state this another way, the accumulative need of the tissues for blood flow is the principal factor that regulates cardiac output.

Yet there are times when the normal unstimulated heart simply cannot pump to all the tissues sufficient blood flow to satisfy the accumulative need. Under these conditions, particularly in exercise, the sympathetic nervous system usually becomes very powerfully stimulated and the parasympathetic system becomes inhibited. These two effects work together to increase the heart rate, to increase the strength of the heart, and to constrict the blood reservoirs in the systemic circulation, all of which increase the cardiac output still more.

Measurement of Cardiac Output

A *flowmeter* can be inserted into the arteries or veins of an experimental animal in such a manner that all the blood flowing through the heart will be recorded. Such procedures, however, are not very useful even in animal experimentation because of the extreme disruption of normal function resulting from the necessary surgery. Obviously, in the human being they could not be used at all. Therefore, for measuring cardiac output in human beings and in intact animals the following indirect procedures have been developed.

THE FICK PRINCIPLE. One indirect method for measuring cardiac output is based on the so-called Fick principle, which is explained by the example of Figure 144. The oxygen in a sample of venous blood is analyzed chemically to be 160 ml. of oxygen per liter of blood, and in the arterial blood, 200 ml. Thus, as each liter of blood passes through the lungs it picks up 40 ml. of oxygen. At the same time, using appropriate respiratory apparatus, the amount of oxygen absorbed from the lungs is measured to be 200 ml. per minute. From these values, one can calculate the cardiac output in the following manner. If the amount of oxygen absorbed per minute is 200 ml. and the amount of oxygen picked up by each liter of blood flowing through the lungs is 40 ml., on dividing 200 by 40 one can conclude that five 1-liter portions of blood must pass through the lungs each minute. Thus, the total cardiac output is calculated to be 5×1 liter or 5 liters per minute.

Expressing the Fick principle mathematically the following formula applies:

Cardiac output in liters per minute =

$$\frac{\text{Total oxygen absorbed per minute}}{\text{Oxygen absorbed by each liter of blood}}$$

THE DYE METHOD. Another simple procedure for measuring cardiac output is to inject a small quantity of foreign substance, such as a brightly colored dye, rapidly into a vein. As the substance passes through the heart and lungs into the arterial system, its concentration in the arterial blood is recorded by a photoelectric "densitometer" or by taking periodic blood samples for the next 20 to 30 seconds and analyzing these for their content of dye. The more rapidly the blood flows through the heart, the more rapidly the dye appears in the arteries and the sooner it will disappear. From the recorded concentration curve of the dye as it passes through the arteries one can calculate the cardiac output quite accurately.

VENOUS PRESSURE

Right Atrial Pressure

The pressure in the venous system is determined mainly by the pressure in the right atrium. Normally, the right atrial pressure is approximately zero—that is, it is almost exactly equal to the pressure of the air surrounding the body. This does not mean, though, that the force distending the walls of the right atrium is zero, because the pressure in the thoracic cavity surrounding the heart is about 4 mm. Hg less than atmospheric pressure. This partial vacuum actually pulls the walls of the atrium outward so that blood is normally sucked from the veins into the atrium.

If the heart becomes weakened, blood begins to dam up in the right atrium, thereby causing the right atrial pressure to rise. Or, if excessive quantities of blood attempt to flow into the heart from the veins, the right atrial pressure rises. Threfore, the right atrial pressure, though normally almost exactly zero, often rises above zero in abnormal conditions.

Peripheral Venous Pressure

The pressure in a systemic vein is determined by five major factors: (1) the right atrial pressure, (2) the resistance to blood flow from the vein to the right atrium, (3) the rate of blood flow along the vein, (4) hydrostatic pressure, and (5) the venous pump. The effect of most of these factors is self-evident, but some of them need special explanation.

Since blood flows always toward the heart, the pressure in any peripheral vein of a person in the lying position must be as great as or greater than the pressure in the right atrium. If considerable resistance exists between the peripheral vein and the heart,

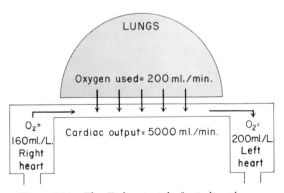

Figure 144. The Fick principle for indirectly measuring cardiac output.

and the flow of blood is of reasonable quantity, then the venous pressure will be considerably higher than right atrial pressure. Many of the veins are compressed where they pass abruptly over ribs, between muscles, between organs of the abdomen, or so forth. At these points the venous resistance is usually very great, impeding blood flow and increasing the pressure in the distal veins. On the average, the pressure in a vein of the arm or leg of a person lying down is approximately 6 to 8 mm. Hg.

EFFECT OF HYDROSTATIC PRESSURE ON PERIPHERAL VENOUS PRESSURE. Hydrostatic pressure is the pressure that results from the weight of the blood itself. Figure 145 illustrates the human being in a standing position, showing that for blood to flow from the lower veins up to the heart, considerable extra pressure must develop in these veins to drive the blood uphill. The weight of the blood from the level of the heart down to the bottom of the foot is great enough that when all other factors affecting venous pressure are non-operative, the venous pressure in the foot will be 90 mm. Hg.

THE VENOUS PUMP. To prevent the extremely high venous pressures that hydrostatic pressure can cause, the venous system is provided with a special mechanism for propelling blood toward the heart. This is called the *venous pump* or sometimes the *muscle pump.* All peripheral veins contain valves which allow blood to flow only toward the heart. Every time a muscle contracts or every time movement of any portion of a limb is effected in any other way, the blood in some of the veins is compressed. Since the valves prevent backward flow, the blood is always pushed toward the heart, in this way emptying the veins and decreasing the peripheral venous pressure.

The hydrostatic pressures shown in Figure 145 occur only when a person is standing completely still or when some disease condition has destroyed the valves of his veins. Ordinarily, the venous pump is so effective that pressures in the leg veins are only 15 to 30 mm. Hg. But when the valves are destroyed, such high pressures (80 to 90 mm. Hg) develop in the leg veins that they become progressively distended to diameters four and five times normal, causing the condition known as *varicose veins.*

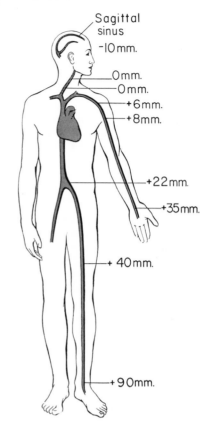

Figure 145. Hydrostatic pressures in various parts of the venous system of a person standing quietly so that the venous pump is inactive.

Measurement of Venous Pressures

The most important of all venous pressures is the right atrial pressure, for one can usually tell from this how well the heart is pumping. Right atrial pressure is usually measured through a cardiac catheter in the manner illustrated in Figure 146A. The catheter is threaded into a vein of the arm, then upward through the veins of the shoulder, into the thorax, and finally into the right atrium. The pressure transmitted through the catheter is recorded by a water manometer located at the patient's side. This same technique can be used for recording pressures in other parts of the heart or pulmonary circulation simply by sliding the catheter through the tricuspid valve into the right ventricle, or on through the pulmonary valve into the pulmonary artery.

For measuring peripheral venous pres-

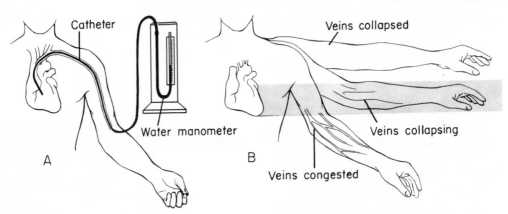

Figure 146. Measurement of (A) right atrial and (B) peripheral venous pressures.

sures, either a catheter or a needle can be inserted directly into a peripheral vein and the pressure measured as just described. However, one can usually estimate peripheral venous pressures quite satisfactorily using the simple technique illustrated in Figure 146B of raising and lowering the arm above or below the level of the heart. When the veins are lower than the heart they become full of blood and stand out beneath the skin. Then, as the arm is raised, the veins normally collapse at a level approximately 9 cm. above the level of the heart. This means that the pressure in the arm veins, when the arm is at heart level, is about 9 cm. of water. If the veins do not collapse until they rise to a level of 20 cm. above the heart, then the peripheral venous pressure is approximately +20 cm. of water.

CARDIAC FAILURE

The term *cardiac failure* means *depressed pumping effectiveness of the heart.* The most frequent cause of cardiac failure is actual damage to the heart itself by some disease process. For instance, the coronary arteries might become blocked because of *atherosclerosis,* or sometimes the valves of the heart are destroyed by rheumatic heart disease. In either case, the effectiveness of the heart as a pump is decreased so that even the normal amount of blood is not pumped as well as normally.

LOW CARDIAC OUTPUT FAILURE. A failing heart often fails to pump adequate quantities of blood to the tissues. This is called *low cardiac output failure.* It can be caused by weakness of any part of the heart or of the whole heart, because failure of any one part can often hinder satisfactory pumping.

PULMONARY CONGESTION RESULTING FROM CARDIAC FAILURE. When it is primarily the left side of the heart that is failing, the right heart continues to pump blood into the lungs with normal vigor while the left heart is unable to move the blood on into the systemic circulation. The resulting accumulation of blood in the lungs increases the pressures in all the pulmonary vessels and engorges them with blood. If the pulmonary capillary pressure becomes high enough, fluid will leak into the lung tissues, resulting in *pulmonary edema.* Indeed, in severe cases of left heart failure, fluid can even leak into the air spaces of the lungs, resulting in *hypostatic pneumonia.* Many patients who die from failure of the left heart do not die because of decreased cardiac output, but because of failure of the water-soaked lungs to aerate the blood.

PERIPHERAL CONGESTION AND EDEMA RESULTING FROM CARDIAC FAILURE. If the right side of the heart fails or if the entire heart fails (which is the usual case), the right atrial pressure rises, causing much of the blood attempting to return to the heart to be dammed in the peripheral veins. As a result, the pressures throughout the entire venous system rise. The veins of the neck become greatly distended, as illustrated in Figure 147, and the venous reservoirs such as the liver and spleen become engorged with blood. Also, the capillary pressure throughout the entire systemic circulatory system may become so great that fluid leaks con-

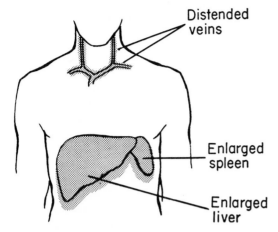

Figure 147. Engorgement of the peripheral venous system and blood reservoirs in cardiac failure.

tinually into the tissue spaces, resulting in extreme *generalized edema.* One of the conditions occasionally encountered in severe heart disease is swelling from head to toe as the result of excessive interstitial fluid, a condition called *dropsy.*

The edema in cardiac failure is caused not only by high pressure in the capillaries but also by two other factors: First, decreased cardiac output decreases the amount of urine formed by the kidneys, causing excessive quantities of water and electrolytes to remain in the body, enhancing the volume of extracellular fluid. Second, diminished blood flow through the body causes the adrenal cortex to secrete large quantities of aldosterone which also makes the kidneys retain water and salt, thus further increasing the total extracellular fluid volume.

One of the most essential features in the treatment of cardiac failure is to control the amount of salt that the person eats and the amount of water that he drinks. Also, very powerful drugs called *diuretics,* which increase the output of urine by the kidneys, are frequently administered.

CIRCULATORY SHOCK

Circulatory shock is the condition resulting when the cardiac output becomes so reduced that the tissues everywhere in the body fail to receive an adequate blood supply. The tissues suffer from inadequate nutrition and inadequate removal of cellular

excretory products because of the reduced blood flow.

Any condition that decreases the cardiac output to a very low level can cause circulatory shock. Therefore, shock can result from weakness of the heart itself or from diminished venous return, for which reason it can be classified into two main types, (1) *cardiac shock,* and (2) *low venous return shock.*

Cardiac Shock

Cardiac shock is caused by decreased effectiveness of the heart as a pump. This type of shock occurs most frequently immediately after a severe heart attack because the heart's ability to pump blood often decreases many-fold in only a few minutes. The person frequently dies because of diminished blood supply to his tissues before the heart can begin to recover. In essence, then, cardiac shock is a severe degree of low cardiac output failure.

Shock Caused by Diminished Venous Return

Any of the factors that decrease the tendency for blood to return to the heart can cause shock. These are (1) decreased blood volume, which causes *hypovolemic shock,* (2) increased size of the vascular bed so that even normal amounts of blood will not fill the vessels adequately—this is called *venous pooling shock,* or (3) obstruction of blood vessels, particularly veins.

HYPOVOLEMIC SHOCK. The blood volume can be decreased as a result of hemorrhage, loss of plasma through exuding wounds, or burns, loss of plasma into severely crushed tissues, or dehydration caused by extreme sweating, lack of water to drink, or excessive loss of fluids through the gut or kidneys. In any of these conditions, the diminished blood volume decreases the mean systemic pressure so greatly that inadequate quantities of blood return to the heart, and the diminished venous return causes shock.

VENOUS POOLING SHOCK. If the blood vessels lose their vasomotor tone, their diameters may increase so greatly that the blood collects, or "pools," in the highly distensible veins. As a result, the pressures in the systemic circulation fall so low that little *pressure gradient for venous return* then exists to make the blood flow toward the heart, and venous return becomes so slight that circulatory shock ensues.

A special type of venous pooling shock is *neurogenic circulatory shock* caused by sudden cessation of sympathetic impulses from the central nervous system to the peripheral vascular system. The result is loss of normal vasomotor tone, diminished pressures everywhere in the systemic circulation, and consequently diminished venous return. Fainting is an example of this type of shock.

Extreme allergic reactions can also cause venous pooling shock. This type of shock is known as *anaphylactic shock*, and its probable mechanism is that illustrated in Figure 148, which may be explained as follows: When a person becomes immune to a disease, his body manufactures *antibodies;* these are special proteins that destroy bacteria or toxins as explained in Chapter 9. Occasionally the reaction of the antibodies with substances entering the circulation damages the body's own tissues, at the same time causing the substance *histamine* to be released into the body fluids. The histamine then circulates to all the peripheral blood vessels and promotes vasodilatation, venous pooling, and diminished venous return, culminating immediately in a state of anaphylactic shock. Indeed, venous pooling can sometimes result so rapidly and to such an extreme extent during anaphylaxis that the person may die before therapy can be started.

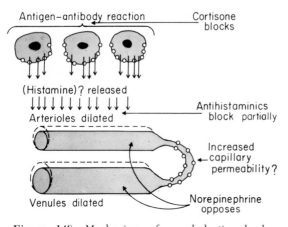

Figure 148. Mechanism of anaphylactic shock.

SHOCK CAUSED BY OBSTRUCTION TO VENOUS RETURN. Very rarely shock is caused by obstruction of blood vessels, which can prevent adequate return of blood to the heart from the systemic circulation. This occurs particularly when large veins are blocked, as when growths press on the inferior vena cava in the region of the liver or in the abdomen.

Stages of Shock

COMPENSATORY STAGE. If the abnormality causing shock is mild, the regulatory mechanisms of the circulatory system can compensate for it, and no damage will result. Figure 149 illustrates this ability of the circulatory system to compensate for hemorrhage. At the bottom of the figure, the small circles represent the changing size of a representative blood vessel. As more and more blood is removed from the circulation, the blood loss is compensated by constriction of the veins and venous reservoirs. Consequently, even though blood is lost, the return of blood to the heart continues as usual, and the cardiac output remains essentially normal until 15 to 20 per cent of the total blood volume (about 900 ml.) is removed. However, beyond this point, the blood loss can no longer be compensated, and cardiac output begins to fall. This early stage of shock is called the *compensatory stage.*

Fortunately, the arterioles supplying the heart do not constrict during the compensatory stage of shock, and those supplying the brain constrict very little, so that blood continues to flow to these essential organs in normal or almost normal quantities. Consequently, during this stage of shock, the person is not in imminent danger of dying, and unless the cause of the shock is intensified, the condition will be corrected and the person will return to normal within a few hours or certainly within a day or two.

PROGRESSIVE STAGE. If the degree of shock becomes very severe, regardless of the initial cause of the shock, *the shock itself promotes more shock.* Some of the reasons for this are shown in Figure 150. First, if the degree of shock is so great that the heart fails to pump enough blood to supply its own coronary vessels, the heart itself becomes progressively weakened. This further diminishes the cardiac output, which

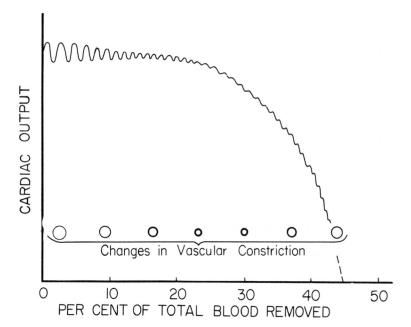

Figure 149. Effect of blood loss on cardiac output, showing compensatory vascular constriction.

weakens the heart even more and diminishes the cardiac output again, leading to a vicious cycle that eventually kills the person.

Second, very poor blood flow to the brain

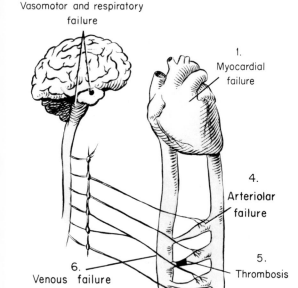

Figure 150. Factors that cause progression of shock and finally cause it to become irreversible.

causes damage to the vasomotor and respiratory centers. Failure of the vasomotor center allows venous pooling, resulting in even more extensive shock, and respiratory failure causes diminished oxygenation of the blood and, therefore, malnutrition of the tissues, which also makes the shock worse.

Peripheral vascular failure is a third cause of a vicious cycle. Ischemia of the blood vessels can probably make the vascular musculature so weak that the vessels dilate, resulting in vascular pooling of blood, diminished venous return, increased shock, increased ischemia, and so forth, the vicious cycle repeating itself again and again until the vessels are fully dilated or the patient is dead.

Fourth, recent experiments have shown that greatly dimished blood flow causes minute blood clots to develop in the small vessels. As a consequence of the plugged vessels, venous return decreases, and the progressively more sluggish flow of blood through the peripheral vessels causes more and more clotting, creating still another vicious cycle.

Fifth, it is believed that greatly prolonged shock might damage the tissue enzymes, which causes all the circulatory functions to diminish further.

Thus, because of several vicious cycles that can develop once shock has reached a certain degree of severity, progressive deterioration of the circulatory system causes the shock to become more and more severe until death ensues.

Irreversible shock. In the early stages of progressive shock, the person's life can usually be saved by appropriate and rapid treatment. For instance, if the shock is initiated by hemorrhage, rapid transfusion of blood can restore normal circulatory dynamics unless the damage to the circulation is already very great. However, if the shock has progressed for a long period of time, such extreme damage may have occurred that no amount of treatment can be successful in restoring the cardiac output enough to sustain life. When this state has been reached, the patient is said to be in *irreversible shock.* The heart, for instance, may have become damaged beyond repair by this time so that it is incapable of pumping more than 1 to 1.5 liters per minute regardless of what therapy is instituted. Or so many peripheral vessels may have become plugged with small clots that blood can no longer flow rapidly enough to repair the damage. In these instances, despite any amount of treatment, the vicious cycles of the progressive stage of shock continue on and on until death of the person.

Treatment of Shock

The treatment of shock depends on the cause. In the case of cardiac shock, the best therapy is to increase the pumping effectiveness of the heart. Usually this is not easily accomplished, but bed rest and the drug digitalis, which often strengthens a weakened heart muscle, are of value. Also, one can sometimes help a person with cardiac shock by judiciously transfusing blood. This will increase the pressures in the systemic circulation and thereby force an increased quantity of blood into the heart. Even a weakened heart can usually pump increased blood when its chambers are stretched a little.

If the shock results from hypovolemia, the blood volume can be increased by administering blood, plasma, or sometimes even isotonic salt solution. Return of the blood volume to normal increases the pressures throughout the systemic vessels, which brings the venous return back to normal and thereby alleviates the shock.

In both neurogenic and anaphylactic shock, the main difficulty is venous pooling. Administration of drugs that act like the sympathetic nervous system in constricting the blood vessels—such as norepinephrine—compresses the blood in the vessels and forces it along the veins toward the heart, bringing the patient out of shock.

REFERENCES

Bevegard, B. S., and Shepherd, J. T.: Regulation of the circulation during exercise in man. *Physiol. Rev.* 47(2):178, 1967.

Chien, S.: Role of the sympathetic nervous system in hemorrhage. *Physiol. Rev.* 47(2):214, 1967.

Crowell, J. W., and Smith, E. E.: Oxygen deficit and irreversible hemorrhagic shock. *Amer. J. Physiol.* 206:313, 1964.

Friedberg, C. K.: *Diseases of the Heart.* 3rd Ed. Philadelphia, W. B. Saunders Company, 1966.

Guyton, A. C.: *Circulatory Physiology: Cardiac Output and Its Regulation.* Philadelphia, W. B. Saunders Company, 1963.

Guyton, A. C.: Determination of cardiac output by equating venous return curves with cardiac response curves. *Physiol. Rev.* 35:123, 1955.

Guyton, A. C.: Venous return. *Handbook of Physiology,* Sec. II, Vol. II, p. 1099. Baltimore, The Williams & Wilkins Company, 1963.

Guyton, A. C., and Crowell, J. W.: Dynamics of the heart in shock. *Fed. Proc.* 20:51, 1961.

Hamilton, W. F.: Measurement of the cardiac output. *Handbook of Physiology,* Sec. II, Vol. I, p. 551. Baltimore, The Williams & Wilkins Company, 1962.

Milstein, B. B.: *Cardiac Arrest and Resuscitation.* Chicago, Year Book Medical Publishers, 1964.

Stephenson, H. E.: *Cardiac Arrest and Resuscitation.* St. Louis, The C. V. Mosby Company, 1963.

Stone, H. L., Bishop, V. S., and Guyton, A. C.: Progressive changes in cardiovascular function after unilateral heart irradiation. *Amer. J. Physiol.* 206:289, 1964.

Symposium on congestive heart failure. *Circulation* 21:95, 1960.

SECTION FOUR

THE BODY FLUIDS AND THE URINARY SYSTEM

CHAPTER 16

BODY FLUIDS, CAPILLARY MEMBRANE DYNAMICS, AND THE LYMPHATIC SYSTEM

In Chapter 5, it was pointed out that there are two major types of fluid in the body, *extracellular fluid* and *intracellular fluid,* each of which has a chemical composition different from that of the other. In the present chapter, we will discuss the relationships of these fluids to overall body function, especially movement of fluids back and forth between the blood and interstitial spaces, regulation of their volumes and constituents, and their role in some of the special fluids, such as cerebrospinal fluid, ocular fluid, and lymph.

BODY WATER, INTRACELLULAR FLUID, AND EXTRACELLULAR FLUID

The total water in the average 70 kilogram man is about 40 liters, or 57 per cent of his total body weight. Of this, about 62 per cent is in the *intracellular compartment* and about 38 per cent in the *extracellular compartment.* Thus, the average 70 kilogram man has approximately 25 liters of intracellular fluid

and 15 liters of extracellular fluid. These relationships are illustrated in Figure 151.

Blood

About 93 per cent of the blood is fluid. However, this fluid, like the total body water,

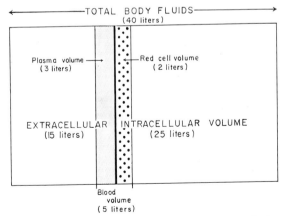

Figure 151. Diagrammatic representation of the body fluids, showing the extracellular fluid volume, blood volume, and total body fluids.

181

comprises intracellular and extracellular fluids. The *plasma* portion of the blood, which is the clear fluid between the cells, is typical extracellular fluid except that it has a higher concentration of protein than do the extracellular fluids elsewhere in the body; this protein is very important for keeping the plasma inside the circulatory system, as is discussed later in the chapter. The *cells* are of two types, *red blood cells* and *white blood cells.* The number of white blood cells is only one five-hundredth the number of red blood cells so that, for practical purposes, essentially all of the intracellular fluid in blood is contained in the red blood cells. The normal total volume of plasma is 3 liters and of red cell fluid 2 liters.

Interstitial Fluid

The interstitial fluid is that portion of the extracellular fluid that lies outside the capillaries and between the cells. The volume of the interstitial fluid, therefore, is equal to the total extracellular fluid volume minus the plasma volume, or 15 − 3 = 12 liters of interstitial fluid volume in the normal, average adult man. Interstitial fluid is usually considered to include such special fluids as those in the cerebrospinal fluid system, the fluid in the eyes, in the minute intrapleural spaces, in the peritoneal cavity, in the pericardial cavity, and in the joint spaces.

MEASUREMENT OF FLUID VOLUMES OF THE BODY

THE DILUTION PRINCIPLE. Frequently it is important for the physiologist to measure the fluid volumes in different compartments of the body. To do this, he utilizes a basic principle called the *dilution principle* as follows:

Figure 152 illustrates a fluid chamber in which a small quantity of dye or other foreign substance is injected. After the substance mixes with the fluid in the chamber, its concentration becomes very dilute and equal in all areas of the chamber (Fig. 152B). Then a sample of the fluid is removed and the concentration of the substance analyzed by chemical, photoelectric, or any other means. The volume of the chamber can then be calculated by the following formula:

$$\text{Volume in ml.} = \frac{\text{Quantity of test substance injected}}{\text{Concentration per ml. of sample fluid}}$$

It should be noted that all one needs to know is (1) *the total quantity of the test substance* injected into the chamber and (2) the *concentration of the substance in the fluid after dispersion.*

MEASUREMENT OF TOTAL BODY WATER. To measure total body water one can inject either *heavy water* or *radioactive water* into a person and then 30 minutes to an hour later remove a sample of blood and measure

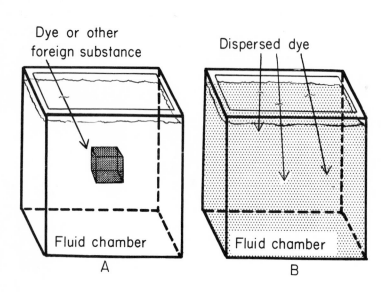

Figure 152. The dilution principle for measuring the volume of a fluid compartment.

the concentration of heavy water or radioactive water in the blood or plasma. Since either one of these types of water disperses throughout the body in exactly the same way as normal water, one can use this measurement to calculate the total body water. Let us assume that 50 ml. of heavy water has been injected into the person and that its concentration in each ml. of plasma is found to be 0.001 ml. Dividing this concentration into the total amount injected (using the above formula), we find that the total body water of this person is 50 liters, which is a value slightly higher than the value of 40 liters in the normal adult man.

MEASUREMENT OF EXTRACELLULAR FLUID VOLUME. To measure the extracellular fluid volume one simply uses a different test substance, a substance that will diffuse everywhere in the extracellular fluid compartment but that will not go through the cell membranes into the intracellular fluid compartment. Substances of this type include *radioactive sodium, thiocyanate ions,* and *inulin,* all of which have been used at one time or another for determining the extracellular fluid volume. The procedure is almost identical with that used for measuring total body water. The test substance is injected intravenously, and a sample of plasma is removed about 30 minutes later, after appropriate mixing throughout the entire extracellular compartment has occurred. The same formula above is used to calculate the extracellular fluid volume.

Calculation of intracellular fluid volume. Once the total body water has been determined using heavy water or radioactive water and the extracellular fluid volume has been determined using radioactive sodium or one of the other test substances, one can calculate intracellular fluid volume by subtracting extracellular fluid volume from total body water.

MEASUREMENT OF PLASMA VOLUME. The plasma volume is measured by injecting some substance that will stay in the plasma compartment of the circulating blood. Such a substance frequently used is a dye called T-1824 or Evans blue which, upon intravenous injection, combines almost immediately with the plasma proteins. Since plasma proteins do not leak readily out of the plasma compartment of the blood, neither will the dye. After appropriate mixing of the blood has occurred, within about 10 minutes, a sample of blood is removed, the plasma separated from the red cells, and the concentration of dye in the plasma measured. Then using the foregoing dilution formula, one determines the plasma volume.

Calculation of interstitial fluid volume. Once the plasma and extracellular fluid volumes have been measured, one can calculate the interstitial fluid volume by subtracting plasma volume from extracellular fluid volume.

Calculation of blood volume. If one knows the relative percentage of the blood that is blood cells, he can calculate the total blood volume from the plasma volume. To determine the percentage of the blood that is plasma, blood is centrifuged at a high speed for 15 to 30 minutes, which separates the plasma from the cells so that the relative percentages of these two can be measured directly. The percentage of blood cells is called the *hematocrit* of the blood, as was explained in Chapter 8. From the hematocrit and the measured plasma volume one calculates total blood volume using the following formula:

$$\text{Blood volume} = \frac{100}{100 - \text{Hematocrit}} \times \text{Plasma volume}$$

CAPILLARY MEMBRANE DYNAMICS

Molecules diffuse continually both inward and outward through the capillary pores. Ordinarily, the rates of diffusion in the two directions are almost exactly equal so that the volumes of plasma and interstitial fluid remain essentially constant. Yet, under special circumstances the rate of fluid diffusion in one direction becomes greater than in the other direction, in which case the plasma and interstitial volumes change accordingly. Therefore, it is important that we consider the dynamics of this net exchange of fluid between the plasma and the interstitial fluid and the mechanisms by which normal volumes are ordinarily maintained.

ANATOMY OF THE CAPILLARY SYSTEM. Figure 153 illustrates a typical capillary bed, showing blood entering the capillaries from the arteriole, passing through the metarteriole, then into the capillaries, and

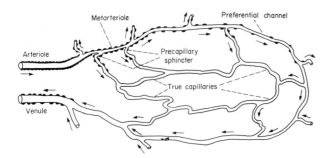

Figure 153. A typical capillary system. (Drawn from a figure by Zweifach: *Factors Regulating Blood Pressure.* Josiah Macy, Jr., Foundation, 1950.)

finally to the venule. The *arteriole* has a muscular coat that allows it to contract or relax in response to stimuli which come mainly from the sympathetic nerves. The *metarteriole* has sparse muscle fibers, and the openings to the *true capillaries* are usually guarded by small *muscular precapillary sphincters* which can open and close the entryway into the capillaries. The muscles of the metarterioles and the precapillary sphincters are controlled mainly in response to local conditions in the tissues. For instance, lack of oxygen allows these muscles to relax, which increases the flow of blood through the capillary bed and in turn increases the amount of oxygen available to the tissues.

PRESSURES IN THE CAPILLARY SYSTEM. Though the pressure at the arterial end of a capillary varies tremendously depending upon the state of contraction or relaxation of the arteriole, metarteriole, and precapillary sphincters, the normal mean pressure at the arterial end of the capillary is probably between 25 and 35 mm. Hg. In the present discussion we will use an average value of about 30 mm. Hg. Likewise, the average pressure at the venous end of the capillary is about 10 mm. Hg, and the average mean pressure in the capillary is probably about 18 mm. Hg. Unfortunately, we do not know these values exactly, for they have never been determined entirely satisfactorily. Pressures have been measured at different points in capillaries with minute pipets, these pressures having been slightly higher than those just given. However, the use of the pipet itself obstructs the capillary and can cause abnormal pressures. The values just given have been estimated from experiments in which the pressures in the arterial and venous ends of a tissue bed have been changed to many different values and the rates of fluid transudation through the capillaries measured by special techniques. From

these data one can calculate the values for capillary pressure.

Tissue Pressure

Tissue pressure is the pressure of the fluid in the spaces between the cells and, therefore, is also the pressure of the fluid outside the capillaries pressing against the capillary membrane.

Tissue pressure is another value that has been difficult to measure because the interstitial spaces generally have a width of less than 1 micron, and introduction of any needle or pipet to measure the pressure can cause an abnormal reading. However, during the past few years we have studied this problem in our own laboratory in a completely different way, by implanting into tissues perforated capsules such as the one illustrated in Figure 154. The tissue grows inward through the holes and lines the inner wall of the capsule. A blood vascular system also develops in the new tissue. Yet, a large fluid space remains in the middle of the capsule. A needle connected to a manometer system can then be inserted through one of the perforations into the cavity and the pressure

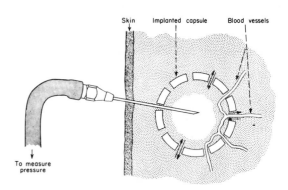

Figure 154. The perforated capsule method for measuring tissue fluid pressure.

measured. Tissue pressures measured in this manner have been about −6 mm. Hg— that is, 6 mm. Hg less than atmospheric pressure — which demonstrates that the tissue spaces actually have a partial vacuum in them.

Measurements of tissue pressure using minute needles inserted in the tissues under special conditions have also given negative values of about −6 mm. Hg.

PRESSURE DIFFERENCE ACROSS THE CAPILLARY MEMBRANE. If the average mean pressure in the capillary is 18 mm. Hg and the average pressure outside the capillary is −6 mm. Hg, the total pressure difference between the two sides of the membrane would be 18 − (−6) or a total of 24 mm. Hg. That is, the pressure inside the capillary is 24 mm. Hg greater than the pressure outside the capillary. This pressure difference between the two sides of the membrane makes fluid tend to diffuse out of the capillary into the tissue spaces more rapidly than in the opposite direction. Fortunately, however, *colloid osmotic pressure*, another mechanism operating at the capillary membrane, opposes this tendency for fluid to leak out of the capillaries, as is discussed in the following section.

Colloid Osmotic Pressure at the Capillary Membrane

THE PRINCIPLE OF OSMOTIC PRESSURE. The principles of osmosis and osmotic pressure were discussed in Chapter 5. To review briefly, if two solutions are placed on either side of a semipermeable membrane so that water molecules can go through the membrane pores but solute molecules cannot, water will move by the process of osmosis from the more dilute solution into the more concentrated solution. However, if pressure is applied to the more concentrated solution, this movement of water by osmosis can be slowed or halted. The amount of pressure that must be applied to stop completely the process of osmosis is called the osmotic pressure. For a much deeper understanding of why these effects take place, the student is referred back to Chapter 5.

COLLOID OSMOTIC PRESSURE. The amount of osmotic pressure that develops at a membrane is determined partly by the size of the pores in the membrane, because

osmosis occurs only when dissolved particles cannot go through the pores. In the case of the cell membrane, such particles include sodium ions, chloride ions, glucose molecules, urea molecules, and others, all of which cause osmotic pressure. However, at the capillary membrane the pore diameter is roughly 10 times that of the cell membrane (80 Angstroms in contrast to 8 Angstroms) so that sodium ions, glucose molecules, and essentially all other constituents of extracellular fluid pass directly through the capillary pores without causing osmotic pressure. However, the proteins in the plasma in the interstitial fluids are an exception, because their molecular sizes are large enough that they fail to penetrate the pores. Therefore, they do cause osmotic effects at the capillary membrane. Because solutions of plasma protein appeared to early chemists to be colloidal suspensions rather than true solutions, this protein osmotic pressure at the capillary membrane is called *colloid osmotic pressure*. However, some physiologists also call this pressure *oncotic pressure*.

Plasma has a normal protein concentration of 7 gm. per cent, while interstitial fluid has a concentration of about 1.5 gm. per cent, making a difference between the two sides of the capillary membrane of about 5.5 gm. per cent. Because the larger concentration is inside the capillary, osmosis of fluid tends to occur always from the interstitial fluid into the capillaries.

Colloid osmotic pressure of plasma and interstitial fluids. If pure plasma were placed on one side of a capillary membrane and pure water on the other side, the colloid osmotic pressure that would develop would be about 28 mm. Hg. If pure interstitial fluid were placed on one side of the membrane and pure water on the opposite side, the colloid osmotic pressure that would develop would be about 4 mm. Hg. Therefore, we say that the colloid osmotic pressure of plasma is 28 mm. Hg and of interstitial fluid 4 mm. Hg.

The *colloid osmotic pressure difference* between the two sides of the membrane is equal to the difference between the colloid osmotic pressures of the two fluids on the respective sides of the membrane. Therefore, the colloid osmotic pressure difference at the capillary membrane is 28 − 4, or 24 mm. Hg.

Equilibration of Pressures at the Capillary Membrane — The Law of the Capillaries

One will note from the foregoing discussions that under normal conditions the fluid pressure difference across the capillary membrane (24 mm. Hg) is exactly equal to the colloid osmotic pressure difference across the membrane (24 mm. Hg). However, the fluid pressure tends to move fluid out of the capillary, while the colloid osmotic pressure tends to move fluid into the capillary. This balance between the two forces explains how it is possible for the circulation to keep its blood volume constant even through the capillary pressure is considerably higher than the tissue pressure. Were it not for the colloid osmotic forces, fluid would be lost continually from the circulation until eventually the blood volume would be insufficient to maintain cardiac output.

The normal state of equilibrium between the forces tending to make fluid leave the capillaries and those tending to return fluid to the capillaires is called the *law of the capillaries*, and it is illustrated mathematically in Figure 155. The mean fluid pressure in the capillary is 18 mm. Hg and the tissue pressure is −6 mm. Hg making a total fluid pressure of 24 mm. Hg tending to move fluid out of the capillary. On the other hand, the colloid osmotic pressure in the capillary is 28 mm. Hg and in the interstitial fluid 4 mm. Hg, making a net difference again of 24 mm. Hg. Thus, the two pressures are mathematically in balance so that the fluid volumes of neither the plasma nor interstitial spaces will be changing.

EFFECT OF NON-EQUILIBRIUM AT THE CAPILLARY MEMBRANE. Occasionally the forces at the capillary membrane lose their state of equilibrium because one or more of them changes to a new value. When this occurs, fluid transudes through the membrane very rapidly and causes a new state of equilibrium to develop, usually within a few minutes to an hour or so.

Let us consider, for example, an increase in capillary pressure from the normal value of 18 up to 25 mm. Hg. This would increase the net force across the capillary membrane by 7 mm. Hg in the outward direction, which would cause rapid transudation of fluid into the tissue spaces. The loss of fluid (but not of plasma proteins) out of the plasma would cause the *capillary pressure to fall* because of decreasing blood volume and would cause the *plasma colloid osmotic pressure to rise* because of increasing concentration of the plasma proteins. In addition, the *tissue pressure would rise* because of the increasing volume of interstitial fluid, and the *tissue colloid osmotic pressure would fall* because of dilution of the tissue proteins. Thus, after a short period of time, we would have a capillary pressure of 22 mm. Hg, a plasma colloid osmotic pressure of 29 mm. Hg, a tissue pressure of −4 mm. Hg, and a tissue colloid osmotic pressure of 3 mm. Hg. If the student will add these pressures, he will see that a new state of equilibrium has been established.

In a similar manner, any other changes in any one of the pressure values on the two sides of the membrane will cause rapid osmosis of fluid through the capillary membrane until equilibrium is reestablished within a few minutes to an hour or more.

Flow of Fluid Through the Tissues

In addition to the very rapid diffusion of water and dissolved substances through the capillary membrane, there is a small amount of actual *flow* of fluid through the tissues. The distinction between diffusion and flow is the following: Diffusion means movement of each molecule along its own pathway as a result of its kinetic motion irrespective of all other molecules, whereas flow means that large quantities of molecules all move in the same direction.

Flow of fluid through the tissues is caused by fluid movement through the arterial and venous ends of the capillaries as follows: If

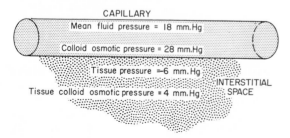

Figure 155.   Mean capillary dynamics.

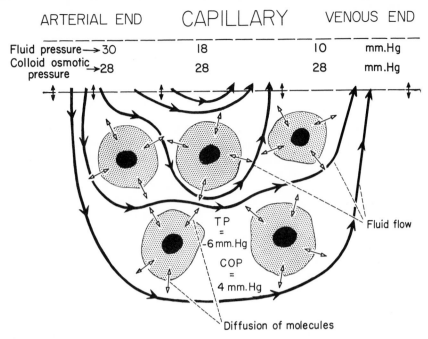

Figure 156.   Flow of fluid through the tissue spaces.

we observe Figure 156, we will see that the fluid pressure in the capillary at the arterial end is 30 mm. Hg, while at the venous end it is 10 mm. Hg. Now, let us study the forces that cause fluid movement through the capillary membrane at both ends of the capillary. A sum of all the forces acting at the arterial end of the capillary gives a net pressure of 12 mm. Hg in favor of movement of fluid out of the capillary, which is called the *filtration pressure* at the arterial ends of the capillaries.

At the venous ends of the capillaries, the greatly decreased fluid pressure in the capillary causes a balance of forces in the opposite direction. This time the sum of the forces is 8 mm. Hg tending to move fluid inward rather than outward. This pressure gradient is called the *absorption pressure.*

Note that the filtration and absorption pressures are not equal, but the difference is counterbalanced by the fact that the venous ends of the capillaries are larger and therefore have about 50 per cent more surface area. As a result, approximately equal amounts of fluid pass out of the arterial end of the capillary and return by the venous end. A small portion of the fluid, however, between $1/10$ and $1/100$ of the total, returns to the circulation by way of the lymphatics rather than by absorption into the venous ends of the capillaries.

THE LYMPHATIC SYSTEM

In addition to the blood vessels, the body is supplied with an entirely separate set of very small and thin-walled vessels called *lymphatics.* These originate in nearly all the tissue spaces as very minute *terminal lymphatics* which are also called *lymphatic capillaries.* The relationship of the terminal lymphatics to the cells and to the capillaries is shown in Figure 157. The terminal

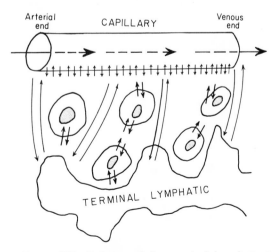

Figure 157.   Relation of the terminal lymphatic to the tissue cells and to a blood capillary.

lymphatics then coalesce into progressively larger and larger lymphatic vessels, finally terminating in the neck as shown in the diagram of the entire lymphatic system in Figure 158. The lymphatics from the entire lower part of the body, from the left arm, and from the left side of the head empty into the venous system at the juncture of the left internal jugular and subclavian veins, while the lymphatics from the right arm, right upper chest, and right side of the head empty into the veins at the juncture of the right internal jugular and subclavian veins.

The lymphatics are an accessory system for flow of fluid from the tissue spaces into the circulation. The terminal lymphatic capillaries are so permeable that even very large particles and protein molecules can pass directly into them along with tissue fluids. Therefore, in effect, the fluid that flows up the lymphatics is actually overflow fluid from the tissue spaces; it is known as *lymph*, and it has identically the same constituents as normal interstitial fluid.

At many points along the lymphatics, particularly where several smaller lymphatics combine to form larger ones, the ves-

sels pass through *lymph nodes*, which are small organs that filter the lymph, taking all particulate matter out of the fluid before it empties into the veins.

Functions of the Lymphatics

RETURN OF PROTEINS TO THE CIRCULATION. The single most important function of the lymphatics is to return proteins to the circulation when they leak out of the capillaries. Some of the pores in the capillaries are so large that there is a continual small rate of protein loss from the plasma, amounting each hour to approximately 1/25 of the total protein in the circulation. If these proteins were not returned to the circulation, the person's plasma colloid osmotic pressure would fall so low and he would lose so much blood volume that he would die within 12 to 24 hours. Furthermore, no other means is available by which proteins can return to the circulation except by way of the lymphatics. Later in the chapter, when we discuss edema, we will see how serious it is for proteins not to be returned to the circulation even from a single part of the body.

Regulation of tissue colloid osmotic pressure by the lymphatics. The transport of proteins from the interstitial fluid spaces by way of the lymphatics keeps the interstitial fluid proteins low in concentration while keeping the plasma proteins high in concentration. Whenever the interstitial fluid protein concentration rises above normal, the tissue colloid osmotic pressure rises correspondingly and causes osmosis of fluid into the tissue spaces. This increase in interstitial fluid volume causes fluid to pass rapidly into the lymphatic capillaries. Thus, any increase in protein in the tissues causes fluid to wash automatically through the tissues into the lymphatics, and this washes the protein back into the circulation, thereby keeping interstitial fluid protein concentration and tissue colloid osmotic pressure at very low levels in comparison with those in the plasma.

FLOW OF LYMPH ALONG THE LYMPHATICS. Two principal factors determine the rate of lymph flow along the lymphatics: (1) the tissue pressure and (2) the degree of activity of the lymphatic pump.

Whenever the *tissue pressure* rises above normal, fluid flows easily into the lym-

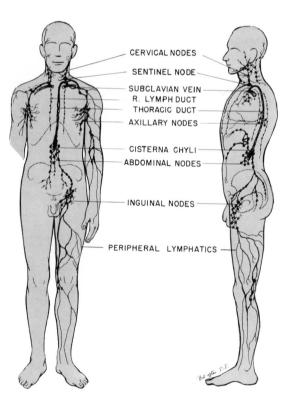

Figure 158.   The lymphatic system.

CERVICAL NODES

SENTINEL NODE

SUBCLAVIAN VEIN
R. LYMPH DUCT
THORACIC DUCT
AXILLARY NODES

CISTERNA CHYLI
ABDOMINAL NODES

INGUINAL NODES

PERIPHERAL LYMPHATICS

phatic capillaries which have wide open communications with these spaces. Therefore, the greater the tissue pressure the greater also is the quantity of lymph formed each minute.

The *lymphatic pump*, the second factor affecting lymph flow, is the pumping of lymph along the lymph vessels by body movements, whih may be explained as follows: Figure 159 illustrates a lymph vessel opened along one side to show a *lymphatic valve*. These valves are present throughout all major lymphatic channels, and every time a muscle contracts, thereby compressing the lymphatics, fluid in the vessel moves out of the compressed portion of the vessel in the direction that the valves are oriented. Since the valves are oriented toward lymph emptying sites into the veins, each compression site of the lymphatic vessel acts as a small individual pump, and the aggregate function of all these individual pumps is called the *lymphatic pump*. When one runs or exercises in any other manner, the total quantity of lymph leaving the tissues increases enormously.

In summary, the amount of lymph flowing along the lymphatics is determined (a) by the tissue pressure, which causes fluid to enter the lymphatic capillaries, and (b) by the intensity of body movements, which activates the lymphatic pump to propel lymph along the vessels.

Role of the lymphatic pump in maintaining negative tissue pressure. Were it not for the lymphatic pump, it would be impossible for the tissues to maintain negative tissue pressure. Each time a lymphatic vessel is compressed, fluid is pushed away from the tissues. This pump operates even in the terminal lymphatic capillaries because the endothelial cells of the terminal lymphatics are arranged so that cells overlap each other to form inlet valves. That is, fluid can flow between cells into the terminal lymphatic but cannot flow backwards because the overlapping cells close the spaces whenever the pressure inside the lymphatic rises.

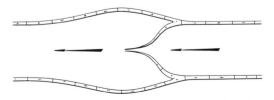

Figure 159.   A lymphatic vessel, showing a lymphatic valve.

Rate of lymph flow.   Lymph flow varies tremendously from time to time, but in the average person, the total lymph flow in all vessels is approximately 100 ml. per hour, or 1 to 2 ml. per minute. One can readily see that this small rate of lymph flow plays hardly any role in returning *fluids* from the tissue spaces to the circulation, but it does play a major role in returning *proteins* to the circulation. The average lymph contains 3 to 4 per cent protein (2 per cent from peripheral tissues and as high as 5 to 6 per cent from the liver where a major share of the lymph is formed).

FILTRATION OF THE LYMPH BY THE LYMPH NODES.   Another function of the lymphatic system is to filter all the lymph as it returns from the tissues before emptying back into the circulation. Since large molecules or particulate matter of any type cannot be absorbed directly into the blood capillaries, it is only through the lymphatic vessels that these substances can leave the tissues. Yet, many of these can be very dangerous to the person if they get into the blood, for instance, bacteria or high molecular weight protein toxins such as tetanus toxin. Fortunately, as the lymph passes through the lymph nodes, special cells, so-called *reticuloendothelial cells*, phagocytize both foreign protein molecules and particles, then digest both of these and release the products into the lymph in the form of amino acids or other small molecular breakdown products. Other cells in the lymph nodes form antibodies which, as we have seen in Chapter 9, play a major role in protecting the body against foreign toxins and bacteria.

THE NORMAL "DRY" STATE OF THE BODY, AND DEVELOPMENT OF EDEMA

IMPORTANCE OF NEGATIVE PRESSURE IN THE TISSUES.   Up to the present point very little has been said about the importance of the negative pressure in the interstitial fluid spaces. It plays two major roles: First, the negative pressure is a partial vacuum between the parts of the body and, therefore, serves to hold parts of the body together. Second, it keeps the amount of fluid in the tissues to a minimum. That is, the colloid osmotic pressure of the plasma keeps pulling fluid out of the tissue spaces until they become almost as small as possible, and it is in this state that the tissues normally operate. Therefore, from a relative point of view, we can say that the interstitial spaces are normally "dry," even though they still have 12 liters of fluid in them. This statement will become more meaningful as we discuss edema in the succeeding paragraphs. This state of minimal fluid in the interstitial spaces is particularly beneficial to the diffusion of nutrients from the capillaries to the cells, for one of the basic principles of diffusion is that the rate of transport of substances by this mechanism is *inversely* proportional to the distance that the substances must diffuse.

When the body fails to maintain the "dry" state in the interstitial spaces, but instead collects large quantities of *extra* fluid, the condition is called *edema*. The transport of nutrients to the tissues then becomes impaired, sometimes seriously enough to cause gangrene of the tissues. For instance, prolonged swelling of the feet frequently causes gangrenous ulcers in the skin, this resulting from too great a distance for the nutrients to diffuse from the capillaries to the tissue cells.

The Basic Causes of Edema

POSITIVE TISSUE PRESSURE IN THE EDEMATOUS STATE. Figure 160 illustrates the compactness of the cells in normal tissues and the spreading apart of the cells when the tissue pressure rises and causes edema. Recent experiments have shown that as long as the tissue pressure remains negative, the cells remain in a compact state, but just *as soon as the tissue pressure rises above the zero level and becomes positive, the cells spread apart*. At tissue pressures of +3 mm. Hg. the interstitial fluid volume is often increased to as much as 3 to 4 times the normal amount, and at +8 mm. Hg the interstitial fluid volume, at least in some tissues, often rises to as much as 20 times normal, causing very great increases in sizes of the interstitial spaces, with resultant increase in dis-

tances for nutrients to diffuse from the capillaries to the cells.

Since it is positive tissue pressure that causes edema, one can readily understand that any change in the dynamics of fluid transfer at the capillary membrane that will increase the tissue pressure from its normal negative value up to a positive value can cause edema. Three basic abnormalities of capillary membrane dynamics often result in edema. These are: (1) elevated fluid pressure in the capillaries, (2) decreased colloid osmotic pressure in the capillaries, and (3) increased tissue colloid osmotic pressure.

ELEVATED FLUID PRESSURE IN THE CAPILLARIES. Though the normal mean capillary pressure is about 18 mm. Hg, this can rise to values as high as 40 to 50 mm. Hg in some abnormal states of the circulation. For instance, when the *heart fails*, blood dams up in the venous system, causing back pressure on the capillaries and thereby raising the capillary pressure. Also, *blockage of the veins* can do the same thing, or *overdilatation of the arterioles* can cause such rapid blood flow into the capillaries that their pressure rises. In any of these conditions, a rise in the mean pressure in the capillaries will cause increased tissue pressure and resultant edema.

LOW PLASMA COLLOID OSMOTIC PRESSURE. Another cause of edema is a decrease in plasma colloid osmotic pressure, which results from diminished plasma protein concentration. This occurs very frequently in persons with *kidney disease* who lose tremendous quantities of plasma proteins into the urine or in persons with *severe burns* who exude large amounts of proteinaceous fluid through their denuded skin. Also, lack of adequate protein in the diet, such as occurs in *famine areas* of the world, can cause failure of formation of adequate plasma proteins. When the plasma colloid osmotic pressure falls below about 3 grams per cent, the normal negative pressure in the tissues becomes lost, and instead the pressure rises to a positive value. Here again, very severe edema can result.

Tissue pressure
= -6 mm.Hg

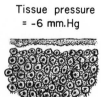

Tissue pressure
= 3 mm.Hg

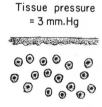

Tissue pressure
= 8 mm.Hg

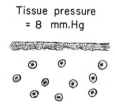

Figure 160. Effect of various tissue pressures on the amount of edema developing in the interstitial spaces.

INCREASED TISSUE COLLOID OSMOTIC PRESSURE. Increased tissue colloid osmotic pressure can also cause high tissue pressure and edema. The most common cause of this is *blockage of the lymphatics,* which prevents return of proteins to the circulation. The proteins that leak through the capillary walls gradually accumulate in the tissue spaces until the tissue colloid osmotic pressure approaches the plasma colloid osmotic pressure. As a result, the capillaries lose their normal osmotic force to hold fluid in the circulation so that fluid in abundance accumulates in the tissues. This condition can cause edema of the greatest proportions.

Lymphatic blockage commonly occurs in the South Sea Island disease called *filariasis,* in which *filariae* become entrapped in the lymph nodes and cause so much growth of fibrous tissue in the nodes that lymph flow through the nodes becomes totally or almost totally blocked. As a result, certain areas of the body, such as a leg or an arm, swell so greatly that the swelling is called "elephantiasis." A single leg with this condition can weigh as much as the entire remainder of the body, all because of the extra fluid in the tissue spaces.

SPECIAL FLUID SYSTEMS OF THE BODY

The Potential Fluid Spaces

There are a few spaces in the body that normally contain only a few milliliters of fluid and yet can collect many liters of fluid under abnormal conditions. These spaces, called the *potential spaces* of the body, include the *intrapleural space,* which is the space between the lungs and the chest wall; the *pericardial space,* which is the space between the heart and the pericardial sac in which it resides; the *peritoneal space,* which is the space between the gut and the abdominal wall; the *joint spaces;* and the *bursae.* Ordinarily, all of these spaces are totally collapsed except for the presence of a highly viscid lubricating fluid that enables the surfaces of the spaces to slip over each other, allowing free movement of the lungs in the chest cavity, free movement of the heart in the pericardial cavity, and so forth.

DYNAMICS OF FLUID EXCHANGE. The dynamics of fluid exchange between these potential spaces and the capillaries are almost identical to the dynamics of fluid exchange at the usual capillary membrane, because the linings of these cavities are almost totally permeable to proteins and fluids so that fluid flows with ease between the cavities and the capillaries in the surrounding tissues.

Figure 161 illustrates the intrapleural space, showing a capillary in the tissue adjacent to the cavity. This figure shows diffusion of fluid back and forth between the cavity and the capillary and also diffusion of fluid through the visceral pleura into the tissue of the lungs. Obviously, with such free diffusion, the pleural space is actually nothing more than an enlarged tissue space. There is negative pressure in the intrapleural space just as there is negative pressure elsewhere in the tissues. Therefore, normally, the intrapleural space is kept "dry" in exactly the same way as the other tissue spaces of the body.

"EDEMA" OF THE POTENTIAL SPACES. Under abnormal conditions, a person can develop "edema" of the potential spaces in exactly the same way that he develops edema in his tissue spaces. Thus, if the capillary pressure rises too high, if the plasma colloid osmotic pressure falls too low, or if the colloid osmotic pressure in the pleural cavity rises too high, fluids will transude into the space in exactly the same way that they transude into the tissue spaces in edema. "Edema" of a potential space is called *effusion,* though physiologically it is exactly the same process as edema.

LYMPHATIC DRAINAGE OF THE POTENTIAL SPACES. All potential spaces have lymphatic drainage systems similar to that of the tissue spaces. Figure 161 illustrates the

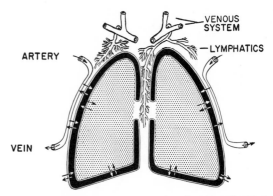

Figure 161. Dynamics of fluid exchange in the intrapleural spaces.

lymphatic drainage from the pleural cavity mainly into the mediastinum. This drainage normally removes the proteins that leak into the intrapleural space and, therefore, keeps the colloid osmotic pressure in the space at a low value.

One of the most serious causes of effusion in a potential space is *infection,* for this usually causes the lymphatics to become plugged with large masses of white blood cells and tissue debris caused by the infection. Also, infection increases the porosity of the surrounding capillaries. Thus, the proteins cannot leave the space even though excessive quantities of proteins are leaking into the space. As a result, the colloid osmotic pressure increases tremendously, and, as one would expect, fluid transudes into the space in massive amounts. Thus, an infected joint swells tremendously, an infected pleural cavity develops terrific amounts of pleural effusion, and an infected abdominal cavity develops large amounts of abdominal effusion which in this case is often called *ascites.*

The Fluid System of the Eye

The fluid system of the eye maintains a constant pressure in the eyeball of almost exactly 20 mm. Hg. This pressure, along with the very strong wall of the eyeball (the sclera) keeps the eyeball in a relatively rigid shape, keeping the distances between

cornea, lens, and *retina* always constant within fractions of a millimeter. This is essential to the functioning of the eye's optical system for focusing images on the retina, as will be discussed in Chapter 24.

The fluids of the eye are divided, as shown in Figure 162, into two separate compartments: the *aqueous humor,* which lies in the *anterior chamber* of the eye between the front of the lens and the cornea, and the *vitreous humor,* which lies in the *posterior chamber* between the lens and the retina. The vitreous humor is a gelatinous mass, and the fluid in it can *diffuse* through the mass but cannot *flow* through it. On the other hand, the aqueous humor in front of the lens is freely flowing. Aqueous humor also spreads backward through the ligaments of the lens along the surface of the *ciliary body* as also illustrated in Figure 162.

FORMATION OF AQUEOUS HUMOR. Fluid is continually formed by the ciliary body from which it passes between the ligaments of the lens and into the anterior chamber of the eye. Then it flows, as illustrated by the arrows in Figure 162, into the angle between the cornea and the iris where it makes its way through minute spaces called the *spaces of Fontana* in the *canal of Schlemm,* a circular vein that passes all the way around the eye at the juncture of the cornea and the sclera.

The ciliary body is made up of a large number of small folds of epithelium which have a total surface area in each eye of about

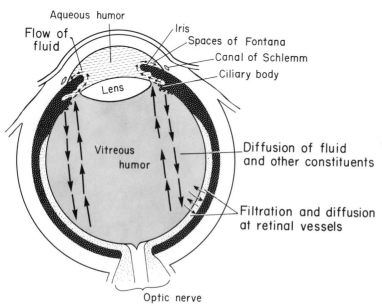

Figure 162. The fluid system of the eye.

6 square centimeters. This epithelium secretes sodium ions continually into the aqueous humor, which in turn sets off the the following effects: (1) The sodium ions cause a positive electrical charge to develop in the aqueous humor. This positive charge in turn pulls negative ions, especially chloride ions, through the epithelium, thus increasing the number of these ions in the aqueous humor as well. (2) The increased quantity of sodium ions and anions in the aqueous humor causes osmosis of water through the epithelium. Thus, indirectly, the secretion of sodium causes a continuous flow of fluid, about *2 cubic millimeters per minute,* into the aqueous humor from the ciliary body.

CONTROL OF PRESSURE IN THE EYEBALL. The pressure in the eye is controlled by the minute structures at the angle between the cornea and the iris, which is where the fluid leaves the eyeball and enters the canal of Schlemm. If the pressure rises too high, increased amounts of fluid leave the eye. On the other hand, if the pressure falls too low, the outflow of fluid from the eyeball automatically decreases. The precise mechanism by which this change in rate of fluid outflow occurs is not known, but it seems to act like a release valve on a steam boiler. When the pressure is above a critical value, the valve structures at the angle of the cornea and iris, perhaps the spaces of Fontana, appear to open up, and when the pressure is below the critical value, the spaces appear to close. Regardless of the precise details by which this mechanism works, it regulates the pressure in the eyeball almost exactly at 20 mm. Hg, rarely allowing it to rise above or to fall below this value more than a few millimeters Hg.

Glaucoma. Glaucoma is a condition in which the pressure of the eye becomes very high, often high enough to cause blindness. The cause of the condition is failure of fluid to flow normally into the canal of Schlemm. When this happens, the continual formation of fluid by the ciliary body, without proper removal of the fluid into the canal of Schlemm, causes the eye pressure to rise sometimes to tremendous levels. For instance, *infection* or *inflammation* can cause debris or white blood cells to plug the openings of the spaces of Fontana. Also, many people develop narrowing of these spaces or pores into the canal of Schlemm for reasons that are unknown.

In rare instances the pressure in the eye rises to as high as 60 to 80 mm. Hg, though in the more usual form it rises to values of 30 to 45 mm. Hg. The very high pressures can cause blindness within a few days by damaging the optic nerve where it enters the eyeball and by compressing the blood vessels to the retina so severely that the nutritive blood flow is cut off. When the eyeball pressure is elevated only moderately, on the other hand, blindness can develop gradually over a period of years.

The Cerebrospinal Fluid System

The cerebrospinal fluid system is similar to that of the eye in that most of the fluid is formed in one area and is reabsorbed in an entirely different area. The fluid of the cerebrospinal fluid cavity plays a special role as a cushion for the brain; the density of the brain is almost the same as that of the cerebrospinal fluid, so that the brain literally *floats* in this fluid. Furthermore, the cranial vault in which the brain and cerebrospinal fluid lie is a very solid structure so that when it is hit on one side the whole vault moves as a unit, both the fluid and the brain being propelled in the same direction at the same time. Therefore, because of the cushioning effect of the cerebrospinal fluid, no damage to the brain results despite the fact that the brain is among the softest tissues of the entire body.

FORMATION, FLOW, AND ABSORPTION OF CEREBROSPINAL FLUID. Essentially all cerebrospinal fluid is formed by the *choroid plexuses,* which are cauliflower-like growths that protrude into all four ventricles shown in Figure 163—the two *lateral ventricles* in

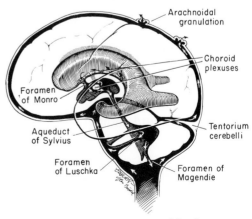

Figure 163.   The cerebrospinal fluid system.

the two cerebral hemispheres, the *third ventricle* in the diencephalic region, and the *fourth ventricle* between the hindbrain and the cerebellum. Fluid formed in the lateral ventricles flows through two small openings called the *foramina of Monro* into the third ventricle, and fluid from the third ventricle flows through the *aqueduct of Sylvius* into the fourth ventricle. From here the fluid flows out of the ventricles, through the *foramina of Luschka* and *foramen of Magendie*, into the *subarachnoid space* which is the fluid space between the brain and the cranial vault. The fluid then flows around the brain and upward over its upper surfaces where it is absorbed into the venous sinuses through the *arachnoidal granulations.*

The choroid plexuses, like the ciliary processes of the eye, continually secrete sodium ions and water by essentially the same mechanism as that described for the eye. Likewise, the arachnoidal granulations function in a manner similar to that of the structures that lie in the angle between the iris and cornea of the eye, for these granulations act as an overflow valvular system. When the pressure in the cerebrospinal fluid is about 10 mm. Hg above that in the veins, fluid flows through the granulations, but when the pressure in the cerebrospinal fluid system is less than this level, fluid does not leave the system.

The only major difference between the fluid system of the eye and the cerebrospinal fluid system is the level of the pressure, for the pressure required to open the overflow valves in the eye is about 20 mm. Hg, while the pressure required to release fluid from the cerebrospinal fluid system into the veins is about 10 mm. Hg.

BLOCKAGE OF FLOW IN THE CEREBRO-SPINAL FLUID SYSTEM. Referring again to Figure 163, one can see by the small white arrows that flow of fluid in the cerebrospinal fluid system can be blocked at many places, especially at the foramina of Monro, at the aqueduct of Sylvius, and at the arachnoidal granulations. Many babies are born with congenital blockage of the aqueduct of Sylvius so that fluid formed in the lateral and third ventricles cannot escape to the sur-

face of the brain to be absorbed. As a result, the ventricles swell larger and larger, compressing the brain against the cranial vault and destroying much of the neuronal tissue. Another common cause of blockage is infection in the cerebrospinal fluid cavity, which causes so much debris and white blood cells in the fluid that the arachnoidal granulations become plugged. In this case, fluid accumulates in the entire cerebrospinal fluid system, and the pressure rises very high. In rare instances the pressure in the cerebrospinal fluid system can rise so high that it actually impedes blood flow in the brain so that the brain cannot receive adequate nutrition.

REFERENCES

Allen, L.: Lymphatics and lymphoid tissues. *Ann. Rev. Physiol.* 29:197, 1967.

Bland, J. H. (ed.): *Clinical Metabolism of Body Water and Electrolytes.* Philadelphia, W. B. Saunders Company, 1963.

Davson, H.: Intracranial and intraocular fluids. *Handbook of Physiology,* Sec. I, Vol. III, p. 1761. Baltimore, The Williams & Wilkins Company, 1960.

Drinker, C. K.: *The Lymphatic System.* Stanford, Calif., Stanford University Press, 1942.

Guyton, A. C.: Concept of negative interstitial pressure based on pressures in implanted perforated capsules. *Circ. Res.* 12:399, 1963.

Guyton, A. C.: Interstitial fluid pressure: II. Pressure-volume curves of interstitial space. *Circ. Res.* 16: 452, 1965.

Landis, E. M., and Pappenheimer, J. R.: Exchange of substances through the capillary walls. *Handbook of Physiology,* Sec. II, Vol. II, p. 961. Baltimore, The Williams & Wilkins Company, 1963.

Mayerson, H. S.: The physiologic importance of lymph. *Handbook of Physiology,* Sec. II, Vol. II, pp. 1035-1073. Baltimore, The Williams & Wilkins Company, 1963.

Millen, J. W., and Wollam, D. H. M.: *The Anatomy of the Cerebrospinal Fluid.* New York, Oxford University Press, 1962.

Moore, F. D., Olesen, K. H., McMurrey, J. D., Parker, H. V., Ball, M. R., and Boyden, C. M.: *The Body Cell Mass and Its Supporting Environment.* Philadelphia, W. B. Saunders Company, 1963.

Pappenheimer, J. R., Heisey, S. R., Jordan, E. F., and Downer, J. D.: Perfusion of the cerebral ventricular system in unanesthetized goats. *Amer. J. Physiol.* 203:763, 1962.

Wolf, A. V., and Crowder, N. A.: *Introduction to Body Fluid Metabolism.* Baltimore, The Williams & Wilkins Company, 1964.

CHAPTER 17

FORMATION OF URINE BY THE KIDNEY, AND MICTURITION

Physiologic Anatomy of the Kidney

The kidney forms urine and in doing so regulates the concentrations of most of the substances in the extracellular fluid. It accomplishes this by removing those materials from the blood plasma that are present in excess, while conserving those substances that are present in normal or subnormal quantities.

Figures 164 and 165 illustrate respectively the gross and microscopic structures of the kidney which are responsible for this fluid purifying function. Figure 164 shows the renal artery entering the substance of the kidney and the renal vein returning from it. Urine is formed from the blood by the *nephrons,* one of which is shown in detail in Figure 165. From these urine flows into the *renal pelvis* and then out through the *ureter* into the *urinary bladder.* The two kidneys contain approximately two million nephrons, and because each nephron operates almost exactly the same as all others, we can characterize most of the functions of the kidney

as a whole by explaining the function of a single nephron.

The nephron is composed of two major parts, the *glomerulus* which filters water and solutes from the blood, and the *tubules* which reabsorb from the filtrate those substances that are needed in the body, while allowing the unneeded substances to flow

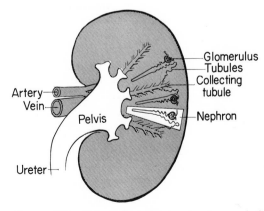

Figure 164. Principal anatomic structures of the kidney.

195

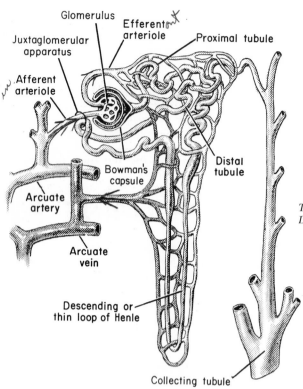

Figure 165. The nephron. (Modified from Smith: *The Kidney: Structure and Function in Health and Disease.* Oxford University Press.)

into the renal pelvis as urine. The glomerulus is a tuft of capillaries surrounded by a capsule called *Bowman's capsule.* Fluid filters out of the capillaries into this capsule and then flows from here, first, into the *proximal tubule,* second, into a long loop called the *loop of Henle,* third, into the *distal tubule,* fourth, into a *collecting tubule,* and, finally, into the *renal pelvis.* As the filtrate passes through the tubules, most of the water and electrolytes are reabsorbed into the blood, but almost all of the end-products of metabolism pass on into the urine. In this way, water and electrolytes are not depleted from the body, though the waste products of metabolism are removed constantly.

FUNCTION OF THE NEPHRON

In discussing the function of the nephron it is desirable to use the simplified diagram shown in Figure 166. This shows the "functional nephron" with an *afferent arteriole* supplying blood to the glomerulus and the blood leaving the glomerulus through an *efferent arteriole,* to flow into the *peritubular*

capillaries and finally into the vein. Also shown are the *glomerular membrane, Bowman's capsule,* the *tubules,* and the *kidney pelvis.*

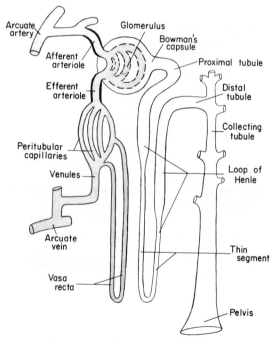

Figure 166. The functional nephron.

Glomerular Filtration

The membranes of the capillaries in the glomerular tuft are collectively called the *glomerular membrane.* This membrane is much more permeable to water and small molecular solutes than is the usual capillary membrane elsewhere in the body, but otherwise the same principles of fluid dynamics apply as to other capillary membranes. Like other capillary membranes, the glomerular membrane is almost completely impermeable to plasma protein.

However, the pressure in the glomerulus is very high, believed to be about 70 mm. Hg, in contrast to the low pressures, between 15 and 20 mm. Hg, in capillaries elsewhere in the body. Because of this high pressure, fluid leaks continually out of all portions of the glomerular membrane into Bowman's capsule. We shall see that most of the fluid that leaks out of the glomerular membrane is later reabsorbed from the renal tubules into the peritubular capillaries. That which is not reabsorbed becomes urine.

FLUID DYNAMICS AT THE GLOMERULAR MEMBRANE, AND THE GLOMERULAR FILTRATION RATE. The fluid pressures in the normal nephron are illustrated in Figure 167A. This figure shows that the *glomerular pressure* is normally 70 mm. Hg, while the colloid pressure in the glomerulus is normally 32 mm. Hg. The pressure in Bowman's capsule is about 14 mm. Hg, and the colloid osmotic pressure in this capsule is essentially zero. Therefore, the pressure tending to force fluid out of the glomerulus is 70 mm. Hg, and the total pressure tending to move fluid in the opposite direction is 32 + 14, or 46 mm. Hg. The difference between these two, 24 mm. Hg, is the net pressure pushing fluid into Bowman's capsule; this is called the *filtration pressure.*

The rate at which fluid flows from the blood into Bowman's capsule, called the *glomerular filtration rate,* is directly proportional to the filtration pressure. Therefore, any factor that changes any one of the pressures on the two sides of the glomerular membrane will also change the glomerular filtration rate. Thus, an increase in glomerular pressure increases the rate of glomerular filtrate formation, while an increase in either the glomerular colloid osmotic pressure or the pressure in Bowman's capsule decreases the rate of filtrate formation.

The normal rate of formation of glomerular filtrate (called simply the *glomerular filtration rate*) is 125 ml. per minute. This is approximately 180 liters each day or 4.5 times the amount of fluid in the entire body, which illustrates the magnitude of the renal mechanism for purifying the body fluids.

Effect of afferent arteriolar constriction on filtration. Figure 167B shows the effect on glomerular filtration of constricting the *afferent arteriole.* The major effect is a drastic decrease in the pressure in the glomerulus. In this figure the glomerular pressure has fallen to 43 mm. Hg, while the total of the colloid osmotic pressure plus the pressure in Bowman's capsule equals 42 mm. Hg. Therefore, the filtration pressure is now only 1 mm., and, as a result, the glomerular filtration rate has been decreased to less than 5 per cent of that occurring in the example of Figure 167A.

The afferent arterioles are controlled partly by sympathetic nerves, and partly by a control mechanism intrinsic in the nephron itself, called *autoregulation,* which will be discussed in detail later in the chapter. Sympathetic stimulation constricts the arterioles and lowers the glomerular pressure, thereby decreasing glomerular filtration. On the other hand, diminished sympathetic stimulation allows afferent arteriolar dilatation and consequently greatly increased glomerular filtration.

Effect of arterial pressure on glomerular filtration. When the systemic arterial pressure rises, increased amounts of blood flow into the glomerulus thereby increasing the glomerular pressure. This in turn increases the glomerular filtration rate and the output of urine. This effect of arterial pressure on kidney output affords a method by which the

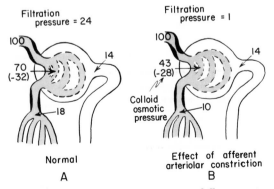

Filtration pressure = 24

Filtration pressure = 1

100   70 (-32)   14   18

100   43 (-28)   14   10   Colloid osmotic pressure

Normal
A

Effect of afferent arteriolar constriction
B

Figure 167. (A) Normal pressures at different points in the nephron, and the normal filtration pressure. (B) Effect of afferent arteriolar constriction on the pressures in the nephron and on the filtration pressure.

arterial pressure itself is automatically regulated. That is, increased pressure causes increased loss of fluid from the blood. This decreases the blood and extracellular fluid volumes and thereby lowers arterial pressure. On the other hand, when the arterial pressure falls, the kidneys stop excreting fluid, thus allowing the blood and extracellular fluid volumes to increase again until the arterial pressure returns to a normal value. This effect was explained in more detail in Chapter 14 in relation to blood pressure regulation.

CHARACTERISTICS OF GLOMERULAR FILTRATE. The filtrate entering Bowman's capsule, the *glomerular filtrate,* is an ultrafiltrate of plasma. The glomerular membrane is porous enough so that water and essentially all of the dissolved constituents of plasma except proteins can filter through. Therefore, glomerular filtrate is almost identical with plasma except that only a very minute quantity of protein (0.03 per cent) is present in the filtrate, while the protein concentration in plasma is approximately 7 per cent. In Chapter 16 it was pointed out that interstitial fluid is also an ultrafiltrate of plasma, having all the constituents of plasma except proteins. Thus, glomerular filtrate is almost identical to interstitial fluid.

Tubular Reabsorption

After the glomerular filtrate enters Bowman's capsule, it passes into the tubular system where each day all but one liter of the 180 liters of glomerular filtrate is reabsorbed into the blood, the remaining one liter passing into the renal pelvis as urine.

Figure 168 illustrates a microscopic cross-section of a tubular region of the kidney,

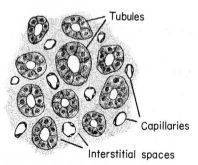

Figure 168. Cross-section of the tubules and adjacent capillaries in the kidney.

showing the close proximity of the tubules to the peritubular capillaries. The tubular fluid is reabsorbed first into the interstitial spaces and then from these spaces into the capillaries. Some of the substances are reabsorbed through the tubular epithelium by the process of *active reabsorption*, while other substances are reabsorbed by the process of *diffusion and osmosis.*

ACTIVE REABSORPTION. The term *active reabsorption,* which was discussed in Chapter 5, means transport of substances through the tubular epithelial cells and into the interstitial spaces by means of special chemical transport mechanisms. From the interstitial spaces, the substances diffuse into the peritubular capillaries. Some of the substances reabsorbed in this manner are glucose, amino acids, proteins, uric acid, and most of the electrolytes—sodium, potassium, magnesium, calcium, chloride, and bicarbonate. Glucose, amino acids, and proteins are reabsorbed almost entirely in the proximal tubules, while electrolytes are reabsorbed in all the tubules. The active reabsorption processes for glucose, amino acids, and proteins are so powerful that ordinarily almost no glucose, proteins, or amino acids pass into the urine. On the other hand, the degree to which the electrolytes are reabsorbed is variable and is regulated by special control systems which are discussed in the following chapter. When an electrolyte is present in the extracellular fluids in excess, it will be reabsorbed to a lesser extent than when it is present in too small an amount. By this process of selective reabsorption, the renal tubules control the concentrations of electrolytes in the body fluids.

The substance that is actively reabsorbed from the tubules to the greatest extent of all is sodium chloride, the total quantity reabsorbed each day being approximately 1200 grams, which is about three fourths of all the substances actively reabsorbed from the glomerular filtrate. The reabsorption of sodium chloride is regulated partially by the hormone *aldosterone* which is secreted by the adrenal cortex. This regulatory mechanism is discussed in detail in the following chapter.

Figure 169 illustrates the mechanism of active sodium reabsorption. The serrated border of the tubular epithelial cell, called the *brush border,* is extremely permeable to sodium and allows sodium to diffuse rapidly from the lumen of the tubule to the in-

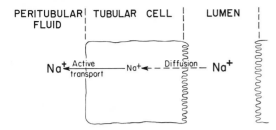

Figure 169. Mechanism of active reabsorption of sodium from the tubule, illustrating diffusion into the tubular epithelial cell from the tubular lumen and then active transport through the base of the cell into the peritubular fluid.

side of the cell. At the base of the cell, on its opposite end from the brush border, the membrane has entirely different properties. Here the membrane is almost completely impermeable to *diffusion* of sodium, but it does *actively transport* sodium in the outward direction from the cell into the peritubular fluid. This active transport probably occurs in the manner explained in Chapter 5 as follows: It is believed that the sodium ion combines with a *carrier* which is dissolved in the cell membrane. The combined sodium-plus-carrier then diffuses to the opposite side of the membrane where the sodium is released into the peritubular fluid. Enzymes in the cell membrane cause the necessary reactions to take place, and high energy phosphate compounds supply the energy required to make the reactions occur.

Active transport of other substances occurs in the same manner. However, each substance has its own specific carrier and its own set of enzymes for catalyzing the reactions.

One of the most important features of active reabsorption is that it can cause absorption of a substance even when its concentration is less in the tubule than in the interstitial fluid. To do this, however, the tubular epithelial cells must expend much energy. Therefore, these cells require tremendous amounts of nutrition, and their metabolic systems are so geared that they can transform the potential energy of their nutrients into the energy required to transport substances against the concentration gradient.

One will readily recognize the similarity between the transport mechanism in the tubular epithelial membrane and that in the cell membrane as was described in Chapter 5. In both instances, specific carriers and enzymes must be available, and large amounts of energy are expended to cause the transport.

REABSORPTION BY DIFFUSION AND OSMOSIS. At this point we need to recall the basic principles of diffusion and osmosis because these play a major role in the reabsorption of water and a few other substances from the tubules. Diffusion means the random movement of molecules in a fluid, and it is caused by kinetic movement of all fluid molecules. In other words, each water molecule or each dissolved molecule in the water is constantly bouncing among all the others, wending its way from place to place, going first in one direction and then another. If a suitable pore is present in a membrane, a molecule can pass through the membrane. The tubular epithelium is permeable to certain types of molecules, including water molecules. Therefore, water can diffuse from the tubules into the interstitial spaces of the kidney.

Osmosis is net diffusion in one direction caused by the presence of a greater concentration of nondiffusible substances on one side of a membrane than on the other side. The basic principles of osmosis were described in Chapter 5. The principal method for reabsorption of water from the tubules is by osmosis, which may be explained as follows:

Active transport of electrolytes and certain non-electrolytes such as glucose and amino acids causes the concentrations of these substances to diminish in the tubular fluids while at the same time to increase in the interstitial fluids surrounding the tubules. All of these substances are relatively nondiffusible through the pores of the membrane; therefore, a large concentration difference develops between the two sides of the membrane, which in turn causes osmosis of water through the membrane.

Thus, in effect, the initiating factor that causes reabsorption of all substances from the tubules is the active reabsorption processes, the osmotic reabsorption of water simply following in the wake of the active reabsorption of solutes.

FAILURE OF REABSORPTION OF UNWANTED SUBSTANCES. Some of the substances in the glomerular filtrate are undesirable in the body fluids; these substances in general are reabsorbed either not at all or very poorly by the tubules. For instance, *urea*, an end-product of protein metabolism, has no functional value to the body and must be removed continually if protein metabolism is to continue. This substance is not actively reabsorbed, and the pores of the tubule are so small that it diffuses through the tubular membrane several hundred times less easily than water. Therefore, while water is being osmotically reabsorbed, only about 50 per cent of the urea diffuses out of the tubules. Instead, the larger part of the urea remains behind and passes on into the urine. Thus, the primary function of the kidney is this separation process in the tubules, the tubules reabsorbing those substances such as amino acids, electrolytes, and water which are needed by the body while at the same time allowing urea, a substance that is not needed by the body, to pass on into the urine.

Other substances that have a fate similar to that of urea include *creatinine, phosphates, sulfates, nitrates, uric acid,* and *phenols,* all substances that are end products of metabolism and would damage the body if they remained in the body fluids.

Active Tubular Secretion

A few substances are actively secreted from the blood into the tubules by the tubular epithelium, including creatinine, potassium, and hydrogen ions. Active secretion presumably occurs by the same mechanism as active reabsorption but in the reverse direction.

In some fishes, tubular secretion is the only means by which waste products are removed. Also, a number of drugs such as penicillin, Diodrast, Hippuran, and phenolsulfonphthalein are removed from the blood primarily by active secretion rather than by glomerular filtration. However, in normal function of the kidneys, tubular secretion is important only to help in regulation of potassium and hydrogen ion concentrations in the body fluids.

Recapitulation of Nephron Function

Now that both glomerular filtration and tubular reabsorption have been discussed, it will be valuable to review again the total function of the nephron as follows: The total blood flow into all the nephrons of both kidneys is approximately 1200 ml. per minute. Approximately 650 ml. of this is plasma, and about one fifth of the plasma filters

through the glomerular membranes of all the nephrons into Bowman's capsules, forming an average of 125 ml. of glomerular filtrate per minute. The glomerular filtrate is actually plasma minus the proteins. The pH of glomerular filtrate is approximately 7.4, which is equal to that of the plasma. As the glomerular filtrate passes downward through the tubules, approximately 80 per cent of the water and electrolytes is reabsorbed in the proximal tubules, while all of the glucose and proteins and a major portion of the amino acids are reabsorbed. As the remaining 20 per cent of the glomerular filtrate passes through the loop of Henle, distal tubules, and collecting tubules variable amounts of the remaining water and electrolytes are absorbed, depending on the need of the body for these substances, as is discussed in the following chapter. The pH of the tubular fluid may rise or fall depending on the relative amounts of acidic and basic ions reabsorbed by the tubular walls. Also, the osmotic pressure of the tubular fluid may rise or fall depending on whether large quantities of electrolytes or large quantities of water are actively reabsorbed. Thus, the pH of the finally formed urine may vary anywhere from 4.5 to 8.2, while the crystalloidal osmotic pressure may be as little as one fourth that of plasma or as great as four times that of plasma.

The final quantity of urine formed is normally about 1 ml. per minute or $\frac{1}{125}$ of the amount of glomerular filtrate filtered each minute. This 1 ml. of urine contains about one half of the urea that is in the original glomerular filtrate, all of the creatinine, and large proportions of the uric acid, phosphate, potassium, sulfates, nitrates, and phenols. Thus, even though almost all the water and salt in the tubular fluid are reabsorbed, a very large proportion of the waste products in the original glomerular filtrate is never reabsorbed, but instead passes into the urine in a highly concentrated form.

Regulation of Glomerular Filtration Rate – The Phenomenon of Autoregulation

The reabsorption of water, salts, and other substances from the tubules depends greatly upon the rate at which glomerular filtrate flows through the tubular system. If the rate is very fast, nothing is reabsorbed adequately before the fluid empties into the urine. On the other hand, when very little glomerular filtrate is formed each minute, essentially everything is reabsorbed, including urea and other end products of metabolism. Therefore, for optimum effectiveness in reabsorbing water and salts while not reabsorbing too much urea and other end products of metabolism, the glomerular filtration rate in each nephron must be very exactly controlled. The mechanism by which this occurs, called *autoregulation of glomerular filtration*, is probably the following:

Referring back to Figure 165 which illustrates the entire nephron, note that the distal tubule, where it comes up from the loop of Henle, lies adjacent to the afferent arteriole. Figure 170 illustrates this juncture, called the *juxtaglomerular apparatus*, showing that at the point of contact the tubular cells are increased in number and are dense, forming a structure called the *macula densa*. Also, the muscle cells of the afferent arteriole at the point of juncture are swollen and filled with granules. These cells are called *juxtaglomerular cells*. When high concentrations of certain substances such as sodium chloride are injected into the distal tubule, the afferent arteriole immediately becomes constricted. This illustrates that the composition of the fluid in the distal tubule controls the degree of constriction of the afferent arteriole and in this way controls the amount of glomerular filtrate that is formed by the nephron.

Therefore, though the story is not yet completely clear, it is almost certain that each nephron monitors the concentrations of substances in the tubular fluid entering the distal tubule from the loop of Henle. This in turn controls the rate of glomerular filtration. If the glomerular filtration rate becomes too great, the concentrations of substances at the macula densa decrease. On the other hand, if the rate becomes too little, these concentrations increase. Yet, by this feedback control system, the concentrations at the macula densa can be controlled at a very exact value, thus allowing precisely the amount of glomerular filtrate to enter the tubular system that can be processed properly in the tubule, no more, no less.

THE CONCEPT OF CLEARANCE

The function of the kidney is actually to *clean* or to *clear* the extracellular fluids of the body. Every time a small portion of plasma filters through the glomerular membrane, passes down the tubules, and then is reabsorbed into the blood, it leaves behind in the tubules a large proportion of the unwanted products. In doing so it is "cleared" of these substances. For instance, out of the 125 ml. of glomerular filtrate formed each minute, approximately 60 ml. of that reabsorbed leaves its urea behind. In other words, 60 ml. of plasma is cleared of urea each min-

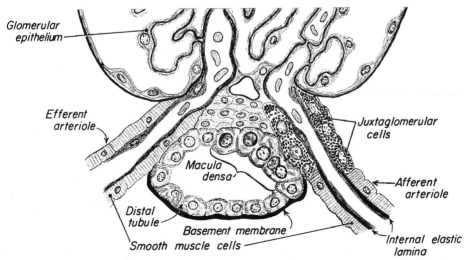

Figure 170. Structure of the juxtaglomerular apparatus, illustrating its possible feedback role in the control of nephron function. (Modified from Ham: *Histology.* J. B. Lippincott Co.)

ute by the kidneys. In the same way, each minute 125 ml. of plasma is cleared of creatinine; 12 ml., of uric acid; 12 ml., of potassium; 25 ml., of sulfate; 25 ml., of phosphate; and so forth.

CALCULATION OF RENAL CLEARANCE. The method by which one determines how much plasma is cleared of a particular substance each minute is to take simultaneous samples of blood and urine. From these samples the quantity of the substance in each milliliter of blood is analyzed chemically, and the quantity of the substance appearing in the urine each minute is determined. By dividing the quantity of the substance in each milliliter of plasma into the quantity passing into the urine during one minute, one can calculate the milliliters of plasma cleared per minute. That is,

Plasma clearance =

$$\frac{\text{Milligrams secreted in urine per minute}}{\text{Milligrams in each milliliter of plasma}}$$

As an example, if the concentration of urea in the plasma is 0.2 mg. in each ml. and the quantity of urea entering the urine per minute is 12 mg., then the amount of plasma that loses its urea during that minute would be 60 ml. To express this another way, the *plasma clearance* of urea is 60 ml. per minute.

RENAL CLEARANCE AS A TEST OF KIDNEY FUNCTION. Since the primary function of the kidneys is to clear the plasma of undesired substances, one of the best means for testing overall kidney function is to measure the clearance of these substances. The clearance of urea has been one of the most widely used tests of renal function. Normal plasma clearance of urea is approximately 60 ml. per minute; the clearance is less than this if the kidneys are damaged, and the amount that it is depressed gives one an approximation of the degree of kidney damage.

MEASUREMENT OF GLOMERULAR FILTRATION RATE BY RENAL CLEARANCE OF INULIN. The clearance of the substance *inulin* by the kidneys is exactly equal to the glomerular filtration rate for the following reasons: Inulin filters through the glomerular membrane as easily as water, so that the concentration of inulin in the glomerular filtrate is exactly equal to that in the plasma. However, inulin is not reabsorbed or secreted even in the minutest degree by the tubules. Therefore, all of the inulin of the glomerular filtrate appears in the urine. In other words, all of the originally formed glomerular filtrate is cleared of inulin, which means that the rate of inulin clearance is equal to the rate of glomerular filtrate formation. As an example, a small quantity of inulin is injected into the blood stream of a person, and after mixing with the plasma its concentration is found to be 0.001 gm. in every milliliter of plasma. The amount of inulin appearing in the urine is 0.125 gm. per minute. On dividing, we find that the plasma clearance of inulin is 125 ml. per minute. This, therefore, is also the amount of glomerular filtrate formed each minute.

ESTIMATION OF RENAL BLOOD FLOW BY CLEARANCE METHODS. The amount of blood flow through the two kidneys can be estimated from the clearance of either Diodrast or para-aminohippuric acid. These two substances, when injected into the blood in small quantities, are almost totally cleared by active tubular secretion. Therefore, if the plasma clearance of Diodrast is found to be 600 ml. per minute, one can be certain that at least 600 ml. of plasma flowed through the kidneys during that minute. This corresponds to a total flow of whole blood through the kidneys of at least 1000 ml. per minute.

ABNORMAL KIDNEY FUNCTION

Almost any type of kidney damage decreases the ability of the kidney to cleanse the blood. Therefore, kidney abnormalities usually cause an excess of unwanted metabolic waste products in the body fluids, and poor regulation of the electrolyte and water composition of the fluids.

KIDNEY SHUTDOWN. Several diseases can cause the kidneys to stop functioning suddenly and completely. Two of the most common of these are, first, poisoning of the nephrons by mercury, uranium, gold, or other heavy metals, and, second, plugging of the kidney tubules with hemoglobin following a transfusion reaction. In addition to these, almost any of the other diseases of the kidneys that are listed below can cause either gradual or rapid shutdown.

Following kidney shutdown the concentrations of urea, uric acid, and creatinine, all of which are metabolic waste products, may reach 10 times normal levels. Also, the body fluids may become extremely acidotic because of failure of the kidneys to excrete sufficient quantities of acid, and if the person continues to drink water he will become very edematous because of failure to rid himself of the ingested fluid. The person passes into coma within a few days, mainly because of acidosis. If the shutdown is complete, he will die in 8 to 14 days.

KIDNEY ABNORMALITIES THAT CAUSE LOSS OF NEPHRONS. Many kidney diseases destroy large numbers of whole nephrons at a time. For instance, infection of the kidney can destroy large areas of the kidneys; trauma can destroy part of or an entire kidney; occasionally a person is born with congenitally abnormal kidneys in which many of the nephrons are already destroyed; or nephron destruction can result from poisons, toxic diseases, and arteriosclerotic blockage of renal blood vessels.

As many as three fourths of the nephrons in the two kidneys can usually be destroyed before the composition of the person's blood becomes excessively abnormal. The reason for this large margin of safety is that the undamaged nephrons can then function much more rapidly than usual. The amount of blood flowing into each nephron becomes greatly increased, and the glomerular filtration rate per nephron can rise to two times normal. This increased activity compensates to a great extent for the lost nephrons, allowing the metabolic waste products to be removed in sufficient quantity to maintain normal body fluid composition. However, these persons usually are treading a thin line of safety because bouts of excess metabolism caused by exercise, fever, or even ingestion of too much food may present the kidneys with far more waste products than they can handle.

Obviously, as the degree of kidney destruction progresses, the derangements of the extracellular fluid become progressively more severe until finally the condition approaches that of kidney shutdown. The person then develops extreme edema, acidosis, and eventually coma and death.

GLOMERULONEPHRITIS. A very common kidney disease is glomerulonephritis. This, like rheumatic heart disease, is an allergic disease caused by toxins of certain types of streptococcic bacteria. Almost all episodes of acute glomerulonephritis occur approximately two weeks after a severe streptococcic sore throat or some other streptococcic infection. The glomeruli become acutely inflamed, swollen, and engorged with blood. Blood flow through the glomeruli almost ceases, and the glomerular membranes become extremely porous, allowing both red blood cells and protein to flow freely into the tubules. A person with glomerulonephritis has decreased kidney function, sometimes to the extent of complete kidney shutdown. If any urine is still formed, it contains large quantities of red blood cells and proteins.

In many instances of acute glomerulonephritis, the inflammation of the glomeruli regresses within two to three weeks, but even so the disease usually permanently destroys a large number of nephrons. Repeated small bouts of glomerulonephritis may destroy more and more nephrons, causing *chronic glomerulonephritis*. This can run a course of a few to many years, leading eventually to edema, depressed metabolism, coma, and death.

Effect of Diuretics to Enhance Kidney Output

A number of different drugs, called *diuretics*, can be given to a person who has either normal or diseased kidneys to increase his urine output. The actions of some of these are:

1. *Xanthines*, such as caffeine in coffee and theophylline in tea, cause dilatation of the afferent arterioles, thereby increasing the glomerular pressure and the glomerular filtration rate. As a result, the amount of urine formed per minute is greatly enhanced.

2. Administration of large quantities of *urea* or non-reabsorbed sugars such as *sucrose* can cause rapid flow of urine in the following way: These substances are filtered through the glomerulus into the tubules, but they are not reabsorbed by the tubules to a major extent. Their presence in the tubules in excessive amounts greatly increases the crystalloidal osmotic pressure of the intratubular fluids, and thereby opposes water reabsorption. As a result, increased quantities of fluid flow on through the tubules and into the urine.

3. Certain types of drugs such as *organic mercurial compounds, chlorothiazide,* and *Diamox* either poison or block some of the specific enzymes in the tubular epithelial cells used for active reabsorption, especially those enzymes that promote reabsorption of sodium and chloride. As a result, larger quantities than usual of these substances

remain in the tubules, and their crystalloidal osmotic pressure opposes the reabsorption of water, promoting increased urine flow.

Use of the Artificial Kidney

When the kidneys are damaged so severely that they can no longer maintain normal composition of the extracellular fluid, it is sometimes desirable to readjust the constituents of the extracellular fluid by use of an artificial kidney, such as the one shown in Figure 171. The artificial kidney is nothing more than a semiporous cellophane membrane arranged so that blood flows over one surface and a *dialyzing solution* flows over the other surface. The membrane is porous to all substances in the blood except the proteins and red blood cells. Almost all the substances can diffuse from the blood into the dialyzing solution, and the substances in the solution can diffuse into the blood. The fluid contains none of the waste products of metabolism. Consequently, the waste products of metabolism diffuse into the bath. On the other hand, the bath contains approximately the same concentrations of electrolytes as those found in normal plasma. Therefore, electrolytes diffuse in both directions, which prevents the blood from losing its normal electrolytes. Occasionally, the artificial kidney is used to supply supplementary nutrition. Large quantities of glucose, for instance, may be placed in the bath so that while the waste products are being removed glucose diffuses into the blood, bolstering the patient's nutritional status.

Unfortunately, the artificial kidney cannot be used with impunity because the person's blood must be rendered incoagulable during its use and because a very large amount of blood must flow through the artificial kidney

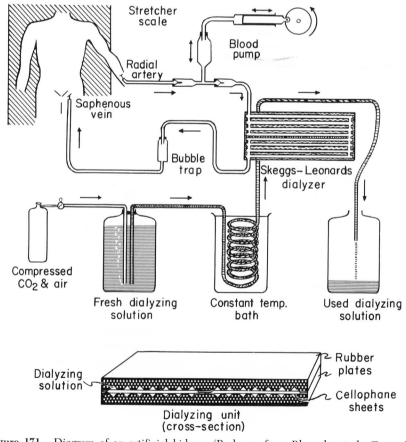

Figure 171. Diagram of an artificial kidney. (Redrawn from Bluemle et al.: *Tr. and Stud. Coll. Physicians* 20:157, 1953.)

to make it work. For these reasons it can be applied only for a few hours out of every several days.

MICTURITION

TRANSMISSION OF URINE TO THE BLADDER. Urine, formed by the kidneys, passes into the *pelves* of the kidneys, shown in Figure 164, and then through the *ureters* to the *urinary bladder*, shown in Figure 172. Urine is forced along the ureters by *peristalsis*, an intermittent wave-like constriction beginning at the pelvis and spreading downward along the ureter toward the bladder. The constriction forces the urine ahead of it. Ordinarily, the urine is transported the entire distance from the pelvis to the bladder in less than one minute.

Occasionally, severe infection or congenital abnormalities destroy the ability of the ureteral wall to contract. As a result, urine begins to collect in the kidney pelvis, causing it to swell and promoting infection that may extend into the kidney. Also, the stagnation of urine may lead to precipitation of crystalline substances, the most prominent of which are various calcium compounds that can grow eventually into large *calculi* or *renal stones* that partially or totally fill the pelvis. These stones in turn often cause extreme pain and obstruction to urine flow.

STORAGE OF URINE IN THE BLADDER. The urinary bladder is a storage reservoir designed to prevent constant dribbling of urine. The exit of urine from the bladder is through the *urethra*, and two muscular sphincters surround the urethra, the *internal*

urethral sphincter which is controlled by the autonomic nervous system and the *external urethral sphincter* which is controlled by the conscious portion of the brain. Ordinarily, both of these sphincters remain contracted so that urine cannot flow out of the bladder except when the person needs to urinate, as is explained below.

The urinary bladder itself can expand from a volume as small as one milliliter to almost a liter. Until the bladder has filled to a volume of 200 to 400 ml., the intrabladder pressure does not increase greatly. This results from the ability of the smooth-muscle bladder wall to stretch tremendously without building up any significant tension in the muscle. However, once the bladder fills beyond 200 to 400 ml., the pressure does begin to rise, and it sometimes reaches as much as 50 mm. Hg when the bladder fills to 600 to 700 ml.

EMPTYING OF THE BLADDER — THE MICTURITION REFLEX. The term *micturition* means emptying of the bladder, and in the human being it is caused by a combination of involuntary and voluntary nervous activity which may be explained as follows: When the volume in the bladder is greater than 200 to 400 ml., special nerve endings in the bladder wall called "stretch receptors" become excited. These transmit nerve impulses through the *visceral afferent nerve pathway* into the spinal cord (see Figure 172), initiating both a conscious desire to urinate and a subconscious reflex called the *micturition reflex.* The nervous centers for the subconscious reflex are in the lower tip of the cord. The efferent impulses are transmitted from here by the parasympathetic nerves to both the bladder wall and the internal urethral sphincter. The bladder wall contracts while the internal sphincter relaxes. Then, the only impediment to urination is the still contracted external urethral sphincter. If the time and place are propitious for urination, the conscious portion of the brain will relax the external sphincter and urination will take place.

If a person wishes to urinate before a micturition reflex has occurred, he can usually initiate the reflex by contracting the abdominal wall, which pushes the abdominal contents down against the bladder and momentarily excites some of the stretch receptors in the bladder wall, thus initiating the reflex.

Many times it is not convenient to urinate when a micturition reflex takes place. In

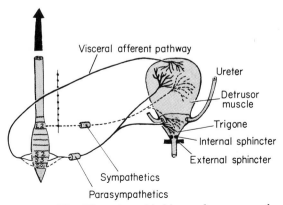

Figure 172. The urinary bladder, and nervous pathways for the micturition reflex.

this case, the reflex usually subsides within a minute or so, and the person loses his desire to urinate. The reflex then remains inhibited for another few minutes to as much as an hour before it returns again. If the reflex is again prevented from causing urination, it will once again become dormant for another few minutes. However, as the bladder becomes more and more filled, the micturition reflex finally becomes so powerful that it is essential to urinate.

Babies, who have not developed voluntary control over the external urethral sphincter, automatically urinate every time the bladder fills. Also, adults whose spinal cords have been severed from the brain cannot keep the external sphincter contracted, so that they, too, urinate automatically when the bladder fills. However, these persons can often initiate this reflex by scratching the genital region, in this way controlling the time that the bladder will empty rather than waiting for the automatic and unannounced emptying.

REFERENCES

Chinard, F. P.: Kidney, water, and electrolytes. *Ann. Rev. Physiol.* 26:187, 1964.

Giebisch, G., and Windhager, E. E.: Characterization of renal tubular transport of sodium chloride and water as studied in single nephrons. *Amer. J. Med.* 34:1, 1963.

Gottschalk, C. W., Lassiter, W. E., Mylle, M., Ullrich, K. J., Schmidt-Nielsen, B., O'Dell, R., and Pehling, G.: Micropuncture study of composition of loop of Henle fluid in desert rodents. *Amer. J. Physiol.* 204:532, 1963.

Kuru, M.: Nervous control of micturition. *Physiol. Rev.* 45(3):425, 1965.

Lotspeich, W. D.: *Metabolic Aspects of Renal Function.* Springfield, Ill., Charles C Thomas, 1959.

Pitts, R. F.: *Physiology of the Kidney and Body Fluids.* Chicago, Year Book Medical Publishers, 1968.

Reubi, F. C.: *Clearance Tests in Clinical Medicine.* Springfield, Ill., Charles C Thomas, 1963.

Smith, H. W.: *The Kidney: Structure and Function in Health and Disease.* New York, Oxford University Press, 1951.

Thurau, K., Valtin, H., and Schnermann, J.: Kidney. *Ann. Rev. Physiol.* 30:441, 1968.

Wesson, L. G., Jr.: *Physiology of the Human Kidney.* New York, Grune & Stratton, 1963.

CHAPTER 18

REGULATION OF BODY FLUID CONSTITUENTS AND VOLUMES

Now that the function of the kidneys has been explained, it is possible to discuss the mechanisms by which most of the body fluid constituents are regulated. The kidneys play a special role in the regulation of (a) electrolyte concentrations of the extracellular fluid, (b) osmotic pressure of all body fluid, and (c) acidity of all the fluid and volumes of both extracellular fluid and blood. In the regulation of acidity, the respiratory system also plays a major role. All of these interrelationships are discussed in this chapter.

Regulation of Sodium Ion Concentration in the Extracellular Fluid

About 90 per cent of the positively charged ions (cations) in the extracellular fluid are sodium. Furthermore, even moderate changes in sodium ion concentration affect such things as strength of cardiac pumping, transmission of nerve impulses, function of the brain, secretion by different glands, and so forth. Therefore, it is extemely important that the sodium ion concentration be regulated very exactly. Ordinarily its concentration does not vary more than 5 milliequivalents above or below a mean value of 142 milliequivalents per liter.

Figure 173 illustrates the overall mechanism by which sodium ion concentration in the extracellular fluids is regulated. The hormone *aldosterone* secreted by the adrenal cortex acts on the renal tubules to cause increased reabsorption of sodium. This obviously conserves sodium in the extracellular fluids so that the extracellular fluid concentration of sodium increases as the person eats salt in his diet. Conversely, if the quantity of aldosterone is reduced, increased amounts of sodium are lost into the urine so that the extracellular fluid concentration of sodium decreases.

The rate of secretion of aldosterone by the adrenal cortex is controlled by the sodium concentration in the extracellular fluids. The exact mechanism by which sodium exerts this effect on aldosterone secretion is unknown, but there are three different theories to explain the effect.

Some physiologists believe that sodium deficiency causes the kidneys to release a humoral factor, *angiotensin,* which then

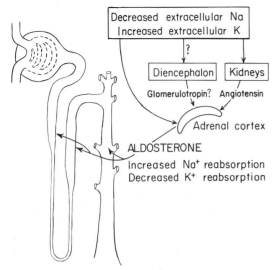

Figure 173. Postulated mechanisms for control of sodium and potassium concentration in the extracellular fluid.

stimulates the adrenal cortex to produce aldosterone.

Others believe that sodium deficiency causes structures in the midbrain near the pineal gland to secrete another humoral substance, *glomerulotropin*, that in turn stimulates the glomerular portion of the adrenal cortex to secrete aldosterone.

Finally, it is known that sodium deficiency has a direct effect on the adrenal cortex to increase aldosterone secretion, but it is still doubtful whether this effect is enough to cause the amount of aldosterone secretion that is known to occur.

Now, let us recapitulate the overall mechanism for control of sodium concentration in the body fluids. If the sodium concentration becomes too low, the adrenal cortex secretes increased quantities of aldosterone. The aldosterone in turn promotes increased reabsorption of sodium so that the extracellular fluid concentration of sodium rises toward normal.

It is immediately recognized that this is a typical feedback regulatory mechanism. That is, the initiating stimulus, the sodium concentration, sets off a series of reactions that returns the sodium concentration to its normal mean level.

Regulation of Other Electrolytes in the Extracellular Fluid

The renal tubules also have the ability to regulate the concentrations of all the other electrolytes in the extracellular fluid as well as sodium, though the mechanisms of these other regulations are not nearly so well understood.

Potassium concentration is mainly regulated as a secondary effect of the regulation of sodium concentration. That is, when aldosterone increases the reabsorption of sodium, it simultaneously decreases potassium reabsorption. Furthermore, increased potassium concentration stimulates the adrenal cortex to secrete aldosterone in the same way that low sodium concentration stimulates it. Thus, at the same time that the aldosterone mechanism helps the body to retain sodium, it simultaneously helps the extracellular fluid rid itself of potassium, thereby keeping the potassium concentration in the extracellular fluid low.

In addition to the aldosterone mechanism for potassium regulation, still another kidney mechanism also helps in potassium regulation: The distal tubules and collecting ducts of the kidneys secrete potassium into the urine whenever the potassium concentration becomes too high in the extracellular fluid; this effect presumably results from a direct stimulatory effect of high potassium concentration on the tubular epithelial cells.

Chloride and *bicarbonate* ion concentrations are also regulated mainly secondarily to the regulation of sodium concentration. When sodium is reabsorbed from the tubules, this transfers positively charged ions out of the tubular fluid into the interstitial fluids of the kidneys, creating a state of electronegativity in the tubules and electropositivity in the interstitial fluids. The negativity in the tubules repels negative ions from the tubules while the positivity of the interstitial fluids attracts the negative ions. Therefore, when sodium ions are reabsorbed from the tubules, chloride and bicarbonate ions are also reabsorbed. Sometimes more chloride than bicarbonate ions are reabsorbed while at other times more bicarbonate ions are reabsorbed; which one it will be is determined by the acid-base balance of the extracellular fluid, as is discussed later in the chapter.

The details of the mechanisms by which calcium, magnesium, and phosphate concentrations are regulated by the kidneys are not very clear. Yet, too high a concentration of any one of these substances in the extracellular fluid causes the tubules to reject it and to pass it on into the urine. On the other hand, a low concentration causes the opposite effect, that is, rapid reabsorption of the substance until its concentration in the extra-cellular fluids returns to normal.

Regulation of Osmotic Pressure of the Extracellular Fluid

THE "COUNTER-CURRENT" MECHANISM OF THE KIDNEY.   Before it is possible for us to explain the mechanism by which osmotic pressure of the extracellular fluid is regulated, we must describe the means by which the kidneys can form either a concentrated or a dilute urine. Figure 174 illustrates the so-called *counter-current mechanism* of the kidney that is responsible for the ability of the kidney to concentrate the urine. This can be explained as follows: The kidneys are divided into two major portions, the *cortex* and the *medulla*. The glomeruli and the proximal and distal tubules are located in the cortex. On the other hand, the *loop of Henle*, which is a portion of the tubule system, extends deeply into the medulla. Also, the *collecting tubule* passes through the medulla before it empties into the renal pelvis. The arrangement of the *peritubular capillaries* in the medulla is also peculiar: These form long loops called the *vasa recta*, so that the blood passes downward into the medulla and then back upward to the cortex before emptying into the veins.

Keeping in mind these anatomic details and referring again to Figure 174, we can now explain the counter-current mechanism. As the tubular fluid passes through the loop of Henle, sodium chloride is actively absorbed into the interstitial fluids of the medulla. The concentration of sodium chloride in the interstitial fluids becomes about three times as great as that of the usual extracellular fluid. The loop arrangement of the vasa recta prevents the blood from carrying large quantities of this sodium chloride away from the interstitial fluids in the following manner: As the blood flows downward in the loop, sodium chloride diffuses rapidly into the blood, from the peritubular fluid. Its concentration rises very rapidly to 3 times normal by the time the blood has reached the bottom of the loop. Then, as the blood flows upward and out of the medulla, most of this sodium chloride diffuses back out of the blood into the peritubular fluid. Thus, the "counter-current" flow of blood in the vasa recta allows most of the sodium chloride that is absorbed into the blood to leave the blood again, thus preventing excessive loss of sodium chloride out of the medulla.

Now, keep in mind the fact that the interstitial fluid in the medulla has a very high concentration of sodium chloride. As the tubular fluid passes through the loop of Henle large quantities of electrolytes are usually actively reabsorbed by the tubular

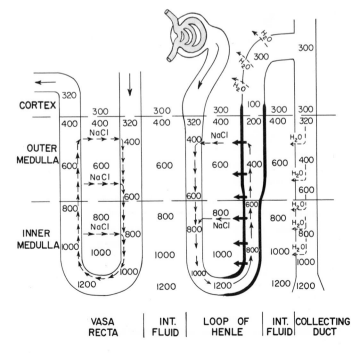

Figure 174. The "counter-current" mechanism by which the kidneys concentrate the urine.

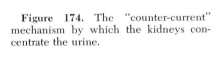

epithelium, so that the tubular fluid is dilute when it empties into the distal tubule. If the fluid passes without further change through the distal tubule and collecting tubule into the renal pelvis, the urine will be dilute. However, if large quantities of water should be reabsorbed from this fluid in the collecting tubule, the urine would become highly concentrated. Thus, if we should simply have a means for increasing or decreasing the amount of water reabsorbed from the collecting tubule, a mechanism would exist for changing the degree of concentration of the urine. Such a mechanism is the following:

The epithelial lining of the distal tubule and collecting tubule has a variable permeability to water. A hormone released from the neurohypophysis, called *antidiuretic hormone*, causes the pores of these tubules to become extremely permeable to water, but, on the other hand, when antidiuretic hormone is not present, the pores become almost entirely impermeable to water. Therefore, in the absence of antidiuretic hormone the dilute fluid entering the distal tubule from the loop of Henle passes all the way through the collecting tubules into the renal pelvis, forming a dilute urine. On the other hand, *in the presence of antidiuretic hormone, most of the water is absorbed by osmosis as the tubular fluid passes down the collecting tubule.* Loss of this water concentrates the fluid so that the urine also becomes highly concentrated.

It is the highly concentrated interstitial fluid of the medulla that causes rapid osmosis of water from the collecting tubule when the epithelial pores are opened by antidiuretic hormone. Therefore, we can now see the importance of the counter-current mechanism which was responsible in the first place for developing a high concentration of sodium chloride in the medulla, thereby creating an osmotic force that can concentrate the urine.

In the absence of antidiuretic hormone the urine is diluted to a concentration about one-fourth that of the extracellular fluid, while in the presence of antidiuretic hormone, the urine is concentrated to about three times the concentration of the extracellular fluid.

FUNCTION OF THE OSMORECEPTOR SYSTEM IN REGULATING THE EXTRACELLULAR OSMOTIC PRESSURE. The complex renal mechanisms for diluting and concentrating the body fluids would not be of any importance if the body did not have some mechanism for increasing or decreasing the rate of secretion of antidiuretic hormone. Figure 175 illustrates the *osmoreceptor control system* that controls antidiuretic hormone secretion; this can be explained as follows: Located in the *supraoptic nuclei* and close thereby in the hypothalamus of the brain are a number of specialized neurons called *osmoreceptors.* These are postulated to have small fluid chambers that swell when the body fluids become too dilute and contract when the body fluids become too concentrated. In the contracted state the osmoreceptors emit impulses that pass downward through nerve fibers into the neurohypophysis and cause release of *antidiuretic hormone.* The antidiuretic hormone then passes by way of the blood to the kidneys where it causes increased reabsorption of water from the collecting tubules.

Now, let us recapitulate the mechanism

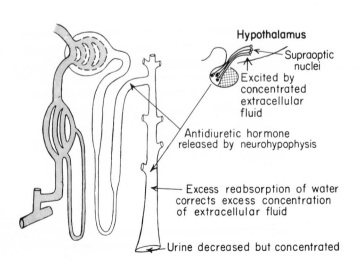

Hypothalamus

Supraoptic nuclei

Excited by concentrated extracellular fluid

Antidiuretic hormone released by neurohypophysis

Excess reabsorption of water corrects excess concentration of extracellular fluid

Urine decreased but concentrated

Figure 175. Osmoreceptor system for regulating crystalloidal osmotic pressure in the extracellular fluid.

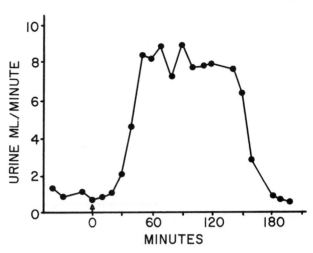

Figure 176. Water diuresis caused by the osmoreceptor system. (Redrawn from Smith: *The Kidney: Structure and Function in Health and Disease.* Oxford University Press.)

by which the osmotic pressure of the extracellular fluids is regulated. If the concentration of the fluids becomes too great, the osmoreceptors contract and emit impulses to the neurohypophysis. As a result, antidiuretic hormone is released which causes the kidneys to reabsorb excessive amounts of water while letting solutes continue to pass into the urine. As a result, solutes are lost while water is retained in the body. Thus, the concentration of the body fluids becomes diluted, returning toward normal. Conversely, if the concentration becomes dilute, the osmoreceptors stop emitting impulses, the neurohypophysis stops releasing antidiuretic hormone, and the kidneys excrete a very dilute urine, thereby causing loss of water from the body and concentration of the fluids back toward normal.

Water diuresis. Figure 176 demonstrates the function of the osmoreceptor system. The curve depicts the rate of urine output per minute. At zero minutes, the person drank several glasses of water. In approximately 20 minutes the water was absorbed into the extracellular fluid, diluting it. At this point the osmoreceptor system caused the kidneys to release excessive quantities of very dilute urine. This effect is called *water diuresis.* After an hour or more of losing the dilute urine, composed mainly of water, the concentration of the extracellular fluid had returned to normal. Therefore, urine flow and urine concentration also returned to normal.

Regulation of Acid-Base Balance

Regulation of acid-base balance actually means regulation of hydrogen ion concentration in the body fluids. When the hydrogen ion concentration is great, the fluids are *acidic;* when the hydrogen ion concentration is slight, the fluids are *basic.* The chemical reactions of the cells depend very greatly on the hydrogen ion concentration, which is the reason the acid-base balance must be regulated very exactly.

The normal concentration of hydrogen ions in the body fluids is 4×10^{-8}. Usually this is expressed in terms of *pH*, which is the logarithm of the reciprocal of the hydrogen ion concentration. The normal pH of extracellular fluid and blood, therefore,

$$\left(\text{the logarithm of } \frac{1}{4 \times 10^{-8}}\right)$$

is 7.4. A pH of less than 7.4 is on the acidic side, and a pH greater than 7.4 is on the basic side. The acid-base regulatory systems of the body are all geared toward maintaining a normal pH of approximately 7.4 in the extracellular fluid. Even in disease conditions it almost never becomes more acidic than 7.0 or more basic than 7.8.

REGULATION OF ACID-BASE BALANCE BY CHEMICAL BUFFERS. All of the body fluids contain *acid-base buffers*. These are chemicals that can combine readily with any acid or base in such a way that they keep the acid or base from changing the pH of the fluids greatly. One of the chemical systems that performs this function is the *bicarbonate buffer,* which is present in all body fluids. The bicarbonate buffer is a mixture of carbonic acid (H_2CO_3) and bicarbonate ion (HCO_3^-). When a strong acid is added to this mixture it combines immediately with the bicarbonate ion to form carbonic acid. Carbonic acid is an extremely weak acid; therefore, the buffer system changes the strong acid into a weak one and keeps the fluids from becoming strongly acid. On the other hand, when a strong base is added to this mixture, the base immediately combines with the carbonic acid to form water and a neutral bicarbonate salt. Loss of the weak acid and the addition of the neutral salt hardly affect the hydrogen ion concentration in the body fluids. Thus, it can be seen that this mixture of carbonic acid and bicarbonate ion protects the body fluids from becoming either too acidic or too basic.

Other important buffers are *phosphate* and *protein buffers.* These are especially important for maintaining normal hydrogen ion concentrations in the intracellular fluids, because their concentrations inside the cells are many times as great as the concentration of the bicarbonate buffer.

In essence, the buffers of the body fluids are the first line of defense against changes in hydrogen ion concentration, for any acid or base added to the fluids immediately reacts with these buffers to prevent marked changes in the acid-base balance.

RESPIRATORY REGULATION OF ACID-BASE BALANCE. Carbon dioxide is continually formed by all cells of the body as one of the end products of metabolism. This carbon dioxide combines with water to form carbonic acid in accordance with the following reaction:

$$CO_2 + H_2O \leftrightarrows H_2CO_3$$

In other words, even the normal metabolic processes are always pouring acids into the body fluids. On the other hand, one of the functions of respiration is to expel carbon dioxide through the lungs into the atmosphere. Normally, respiration removes carbon dioxide at the same rate it is being formed. If, however, pulmonary ventilation decreases below normal, carbon dioxide will not be expelled normally but instead will pile up in the body fluids, causing the concentration of carbonic acid to increase also. As a result, the hydrogen ion concentration rises. On the other hand, if the rate of pulmonary ventilation rises above normal, the opposite effect occurs; carbon dioxide is blown off at a more rapid rate than it is being formed, thereby decreasing the carbon dioxide and carbonic acid concentrations. Complete lack of breathing for a minute reduces the pH of the extracellular fluid from the normal level of 7.4 down to about 7.1, while very active overbreathing can increase it to 7.7 in about

one minute. Thus the acid-base balance of the body can be changed greatly by over- or under-ventilation of the lungs.

Control of respiration by the hydrogen ion concentration of the body fluid. In the preceding paragraph the effect of respiration on the acid-base balance was discussed. In this paragraph the opposite effect, that of acid-base balance on respiration, is discussed. A high hydrogen ion concentration stimulates the respiratory center in the medulla of the brain, greatly enhancing the rate of ventilation. Conversely, a low hydrogen ion concentration depresses the rate of ventilation. This effect of hydrogen ion concentration on the activity of the respiratory center affords an automatic mechanism for maintaining a fairly constant pH of the body fluids. That is, an increase in hydrogen ion concentration increases the rate of ventilation, which in turn removes carbonic acid from the fluids. Loss of the carbonic acid decreases the hydrogen ion concentration back toward normal.

This respiratory mechanism for regulating acid-base balance reacts almost immediately when the extracellular fluids become either too acidic or too basic. Figure 177 illustrates the effect on the respiration of adding first a small amount of acid to the blood, and then later a small amount of alkali. The up and down curves represent the depth of respiration, and the rapidity of the curves represents the frequency of respiration. Note that acidosis greatly increases both the depth and rate of respiration, while alkalosis depresses respiratory function greatly so that the depth of respiration becomes very slight and the rate very slow. This respiratory mechanism is so effective for regulating acid-base balance that it usually can return the pH of the body fluids almost to normal within a few minutes after an acid or alkali has been administered.

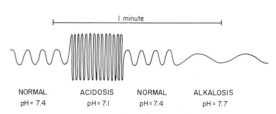

| NORMAL | ACIDOSIS | NORMAL | ALKALOSIS |
| pH = 7.4 | pH = 7.1 | pH = 7.4 | pH = 7.7 |

Figure 177.   Effect of acidosis and alkalosis on respiration.

RENAL REGULATION OF ACID-BASE BALANCE. A number of other acids besides carbonic acid are continually being formed by the metabolic processes of the cells, and can be eliminated from the body only by the kidneys. These include large quantities of phosphoric acid and small quantities of sulfuric, uric, and keto acids, all of which, on entering the extracellular fluids, can cause acidosis. The kidneys normally rid the body of these excess acids as rapidly as they are formed, preventing an excessive build-up of hydrogen ions. These acids are generally called *metabolic acids,* in contradistinction to carbonic acid, which is called a *respiratory acid.*

In rare instances too many basic compounds enter the body fluids rather than too many acidic compounds. In general, this occurs when basic compounds are injected intravenously or when the person ingests a large quantity of alkaline foods or drugs.

Mechanisms by which the kidneys regulate acid-base balance. The kidneys regulate acid-base balance by (1) excreting hydrogen ions into the urine when the extracellular fluids are too acidic and (2) excreting basic substances, particularly sodium bicarbonate, into the urine when the extracellular fluids become too alkaline.

Excretion of hydrogen ions and reabsorption of sodium. The distal tubular epithelium continually secretes hydrogen ions into the tubular fluid. These ions combine mainly with sodium salts in the tubular fluid to form weak acids and to free the sodium ion that is bound in the salt. The sodium in turn is absorbed through the tubular wall into the body fluids. Thus, there is a net exchange of hydrogen for sodium ions, the hydrogen ions passing into the urine and the sodium being reabsorbed. The precise manner in which this mechanism is used to regulate acid-base balance is quite complex, but it can be paraphrased in a few simple sentences as follows:

Ordinarily, the amount of hydrogen ions secreted into the tubules is almost exactly equal to the quantity of bicarbonate ions in the tubular fluid. The hydrogen ions can be considered to be an acidic substance and the bicarbonate ions an alkaline substance. The combination of the hydrogen ions and the bicarbonate ions forms carbonic acid which in turn splits into carbon dioxide and water. The carbon dioxide is reabsorbed from the tubules into the body fluids and then is expelled through the lungs. Thus, we have an acidic substance and a basic substance both being destroyed, and the overall acid-base balance does not change when equal amounts of acidic and basic substance are destroyed. However, if the extracellular fluids become very acidic, the quantity of bicarbonate ions in the glomerular filtrate decreases. Therefore, now, far more hydrogen ions are secreted into the tubules than there are bicarbonate ions available to react with them. As a result, the hydrogen ions combine with the buffers of the tubular fluid and are carried into the urine. Thus, the acidic hydrogen ions are lost from the body fluids, returning the fluids to normal acid-base balance.

On the other hand, if the extracellular fluids become too alkaline, the quantity of bicarbonate ions in the glomerular filtrate becomes far greater than the quantity of hydrogen ions secreted by the tubules. The excess bicarbonate ions pass on into the urine combined mainly with sodium in the form of sodium bicarbonate. Therefore, a large amount of base is lost so that the body fluids become more acidic, returning the acid-base balance again back toward normal.

Ammonia secretion. Sometimes the body fluids become so acidic that it is impossible for the buffers in the tubules to carry all of the secreted hydrogen ions into the urine. When this happens, the distal tubules begin to secrete ammonia. The ammonia combines with the hydrogen ions to form ammonium ions, which displace sodium from the buffers and allow the sodium to be reabsorbed into the extracellular fluids. The ammonium ions then pass on into the urine combined in the form of a salt.

In summary, the renal mechanisms for regulating acid-base balance remove hydrogen ions from the extracellular fluids when the hydrogen ion concentration becomes too great, and they add hydrogen ions while removing sodium ions when the hydrogen concentration becomes too low. This principle is illustrated in Figure 178 which shows at point A a pH of about 7.5 in the extracellular fluid. Because this is on the alkaline side, the pH of the urine becomes alkaline because of loss of alkaline substances from the body fluids. On the other hand, at point B the pH has fallen to 7.3, and the pH of the urine has become very acidic because of loss of large quantities of acidic substances from the body fluids. In both of these instances

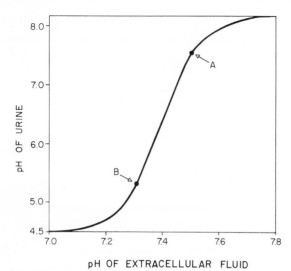

Figure 178. Effect of extracellular fluid pH on urine pH.

the loss of alkaline or acidic substances returns the pH toward normal.

ABNORMALITIES OF ACID-BASE BALANCE. Many disorders of the respiratory system, the kidneys, or the metabolic systems for forming acids and bases can cause serious derangement of the acid-base balance. Some of the effects of these conditions are shown in Figure 179. The normal pH of the blood is 7.4, but intense over-ventilation can cause the pH to rise sometimes to as high as 7.8. On the other hand, asphyxia, which means extreme decrease in ventilation, causes a build-up of carbon dioxide and carbonic acid in the body fluids, thereby promoting

acidosis in which the pH of the blood sometimes falls to as low as 7.0.

A common cause of alkalosis is ingestion of alkaline drugs used for treatment of gastritis or stomach ulcers. The drugs sometimes are absorbed into the body fluids in quantities greater than the kidneys can remove, resulting in alkalosis.

Loss of large quantities of fluids from the intestinal tract at times causes acidosis but at other times alkalosis. For instance, vomiting large quantities of hydrochloric acid from the stomach gradually depletes the acid reserves of the extracellular fluid and causes alkalosis. On the other hand, vomiting of fluids from the lower intestine or loss of fluids as a result of diarrhea usually causes acidosis. This is because secretions from the lower intestine contain large quantities of sodium bicarbonate. When this basic salt is lost from the body it is immediately replaced by carbonic acid formed by the reaction of carbon dioxide with water. Therefore, the net effect is loss of sodium ions and gain of hydrogen ions, causing acidosis.

Finally, Figure 179 shows extreme acidosis that sometimes results from severe diabetes mellitus. In diabetes many of the fats normally used by the body for energy are not completely metabolized but instead are broken into substances called *keto acids* that build up in the body fluids, producing extreme acidosis.

Effects of acidosis and alkalosis on bodily functions. Acidosis generally promotes depressed mental activity culminating in coma and death. Usually the afflicted person will

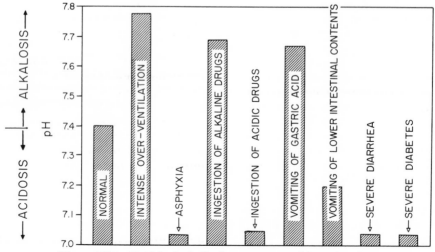

Figure 179. pH of the body fluids in various acid-base disorders.

pass into coma when the pH of his extracellular fluid falls below approximately 6.9.

On the other hand, alkalosis causes overexcitability of the nervous system, often resulting in excessive initiation of impulses. These can cause tetanic contraction of all the muscles or they can actually kill a person by causing convulsions or other derangements of nervous activity.

Regulation of Blood Volume

The normal blood volume is almost exactly 5000 ml., and it rarely rises or falls more than a few hundred milliliters from this value. Two principal mechanisms for maintaining this constancy are (1) the capillary fluid shift mechanism and (2) the kidney mechanism.

THE CAPILLARY FLUID SHIFT MECHANISM. When the blood volume becomes too great, the pressures increase in all the vessels throughout the body, including the capillaries. The normal pressure in the capillaries is about 18 mm. Hg, as illustrated in Figure 180. When capillary pressure rises above this value, fluid automatically leaks into the tissue spaces as we explained in Chapter 16. Thus, the blood volume decreases back toward normal. When the capillary pressure has returned to normal, further loss of fluid into the tissue spaces ceases. Conversely, when the blood volume falls too low, the capillary pressure falls, and fluid is then absorbed from the interstitial spaces, increasing the blood volume again back toward normal.

THE KIDNEY MECHANISM. As was explained in the previous chapter, when the glomerular pressure in the kidneys rises, the amount of glomerular filtrate and consequently the amount of urine formed by the kidneys greatly increase. Figure 180 also shows that the normal glomerular pressure is about 70 mm. Hg. An increased blood volume raises this to a higher level by two different mechanisms: First, the increased volume increases the arterial pressure, which causes increased flow of blood through the afferent arterioles into the glomeruli, thus raising their pressure. Second, the increased vascular pressure stretches the blood vessels in the thorax and neck that are supplied with stretch receptors called *baroreceptors.* Stretching these initiates a nervous reflex, as explained in Chapter 14, that allows the renal afferent arterioles to dilate, thus increasing the blood flow into the kidney and correspondingly increasing the amount of urine formed.

One will immediately recognize both of these effects to be feedback mechanisms by which the blood volume can be regulated; that is, an increase in blood volume initiates an increase in urinary output, which automatically decreases the blood volume back toward normal.

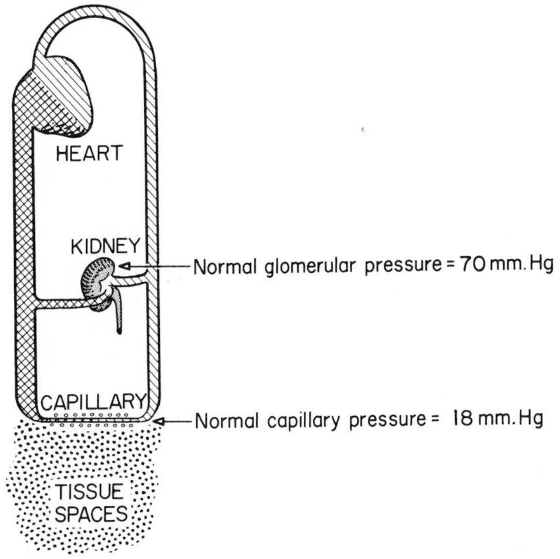

Regulation of Extracellular Fluid Volume

One can understand the regulation of extracellular fluid volume best by first recalling from Chapter 16 the mechanism by which negative interstitial fluid pressures develop throughout the body. It will be recalled that the colloid osmotic pressure of the blood is ordinarily considerably greater than the capillary pressure, which causes a tendency for fluid always to be absorbed from the interstitial spaces. This creates a negative pressure in the interstitial spaces of about −6 mm. Hg.

Therefore, in effect, the interstitial fluid volume is normally regulated to just that minimum volume that is necessary to fill the interstitial spaces. If ever the interstitial fluid volume becomes too great, the inter-

Figure 180. The capillary and kidney systems for regulating blood volume.

HEART

KIDNEY

—Normal glomerular pressure = 70 mm. Hg

CAPILLARY

—Normal capillary pressure = 18 mm. Hg

TISSUE SPACES

stitial pressure rises markedly so that fluid is absorbed into the blood far more easily than usual, which increases the blood volume. The increased blood volume in turn causes the kidneys to form increased amounts of urine. Thus, in effect, the interstitial fluid volume is lost into the urine by way of the blood.

Conversely, if the interstitial fluid volume falls too low, which decreases the interstitial fluid pressure below the normal value of −6 mm. Hg, fluid then leaks out of the blood to replenish the interstitial fluids. As a result, the blood volume falls too low, the rate of urine formation becomes greatly decreased, and increased amounts of water and salts are retained to build up the blood volume back to normal.

Regulation of Intracellular Fluid Volume

Each cell is autonomous in regulating its own fluid volume. It automatically transports those electrolytes from the extracellular fluid that are needed for intracellular function and these in turn pull water into the cell by osmosis. Thus, because osmotic equilibrium between the intracellular and extracellular fluid compartments is maintained at all times, as was explained in Chapter 16, the fluid volume of each cell is directly proportional to the quantity of electrolytes pulled into the cell by its metabolic processes. Therefore, the regulation of intracellular fluid volume, unlike the regulation of the extracellular fluid volume, is not a direct function of the kidneys; yet, indirectly, the kidneys help in this regulation by maintaining appropriate constituents in the extracellular fluids.

Thirst, and Appetite For Electrolytes

In our discussion thus far, we have assumed that there is a constant intake of fluid and electrolytes and that regulation of body fluid constituents is carried out by increasing or decreasing the quantity of water or electrolytes lost into the urine. However, this is not entirely true, because a person's intake of both water and electrolytes is regulated also to a very great extent by the thirst and appetite mechanisms. Both of these are controlled by special centers in the hypothalamus of the brain. We know much about the mechanism that controls thirst but unfortunately very little about those that control appetite for the electrolytes, except for the following well known facts: First, animals or human beings living far from regions where salt is plentiful have a definite craving for salt. Indeed, animals will seek out salt deposits called *salt licks* and will even fight for the salt just as they will fight for food. We know also that human beings, under special conditions, such as the pregnant mother, crave excessive amounts of salt and at times even other types of electrolytic substances such as sour pickles (acidic), or even calcium salts. Yet, the usual human being, living in our modern society of plenty, has such an abundance of the different types of salt in his diet that his appetite for a specific salt is almost never a significant factor in the regulation of the electrolyte content of his body fluids. On the other hand, the thirst mechanism does play an exceedingly important role in the hour by hour regulation of body fluid volumes and osmotic pressure.

THE DRINKING CENTER OF THE BRAIN. Figure 181 illustrates a small area located in the hypothalamus a few millimeters behind the supraoptic nuclei, an area called the *drinking center*. When this center is stimulated electrically, the animal searches for the nearest water and begins to drink profusely. Also, injection of a concentrated solution of electrolytes into this center will likewise cause the animal to begin drinking.

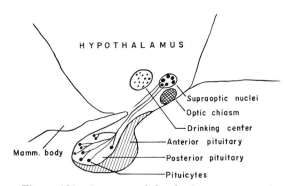

HYPOTHALAMUS

Supraoptic nuclei
Optic chiasm
Drinking center
Anterior pituitary
Mamm. body
Posterior pituitary
Pituicytes

Figure 181. Location of the drinking center in the hypothalamus in relation to the supraoptic nuclei which control the antidiuretic hormonal system.

One will note the close proximity of the drinking center to the osmoreceptors located in the supraoptic nuclei. When the body fluids become too concentrated, they excite both the drinking center and the osmoreceptors at the same time. The drinking center causes the animal to drink water, and the osmoreceptors cause the kidneys to conserve water.

REFERENCES

Albert, S. N.: *Blood Volume.* Springfield, Ill., Charles C Thomas, 1962.

Berliner, R . W.: Outline of renal physiology. In Strauss, M. B., and Welt, L. G. (eds.): *Diseases of the Kidney.* Boston, Little, Brown & Company, 1963.

Brown, E. B., Jr., and Goott, B.: Intracellular hydrogen ion changes and potassium movement. *Amer. J. Physiol.* 204:765, 1963.

Christensen, H. N.: *Body fluids and the Acid-Base Balance.* Philadelphia, W. B. Saunders Company, 1964.

Davenport, H. W.: *The ABC of Acid-Base Chemistry.* 4th Ed. Chicago, University of Chicago Press, 1958.

Merrill, J. P.: The artificial kidney. *Sci. Amer.* 205(1): 56, 1961.

Reeve, E. B., Allen, T. H., and Roberts, J. E.: Blood volume regulation. *Ann. Rev. Physiol.* 22:349, 1960.

Schmidt-Nielsen, B., and Laws, D. F.: Invertebrate mechanisms for diluting and concentrating the urine. *Ann. Rev. Physiol.* 25:631, 1963.

Selkurt, E. E.: The renal circulation. *Handbook of Physiology,* Sec. II, Vol. II, p. 1457. Baltimore, The Williams & Wilkins Company, 1963.

Weisberg, H. F.: *Water, Electrolyte, and Acid-Base Balance.* 2nd Ed. Baltimore, The Williams & Wilkins Company, 1962.

Wolf, A. V.: *Thirst: Physiology of the Urge to Drink and Problems of Water Lack.* Springfield, Ill., Charles C Thomas, 1958.

SECTION FIVE

RESPIRATION

CHAPTER 19

MECHANICS OF RESPIRATION AND TRANSPORT OF OXYGEN AND CARBON DIOXIDE

The function of the respiratory system is, first, to supply oxygen to the blood and, second, to remove carbon dioxide. Figure 182 illustrates the principal structures of this system, showing the lungs, trachea, glottis, and nose. The lungs contain millions of small air sacs, called *alveoli,* connected by the bronchioles and trachea with the nose and mouth. With each intake of breath the alveoli expand, and during expiration air is forced out of the alveoli again to the exterior. Thus, there is continual renewal of air in the alveoli, a process called *pulmonary ventilation.* Later in the Chapter, in Figure 191, we will see more details of the structure of these terminal air sacs in the lungs.

To the upper right in Figure 182 is illustrated the functional relationship of an alveolus to a pulmonary capillary. Each alveolus has on all its sides a network of capillaries, and the membrane between the air in the alveolus and the blood in the pulmonary capillary is so thin that oxygen can diffuse into the blood with extreme ease

and carbon dioxide out with even greater ease. Therefore, the role of the basic structure of the lungs is simply to aerate the blood, and to allow replenishment of oxygen and removal of carbon dioxide. And the object of breathing is to move air continually into and out of the alveoli.

FUNCTIONS OF THE RESPIRATORY PASSAGEWAYS

Function of the Nose

The nose is not merely a passageway for movement of air into the lungs, but it also preconditions the air in several ways, including (a) warming the air, (b) humidifying the air, and (c) cleansing the air. These functions may be explained as follows:

The inside surface of the nose is extensive. The nasal cavity is divided by a central

221

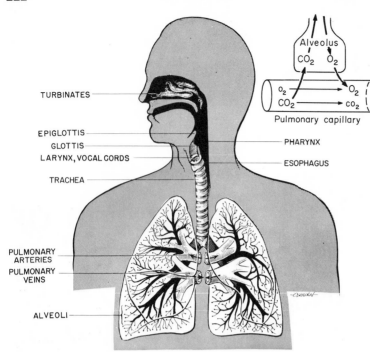

TURBINATES

EPIGLOTTIS
GLOTTIS
LARYNX, VOCAL CORDS

TRACHEA

PULMONARY
ARTERIES
PULMONARY
VEINS

ALVEOLI

Alveolus
CO_2 O_2

Pulmonary capillary
O_2
CO_2
O_2
CO_2

PHARYNX

ESOPHAGUS

Figure 182. The respiratory system, showing the respiratory passages and function of the alveolus to oxygenate the blood and to remove carbon dioxide.

septum, and several projections called *turbinates* extend into each cavity from the lateral side. Air passing through the nose, on coming in contact with all of the nasal surfaces, becomes warmed and humidified. The turbinates also cause many eddies in the flowing air, forcing it to rebound in many different directions before finally completing its passage through the nose. This causes small dust or other suspended particles in the air to precipitate against the nasal surfaces by the following mechanism: When air containing a foreign particle travels toward a surface and then suddenly changes its direction of movement, the momentum of the particle makes it continue to travel in the original direction, while the air, which has little momentum because of its low mass, takes off in the new direction. The particles impinge on a turbinate or on another surface of the nasal passageways and are entrapped in the layer of mucus covering the surface. Then ciliated epithelial cells, whose cilia protrude into the mucus and beat toward the pharynx, slowly move the mucus and entrapped particles into the throat to be swallowed. This method of removing foreign particles from the air is so efficient that very rarely do particles of greater size than 4 to 6 microns pass through the nose into the lower respiratory passageways.

Functions of the Pharynx and Larynx

The *pharynx*, which is commonly called the throat, separates into the trachea and esophagus immediately above the larynx. Here food is separated from air, the air passing into the *trachea* while the food passes into the *esophagus*. This separation of food and air is controlled by local nerve reflexes. Whenever food touches the surface of the *pharynx*, the *vocal cords* close together and the *epiglottis* automatically closes over the opening of the trachea, allowing the food to slide on into the esophagus. When air passes into the pharynx, the epiglottis and vocal cords remain open, and the air takes the course of least resistance into the trachea and lungs.

FUNCTION OF THE VOCAL CORDS. The vocal cords are the portion of the larynx that makes sound. The *vocal cords* are two small vanes located on either side of the air passageway, as shown in Figure 183. Contraction of muscles in the larynx can bring these vanes close to each other or can spread them apart. They can also be stretched or relaxed, and their edges can be flattened or thickened by muscles actually in the cords themselves. When the cords are together and air is forced between them, they vibrate

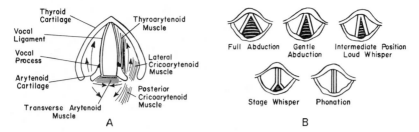

Figure 183. Laryngeal function in phonation. (Modified from Greene: *The Voice and Its Disorders.* Pitman Medical Publishing Co.)

to generate sound, and the different *pitches* of sound are controlled by the degree to which the cords are stretched and by the degree of flattening or thickening of the vocal cord edges. The formation of words or other complicated sounds is a function of the mouth as well as the larynx because the *quality* of a sound depends upon the momentary position of the lips, cheeks, teeth, tongue, and palate.

For speech or other sounds to be emitted, the respiration, the vocal cords, and the mouth must all be controlled at the same time. This is effected by a special brain center, called *Broca's area*, located in the left frontal lobe of the brain. The organization and function of this center will be discussed in Chapter 27.

The Cough and Sneeze Reflexes

A means for keeping the respiratory passages clean is to force air very rapidly outward by either coughing or sneezing. The cough reflex is initiated by any irritant touching a surface of the glottis, the trachea, or a bronchus. Sensory impulses are transmitted to the medulla of the brain, and, in turn, appropriate impulses are transmitted back to the respiratory system and larynx to cause the cough. The respiratory muscles first contract very strongly, building up high pressure in the lungs, while at the same time the vocal cords remain clamped tightly closed. Then suddenly the vocal cords open, allowing the pressurized air to flow with a blast out of the lungs. Sometimes the air flows as rapidly as 70 miles per hour. In this way unwanted foreign matter such as particles, mucus, or other substances are expelled from the respiratory passageways.

The sneeze reflex is very similar to the cough reflex except that it is initiated by irritants in the nose. Impulses pass from the nose to the medulla and then back to the respiratory system. A sudden forceful expiration blows air outward while the soft palate varies its position to allow rapid flow of air successively through the nose and mouth. In this way the sneeze is capable of clearing the nasal passageways in the same manner that the cough reflex clears many of the lower passageways.

FLOW OF AIR INTO AND OUT OF THE LUNGS

The lungs are enclosed in the *thoracic cage,* which is comprised of the *sternum* in front, the *spinal column* in back, the *ribs* encircling the chest, and the *diaphragm* below. The act of breathing is performed by enlarging and contracting the thoracic cage. The cavity formed by the thoracic cage is called the *pleural cavity,* and the lungs normally fill this cavity entirely. The lungs are covered with a lubricated membrane called the *visceral pleura,* and the inside of the pleural cavity is lined with a similar membrane called the *parietal pleura.* The lungs slide freely inside the pleural cavity, so that any time the cavity enlarges, the lungs must also enlarge. In other words, any change in the volume of the thoracic cage is immediately reflected by a similar change in the volume of the lungs.

THE MUSCLES OF RESPIRATION. *The inspiratory muscles.* Figure 184 illustrates several of the muscles of respiration, and also the changes in shape of the thoracic cage during inspiration and expiration. The major muscles of inspiration are the *diaphragm,* the *external intercostals,* and a number of small muscles in the neck which pull upward on the front of the thoracic cage. The inspiratory muscles cause the pleural cavity to en-

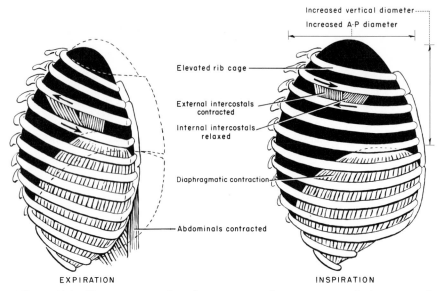

Increased vertical diameter
Increased A·P diameter
Elevated rib cage
External intercostals contracted
Internal intercostals relaxed
Diaphragmatic contraction
Abdominals contracted

EXPIRATION   INSPIRATION

Figure 184. The major muscles of expiration and inspiration, and changes in the thoracic cage during expiration and inspiration.

large in two ways. First, downward movement of the diaphragm pulls the bottom of the pleural cavity downward, thus elongating it, which is shown to the right in Figure 184. Second, the external intercostals and neck muscles lift the front of the thoracic cage, causing the ribs to angulate more directly forward than previously, increasing the thickness of the cage, which is also shown in the right half of the figure.

The expiratory muscles. The major muscles of expiration are the *abdominals* and, to a lesser extent, the *internal intercostals.* The abdominal muscles cause expiration in two ways. First, they *pull downward on the chest cage* thereby decreasing the thoracic thickness. Second, they *force the abdominal contents upward* against the diaphragm, decreasing the longitudinal dimension of the pleural cavity. The internal intercostals help in the process of expiration to a slight extent by pulling the ribs downward. When they are in this downward position, the thickness of the chest is considerably decreased, as can be seen in the left half of Figure 184.

Pulmonary Pressures

ALVEOLAR PRESSURE. During inspiration the thoracic cage enlarges, which also enlarges the lungs. One will recall from the basic laws of physics that when a volume of gas is suddenly enlarged, its pressure falls. Thus, during inspiration the enlargement of the thoracic cage decreases the pressure in the alveoli to about −3 mm. Hg, and it is this negative pressure that pulls air through the respiratory passageways into the alveoli.

During expiration exactly the opposite effects occur; compression of the thoracic cage around the lungs increases the alveolar pressure to approximately +3 mm. Hg, which obviously pushes air out of the alveoli to the atmosphere.

If a person breathes with maximal effort but with his nose and mouth closed so that air cannot flow into or out of his lungs, the alveolar pressure can be decreased to as low as −80 mm. Hg or increased to as high as +100 mm. Hg. Thus, the maximum power of the muscles of respiration is far greater than that needed for normal quiet respiration. This provides a tremendous reserve respiratory ability which can be called upon in times of need for maximal respiratory activity, such as during heavy exercise.

INTRAPLEURAL PRESSURE. The space between the lungs and the thoracic cage is called the *intrapleural space*, and the pressure in this space is called the *intrapleural pressure*. This pressure is always a few millimeters of mercury less than that in the alveoli. Figure 185 illustrates that during inspiration the intra-alveolar pressure is

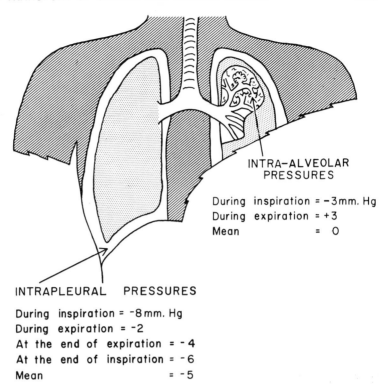

Figure 185. Alveolar and intrapleural pressures during normal breathing.

INTRA-ALVEOLAR PRESSURES

During inspiration = −3mm. Hg
During expiration = +3
Mean   = 0

INTRAPLEURAL   PRESSURES

During inspiration = −8mm. Hg
During expiration = −2
At the end of expiration = −4
At the end of inspiration = −6
Mean   = −5

about −3 mm. Hg while the intrapleural pressure is about −8 mm. Hg. During normal expiration, the intra-alveolar pressure rises to +3 mm. Hg while the intrapleural pressure rises to −2 mm. Hg. It will be noted, therefore, that the intrapleural pressure remains about 5 mm. Hg less than the intra-alveolar pressure all the time. The reason for this is that the lungs are always pulling away from the chest wall because of two effects: First, the surface tension of the fluid lining the inside of the alveoli makes the alveoli try to collapse. Second, elastic fibers spread in all directions through the tissues of the lungs, and these also tend to contract the lungs at all times. The effect of both of these is to pull the lungs away from the thoracic cage, creating an average negative pressure in the intrapleural space of about −5 mm. Hg.

Pneumothorax. When an opening is made in the chest wall the elastic forces in the lungs cause them to collapse immediately, sucking air through the opening into the chest cavity. Then, when the person tries to breathe, instead of the lungs expanding and contracting, air flows in and out of the hole in the chest. Thus, a wound of the chest can kill a person by suffocation; yet the condition can be treated very easily simply by sucking air out of the pleural cavity and plugging the hole.

Surfactant in the alveoli. If the alveoli were lined with water instead of the normal alveolar fluid, the surface tension would be so great that it would cause the alveoli to remain collapsed almost all the time. However, a substance called a *surface active agent* or simply *surfactant* is secreted into the alveoli by the alveolar membrane. This substance is a detergent that greatly decreases the surface tension of the fluid lining the alveoli.

Artificial Respiration

When the respiratory muscles fail, a person can be kept alive only by some means of artificial respiration. The simplest method is to force air into and out of his mouth or nose. Several devices have been made for this purpose, one of which is called the *resuscitator.* To use this apparatus a mask is placed over the mouth and nose and intermittent blasts of air fill the lungs. Between blasts, air is pulled back out of the lungs into the atmosphere.

For prolonged artificial respiration the person is usually placed inside an artificial respirator of the type shown in Figure 186. This is a large tank provided with some means for repetitively increasing and de-

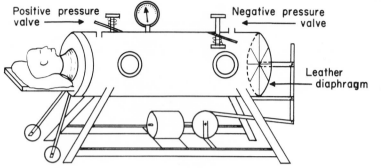

Figure 186. The artificial respirator.

creasing the pressure and for controlling the extent of the rise and fall in pressure. In the respirator of Figure 186 the air pressure is alternated by inward and outward movement of a large leather diaphragm, and the positive and negative pressures are controlled by relief valves shown on top of the tank. Because respiration normally is accomplished by the inspiratory muscles rather than by the expiratory muscles, the pressure inside the tank is usually adjusted to give more of a vacuum cycle than a pressure cycle. A vacuum of about −10 mm. Hg causes approximately normal inspiration by pulling outward on the chest cage and abdominal wall, thereby sucking air into the lungs. Then, the respirator changes from the vacuum to the pressure cycle and applies 2 to 3 mm. Hg of positive pressure. This causes expiration by pushing against the abdomen and chest.

Spirometry and the Divisions of the Respiratory Air

Figure 187 represents a *spirometer,* an apparatus that can be used to record the flow of air into and out of the lungs. It is com-

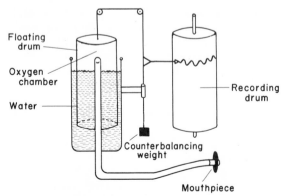

Figure 187. The spirometer.

posed of a drum inverted in a tank of water, and with a tube extending from the air space in the drum to the mouth of the subject. The drum is suspended from pulleys and counterbalanced by a weight. As the person breathes in and out, the drum moves up and down, and the counterweight balancing the drum also rides up and down, recording on a moving paper the changing volume of the chamber. Figure 188 illustrates a typical spirometer recording.

THE TIDAL VOLUME, RESPIRATORY RATE, AND THE MINUTE RESPIRATORY VOLUME. The air that passes into and out of the lungs with each respiration is called the *tidal air,* and the volume of this air in each breath is called the *tidal volume.* The normal tidal volume is about 500 ml., and the normal rate of respiration for an adult is usually about 12 times per minute. Therefore, a total of about 6 liters of air normally passes into and out of the respiratory passageways each minute. This amount is called the *minute respiratory volume.*

THE INSPIRATORY CAPACITY. In Figure 188, after three normal breaths, the subject breathes inward as deeply as possible. The amount of air that he can pull into his lungs beyond that already in his lungs at the beginning of the breath is called the *inspiratory capacity.* This quantity is approximately 3000 ml. in the normal person.

EXPIRATORY RESERVE VOLUME. After several more normal respirations, the subject expires as much as possible (see Figure 188). The amount of air that he is capable of expiring beyond that which he normally expires is called the *expiratory reserve volume.* This is usually about 1100 ml.

Residual volume and the functional residual capacity. In addition to the expiratory reserve volume, there is air in the lungs that cannot be expired even by the

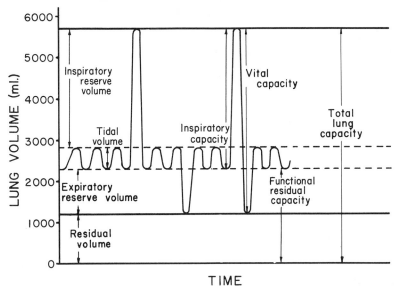

Figure 188. A spirogram, showing the divisions of the respiratory air.

most forceful exhalation. The volume of this air is about 1200 ml., and it is called the *residual volume*. The sum of the expiratory reserve volume plus the residual volume is called the *functional residual capcity;* this is the amount of air remaining in the respiratory system at the end of a normal expiration. It is this air that allows oxygen and carbon dioxide transfer into and out of the blood to continue even between periods of inspiration.

VITAL CAPACITY. Finally, Figure 188 shows the spirogram of a maximal inspiratory effort followed immediately by a maximal expiratory effort. The total change in pulmonary volume between these two extremes is called the *vital capacity*. The vital capacity of the normal person is approximately 4500 ml. A well trained male athlete may have a vital capacity as great as 6500 ml., while an asthenic female frequently has a vital capacity no greater than 3000 ml.

The vital capacity is a measure of the person's overall ability to inspire and expire air, and it is determined by two factors: (1) the strength of the respiratory muscles, and (2) the resistance of the thoracic cage and lungs to expansion and contraction. Disease processes that either weaken the muscles—such as poliomyelitis—or decrease the expansibility of the lungs—such as tuberculosis—can decrease the vital capacity. For this reason vital capacity measurements are an invaluable tool in assessing the functional ability of the breathing system.

Dead Space

Much of the air pulled into the respiratory passages with each breath is expired again before it ever reaches the alveoli. This air is useless from the point of view of oxygenating the blood. Consequently the respiratory passageways are called *dead space*. The total volume of this space is normally about 150 ml., which means that during inspiration of a normal tidal volume of 500 ml., only 350 ml. of new air actually enters the alveoli.

ALVEOLAR VENTILATION. The most important measure of the effectiveness of a person's respiration is his *alveolar ventilation*, which is the total quantity of new air that enters his alveoli each minute. With only 350 ml. of air reaching the alveoli with each breath and with a normal respiratory rate of 12 times per minute, the alveolar ventilation is approximately 4200 ml. per minute. During maximal respiratory effort this can be increased to as high as 100 liters per minute, and at the opposite extreme a person can remain alive at least for a few hours with an alveolar ventilation as low as 1200 ml. per minute.

Exchange of Alveolar Air with Atmospheric Air

At the end of each expiration approximately 2300 ml. of air still remains in the

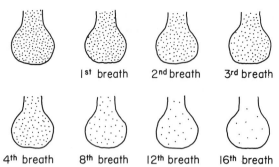

1st breath   2nd breath   3rd breath

4th breath   8th breath   12th breath   16th breath

Figure 189. Expiration of a foreign gas from the alveoli with successive breaths.

lungs. This is called the *functional residual capacity*, as was explained previously. With each breath, about 350 ml. of new air is brought into this air, and with each expiration this same amount of air is removed. It is obvious, then, that each breath does not change the total content of the alveolar air, but instead *changes only about one seventh of the air.* This effect is illustrated in Figure 189, which shows an alveolus containing a quantity of some foreign gas at the beginning of the series of breaths. After the first breath a small portion of this gas has been removed, after the second breath a little more, after the third a little more, and so forth. Note that even after the sixteenth breath a small portion of the foreign gas is still in the alveolus. At a normal alveolar ventilation of 4200 ml. per minute, approximately one-half the alveolar gases are replaced by new air every 23 seconds. This slow turnover of air in the alveoli keeps the

alveolar gaseous concentrations from rising and falling greatly with each individual breath.

ALVEOLAR AIR

The air in the alveoli is separated from the blood in the capillaries by the *pulmonary membrane* which has a thickness of only 0.2 to 0.4 micron. Therefore, oxygen can diffuse readily from the alveolar air into the blood, and carbon dioxide can diffuse from the blood into the alveolar air. One of the principal factors that determine the rate at which a gas will diffuse through this membrane is the *difference between the partial pressures* of the gas on the two sides of the membrane. Because many students may not be familiar with the concept of partial pressure, it is essential that this be explained at this point.

Partial Pressures

Figure 190 shows four chambers of the same size. Chamber B contains two separate compartments, the upper one storing 79 volumes of nitrogen and the lower one 21 volumes of oxygen. The pressure in each of these compartments is normal atmospheric pressure, 760 mm. Hg. The nitrogen is moved to chamber A which holds 100 volumes. In other words, the nitrogen expands from 79 to 100 volumes. In this process the pressure falls to 79/100 its

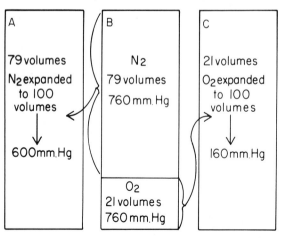

A

79 volumes

N_2 expanded to 100 volumes

↓

600 mm. Hg

B

N_2
79 volumes
760 mm. Hg

O_2
21 volumes
760 mm. Hg

C

21 volumes

O_2 expanded to 100 volumes

↓

160 mm. Hg

D

$N_2 = 79\%$
(600 mm. Hg partial pressure)
+
$O_2 = 21\%$
(160 mm. Hg partial pressure)

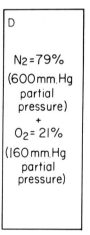

Figure 190. Partial pressures of oxygen and nitrogen in different chambers (explained in the text).

760 mm. Hg TOTAL PRESSURE

original value, or from 760 mm. to 600 mm. Similarly, the oxygen expands when moved to chamber C, and its pressure falls from 760 to 160. Finally, in chamber D, the nitrogn and oxygen in chambers A and C are mixed together. Now it is found that the total pressure in this chamber is once again 760 mm. Hg. Reasoning from the pressure and volume changes in these chambers, it is evident that 600 mm. of the 760 mm. Hg pressure in the fourth chamber is caused by nitrogen and 160 mm. is caused by oxygen. Therefore, the *partial pressure of nitrogen* (PN_2) in this mixture is said to be 600 mm. Hg, while the partial pressure of oxygen (PO_2) is 160 mm. Hg.

The partial pressure of a gas is a measure of the total force that it exerts against the walls surrounding it. Consequently, the penetrating power of a gas is directly proportional to its partial pressure. The greater the partial pressure of a gas in the alveolus, the greater is its tendency to pass through the pulmonary membrane into the blood.

Table 5 shows the relative pressures of nitrogen, oxygen, carbon dioxide, and water vapor in atmospheric and alveolar airs. This table shows that normal atmospheric air contains abut four-fifths nitrogen and one-fifth oxygen, with almost negligible quantities of carbon dioxide and water vapor. On the other hand, alveolar air contains considerable quantities of both carbon dioxide and water vapor, while the oxygen content is considerably less than that of atmospheric air. These differences can be explained as follows:

HUMIDIFICATION OF AIR AS IT ENTERS THE RESPIRATORY PASSAGES. When air is inspired it is humidified immediately by the moisture on the linings of the respiratory passages. At normal body temperature the partial pressure of water vapor in the lungs is 47 mm. Hg, and as long as the body temperature remains constant, this partial pressure also remains constant.

The mixing of water vapor with the in-coming atmospheric air dilutes the air so that the pressure of the other gases becomes slightly decreased. This explains why the nitrogen partial pressure in the alveoli is slightly less than its partial pressure in atmospheric air

OXYGEN AND CARBON DIOXIDE PARTIAL PRESSURE IN THE ALVEOLI. Alveolar air continually loses oxygen to the blood, and this oxygen is replaced by carbon dioxide diffusing out of the blood into the alveoli. This explains why the oxygen pressure in alveolar air is much less than in atmospheric air, and it also explains why the carbon dioxide pressure is much greater in the alveolar air than in the atmospheric air. The normal partial pressure of oxygen in the alveoli is approximately 104 mm. Hg, and that of carbon dioxide is 40 mm. Hg. However, these values change greatly from time to time depending on the rate of alveolar ventilation and the rate of oxygen and carbon dioxide transfer into and out of the blood. For instance, a high alveolar ventilation provides increased amounts of oxygen to the alveoli and increases the alveolar oxygen pressure. Also, a high alveolar ventilation removes carbon dioxide from the alveoli more rapidly than usual, thereby decreasing the carbon dioxide pressure.

TRANSPORT OF GASES THROUGH THE PULMONARY MEMBRANE

The Pulmonary Membrane

The pulmonary membrane is comprised of all the pulmonary surfaces that are thin enough to allow gases to diffuse into the pulmonary blood. These include, as illustrated in Figure 191, the membranes of the respiratory bronchioles, the alveolar ducts, the atria, and the alveolar sacs.

The total area of the pulmonary membrane is approximately 60 sq. m., which is about equal to the floor area of a moderate sized classroom. Only about 60 ml. of blood is in the capillaries at any one time. If one will think for a moment of this small amount of blood spread out evenly over the entire floor of a classroom, he can quite readily understand that very large quantities of gas can enter and leave this blood in only a fraction of a second.

TABLE 5. *Partial Pressures of Respiratory Gases in the Atmosphere and in the Alveoli*

GAS	ATMOSPHERIC AIR	ALVEOLAR AIR
N_2	597.0 (78.62%)	569.0 (74.9%)
O_2	159.0 (20.84%)	104.0 (13.6%)
CO_2	0.15 (0.04%)	40.0 (5.3%)
H_2O	3.85 (0.5%)	47.0 (6.2%)

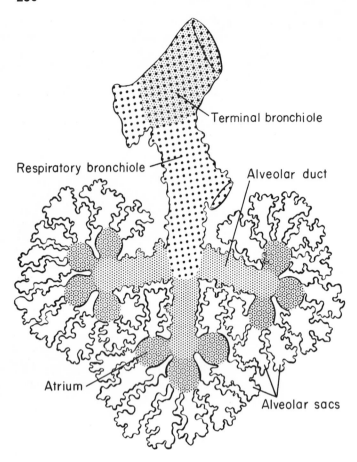

Terminal bronchiole

Respiratory bronchiole

Alveolar duct

Atrium

Alveolar sacs

Figure 191. The respiratory unit, illustrating the anatomy of the respiratory surfaces. (From Miller: *The Lung*. Charles C Thomas.)

The thickness of the pulmonary membrane is only 0.2 to 0.4 micron, which is several times less than the thickness of a red blood cell. This, also, is one of the features of the pulmonary membrane that allows extremely rapid diffusion between the alveolar air and the blood.

Solution and Diffusion of Gases in Water

To understand the transport of gases through the pulmonary membrane, one must first be familiar with the physical principles of gaseous solution and diffusion in water. Figure 192 shows a chamber containing water at the bottom and normal air at the top. Molecules of nitrogen and oxygen are continually bouncing against the surface of the water, and some of the molecules enter the water to become dissolved. The dissolved molecules bounce among the water molecules in all directions, some of them eventually reaching the surface again to bounce

back into the gaseous space. After the air has remained in contact with the water for a long time, the number of molecules passing out of the solution equals exactly the number of molecules passing into the solution. When this state has been reached, the gases in the gaseous phase are said to be in equilibrium with the dissolved gases.

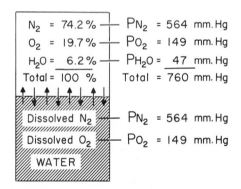

N_2	= 74.2%	PN_2	= 564 mm. Hg
O_2	= 19.7%	PO_2	= 149 mm. Hg
H_2O	= 6.2%	PH_2O	= 47 mm. Hg
Total	= 100 %	Total	= 760 mm. Hg

Dissolved N_2 —— PN_2 = 564 mm. Hg

Dissolved O_2 —— PO_2 = 149 mm. Hg

WATER

Figure 192. A sample of air exposed to water, showing equilibration between gases in the gaseous phase and in the dissolved state.

In the equilibrium state, the partial pressure tending to force nitrogen molecules into the water is 564 mm. Hg, and the pressure of nitrogen molecules pushing outward from the water is also 564 mm. Hg. If the nitrogen pressure becomes greater in the gaseous phase than in the dissolved phase, more nitrogen molecules will move into the solution than out. If the pressure becomes less in the gas than in the solution, more nitrogen molecules will leave the solution than will enter it. In other words, *the pressure of a gas is the force with which it attempts to move from its present surroundings.* Whenever the pressure is greater at one point than at another point, whether this be in a solution or in a gaseous mixture, more molecules will move toward the low pressure area than toward the high pressure area.

DIFFUSION OF GASES THROUGH TISSUES AND FLUIDS. Figure 193 shows a chamber containing a solution. At end A is a high concentration of the dissolved molecules, while at end B is a low concentration. The molecules are bouncing at random in all directions. Obviously, because of the higher concentration at A than at B, it is easier for large numbers of molecules to move from A to B than in the opposite direction, which is indicated by the lengths of the arrows, and eventually the numbers of molecules at both ends of the chamber will be approximately equal. Movement of molecules in this manner from areas of high concentration toward areas of low concentration is called *diffusion.*

The rate of diffusion of molecules in the body fluids and in the body tissues is determined by the pressure differences between the different points. Far more molecules are at point A in Figure 193 than at point B. Consequently, the pressure of the gas at point A is also far greater than at point B.

The difference between the two pressures is called simply the *pressure difference,* and the rate of diffusion between the two points is directly proportional to the pressure difference.

DIFFUSION OF GASES THROUGH THE PULMONARY MEMBRANE. The factors that determine the rate of gas diffusion through the pulmonary membrane are the following:

1. The greater the *pressure difference* between one side of the membrane and the other, the greater will be the rate of gaseous flow. If the pressure of a gas in the alveolus is 100 mm. Hg while that in the blood is 99 mm. Hg, the pressure difference will be only 1 mm., and the rate of gaseous flow will be very slight. If the pressure in the blood suddenly falls to 0 mm. Hg, the pressure difference rises to 100 mm., and the gaseous diffusion increases to 100 times its former rate.

2. The greater the *area of the pulmonary membrane,* the greater will be the quantity of gas that can diffuse in a given period of time. In some pulmonary diseases, such as *emphysema,* large portions of the lungs are destroyed and the total area of the pulmonary membrane is greatly decreased. Sometimes the decrease is enough to cause continual respiratory embarrassment.

3. The *thinner the membrane,* the greater will be the rate of gaseous diffusion. The normal pulmonary membrane is sufficiently thin so that venous blood entering a pulmonary capillary can be brought almost to complete gaseous equilibrium with the alveolar air within approximately one fifth of a second. Occasionally, though, the thickness of the membrane increases many-fold, especially when the lungs become edematous because of pulmonary congestion, pneumonia, or some other lung disease. The person may die simply because gases cannot diffuse through the thick membrane rapidly enough.

4. The diffusion coefficient of a gas is proportional to the *amount of gas that dissolves in the membrane* and is inversely proportional to the *square root of the molecular weight.* Therefore, highly soluble gases and gases with small molecular weights diffuse more rapidly than less soluble gases and gases with large molecular weights. If the diffusion coefficient of oxygen is considered to be 1, the relative rates of diffusion of the three important respiratory gases are the following:

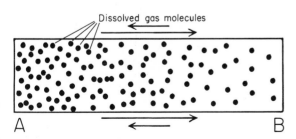

Figure 193. Diffusion of gas molecules through liquids.

Oxygen = 1
Carbon dioxide = 20.3
Nitrogen = 0.53

Thus carbon dioxide diffuses through the pulmonary membrane far more easily than either oxygen or nitrogen. The main reason for the differences in rate of diffusion is that carbon dioxide is about 20 times as soluble as oxygen in the fluids of the pulmonary membrane, while nitrogen is only about half as soluble as oxygen.

TRANSPORT OF OXYGEN THROUGH THE TISSUES

Diffusion of Oxygen from the Alveoli into the Blood and from the Blood into the Tissues

Figure 194 illustrates the overall transport of oxygen from the alveolus to the tissue cell, showing diffusion of oxygen from the alveolus into the pulmonary blood, transport of the blood through the arteries to the tissue capillaries, and, finally, diffusion of oxygen from the capillaries to the cell.

The data of Table 6 explain why oxygen diffuses from the alveoli into the pulmonary blood. The oxygen pressure (PO_2) of "venous" blood entering the lungs is only 40 mm. Hg in comparison with the alveolar air PO_2 of 104 mm. Hg. The pressure difference, therefore, is 64 mm. Hg, which causes extremely rapid diffusion of oxygen into the blood. During the very short time that the

TABLE 6. *Relative Gaseous Pressures in the Alveoli and the Blood*

GAS	ALVEOLAR AIR	VENOUS BLOOD	ARTERIAL BLOOD
O_2	104	40	100
CO_2	40	45	40
N_2	569	569	569

blood remains in the capillary, only about one second, it attains a PO_2 of approximately 100 mm. Hg.

The aerated blood then flows from the lungs to the tissue capillaries, where oxygen diffuses from the blood into the tissue cells, in which the PO_2 always remains very low, the reason for which may be explained as follows: When oxygen enters a cell it reacts very readily with sugars, fat, and proteins to form carbon dioxide and water, as is illustrated in Figure 195. As a result, the concentration of oxygen in the cell rarely rises above 30 mm. Hg. The high pressure of the oxygen in the capillaries, 100 mm. Hg, causes it to diffuse into the interstitial fluid and through the cellular membrane to the interior of the cells.

In summary, oxygen movement, both into the blood of the lungs and from the blood to the tissues, is caused by diffusion. The oxygen pressure difference from the lungs to the pulmonary blood is sufficient to keep oxygen always diffusing into the blood, and the pressure difference from the capillary

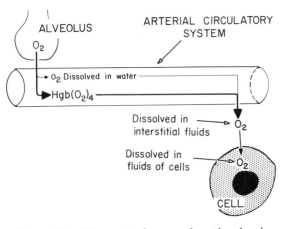

Figure 194. Transport of oxygen from the alveolus to the tissue cell.

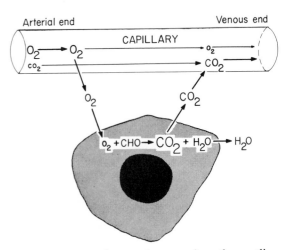

Figure 195. Diffusion of oxygen from the capillary to the tissue cell, formation of carbon dioxide inside the cell, and diffusion of carbon dioxide back to the tissue capillary.

blood to the tissue cells is always sufficient to keep oxygen diffusing from the blood to the cells.

Transport of Oxygen in the Blood by Hemoglobin

When oxygen diffuses from the lungs into the blood a small proportion of it becomes dissolved in the fluids of the plasma and red cells, but approximately 60 times as much combines immediately with the hemoglobin of the red blood cells and is carried in this combination to the tissues. Indeed, without the hemoglobin, the amount of oxygen that can be carried to the tissues is only a fraction of that required to maintain life. When the blood passes through the tissue capillaries, oxygen splits away from hemoglobin and diffuses into the tissues. Thus, hemoblobin acts as a carrier of oxygen, increasing the amount of oxygen that can be transported from the lungs to the tissues to some 60 times more than that which could possibly be transported only in the dissolved state.

THE OXYGEN-HEMOGLOBIN DISSOCIATION CURVE. Figure 196 illustrates the so-called *oxygen-hemoglobin dissociation* curve. This curve shows the per cent of the hemoglobin that is combined, or that is *saturated*, with oxygen at each oxygen pressure level. Aerated blood leaving the lungs usually has an oxygen pressure of about 100 mm. Hg. Referring to the curve, it is seen that at this pressure *approximately 97 per*

cent of the hemoglobin is combined with oxygen. As the blood passes through the tissue capillaries, the oxygen pressure falls normally to about 40 mm. Hg. At this pressure *only about 70 per cent* of the hemoglobin is combined with oxygen. Thus, about 27 per cent of the hemoglobin normally loses its oxygen to the tissue cells. Then, on returning to the lungs it combines with new oxygen and transports this once again to the tissue cells.

UTILIZATION COEFFICIENT AND THE HEMOGLOBIN RESERVE CAPACITY. The proportion of the hemoglobin that loses its oxygen to the tissues during each passage through the capillaries is called the *utilization coefficient.* In the normal person the utilization coefficient is 27 per cent, or expressed another way, approximately one fourth of the hemoglobin is used to transport oxygen to the tissues under normal conditions. When the tissues are in extreme need of oxygen, the oxygen pressure in the tissues can fall to extremely low values, allowing oxygen to diffuse from the capillary blood much more rapidly than usual. As a result, the saturation of the hemoglobin can fall to as low as 10 to 20 per cent instead of the normal level of 70 per cent, and the utilization coefficient rises to 77 to 87 per cent. Therefore, without even increasing the rate of blood flow, the amount of oxygen transported to the tissues in times of serious need can be increased more than three-fold. If one also remembers that the cardiac output can increase as much as five-fold in times of stress, then it is seen that the amount of oxygen that can be transported to the tissues can be

Figure 196. The oxygen-hemoglobin dissociation curve.

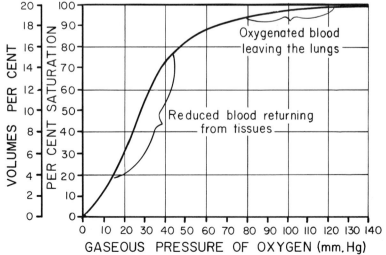

increased to as much as 15 to 20 times normal, part of the increase being caused by an increase in the utilization coefficient and even more by the increased cardiac output.

HEMOGLOBIN AS AN OXYGEN BUFFER. For cellular function to continue at a normal pace, the concentrations of all substances in the extracellular fluid must remain relatively constant at all times. One of the functions of hemoglobin is to keep the oxygen pressure in the tissues almost always between the limits of 20 and 45 mm. Hg. This may be explained as follows: As blood flows through the capillaries, 27 per cent of the oxygen is usually removed from the hemoglobin, making the hemoglobin saturation fall to about 70 per cent. Referring again to Figure 196, it can be seen that to remove this much oxygen the oxygen pressure in the tissues can never rise above 45 mm. Hg. On the other hand, whenever the oxygen pressure falls to 20 mm. Hg, more than three fourths of the oxygen is released from the hemoglobin, and this rapid release usually keeps the tissue oxygen pressure from falling any lower. Thus, hemoglobin automatically releases oxygen to the tissues so that the tissue PO_2 remains almost always between the limits of 20 and 45 mm. Hg.

CARBON MONOXIDE AS A HEMOGLOBIN POISON. Carbon monoxide combines with hemoglobin in an almost identical manner as oxygen, except that the combination of carbon monoxide with hemoglobin is approximately 210 times as tenacious as the combination of oxygen with hemoglobin. Also, because the two substances combine with hemoglobin at the same point on the molecule, they both cannot be absorbed by the hemoglobin at the same time. Carbon monoxide mixed with air in a concentration of 0.1 per cent, 210 times less than that of oxygen, will cause half of the hemoglobin to combine with carbon monoxide and therefore leave only half of the hemoglobin to combine with oxygen. When the concentration of carbon monoxide rises above approximately 0.2 per cent, which is still 100 times less than the normal concentration of oxygen, the quantity of hemoglobin available to transport oxygen becomes so slight that death ensues.

When death is about to occur from carbon monoxide poisoning, it can frequently be prevented by administering pure oxygen. The pure oxygen, on entering the alveoli, has a partial pressure of approximately 600 mm. Hg, six times the normal alveolar oxygen pressure. The force with which oxygen can combine with hemoglobin is therefore increased six-fold, which forces the carbon monoxide off the hemoglobin molecule six times as rapidly as would occur without treatment.

TRANSPORT OF CARBON DIOXIDE

Referring once again to Figure 195, it can be seen that when oxygen is used by the cells for metabolism carbon dioxide is formed. The pressure of carbon dioxide (PCO_2) in the cells becomes very high, about 55 mm. Hg, and a pressure difference develops between the cells and the blood in the capillary. This causes carbon dioxide to diffuse out of the cell, into the interstitial fluid, and from there into the capillary. As shown in Figure 197, it is transported in the blood to the lungs. The PCO_2 of the blood entering the tissue capillaries is about 40 mm. Hg, but this rises to 45 mm. Hg as carbon dioxide enters the blood from the cells. Therefore, the PCO_2 of the blood entering the lungs is approximately 45 mm. Hg, while the PCO_2 in the air of the alveolus is only 40 mm. Hg. A pressure difference of about 5 mm. Hg exists between the blood and the alveolus, causing carbon dioxide to diffuse out of the blood. Furthermore, because of the extremely rapid diffusion coefficient of carbon dioxide, the PCO_2 of the pulmonary blood falls to almost complete equilibrium with the

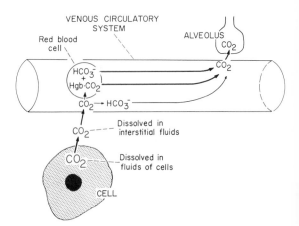

Figure 197. Transport of carbon dioxide from the tissue cell to the alveolus, illustrating some of the chemical combinations of carbon dioxide with blood.

alveolar P_{CO_2}, that is, to 40 mm. Hg, before the blood leaves the capillary.

Chemical Combinations of Carbon Dioxide with the Blood

Only about 5 per cent of the carbon dioxide is transported in the blood in the dissolved state. About 95 per cent of the carbon dioxide diffuses from the plasma into the red cell where it undergoes two reactions: First, carbon dioxide reacts with water to form carbonic acid. Red blood cells contain an enzyme called *carbonic anhydrase* which speeds this reaction approximately 250-fold. Therefore, very large quantities of carbon dioxide combine with water to form carbonic acid inside the red blood cells while only small quantities are formed in the plasma. The carbonic acid in the cells immediately reacts with the acid-base buffers of the cells and becomes mainly bicarbonate ion. This reaction prevents the acidity of the cell from becoming too great.

Second, a few per cent of the total carbon dioxide entering the blood combines directly with hemoglobin to form a compound called *carbaminohemoglobin*. This reaction does not occur at the same point on the hemoglobin molecule as the reaction between oxygen and hemoglobin. Therefore, hemoglobin can combine with both carbon dioxide and oxygen at the same time. In this way hemoglobin acts as a carrier not only of oxygen but also of carbon dioxide. However, the reaction occurs relatively slowly so that this method of carbon dioxide transport is probably unimportant.

To recapitulate, carbon dioxide is transported in the blood in three principal forms. First, only a very small quantity is transported as dissolved carbon dioxide. Second, by far the largest proportion is transported in the form of bicarbonate ion. And, third, a few per cent is transported in the form of carbaminohemoglobin.

When the blood enters the pulmonary capillaries all the chemical combinations of carbon dioxide with blood are reversed, and the carbon dioxide is released into the alveoli.

REFERENCES

Campbell, E. J. M.: Respiration. *Ann. Rev. Physiol.* 30:105, 1968.

Ciba Foundation Symposium: *Pulmonary Structure and Function.* Boston, Little, Brown & Company, 1962.

Clements, J. A.: Surface phenomena in relation to pulmonary function. *Physiologist* 5:11, 1962.

Comroe, J. H., Jr., et al: *The Lung: Clinical Physiology and Pulmonary Function Tests*, 2nd Ed. Chicago, Year Book Medical Publishers, 1963.

Dubois, A. B.: Respiration. *Ann. Rev. Physiol.* 26:421, 1964.

Fenn, W. O.: Carbon dioxide and intracellular homeostasis. *Ann. N.Y. Acad. Sci.* 92:547, 1961.

Greene, M.: *The Voice and Its Disorders.* New York, The Macmillan Co., 1957.

Haldane, J. S., and Priestley, J. G.: *Respiration.* New Haven, Yale University Press, 1935.

Mead, J.: Mechanical properties of lungs. *Physiol. Rev.* 41:281, 1961.

Pattle, R. E.: Surface lining of lung alveoli. *Physiol. Rev.* 45(1):48, 1965.

CHAPTER 20

REGULATION OF RESPIRATION AND THE PHYSIOLOGY OF RESPIRATORY ABNORMALITIES

THE BASIC RHYTHM OF RESPIRATION

Respiration is controlled by the *respiratory center* located on both sides of the brain stem in the lateral reticular substance of the medulla and lower pons, as illustrated in Figure 198A. Nerve signals are transmitted from this center to the muscles of respiration. The most important muscle of respiration, the diaphragm, receives its respiratory signals through a special nerve, the *phrenic nerve,* that leaves the spinal cord in the upper half of the neck and passes downward through the chest to reach the diaphragm. Signals to the expiratory muscles, especially the abdominal muscles, are transmitted down the spinal cord to the spinal nerves that innervate these muscles.

Rhythmic Oscillation in the Respiratory Center

The rhythmic respiratory cycle is caused by oscillation back and forth between *in-*

spiratory and *expiratory* neurons intermingled together in the respiratory center. Figure 198B illustrates the basic mechanism of this oscillating mechanism, showing to the left four expiratory neurons and to the right four inspiratory neurons. It is believed that the expiratory neurons are organized in a loop in such a way that when one of these neurons becomes excited it sends a signal around and around the circuit, thus causing oscillation. As illustrated in the figure, fibers pass from this loop down the spinal cord to carry expiratory impulses to the expiratory muscles.

To the right in Figure 198B is shown a comparable group of inspiratory neurons, also capable of oscillating in a manner similar to that for the expiratory neurons. Each time oscillation occurs in this loop, impulses are sent to the inspiratory muscles.

Now let us see how the expiratory neurons interact with the inspiratory neurons. It is believed that when inspiratory oscillation occurs, this not only sends impulses down to the inspiratory muscles but simultaneously sends *inhibitory* impulses into the

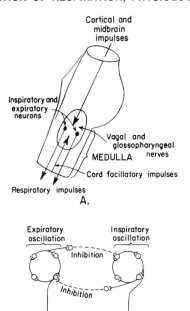

Figure 198. (A) The respiratory center located bilaterally in the lateral reticular substance of the medulla and lower pons. (B) Theoretical mechanism for the rhythmical control of respiration.

expiratory system to reduce or stop expiratory activity. Likewise, when the expiratory system oscillates it sends collateral impulses into the inspiratory system to inhibit it. Thus, the two oscillating groups of neurons mutually inhibit each other. For this reason, when one is oscillating the other stops oscillating. Each of these sets of neurons can oscillate for only 1 to 3 seconds, possibly because of neuronal fatigue. That is, when oscillation first begins the neurons can relay signals with ease, but after a second or more they supposedly become fatigued so that their transmission fades. Thus, at the end of inspiratory oscillation the expiratory neurons, having rested for a few seconds, begin to oscillate. At the end of another 1 to 3 seconds expiratory oscillation ceases and inspiratory oscillation begins again.

The arrows of Figure 198A illustrate several nerve pathways entering the respiratory center. Nerve impulses from the vagal and glossopharyngeal nerves cause reflex excitation or depression of the respiratory center, as will be discussed later. Signals from the spinal cord, including sensory impulses from the limbs during exercise and

impulses from the skin, stimulate respiration. And signals from upper levels of the brain can affect respiration in many different ways. For instance, during speech the cerebral cortex controls respiration to coordinate it with the speech.

EFFECT OF THE HERING-BREUER REFLEX ON THE BASIC RHYTHM OF RESPIRATION. Located throughout the lungs are nerve endings that become stimulated when the lungs are distended. Impulses from these pass through the vagus nerve to the medulla, where they inhibit inspiration and excite expiration. One of the major functions of this reflex, called the *Hering-Breuer reflex*, is to prevent overinflation of the lungs. Excessive distention of the lungs can damage them by tearing the tissues, or, more importantly, it temporarily blocks the flow of blood into the left heart because of dilatation of the lung vessels which take up the blood instead.

The Hering-Breuer reflex also aids in maintaining the basic rhythm of respiration. As the lungs distend during inspiration this reflex inhibits inspiration, and it simultaneously initiates expiration. Then, as the lungs deflate, the Hering-Breuer reflex ceases, allowing the inspiratory center to become excited again and the expiratory center to become inhibited. Therefore, this reflex helps to maintain the alternating rhythmic stimulation of the inspiratory and expiratory centers.

It is evident, then, that at least two different feedback mechanisms help to maintain the respiratory cycle, one intrinsic in the medulla and one operating extrinsically through the Hering-Breuer reflex. If either one of them becomes nonfunctional, the other is still capable of continuing the cycle. This duplication of neurogenic mechanisms is characteristic of most of the important functions of the brain, for any system that must continue to operate without a single instance of failure for up to 100 years must have a large margin of safety should some portion of the system become temporarily damaged.

FAILURE OF THE RESPIRATORY CENTER. Occasionally all the oscillating mechanisms of the respiratory center fail at the same time. One of the more frequent causes of this is cerebral concussion or some other intracerebral abnormality that causes excess pressure on the medulla. The pressure collapses the blood vessels supplying the respiratory center, which blocks all activity of the medulla and as a result stops the respiratory rhythm. Another common cause of failure is acute poliomyelitis, which sometimes destroys neuronal cells in the reticular substance of the hindbrain, thereby depressing the respiratory center. Finally, the most frequent of all causes of respiratory failure is attempted suicide with sleeping drugs. These anesthetize the respiratory neurons and in this way stop the rhythm of respiration.

Failure of the respiratory rhythm is one of the most difficult of all abnormalities of the body to overcome. In general, very few drugs can be used to excite the respiratory center, and those that are available — caffeine, picrotoxin, and a few others — are so weak in their effects on the respiratory center that they are of little value. In general, the only effective treatment is to give prolonged artificial respiration in the hope that the cause of the respiratory depression will be rectified by the body itself. For instance, respiratory depression caused by poliomyelitis, pressure on the brain, or sleeping drugs is usually reversible provided artificial respiration is maintained long enough.

REGULATION OF ALVEOLAR VENTILATION

Even though the basic rhythm of respiration is continuous, the rate of rhythm and the depth of respiration vary tremendously in response to varying physiologic conditions in the body. Obviously, the more oxygen needed by the body and the more carbon dioxide the body needs to expel, the greater must be the alveolar ventilation.

Effect of Carbon Dioxide on Alveolar Ventilation

The most powerful stimulus known to affect the respiratory center is *carbon dioxide*. When the carbon dioxide content of the blood increases above normal, all portions of the respiratory center become excited. The intensity of both inspiratory and expiratory signals becomes greatly enhanced, and the rapidity of the oscillation back and forth between the inspiratory and expiratory centers is also speeded up. Therefore, *both the rate and depth of respiration are increased.* This effect is illustrated in Figure 199, which shows that increased carbon dioxide in the blood can increase the rate of alveolar ventilation to as much as 10 times normal.

About one-half of the carbon dioxide effect on the respiratory center is caused by a direct action of carbon dioxide on the respiratory neurons to increase their excitability. However, the other half of the effect is caused indirectly in the following manner: When carbon dioxide increases in the blood some of it diffuses into the cerebrospinal fluid in which the brain is bathed. This carbon dioxide combines with water to

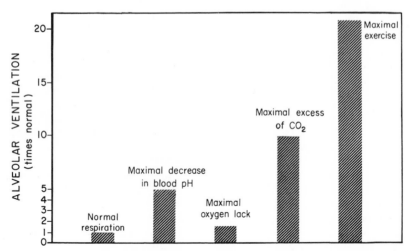

Figure 199. Effect on alveolar ventilation of maximal decrease in blood pH, maximal oxygen lack, maximal excess of CO_2, and maximal exercise.

form carbonic acid, thereby increasing the hydrogen ion concentration in the fluid. Part of the respiratory center lies on the anterior surface of the medulla in immediate contact with the fluid, and this portion of the center is specifically excited by hydrogen ions. Thus, indirectly, the appearance of carbon dioxide in the cerebrospinal fluid excites the respiratory center as a result of increased hydrogen ions in the cerebrospinal fluid.

CONTROL OF CARBON DIOXIDE CONCENTRATION IN THE BODY FLUIDS BY THE CARBON DIOXIDE FEEDBACK MECHANISM. It is extemely important that the respiratory system is controlled by blood concentration of carbon dioxide because this provides a method by which the carbon dioxide concentration can in turn be controlled in all fluids of the body. This can be explained as follows: The concentration of carbon dioxide in the blood is controlled by the rate of alveolar ventilation. That is, increased ventilation causes the lungs to blow off increased amounts of carbon dioxide from the blood. Therefore, when excess carbon dioxide excites the respiratory center, the resulting increase in ventilation causes the excess carbon dioxide concentration to return once more to near normal. Conversely, when carbon dioxide concentration falls too low, the resulting decrease in ventilation allows the carbon dioxide concentration to rise once again back toward normal.

The body has no other means for controlling carbon dioxide concentration in the blood and body fluids, which makes this mechanism all important. Should carbon dioxide concentration rise too high, it would stop essentially all chemical reactions in the body, because carbon dioxide is one of the end products of these reactions. On the other hand, if carbon dioxide concentration should fall too low, other dire consequences would occur, such as development of alkalosis caused by loss of carbonic acid in the body fluids. The alkalosis then would cause increased irritability of the nervous system, sometimes resulting in tetany or even epileptic convulsions.

Regulation of Alveolar Ventilation by the Hydrogen Ion Concentration

The second most powerful influence on alveolar ventilation is the hydrogen ion concentration in the body fluids. When the hydrogen ion concentration becomes high—that is, when the pH becomes low—the respiratory neurons are excited, and alveolar ventilation increases. However, the maximum amount that ventilation can be increased by increasing the hydrogen ion concentration is about five-fold. Therefore, an increase in the hydrogen ion concentration is only one half as potent a stimulus of respiration as an increase in carbon dioxide concentration.

IMPORTANCE OF THE HYDROGEN ION FEEDBACK MECHANISM FOR REGULATING RESPIRATION. Regulation of respiration by the hydrogen ion concentration helps to maintain normal acid-base balance in the body fluids, which was discussed in more detail in Chapter 18. When carbon dioxide is blown off through the lungs, loss of this carbon dioxide from the body fluids shifts the chemical equilibria of the acid-base buffers in such a way that much of the carbonic acid of the body fluids dissociates into water and carbon dioxide. This removes some of the body's acid and thereby decreases the hydrogen ion concentration. Therefore, stimulation of the respiratory center by increased hydrogen ion concentration sets off a feedback mechanism that automatically decreases the hydrogen ion concentration back toward normal. Conversely, decreased hydrogen ion concentration decreases alveolar ventilation, thus increasing the acidity of the fluids back toward normal.

Regulation of Alveolar Ventilation by Oxygen Deficiency

On first thought most students have the idea that respiration should be regulated mainly by the needs of the body for oxygen. However, at ordinary ventilatory rates the hemoglobin always becomes almost completely saturated with oxygen, and extreme increases or moderate decreases in alveolar ventilation make very little difference in the amount of oxygen carried away from the lungs by the hemoglobin. Therefore, there is no need for very acute regulation of respiration to maintain constant oxygen concentration in the blood.

On rare occasions, however, the oxygen concentration in the alveoli does fall too low to supply adequate quantities of oxygen to the hemoglobin. This occurs especially when one ascends to high altitudes where the oxygen concentration in the atmosphere is very low. Or it occurs when a person contracts pneumonia or some disease that reduces the oxygen in the alveoli. Under these conditions the respiratory system needs to be stimulated by *oxygen deficiency.* A mechanism for this is the *chemoreceptor system,* shown in Figure 200. Small *aortic* and *carotid bodies* lie adjacent to the aorta and carotid arteries in the chest and neck, each of these having an abundant arterial blood supply, and containing neuron-like cells called *chemoreceptors,* which are sensitive to the lack of oxygen in the blood. When stimulated, these receptors send impulses along the vagus and glossopharyngeal nerves to the medulla, where they stimulate the respiratory center to increase alveolar ventilation.

The chemoreceptor system is not a powerful stimulator of respiration in comparison with the stimulation of the respiratory center by excess carbon dioxide or hydrogen ions. As illustrated in Figure 199, maximum oxygen lack can increase alveolar ventilation to approximately $1\frac{2}{3}$ times its normal value in comparison with an increase of 10 times caused by excess carbon dioxide and 5 times caused by excess hydrogen ions.

Effect of Exercise on Alveolar Ventilation

Alveolar ventilation increases almost directly in proportion to the amount of work

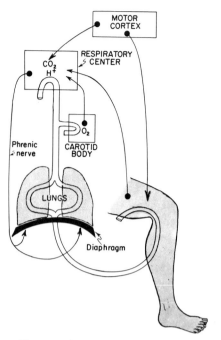

Figure 201. Mechanisms by which muscle activity stimulates respiration.

performed by the body during exercise, reaching as high as 100 to 120 liters per minute in the most strenuous exercise. This value is about 20 times that during normal quiet respiration, as illustrated in Figure 199.

Physiologists still have difficulty explaining the extreme increase in pulmonary ventilation that occurs during exercise. Indeed, despite the extremely rapid production of carbon dioxide during exercise and simultaneous depletion of oxygen from the blood, alveolar ventilation increases so greatly that it prevents the blood concentrations of these gases from changing significantly from normal. Therefore, most physiologists believe that the increase in respiration during exercise is not caused by either or both of these two chemical factors.

If it is not chemical factors that increase ventilation during exercise, it must be some stimulus entering the respiratory center by way of nerve pathways. Two such nerve pathways illustrated in Figure 201 have been discovered. (1) At the same time that the cerebral cortex transmits signals to the exercising muscles it also transmits parallel signals into the respiratory center to increase the rate and depth of respiration. (2) Movement of the limbs and other parts of the body sends sensory signals up the spinal cord to excite the respiratory center. Therefore, it is believed that these two

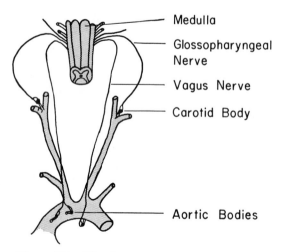

Figure 200. The chemoreceptor system for stimulating respiration.

signals, one from the cerebral cortex and the other from the moving parts of the body, are the factors that increase respiration during exercise. If these two factors fail to increase respiration adequately, carbon dioxide and hydrogen ions begin to collect in the body fluids and oxygen begins to diminish, and these then excite the respiratory center as a second line of defense to increase ventilation.

Other Factors that Affect Alveolar Ventilation

EFFECT OF ARTERIAL PRESSURE. Arterial pressure is another factor that helps to regulate alveolar ventilation. This operates through the *baroreceptor system* which was described in Chapter 14. When the arterial pressure is high, the respiratory center is depressed, and ventilation is reduced correspondingly. Conversely, when the pressure is low, and as a consequence blood flow to the tissues is poor, ventilation is increased; this partially compensates for the poor blood flow by allowing better oxygenation and better removal of carbon dioxide in the lungs.

EFFECT OF PSYCHIC STIMULATION. Impulses initiated by psychic stimulation of the cerebral cortex can also affect respiration. For instance, anxiety states frequently can lead to intense hyperventilation, sometimes so much so that the person makes his body fluids alkalotic, thus precipitating tetany.

EFFECT OF SENSORY IMPULSES. Sensory impulses from all parts of the body can affect respiration. The effect of entering a cold shower is well known; this causes an intense inspiratory gasp followed by a period of prolonged inspiration and then by rapid, forceful breathing. Even a pin prick can cause sudden changes in the rate and depth of respiration. However, these factors affect alveolar ventilation only temporarily, for the chemical factors mentioned previously soon become dominant over these aberrant effects and suppress them.

EFFECT OF SPEECH. The speech centers of the brain also control respiration at times. When one talks it is as important to control the flow of air between the vocal cords as to control the vocal cords themselves. Therefore, whenever nerve impulses are transmitted from the brain to the vocal cords, collateral impulses are sent simultaneously into the respiratory system.

The Multiple Factor Theory for Control of Respiration

It is obvious from the preceding discussions that no single factor regulates the respiratory centers, but that many different factors, including carbon dioxide concentration, hydrogen ion concentration, oxygen availability, exercise, arterial pressure, sensory phenomena, and speech all help to regulate respiration. The most generally accepted theory for the regulation of respiration, then, is the so-called *multiple factor theory.* Some factors are known to be more important than others—carbon dioxide or exercise, for example. Others are of intermediate importance, such as hydrogen ion concentration, and others are of relatively minor importance, such as oxygen deficiency and arterial pressure.

PHYSIOLOGY OF RESPIRATORY ABNORMALITIES

Anoxia

One of the most important effects of most respiratory diseases is anoxia, which means diminished availability of oxygen to the cells of the body. The different types of anoxia that can occur are anoxic anoxia, stagnant anoxia, anemic anoxia, and histotoxic anoxia, the principal mechanisms of which are shown in Figure 202.

ANOXIC ANOXIA. Anoxic anoxia means failure of oxygen to reach the blood of the lungs. The obvious causes of anoxic anoxia are (1) *too little oxygen in the atmosphere,* (2) *obstruction of the respiratory passages,* (3) *thickening of the pulmonary membrane,* and (4) *decreased area of the pulmonary membrane.*

STAGNANT ANOXIA. Stagnant anoxia means failure to transport adequate oxygen to the tissues because of slow blood flow. The most common cause of stagnant anoxia is low cardiac output caused by *heart failure.* Immediately after a heart attack the blood flow to the tissues may be so slight that the person may die of the stagnant anoxia itself.

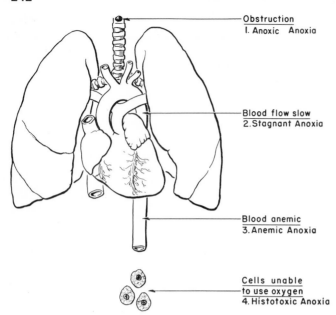

Obstruction
1. Anoxic Anoxia

Blood flow slow
2. Stagnant Anoxia

Blood anemic
3. Anemic Anoxia

Cells unable
to use oxygen
4. Histotoxic Anoxia

Figure 202. The types of anoxia.

At other times the transport of oxygen will be barely enough to keep him alive but his tissues nevertheless will suffer severely from oxygen deficiency.

ANEMIC ANOXIA. Anemic anoxia means too little hemoglobin in the blood to transport oxygen to the tissues. This may be caused by any one of three different abnormalities: First, the person may have *anemia,* which means too few red blood cells. Second, he may have a sufficient number of red blood cells but *too little hemoglobin in the cells.* Third, he may have plenty of hemoglobin, but much of it may have been *poisoned by carbon monoxide* or some other poison so that it cannot transport oxygen; for instance, breathing air with a content of only 0.2 per cent carbon monoxide can decrease the hemoglobin available to transport oxygen to as low as one-third normal.

HISTOTOXIC ANOXIA. This means failure of the tissues to utilize oxygen even though adequate quantities are transported to them. The classic cause of serious histotoxic anoxia is *cyanide poisoning;* this blocks the enzymes responsible for use of oxygen in the cell. Mild forms of histotoxic anoxia frequently occur with *vitamin deficiencies,* for lack of certain of the vitamins often results in diminished quantities of oxidative enzymes in the cells.

Oxygen Therapy

In most types of anoxia, therapy with purified oxygen can be tremendously beneficial, the benefit occurring in three separate ways, which may be described as follows:

First, the normal partial pressure of oxygen in the alveoli is about 104 mm. Hg, but when a person breathes pure oxygen, this can rise to as high as 600 mm. Hg. Such an increase can increase as much as six-fold the pressure gradient for oxygen diffusion through the pulmonary membrane.

Second, even when there is insufficient hemoglobin in the blood to carry enough oxygen to the tissues, oxygen therapy can still be beneficial. The reason for this is that when the partial pressure of oxygen in the alveoli becomes 600 mm. Hg, 2 ml. of oxygen dissolves in the fluid of every 100 ml. of blood and is transported in this form to the tissues. This effect is illustrated in Figure 203. The lower solid curve of this figure shows the total quantity of oxygen combined with hemoglobin at different oxygen partial pressures, and the upper curve is the total oxygen in the blood including that in the dissolved state. At a normal alveolar oxygen partial pressure of about 100 mm. Hg, almost no oxygen is dissolved in the fluids. Yet as the alveolar oxygen partial pressure rises to 600 mm. Hg, one sees by the shaded area of the figure that reasonable quantities of oxygen begin to dissolve in the fluids. This oxygen is the difference between life and death in some anoxic persons.

The third means by which oxygen therapy can benefit some types of anoxia, is by decreasing the volume of gases that must flow in and out of the lungs through the trachea.

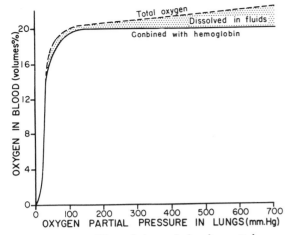

Figure 203. Effect of elevated alveolar partial pressure of oxygen on the total oxygen transported by each 100 ml. of blood.

If the respiratory passageways are partially obstructed, the blood can still be adequately oxygenated with only one-fifth normal alveolar ventilation if the person breathes pure oxygen rather than air because the concentration of oxygen in air is only 20 per cent. Yet the decreased alveolar ventilation will still cause a buildup of carbon dioxide in the body fluids, sometimes enough to produce serious acidosis even though the tissues are adequately oxygenated.

Dyspnea

Dyspnea means *air hunger* or, in other words, a psychic feeling that more ventilation of the blood is needed than is being attained. Most instances of dyspnea occur when some respiratory abnormality causes the blood to become anoxic and to collect too much carbon dioxide. The respiratory center becomes overly excited, and impulses are transmitted to the conscious portions of the brain apprising the psyche of the need for increased ventilation. This causes a feeling of air hunger or dyspnea.

However, an occasional person develops *psychic dyspnea* because of a neurosis. His blood has normal carbon dioxide and is adequately oxygenated, but the neurosis causes him to become unduly conscious of his respiration, making him feel that he is not receiving adequate quantities of air. One of the neuroses that most frequently causes this type of dyspnea is cardiac neurosis, for many who fear cardiac disease have known cardiac patients who have had continuous dyspnea because of stagnant anoxia. Therefore, the neurotic who believes that he has heart disease often develops psychic dyspnea.

Pneumonia

Pneumonia is caused by infection of the lung with bacteria such as the pneumococcus bacterium or with viruses. The infection causes the walls of the alveoli to become inflamed and edematous and the spaces in the alveoli to become filled with fluid and blood cells. This effect is illustrated in the center of Figure 204. Pneumonia causes anoxic anoxia for two reasons: first, because fluid and blood cells fill many alveoli, and, second, because the membranes of those alveoli that are still aerated are frequently so thickened with edema that oxygen cannot diffuse with ease.

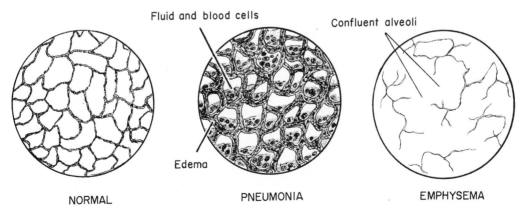

Figure 204. Histologic appearance of the normal lung, of the lung with pneumonia, and of the emphysematous lung.

Pulmonary Edema

Pulmonary edema means collection of fluid in the interstitial spaces of the lungs and in the alveoli; this affects respiration in the same way pneumonia affects it. Generalized pulmonary edema is usually caused by failure of the left heart to pump blood from the pulmonary circulation into the systemic circulation. This can result from mitral or aortic valvular disease or from failure of the left ventricular muscle. In any event, blood is dammed up in the pulmonary circulation, and the pulmonary capillary pressure rises. When this pressure becomes greater than the colloid osmotic pressure of the blood, 28 mm. Hg, fluid transudes very rapidly from the plasma into the alveoli and interstitial spaces of the lungs, causing anoxic anoxia. Sometimes acute failure of the left heart causes pulmonary edema to occur so rapidly that the person dies of anoxia in only 20 to 40 minutes.

Emphysema

Emphysema is a disease caused most frequently by chronic bronchial infection resulting from smoking. In this condition, large portions of the alveolar walls are destroyed; this is illustrated in Figure 204. Therefore, the total surface area of the pulmonary membrane becomes greatly diminished, which also diminishes the aeration of the blood. Anoxic anoxia results, and the quantity of carbon dioxide in the body fluids becomes mildly increased.

The emphysematous person also usually develops pulmonary hypertension, for every time the wall of an alveolus is destroyed the blood vessels in the wall are also destroyed. This increases the pulmonary resistance, which in turn elevates the pulmonary arterial pressure.

Atelectasis

Atelectasis means collapse of a portion of the lung or of an entire lung. A common cause of atelectasis is a chest wound that allows air to leak into the pleural cavity. The surface tension of the fluid in the alveoli and the elastic fibers in the interstitial spaces of the lungs cause the lungs to collapse to a small size, the alveoli losing all their air. Another common cause of atelectasis is plugging of a bronchus, in which case the air in the alveoli is absorbed into the blood, thus collapsing the alveoli.

Not only do the alveoli collapse in atelectasis, but the blood vessels also collapse simultaneously. Consequently, blood flow through the collapsed lung decreases greatly, allowing the major portion of the pulmonary blood to flow through areas of the lungs that are still aerated. Because of this shift of blood flow, atelectasis of a whole lung often does not greatly diminish the aeration of the blood.

Asthma

Asthma is usually caused by an allergic reaction to pollens in the air. The reaction causes spasm of the bronchioles, thus impeding air flow into and out of the lungs. For reasons that are not too clearly understood, the outward flow of air through the bronchioles is obstructed more than the inflow, which allows the asthmatic person to inspire with ease but to expire with difficulty. As a result, the lungs become progressively more and more distended, and with repeated asthmatic attacks year in and year out, the prolonged distention of the thoracic cage causes the chest to become barrel shaped. Asthma rarely becomes severe enough to cause serious anoxia, but it does cause severe dyspnea. Asthma can usually be treated by administering drugs, especially epinephrine, that relax the bronchiolar musculature. Also, it can often be prevented by desensitization of the person to the causative pollens, as discussed in Chapter 9.

REFERENCES

American Physiological Society: *Handbook of Physiology*, Sec. III, Respiration. Vol. I, 1964; Vol. II, 1965. Baltimore, The Williams & Wilkins Company.

Bernstein, L.: Respiration. *Ann. Rev. Physiol.* 29:113, 1967.

Burns, B. D.: The central control of respiratory movements. *Brit. Med. Bull.* 19:7, 1963.

Cunningham, D. J. C. and Lloyd, B. B. (eds.): *The Regulation of Human Respiration.* Philadelphia, F. A. Davis Company, 1963.

Lloyd, B. B.: The chemical stimulus to breathing. *Brit. Med. Bull.* 19:10, 1963.

Milhorn, H. T., Jr., Benton, R., Ross, R., and Guyton, A. C.: A mathematical model of the human respiratory control system. *Biophys. J.* 5:27, 1965.

Pappenheimer, J. R., Fencl, V., Heisey, S. R., and Held, D.: Role of cerebral fluids in control of respiration as studied in unanesthetized goats. *Amer. J. Physiol.* 208:436, 1965.

CHAPTER 21

AVIATION, SPACE, AND DEEP SEA DIVING PHYSIOLOGY

One of the most important problems in aviation, space, and deep sea diving physiology is the altered barometric pressure to which the aviator, spaceman, or diver is exposed. Therefore, many of the physical principles applicable to respiration are also applicable to aviation, space flights, or dives deep beneath the sea.

Problems also arise in aviation, space, and diving physiology because of mechanical stresses on the body or because of extremes of climate. For instance, extreme acceleration during takeoff of a spaceship can cause the spaceman to weigh as much as 1500 pounds, a force that the body can withstand only when in the proper position. Also, in the upper atmosphere the air temperature is 20° to 50° C. below zero, and deep beneath the sea, pressures on the outside of the body can sometimes become so intense that the entire chest wall actually collapses.

EFFECTS OF LOW BAROMETRIC PRESSURE

Oxygen Deficiency at High Altitudes

At high altitudes the barometric pressure is low, as shown in Table 7, and the partial pressure of oxygen is reduced correspondingly. This decreases the amount of oxygen absorbed into the blood, and leads to anoxia. Table 7 also shows the effect of low oxygen partial pressure on the saturation of arterial hemoglobin in the blood. For instance, at an altitude of approximately 23,000 feet only half of the arterial hemoglobin is saturated with oxygen. Obviously, this decreases the effectiveness of oxygen transport to the tissues at least 50 per cent, so that a person at this altitude will be debilitated because of tissue anoxia.

It will be noted also from Table 7 that the alveolar oxygen partial pressure at an altitude of 50,000 feet is only 1 mm. Hg and that the arterial oxygen saturation is only 1 per cent. Therefore, at this altitude, the small quantities of oxygen stored in the tissues actually diffuse backward into the blood and then from the blood into the lungs. Obviously, a person would have no more than a few minutes to live under these conditions.

EFFECTS OF OXYGEN DEFICIENCY. Figure 205 shows graphically the time required at various altitudes for oxygen deficiency to cause either collapse or coma in a normal *unacclimatized* person. Above 30,000 feet an unacclimatized person lapses into coma in about one minute. At 20,000 feet the person usually does not go into coma, but after 10 or more minutes he exhibits signs of col-

TABLE 7. *Effects of Low Atmospheric Pressures on Alveolar Oxygen Concentrations and on Arterial Oxygen Saturation*

Altitude (ft.)	Barometric Pressure (mm. Hg)	P_{O_2} in Air (mm. Hg)	BREATHING AIR		BREATHING PURE OXYGEN	
			P_{O_2} in Alveoli (mm. Hg)	Arterial Oxygen Saturation (%)	P_{O_2} in Alveoli (mm. Hg)	Arterial Oxygen Saturation (%)
0	760	159	104	97	673	100
10,000	523	110	67	90	436	100
20,000	349	73	40	70	262	100
30,000	226	47	21	20	139	99
40,000	141	29	8	5	58	87
50,000	87	18	1	1	16	15

lapse such as weakness, mental haziness, and other deficiencies. He usually does not pass into coma until he ascends to altitudes of 22,000 to 24,000 feet, which is called his *ceiling.*

The first symptom of oxygen deficiency occurs in vision. Even at altitudes as low as 6000 to 10,000 feet, the sensitivity of the eyes to light begins to decrease — not enough to be noticed usually, but nevertheless enough to be measured by instruments. When the aviator rises to altitudes of 12,000 to 15,000 feet he is likely to experience alterations in psychic behavior. Some aviators tend to go to sleep while others develop euphoric exaltation. In almost all instances the acuity of reasoning becomes greatly depressed. In short, the aviator at this altitude is likely to

develop a sleepy slap-happiness, and as he rises to 18,000 to 24,000 feet these symptoms often become so severe that control of the aircraft is endangered. For this reason safety regulations require that an aviator breathe oxygen when he ascends to dangerous altitudes. On occasion, he rises too high before applying his oxygen, and by that time his mental acuity might have become so depressed that he no longer knows his need for oxygen. This often results in a vicious cycle of depressed mentality, for he rises still higher without knowing what he is doing; his mentality becomes further depressed, he rises higher, and so forth until coma develops.

EFFECT OF WATER VAPOR AND CARBON DIOXIDE ON ALVEOLAR OXYGEN PARTIAL PRESSURE. Were it not for water vapor and carbon dioxide in the alveoli, the aviator could go to considerably higher heights than usual before developing oxygen deficiency. Regardless of the altitude, the alveolar pressure of water vapor does not change at all, and the pressure of carbon dioxide does not change greatly. Therefore, unless the barometric pressure is considerably greater than the combined pressures of these two gases, little space will be left in the alveoli for other gases. Figure 206 shows this effect. At sea level the combined pressures of all the gases in the alveoli is 760 mm. Hg. Carbon dioxide is continually excreted from the blood into the alveolar air, keeping the P_{CO_2} at a level of 40 mm. Hg, and water vapor continually evaporates from the surfaces of the alveoli creating a water vapor pressure of 47 mm. Hg, which remains constant as long as the body temperature is normal regardless of changes in barometric pressure. Thus, the combined pressure of

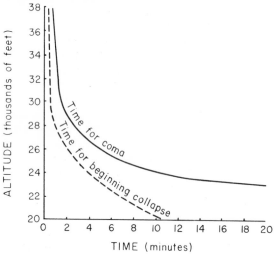

Figure 205. Time required at different altitudes for oxygen deficiency to cause collapse or coma. (From Armstrong: *Principles and Practice of Aviation Medicine.* The Williams & Wilkins Company.)

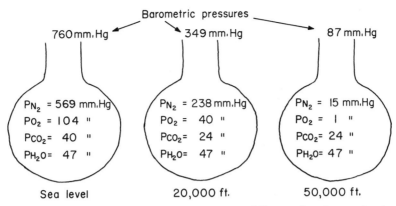

Figure 206. Effects of low barometric pressures at different altitudes on alveolar gas concentrations.

water vapor and carbon dioxide at sea level is 40 plus 47 or 87 mm. Hg. Subtracting this from 760 mm. Hg, the remaining partial pressure available in the alveolus for both nitrogen and oxygen is 673 mm. Hg. This is more than adequate, so that the oxygen partial pressure is high enough to keep the arterial blood saturated with oxygen.

Now, let us see what effects the water vapor pressure and carbon dioxide partial pressure have on alveolar function at an altitude of 50,000 feet. Here, the total barometric pressure is only 87 mm. Hg. The water vapor pressure remains 47 mm. Hg because the body temperature does not change. The P_{CO_2} has fallen to about 24 mm. Hg because of increased respiration. Therefore, the combined pressure of water vapor and carbon dioxide is now 71 mm. Hg, which when subtracted from 87 mm. Hg leaves a total partial pressure for both nitrogen and oxygen of only 16 mm. Hg. Furthermore, the person is so anoxic that his blood absorbs oxygen out of the alveoli almost as rapidly as it enters. Therefore, the P_{O_2} falls to about 1 mm. Hg which is so slight that the person will become unconscious in only a few seconds and will die within another minute or so. Were it not for the carbon dioxide and water vapor, it would be possible at 50,000 feet to supply far more oxygen to the blood.

RESPIRATORY COMPENSATION FOR OXYGEN DEFICIENCY — THE PROCESS OF ACCLIMATIZATION. The chemoreceptor mechanism described in the previous chapter increases alveolar ventilation when a person develops oxygen deficiency at high altitudes. This mechanism unfortunately is not very powerful, for it normally can increase alveolar ventilation only about 60 per cent.

Yet even this amount allows the aviator to ascend several thousand feet higher than he could otherwise.

When a person is exposed to high altitudes for several days at a time, his chemoreceptor mechanism, for reasons not well understood, becomes progressively more active until his alveolar ventilation increases to as much as five times normal. This slow process of developing more and more ventilation is called *acclimatization* to high altitudes.

Another factor that favors acclimatization is an increase in the number of red blood cells, for oxygen deficiency causes rapid production of these cells by the bone marrow, as explained in Chapter 8. Unfortunately, the formation of many new cells is a slow process, requiring several weeks to help acclimate a person to high altitudes. Persons who live at high altitudes all the time often develop red blood cell counts as high as 7 to 8 million per cu. mm., which is 50 per cent or more greater than normal.

Acclimatization to high altitudes is not of special importance in aviation because the aviator rarely remains aloft long enough for acclimatization to occur. It is far more important to the mountain climber, who must become slowly acclimatized if he is to succeed in ascending to the top of the highest mountains. This explains why ascension is normally accomplished in slow stages over a period of weeks, thereby allowing the body to become progressively acclimatized. By using this procedure of acclimatization conquerors of Mt. Everest have been able to remove their masks atop this highest mountain of the world at an altitude over 29,000 feet, though this would cause coma in a normal person in a minute or two.

Oxygen Breathing at High Altitudes

A person can ascend to far higher altitudes when he breathes pure oxygen than when he breathes air, because oxygen then occupies the space in the alveoli normally occupied by nitrogen, in addition to the usual oxygen. This allows the alveolar partial pressure of oxygen to remain quite high even when the barometric pressure falls to a low value. Observing Figure 206 once again, it will be noted that at 20,000 feet the combined pressure of oxygen and nitrogen is 278 mm. Hg. If the nitrogen were replaced with oxygen, the oxygen pressure would also be 278 mm. Hg, but because of the presence of the nitrogen the oxygen pressure is only 40 mm. Hg, which is low enough to cause considerable desaturation of the blood. When breathing pure oxygen at 20,000 feet the arterial hemoglobin is 100 per cent saturated with oxygen, though when breathing air the saturation is only 67 per cent.

Figure 207 presents graphically the percentage saturation of hemoglobin in the arterial blood at different altitudes when breathing air as opposed to breathing pure oxygen. It is evident from this figure that essentially no desaturation occurs when breathing pure oxygen below an altitude of approximately 33,000 feet. Above that point, however, the barometric pressure becomes so low that even with nitrogen eliminated from the alveoli, the pressure of oxygen is still not sufficient to keep the blood saturated. The lowest level of arterial oxygen saturation at which a non-acclimatized person can remain alive is about 50 per cent. Therefore,

the ceiling for a person breathing pure oxygen is approximately 47,000 feet, in comparison with a ceiling of 23,000 feet for one breathing air.

Pressurized Cabins

Obviously, all the problems of low barometric pressure at high altitudes can be overcome by pressurizing the airplane. Usually the air inside the cabin of passenger planes is pressurized to maintain approximately the same pressure as that at 5000 feet.

EXPLOSIVE DECOMPRESSION. One of the major problems of pressurizing high altitude equipment is the possibility that the chamber might explode. Experiments have shown, fortunately, that sudden decompression does not cause any significant damage to the body because of the decompression itself. The danger lies in exposure to the low partial pressures of oxygen in the rarefied atmosphere. Referring back to Figure 205, one sees that at altitudes above 30,000 feet the person has only a minute or more of consciousness, during which time he must institute appropriate life-saving measures, such as putting on an oxygen mask or parachuting to safety. If he parachutes without the aid of a special automatic opening device it is important to open the chute as soon as possible, because he is likely to lapse into coma and fall to earth without benefit of the parachute. On the other hand, if he opens the chute he is likely to lapse into coma and die because of the altitude. Fortunately, modern high-flying jets are equipped with special devices that eject the aviator out of the plane when need be and also automatically delay the opening of the chute until an appropriate altitude is reached.

ACCELERATION EFFECTS OF AVIATION

Airplanes move so rapidly and change their direction of motion so frequently that the body is often subjected to severe physical stress caused by the sudden changes in motion. When the *velocity* of motion changes the effect is called *linear acceleration*. On the other hand, when the *direction* of motion is changed the effect is called *angular acceleration*. In general, the forces induced

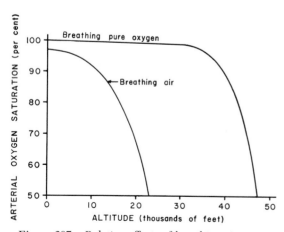

Figure 207. Relative effects of breathing air or pure oxygen on the saturation of arterial hemoglobin with oxygen at different altitudes.

by linear acceleration during normal flight of an airplane are not sufficient to cause major physiologic effects. But when an airplane turns, dives, or loops, the centrifugal forces caused by angular acceleration are frequently sufficient to promote serious derangements of bodily function.

Angular Acceleration

POSITIVE G. Figure 208 illustrates an airplane going into a dive and then pulling out. While the airplane is flying level, the downward force of the aviator against his seat is exactly equal to his weight. However, as he begins to come out of the dive, he is pushed against the seat with much greater force than his weight because of centrifugal action. At the lowest point of the dive the force is six times that which would be caused by normal gravitational pull. This effect is called *positive angular acceleration,* and the person is said to be under the influence of +6 g force, or, in other words, a force six times that of gravity.

NEGATIVE G. At the beginning of the dive in Figure 208 the airplane changes from level flight to a downward direction, which throws the pilot upward against his seat belt. He then is not exerting any force at all against his seat but instead is being held down by the seat belt with a force equal to three times his weight. This effect is called *negative angular acceleration,* and the amount of force exerted is said to be −3 g.

EFFECT OF ANGULAR ACCELERATION ON THE CIRCULATORY SYSTEM. The most important effects of angular acceleration on the body occur in the circulation. Positive g causes the blood in the vascular system to be centrifuged toward the lower part of the body. The veins of the abdomen and legs distend greatly, storing far more blood than usual, so that little or none of it returns uphill to the heart. The cardiac output falls to very low values or even to zero. The arterial pressure falls and the person lapses into coma.

Negative g causes the opposite effects; so much blood is centrifuged into the upper part of the body that the cardiac output rises and the arterial pressure rises. Occasionally the person develops such high pressure in his cerebral vessels that brain edema results, and, rarely, vessels even rupture, causing serious mental impairment.

Figure 209 depicts graphically the time required for blackout (coma) to ensue at various degrees of positive acceleration. A person can usually withstand +4 g almost indefinitely; he might develop a certain amount of dizziness, but, in general, his circulatory control systems can maintain sufficient cardiac output to prevent blackout. However, at acceleration values above +4 g, blackout usually occurs quite rapidly—at 5 g in about 8 seconds, at 12 g in about 3 seconds, and at 20 g in about 2½ seconds. Obviously, the amount of time that an aviator can remain conscious at high rates of acceleration determines how rapidly he can come out of a dive, and it also limits the sharpness of turns

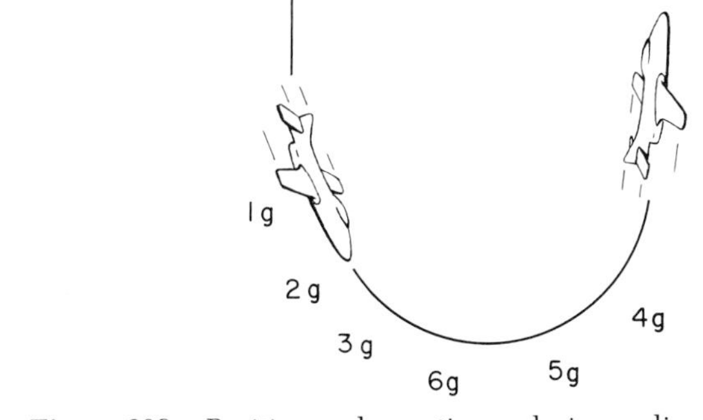

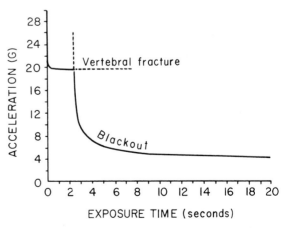

Figure 208. Positive and negative g during a dive and pullout of an airplane.

Figure 209. Effect of exposure time to different degrees of positive acceleration in producing blackout or vertebral fracture. (From Armstrong: *Principles and Practice of Aviation Medicine.* The Williams & Wilkins Company.)

that he can make. Most airplanes are designed to withstand 20 or more positive g. Therefore, the maneuverability of an airplane is dependent more on what the aviator himself can stand than upon the strength of the plane.

ANTI-BLACKOUT MEASURES. Application of pressure to the outside of the legs and lower abdomen can prevent pooling of blood during positive acceleration.

Another means for partially accomplishing the same effect is for the aviator himself to tighten his abdominal muscles as tightly as possible during the pullout from a dive. This allows him to withstand a little more acceleration than he otherwise could and allows him several seconds extra exposure before blacking out.

Different types of "anti-g" suits for accomplishing this have also been devised. One of these utilizes air bags applied to the abdomen and legs. When the airplane pulls out of the dive the bags inflate so that positive pressure is applied to the legs and abdomen. In this way the aviator is not uncomfortably subjected to pressure all the time but only when it is needed.

Because of the speed of the modern jet airplane, the tremendous acceleratory forces that can develop make it impossible for an aviator to turn sharply without blacking out. One of the solutions to this problem will probably be for the aviator to lie prone while flying the plane, for the body can withstand +15 g in the horizontal position in comparison with only +4 g in the sitting position.

EFFECT OF POSITIVE ACCELERATION ON THE SKELETON. Also illustrated in Figure 209 is the effect of positive acceleration in causing vertebral fracture. This graph shows that with the aviator in the sitting position, positive acceleration greater than 20 g is likely to rupture a vertebra because of the intense force of the upper body pressing downward. Only a split second of the intense force can cause the fracture, for the stress that a bone can withstand does not depend on how long the stress is applied, but on how much stress is applied.

DECELERATORY PARACHUTE DESCENT. When a person jumps from an airplane, gravity causes him to begin falling toward earth. Figure 210 illustrates the velocity of fall at different distances (if the person has not opened his parachute). By the time he has fallen 1500 feet his velocity will have reached approximately 175 feet per second.

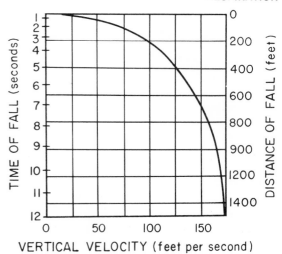

Figure 210. Effect of distance of fall on the velocity attained. (From Armstrong: *Principles and Practice of Aviation Medicine*. The Williams & Wilkins Company.)

At this speed, the air resistance exactly opposes the tendency for gravity to increase his velocity even more. Therefore, this velocity of fall, 175 feet per second, is the maximum that will be attained, and it is called the *terminal velocity*. (At very high altitudes the terminal velocity is considerably greater than 175 feet per second because of the rarefied atmosphere, but it will slow down to 175 feet per second when the person falls into the heavier atmosphere.)

One of the deceleratory forces caused by parachute descent is the *opening shock* that occurs when the parachute opens. The force of the shock is roughly proportional to the square of the velocity of fall at the time of opening. Parachutes are designed so that even after the person has reached the terminal velocity of fall the shock will not be sufficient to harm the body.

On landing, the parachutist approaches earth at a velocity of about 15 miles per hour. This is equivalent to jumping without a parachute from a wall approximately 8 feet high. Even this velocity can be dangerous if the parachutist is not ready to cope with the landing jolt. If he lands stiff-legged, he will almost certainly suffer a fracture of a leg, the pelvis, or a vertebra. On the other hand, if he does not tense his muscles when landing, he will pancake on the ground and likely suffer other injuries. Therefore, it is important that he land with his legs flexed but tense.

SPACE PHYSIOLOGY

The physiologic principles of traveling in spaceships are much the same as those applicable to aviation physiology, except that the importance of some of the factors is intensified. The special problems in space physiology include: (a) weightlessness, (b) intense linear acceleration forces, (c) supply of oxygen and other nutrients, and (d) special environmental hazards, especially radiation.

WEIGHTLESSNESS. Prior to the advent of the spaceship, persons had experienced the phenomenon of weightlessness for not more than a few seconds at a time, this state having been created in the cabins of airplanes undergoing a trajectory path, upward and over a hump, calculated for exact course and velocity to provide zero g forces for a period of 10 to 15 seconds. Yet it was known that once a spaceship was in orbit or was traveling from planet to planet, the spaceship itself would be subjected to exactly the same gravitational forces as the spaceman inside the ship. Therefore, the person would experience the phenomenon of prolonged weightlessness; that is, there would be no force pulling him in any direction toward any part of the spaceship.

Prior to the first space flight, many physiologists were concerned about how a human being would react to weightlessness of long duration. Fortunately, however, now that man has experienced this phenomenon, it has been proved to have little significant effect on the physiology of the body, except that the spaceman frequently has difficulty readjusting to the "weightful" state on returning to earth. However, weightlessness does necessitate the use of a few operational devices for flying the spaceship and for other activities within the spaceship. First, since the spaceman "floats" inside the spaceship, he must be either strapped in place or have appropriate hand holds. Second, his food must be in special containers that have closed tops, because the food will literally float off any plate and fluids will float out of any glass. Ordinarily, foods are prepared in tubes of the toothpaste type, and the food is simply squeezed into the mouth. Likewise, water is sucked from a tube. Excreta must also be forced into a container rather than left to float freely in the spaceship. Otherwise, one can imagine what the atmosphere inside the spaceship would be like after a few days of flying. Except for these simple problems, the state of weightlessness is actually very peaceful and pleasant, which is what one might suspect from his own experience of floating in water.

LINEAR ACCELERATION OF A SPACESHIP. At takeoff, a high velocity is reached in only a few minutes while putting the spaceship into orbit. This is achieved by using rocket propulsion from one or more boosters. Figure 211 shows the intensity of forward acceleration at different times during the five minutes required to put a spaceship into orbit, illustrating that linear acceleration increases to 2 g when the first booster begins to fire. As that booster becomes lighter and lighter because of decreasing fuel, the acceleration increases to as much as 9 to 10 g. Then after the first booster has dropped, the second booster goes through a similar pattern of acceleratory changes. Also, the final stage, which carries the space capsule, accelerates for another minute or so before its rocket propulsion completely ceases. Upon cessation of all propulsion at the end of 5 minutes, the spaceman enters the state of weightlessness, which is designated in the figure by zero g acceleration.

In the sitting position the spaceman cannot withstand more than about 4 g, but in the reclining position, the position assumed in a reclining easy chair, the spaceman can withstand 12 to 15 g, which is greater than any of the forces illustrated in Figure 211. Therefore, this is the position that the spaceman assumes during takeoff.

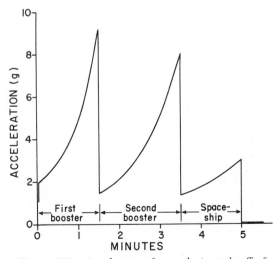

Figure 211. Acceleratory forces during takeoff of a spaceship.

Deceleration. One of the greatest problems in all of space physiology is the effect of slowing the spaceship as it reenters the atmosphere. The tremendous amount of kinetic energy stored in the moving spaceship must be dissipated as the spaceship decelerates. This energy becomes heat and obviously increases the temperature inside the space capsule to very high levels. Therefore, special precautions are taken to provide a heat dissipation shield at the front of the spaceship which absorbs most of this heat rather than allowing it to be transmitted to the cabin of the ship. A spaceship traveling at a speed of Mach 100, a speed about three times that required for orbiting the earth, but which might well be used in interplanetary space flights, requires a distance of about 10,000 miles for safe deceleration. Any more rapid deceleration than this would create more g forces than the spaceman could stand.

REPLENISHMENT OF OXYGEN AND OTHER NUTRIENTS. It is a simple matter to provide all the oxygen and nutrients needed by the spaceman if he is to remain in space for not more than a few days to a few weeks. However, interplanetary travel will require that a spaceman remain in space for many months or even years, for which reason it becomes physically impossible to carry sufficient oxygen and other nutrients within the confines allowable in the spaceship. There-

fore, many different procedures have been attempted to create a complete *life cycle* within the spaceship that will continually replenish the spaceship's oxygen and also supply adequate nutrients to the spaceman. For instance, algae are capable of living on human excreta, utilizing both carbon dioxide and fecal excreta to form oxygen, carbohydrates, proteins, and fats. Theoretically, therefore, algae can be used as food and the oxygen can be breathed. Unfortunately, though, the amount of algae necessary to provide this complete life cycle for spacemen is far greater than space limitations in the spaceship will allow. Furthermore, the foods developed from algae have proved to be neither palatable nor totally life sustaining. Yet future research might provide an adequate means for finally completing this life cycle.

Other procedures for completing the life cycle are based on chemical or electrochemical procedures for separating oxygen from carbon dioxide and for resynthesizing certain types of foodstuffs. Here again, the success has not been completely rewarding, so that today's space flights are still limited by the amount of oxygen and nutrients that can be carried from earth at the time of takeoff.

RADIATION HAZARDS IN SPACE PHYSIOLOGY. Early in the space program, scientists began to realize that enough radia-

Figure 212. The hazardous Van Allen radiation belts around earth. (From Newell, H. E.: *Science* 131:385, 1960.)

tion of different types—gamma rays, x-rays, electrons, cosmic rays, and so forth—exists at certain points in space that these rays alone could be lethal. Figure 212 illustrates two major belts of such radiation, called *Van Allen radiation belts*, that extend entirely around the earth and that are especially prominent in the equatorial plane of the earth. One of these belts begins at an altitude of about 300 miles and extends to about 3000 miles. The outer begins at about 6000 miles and extends to 20,000 miles.

The intensity of radiation in either one of the two Van Allen belts is great enough that a person in a space ship orbiting the earth could receive enough radiation in only a few hours to cause death. Indeed, even when the space ship goes through these two belts on an interplanetary voyage, lack of adequate radiation shielding or going through the two belts slowly could also cause death. However, Figure 212 shows that the radiation does not occur to a significant extent at the two poles of the earth, which means that space travel in the future will perhaps take place mainly from one or the other of the two poles. Since significant radiation does not begin until an altitude of about 300 miles is reached, one can also understand why space travel around the earth has thus far been confined to altitudes below 300 miles.

OTHER PROBLEMS OF SPACE FLIGHTS. Other hazards at high altitudes include some of the same hazards of aviation but to a much severer degree, such as (1) exposure of the spaceman to ultraviolet radiation, which can cause severe sunburn or can blind him, (2) exposure to extreme heat when in the direct pathway of sunrays and to extreme cold when on the opposite side of a spaceship from the sunrays, and (3) exposure to very low barometric pressures in case of decompression of the spaceship. All of these problems require special engineering such as pressurized suits, protective filters for the eyes, and very intricate heat control systems for the space capsule.

DEEP SEA DIVING AND HIGH PRESSURE PHYSIOLOGY

The human body is occasionally exposed to extremely high barometric pressure such as when diving deep beneath the sea or when working in a compression chamber—for instance, in a tunnel beneath a river filled with compressed air to prevent cave-in. The very high barometric pressure increases the amount of gases that pass from the alveoli into the blood and become dissolved in all the body fluids. This extra dissolved gas, unfortunately, often leads to serious physiological disturbances.

PRESSURES AT DIFFERENT DEPTHS OF THE SEA. A person on the surface of the sea is exposed to the pressure of the air above the earth. This pressure is approximately 760 mm. Hg; or it is also stated to be 1 atmosphere pressure. The weight of 33 feet of water is equal to the weight of the atmosphere, so that at a depth of 33 feet beneath the sea the pressure is double that at the surface, or 2 atmospheres of pressure. Thus, the pressure increases as one descends beneath the sea in accordance with the following table:

Feet	Atmosphere(s) pressure
0	1
33	2
67	3
100	4
200	7
300	10

Effect of High Gas Pressures on the Body

Figure 213 illustrates the pressures of each of the alveolar gases when a diver is breathing compressed air at 33 feet below sea level, and also at 100 feet below sea level. It will be noted that the pressures of water vapor and carbon dioxide remain the same regardless of how far below the surface the person descends. The reason for this is that water vapor pressure is directly dependent on the temperature of the body, as explained in Chapter 19, and the pressure of carbon dioxide is dependent on the rate of carbon dioxide release from the blood into the lungs, which continues to be the same regardless of depth. On the other hand, the pressures of nitrogen and oxygen increase almost directly in proportion to the depth below sea level to which the person descends.

OXYGEN POISONING. Ordinarily a person breathing compressed air can descend to

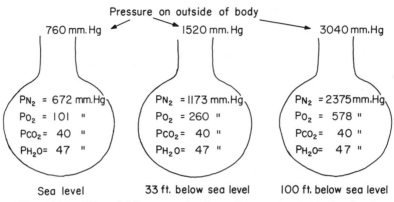

Figure 213.   Effect of different sea depths on the gas pressures in the alveoli.

about 200 feet below the sea without any danger from excess oxygen absorption. However, at levels deeper than this, his tissues begin to be damaged very rapidly by the oxygen. The reason for this is the following: Hemoglobin normally releases oxygen to the tissues at pressures ranging between 20 and 45 mm. Hg. In this way the hemoglobin automatically "buffers" the oxygen in the tissues, maintaining a relatively constant oxygen pressure that also allows a relatively constant rate of cellular oxidation. When there are extreme oxygen pressures in the lungs, however, large quantities of oxygen dissolve in the fluids of the blood, and the oxygen transported in this manner is released to the tissues at pressures that are much greater than 45 mm. Hg, supplying far more than normal amounts of oxygen to the cells. This causes deranged cellular metabolism, often damaging the cells themselves or leading to abnormal cellular function. The most debilitating effect of oxygen poisoning occurs in the brain, and is usually manifested by nervous twitchings, convulsions, and coma. Obviously, the person often loses control of himself, which might make him descend too deeply or fail to operate his breathing apparatus correctly, which will lead to his death.

The dangers of oxygen poisoning can be eliminated by supplying a progressively less and less percentage of oxygen in the diver's breathing mixture as he goes to greater depths.

NITROGEN POISONING. High pressures of nitrogen can also seriously affect a diver's mental functions even though this gas does not enter into any metabolic reactions in the body. When it dissolves in the body fluids in high concentrations, it exerts an anesthetic effect on the central nervous system. Ordinarily, a person can descend to approximately 200 feet below sea level before this anesthetic effect becomes serious, which is about the same safe depth as that for the prevention of oxygen poisoning. However, at depths slightly lower than this the diver will actually fall asleep because of the increasing amounts of nitrogen dissolved in his fluids. Also, in the early stages of this "nitrogen narcosis," the person frequently develops a sense of extreme exhilaration associated with seriously depressed mental acuity, a condition called "raptures of the depths." This state is comparable to severe drunkenness from alcohol, and it can cause the diver to descend much deeper than he should or to perform other dangerous acts that will lead to his death.

To avoid the dangers of nitrogen poisoning, the nitrogen is frequently replaced by helium in the breathing mixture. Helium does not cause an anesthetic effect, and it has an additional advantage of diffusing out of the body fluids more rapidly than nitrogen will when the diver ascends.

THE PROBLEM OF CARBON DIOXIDE WASHOUT FROM THE DIVING HELMET. If high concentrations of carbon dioxide are allowed to collect in a diver's helmet, he will soon develop respiratory acidosis and perhaps even coma. Therefore, carbon dioxide must be washed out of the helmet rapidly enough so that its concentration will never rise to any significant level. To do this, the same volume of gas must flow through the helmet at all times. When the diver descends to 200 feet below the sea, his air must be compressed seven times to withstand the pressure. This also reduces the volume seven times, and requires that the compressor pump seven times as much air for carbon dioxide washout as is required at sea level. It can be seen, therefore, that the rate at which air is supplied to the diver must be increased directly in proportion to the increase in pressure at which he is working. Indeed, it is this factor of carbon dioxide washout that determines how much air must be pumped to the diver rather than the amount of oxygen that he needs.

SCUBA diving. The term SCUBA means "self-contained underwater breathing apparatus." The major factor that limits the time a person can remain under water using a SCUBA system is the problem of carbon dioxide washout from the alveoli. The deeper the diver goes, the less becomes the *volume* of each unit mass of gas liberated from his tank. Therefore, the greater must be the *mass* of gas that flows out of his tank to maintain carbon dioxide washout from his lungs. Thus, a tank of compressed air will last only one seventh as long at 200 feet as at a few feet below the surface of the sea. For this reason, a SCUBA diver can descend to a 200-foot depth for only 15 to 20 minutes at a time. Other than for this limitation, the physical principles of SCUBA diving are essentially the same as those of helmet diving.

Decompression Sickness

SOLUTION OF NITROGEN IN THE BODY FLUIDS. In addition to the anesthetic effect of nitrogen, this gas can also cause very serious damage because of bubble formation in the body fluids when the diver returns toward the surface. The amount of nitrogen normally dissolved in the entire body when a person is exposed to normal atmospheric pressure is about 1 liter. If he remains 100 feet below the sea for several hours, that is, at 4 atmospheres pressure, the amount of dissolved nitrogen increases to 4 times as much, and at 7 atmospheres pressure to 7 times as much.

BUBBLE FORMATION DURING DECOMPRESSION. Figure 214 illustrates to the left the gases dissolved in the body fluids when a diver is breathing air at a pressure of 5000 mm. Hg, which corresponds to a depth

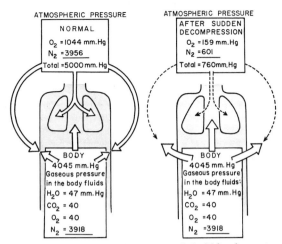

Figure 214. The dynamics of bubble formation during sudden decompression.

below the sea of approximately 200 feet. The pressure on the outside of the body is 5000 mm. Hg, and the pressure of all the dissolved gases in the body fluids is only 4045 mm. Hg. Therefore, the pressure in this instance is great enough on the outside of the body to keep the dissolved gases from ever becoming bubbles. To the right in this same figure, however, the person has suddenly been brought to the surface of the sea where the pressure on the outside of his body and in his lungs is only 760 mm. Hg. Immediately, the gases begin to expand inside the fluids because the pressure on the outside of the body is now much less than the pressure of the gases dissolved in the fluids. The expanding gases form bubbles throughout the body—in the cells, in the blood, in the spinal fluid, and almost everywhere else.

EFFECT OF BUBBLE FORMATION ON THE BODY. The effects of bubble formation in the body are known by many different names: *decompression sickness*, the *bends*, *diver's paralysis*, *caisson disease*, and others. The most distressing effects occur when bubbles develop in the central nervous system. They sometimes occur inside neuronal cells, but mostly in extraneuronal tissues of the brain and spinal cord, causing mechanical rupture of fiber pathways. The damage can lead to serious mental disorders and very frequently to permanent paralysis because of ruptured fiber pathways to the muscles.

Decompression sickness usually causes severe pain, which may result from bubble damage in the central nervous system or in peripheral nerves. Sometimes the pain is caused by bubble formation in pain-sensitive tissues such as the joints and bones, or it can result from sudden distention of the gastrointestinal gases, causing severe bloating of the gut.

In short, many different symptoms are produced by bubble formation and gas expansion in the body, though the most distressing and permanent of these result from damage to the nervous system.

PREVENTION OF DECOMPRESSION SICKNESS. A total of approximately 2 liters of bubbles can usually develop in the body fluids without dangerous consequences, but more than this almost always causes symptoms. These can be prevented, however, by allowing the diver to come to the surface very slowly or by decompressing him in a decompression chamber. In either event, the pressure around his body must be decreased so gradually that the excess gases dissolved in his fluids can leave through his lungs before the quantity of bubbles rises above 2 liters.

REFERENCES

Armstrong, H. G.: *Aerospace Medicine.* Baltimore, The Williams & Wilkins Company, 1961.

Brown, J. H. U.: *Physiology of Man in Space.* New York, Academic Press, 1963.

Haugaard, N.: Cellular mechanisms of oxygen toxicity. *Physiol. Rev.* 48(2):311, 1968.

Lambertsen, C. J., and Greenbaum, L. J., Jr. (eds.): *Proceedings, Second Symposium on Underwater Physiology.* Natl. Acad. Sci.—Natl. Res. Council, Publication 1181, 1963.

Miles, S.: *Underwater Medicine.* Philadelphia, J. B. Lippincott Company, 1962.

Schade, J. P., and McMenemey, W. H. (eds.): *Selective Vulnerability of the Brain in Hypoxaemia.* Philadelphia, F. A. Davis Company, 1963.

Schaefer, K. E. (ed.): *Environmental Effects on Consciousness.* New York, The Macmillan Company, 1962.

Scholander, P. F.: Physiological adaptation to diving in animals and man. *Harvey Lect.* 67:93, 1961-1962.

Sells, S. B., and Berry, A. A. (eds.): *Human Factors in Jet and Space Travel.* New York, The Ronald Press Company, 1961.

Submarine Medicine Practice, Department of the Navy. Washington, U. S. Government Printing Office, 1956.

SECTION SIX

THE NERVOUS SYSTEM

CHAPTER 22

DESIGN OF THE NERVOUS SYSTEM, AND BASIC NEURONAL CIRCUITS

Basic Organization of the Nervous System

The basic functional units of the nervous system are the *neurons*. The *nerve fibers* are filamentous outgrowths of the neuron cell bodies, and it is these fibers that transmit information from one part of the nervous system to another. Also, it is the neurons in the central nervous system that process the incoming information from the periphery and determine the signals to be transmitted back to the different parts of the body to initiate various bodily activities. To perform these functions the nervous system is composed of three major parts: (1) the sensory system, (2) the motor system, and (3) the integrative system, each of which may be described as follows:

THE SENSORY SYSTEM. Most activities of the nervous system originate with sensory experience, whether this be visual experience, auditory experience, tactile sensations from the surface of the body, or other sensations. The sensory experience can cause either an immediate reaction, or its memory can be stored in the brain for many weeks or

years and can then help to determine bodily actions at some future date.

Figure 215 illustrates the general plan of the sensory system, which transmits sensory information from the entire surface of the body and deep structures into the nervous system through the spinal nerves. This information is conducted into (a) the spinal cord at all levels, (b) the basal regions of the brain, including the medulla and pons, and (c) finally into the higher regions of the brain, including the thalamus and the cerebral cortex. In addition to these primary "sensory" areas, signals are secondarily relayed to essentially all other parts of the nervous system.

THE MOTOR SYSTEM. The most important ultimate role of the nervous system is to control the bodily activities. This is achieved by controlling (a) contraction of skeletal muscles throughout the body, (b) contraction of smooth muscle in the internal organs, and (c) secretion by both exocrine and endocrine glands in many parts of the body. These activities are collectively called *motor functions* of the nervous system, and the portion of the nervous system

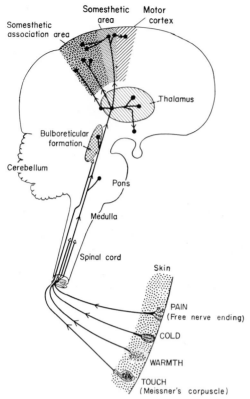

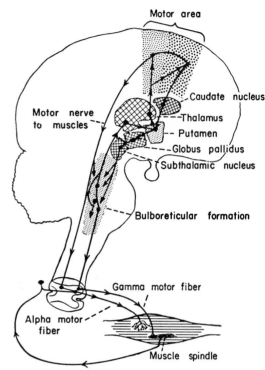

Figure 215. The sensory division of the nervous system.

Figure 216. The motor division of the nervous system.

that is directly concerned with transmitting signals to the muscles and glands is called the *motor division* of the nervous system.

Figure 216 illustrates the general plan of the motor axis for controlling skeletal muscle contraction. Signals originate in (a) the motor area of the cerebral cortex, (b) the basal regions of the brain, or (c) the spinal cord and are transmitted through motor nerves to the muscles. Each specific level of the nervous system plays its own special role in controlling bodily movements, the spinal cord and basal regions of the brain being concerned primarily with automatic responses of the body to sensory stimuli and the higher regions with deliberate movements controlled by the thought processes of the cerebrum.

THE INTEGRATIVE SYSTEM. The term *integrative* means processing of information to determine the correct and appropriate motor action of the body or to provide abstract thinking. Located immediately adjacent to all sensory or motor centers in both the spinal cord and brain are numerous cen-

ters concerned almost entirely with integrative processes. Some of these areas are concerned with the storage of information, which is called *memory*, while others assess sensory information to determine whether or not it is pleasant or unpleasant, painful or soothing, intense or weak, and so forth. It is in these regions that the appropriate motor responses to incoming sensory information are determined; once the determination is made, signals are then transmitted into the motor centers to cause the motor movements.

The Neuron as the Basic Functional Unit of the Nervous System

Transmission of signals by nerve fibers was discussed in detail in Chapter 6. There are twelve billion neurons in the brain and spinal cord; nine billion of these are located in the cerebral cortex, which is the major storehouse for information, that is, for storing memories and for storing the different patterns of thinking, motor responses, and so forth. The neurons, by way of their filamentous nerve fibers, transmit signals one to another, in this way performing the different functions of the nervous system. There are many different types of neurons, some of which make many connections with other neurons and some of which make relatively few connections. Some neurons are very large, and others are very small. Some give rise to extremely large nerve fibers which can transmit signals at velocities as great as 100 meters per second. Others give rise to very minute fibers that transmit nerve signals at velocities as slow as 1 meter per second. It is the goal of this chapter to explain how these neurons are organized into different types of nerve networks that perform the myriad functions of the nervous system.

Reflexes

Many functions of the nervous system are the result of reflexes. A reflex is a motor response that occurs following a sensory stimulus, the response taking place through a *reflex arc* consisting of a *receptor*, a *transmitter*, and an *effector*. A receptor is any type of sensory nerve ending that is capable of detecting one of the usual body sensations such as touch, pressure, smell, sight, and so forth. Once the sensation is detected, a signal is transmitted by the transmitter, which is either a single neuron or several successive neurons connected in series with each other. Finally, the effector is a skeletal muscle or one of the internal organs such as the heart, gut, or a gland that can be controlled by the nerves.

Let us refer back to Figures 215 and 216 and see how a simple reflex could occur. Assume, for instance, that one of the free nerve endings in the skin is stimulated by a pain stimulus and that the pain signal is transmitted into the spinal cord. On entering the cord it excites other neurons that eventually send signals back to appropriate muscles to cause withdrawal of that part of a body contacting the painful stimulus. Thus, if a person steps on a nail, his foot is automatically withdrawn as soon as a sensory signal can be transmitted to the spinal cord and then back to the muscles of the leg. This is called, very simply, the *withdrawal reflex*, and it can occur even after the spinal cord has been completely separated from the brain. In other words, many of our automatic muscle functions are controlled primarily by the spinal cord and not by the conscious portions of our brain.

A much more complex reflex would be one in which many different sensory signals pass into the central nervous system from the eyes, ears, skin, and other portions of the sensory nervous system to apprise the person of danger. After a few seconds of integration, an automatic signal is transmitted back to the muscles to make the person run away. This is basically the same type of response as the simple withdrawal reflex except that it involves far more sensory elements, far more integrative elements, and far more motor elements. It also involves memories stored from previous learning, which make one realize the danger.

Thus, if one stretches his imagination far enough, he can explain almost any function of the nervous system on the basis of progressively more complex reflexes. However, most physiologists like to reserve the term reflex for an almost instantaneous automatic motor response to a sensory input signal, and to call the more complex activities of the nervous system the *higher functions*.

Three Major Levels of Nervous System Function

In phylogenetic development of man, the nervous system has progressed from simple nerve fibers connecting different parts of the body, as occur in primitive multicellular animals, up to the very complex nervous system containing twelve billion neurons in the human being. As animalhood reached the multisegmental stage, nerve fibers and neuronal cell bodies aggregated together to form a *neural axis* along the body, this neural axis providing the basis for the spinal cord of the human being. Then there developed in the head end of the axis enlarged aggregates of neurons which transmitted control signals down the neural axis to all parts of the body. These portions of the nervous system became highly developed in fishes, reptiles, and birds, and they are comparable to the basal regions of the human brain. Finally, there burst forth an overgrowth of nervous tissue surrounding the basal regions of the brain, forming the cerebral cortex. This occurred primarily in mammals and has reached a tremendous level of development in the human being.

During each of these stages of development, new functions of the nervous system were established, and many of these have been inherited all the way from the very lowest animals to the human being. For instance, in the multisegmental worm the most important nervous achievement was reflex response of the worm to noxious stimuli causing the worm's body to withdraw from any damaging influence. The human being has this same type of reflex, as previously described, controlled almost entirely by the spinal cord, an inheritance all the way from the worm to the human being.

In the more complex animal, such as fishes and reptiles, it became important for the animal to be oriented properly in space, which initiated the process of equilibrium, and the human being has inherited this same ability. Furthermore, it is in the basal regions of the human brain, comparable to the highest levels of the fish's brain, where equilibrium is controlled.

Finally, mammals have certain mental abilities over and above those of other animals, including especially the ability to store tremendous quantities of information in the form of memories and to utilize this information throughout life. Associated with this vast storage has also come the thinking process, a faculty that has reached its highest level of expression in the human brain. Along with the thinking process has come the art of communication, by voice, by sight, or other ways. All these functions are primarily controlled by the cerebral cortex.

Thus, in summary, the three major levels of organization in the central nervous system are (1) the *spinal cord*, which controls many of the basic reflex patterns of the body, (2) the *basal regions of the brain*, which control most of the vegetative functions of the body such as equilibrium, eating, gross body movements, walking, and so forth, and finally (3) the *cerebral cortex*, which is the harbinger of our higher thought processes and the controller of our voluntary discrete motor activities.

TRANSMISSION OF NERVE SIGNALS FROM NEURON TO NEURON: FUNCTION OF THE SYNAPSE

The *synapse* is the juncture between two neurons. It is through the synapse that sig-

nals are transmitted from one neuron to another. However, the synapse has the capability of transmitting some signals and refusing other signals, thereby making it a valuable tool of the nervous system for choosing which course of events to follow. It is because of this variable transmission of signals that the synapse is perhaps the most important single determinant of central nervous system function.

Physiologic Anatomy of the Synapse

Figure 217 illustrates a neuron, which is composed of three major parts: the *soma*, or main body of the neuron, and two types of projections, the *dendrites* and the *axon*. Each neuron has only one axon, and signals are transmitted *outward* from the neuron cell body through this axon. On the other hand, most neurons have many dendrites, sometimes as many as several hundred, and these transmit signals into the neuron cell body. They are usually short projections that extend only a few millimeters in the nervous system. However, an exception is the sensory nerve fiber, which is a single long dendrite projecting into a peripheral nerve, and transmits sensory information from a peripheral receptor. Thus, the peripheral nerves are composed of motor nerve axons going outward from the cord to the muscles, the heart, the gut, and so forth, and of sensory nerve dendrites carrying information inward to the cord.

Figure 217 also shows hundreds of small fibers leading to the neuron and terminating in small knobs called *presynaptic termi-*

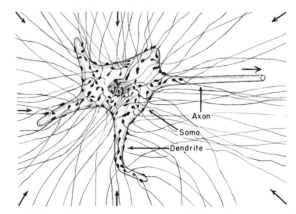

Figure 217. A typical neuron, showing hundreds of presynaptic terminals that originate from other neurons.

nals. These lie on the surfaces of the soma and dendrites. It is this juncture between the presynaptic terminals and dendrites or soma that is called the *synapse*. The small fibers are many branches of axons that come from other neurons.

Transmission of the Impulse at the Synapse

Transmission of the impulse from the presynaptic terminals to the dendrites or soma of the neuron occurs in very much the same way that transmission occurs at the neuromuscular junction as explained in Chapter 6. However, there is one major difference: at the neuromuscular junction the nerve endings always secrete acetylcholine which in turn excites the muscle fiber, while at the synapse some presynaptic terminals secrete an *excitatory transmitter substance* and others secrete an *inhibitory transmitter substance;* therefore some of these terminals excite the neuron and some inhibit it.

EXCITATION OF THE NEURON — "THE EXCITATORY TRANSMITTER." Figure 218 illustrates the structure of a typical presynaptic terminal lying adjacent to the membrane of either the soma or dendrite of a neuron. The terminal contains many small globules of transmitter substance, and when a nerve impulse reaches the presynaptic terminal, momentary changes in the membrane structure of the terminal allow a few of these globules to escape into the synaptic space shown in the figure. The transmitter substance then acts on the surface of the neuron to cause either excitation or inhibition.

The exact chemical nature of neither the excitatory nor inhibitory substance has yet been proved. However, it is believed that the excitatory transmitter, at least in part of the central nervous system, is *acetylcholine,* the same excitatory transmitter that transmits signals from motor nerves into muscle fibers. There are two reasons for this belief. First, in the ganglia of the sympathetic chain which lie outside the central nervous system but are similar in structure and function to the central nervous system, large quantities of acetylcholine are known to be secreted during the transmission of signals through the synapses. Second, certain "cholinergic" nerve fibers are known to enter the substance of the spinal cord and to excite neurons inside the spinal cord. Since it is known that these neurons secrete acetylcholine, it is presumed that acetylcholine is the excitatory transmitter.

EXCITATORY POSTSYNAPTIC POTENTIAL. The manner in which the excitatory transmitter excites the neuron can be explained by reference to Figure 219. The excitatory transmitter increases the permeability of the neuronal membrane immediately beneath the presynaptic terminal. This allows sodium ions to flow rapidly to the inside of the cell, and since sodium ions carry positive charges, the net result is an increase in positive charges inside the cell, which is called the *excitatory postsynaptic potential.* As a result, an electrical current, illustrated by the arrows in Figure 219, is immediately set up around the soma of the cell. If the potential becomes great enough, it will initiate an action potential in the axon that will travel over the nerve fiber leading from the neuron. However, if the excitatory postsynaptic potential is less than a certain threshold value, nothing will happen — that is, no action potential.

SUMMATION OF EXCITATORY POSTSYNAPTIC POTENTIALS. Almost invariably, stimulation of a single presynaptic terminal will not initiate an impulse in the axon. Instead, large numbers of presynaptic terminals must become excited at the same time. Referring once again to Figure 217, it is readily apparent that hundreds of presynaptic terminals could easily be excited simultaneously and that these operating in unison could cause neuronal discharge.

That is, if two terminals release their

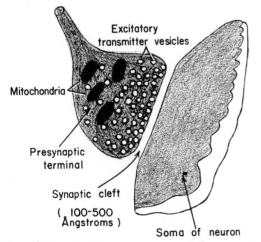

Figure 218. Physiologic anatomy of the synapse.

Excitatory transmitter vesicles

Mitochondria

Presynaptic terminal

Synaptic cleft
(100-500 Ångstroms)

Soma of neuron

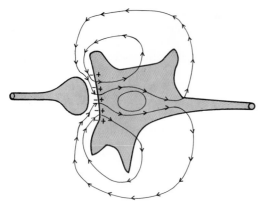

Figure 219. Flow of current beneath the synapse and around the neuron, showing that a synaptic discharge causes increased current flow over the entire body of the neuron.

excitatory transmitter substances simultaneously, twice as much sodium enters the cell body and twice as much excitatory postsynaptic potential develops. This is called *summation.* As more and more terminals summate, the potential becomes greater and greater, until finally it is great enough to excite the axon.

Spatial and temporal summation. Two different types of summation occur at the synapse in the same way that two types of summation occur in nerve and muscle fibers. These are spatial and temporal summation. *Spatial summation* means that two or more presynaptic terminals fire simultaneously, thereby "summing" their individual effects on the excitatory postsynaptic potential. *Temporal summation* means that the same presynaptic terminal fires two or more times in rapid succession, thus adding the effect of the second firing to that of the first. The effect of the first discharge usually lasts for about 15 milliseconds. Therefore, if two successive discharges of the same presynaptic terminal occur within less than $1/70$ second of each other, temporal summation occurs, and the closer together these discharges, the greater will be the degree of summation.

REPETITIVE DISCHARGE OF THE AXON — THRESHOLD FOR FIRING. Once sufficient excitatory transmitter has been secreted by the presynaptic terminals to raise the postsynaptic potential above a critical value called the *threshold* of the neuron, the axon will fire repetitively and will continue to do so as long as the potential remains above

this threshold. Furthermore, as the postsynaptic potential rises higher and higher above the threshold, the more rapidly will the axon fire. For instance, let us assume that an excitatory postsynaptic potential of 15 millivolts is the threshold value. If the potential rises slightly above this, to 16 millivolts, we would expect the axon to discharge perhaps 5 to 10 times per second. But, if the postsynaptic potential rises to as high as 30 millivolts, twice the threshold value, then the axon perhaps would fire as many as 200 to 300 times per second.

Facilitation at the synapse. When presynaptic terminals discharge but fail to cause an action potential in the axon, the neuron still becomes *facilitated;* that is, even though an action potential does not result, the neuron becomes more excitable to impulses from other presynaptic terminals. Let us assume, for instance, that 25 presynaptic terminals must fire simultaneously to discharge a neuron. If 20 presynaptic terminals fire simultaneously, the neuron will not discharge, but it does become sufficiently "facilitated" so that any 5 additional presynaptic terminals anywhere on the surface of the cell could now elicit an impulse. This is the situation that develops in a "nervous" person, for large numbers of his neurons become facilitated though not excited; yet very few extraneous impulses can then cause terrific reactions.

Response characteristics of different neurons. The nervous system is composed of hundreds of different types of neurons. Some of the cell bodies are tremendously large in size while others are very minute. Some are capable of transmitting as many as 1000 impulses per second over their axons and some are capable of transmitting no more than 25 to 50 per second. And, different neurons have different thresholds. All of these differences are fortunate, because neurons in different parts of the brain perform different functions.

Figure 220 illustrates typical *response patterns* of three different neurons. Neuron 1 requires a postsynaptic potential of 5 millivolts to discharge it, and it reaches a maximum discharge rate of 100 per second. The threshold of neuron 2 is 8 millivolts, and the maximum frequency of discharge is 35. Finally, neuron 3 has a high threshold of 18 millivolts, but once it becomes excited, its maximum discharge rate is 120 per second.

INHIBITION AT THE SYNAPSE. Some of

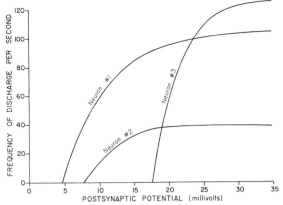

Figure 220. Response characteristics of different types of neurons.

the presynaptic terminals secrete an *inhibitory transmitter* instead of an excitatory transmitter. Unfortunately, the inhibitory transmitter, like the excitatory transmitter, has not been isolated with certainty. However, a substance called *gamma aminobutyric acid* (GABA), found in large quantities in inhibitory nerve fibers, has an inhibitory effect on the synapse. Therefore, it is now believed to be the inhibitory transmitter.

The inhibitory transmitter has an opposite effect on the synapse to that caused by the excitatory transmitter, usually causing a negative potential called the *inhibitory postsynaptic potential.* An example of the manner in which the inhibitory transmitter operates at the synapse is the following: Let us assume that the threshold for stimulation of a neuron is +10 millivolts and that excitatory presynaptic terminals are generating a postsynaptic potential of +20 millivolts. The neuron will be firing repetitively at a rate of perhaps 50 to 200 times per second. At this time, stimulation of inhibitory presynaptic terminals releases inhibitory transmitter which creates an inhibitory postsynaptic potential of −15 millivolts. When this summates with the excitatory potential of +20 millivolts, the net value is only +5 millivolts, a level far below the threshold of the neuron. Therefore the neuron stops discharging.

EXCITATORY AND INHIBITORY NEURONS. The central nervous system is made up of *excitatory neurons* that secrete excitatory transmitter at their nerve endings, and of *inhibitory neurons* that secrete inhibitory transmitter at their nerve endings. Certain neuronal centers of the central nervous system are comprised entirely of excitatory neurons while others are both excitatory and inhibitory, comprised of both types of cells. Therefore, the central nervous system, unlike the peripheral sensory and motor systems, has two modes of activity, either excitation or inhibition instead of simply excitation alone.

Some Special Characteristics of Synaptic Transmission

ONE-WAY CONDUCTION AT THE SYNAPSE. Impulses traveling over the soma or dendrites of the neuron cannot be transmitted backward through the synapses into the presynaptic terminals. Thus, only one-way conduction occurs at the synapse. This is extremely important to the function of the nervous system, for it allows impulses to be channeled in the desired direction.

FATIGUE OF THE SYNAPSE. Transmission of impulses at the synapse is different from transmission in nerve fibers in a very important respect: the synapse *fatigues* very rapidly while nerve fibers fatigue hardly at all. Figure 221 illustrates this effect, showing that at the onset of the input signal, the neuron discharges very rapidly at first but then more and more slowly the longer it is stimulated. Some synapses fatigue very rapidly, while others fatigue very slowly.

One might expect this phenomenon of fatigue to be an impediment to the action of the central nervous system, but, on the

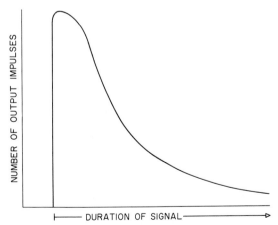

Figure 221. Effect of fatigue on the number of output impulses from the neuron after it begins to be stimulated.

contrary, it is a necessary feature. Were it not for synaptic fatigue, a person could never stop a thought, a rhythmic muscular activity, or any other prolonged repetitive activity of the nervous system once it had begun. Fatigue of synapses is a means by which the central nervous system allows a nervous reaction to fade away to make way for others. Later in the chapter, specific instances of the importance of fatigue are pointed out.

THE "MEMORY" FUNCTION OF THE SYN-APSE. When large numbers of impulses pass through a synapse, the synapse becomes "permanently" facilitated so that impulses from the same origin can pass through the synapse with greater ease at a later time. It is believed that this is the means by which memory occurs in the central nervous system. For instance, a given thought initiated by visual, auditory, or any other type of signal causes impulse transmission through given pathways in the brain. If the same thought is repeated over and over again, such as the thought caused by seeing the same view again and again, then the pathways for the thought become permanently facilitated so that subsequent impulses pass through these pathways with the greatest of ease. Therefore, impulses may later enter the same pathways from some other source besides the visual apparatus, and the person "remembers" the scene rather than actually seeing it. This process of memory will be discussed in greater detail in Chapter 29.

OTHER FACTORS THAT AFFECT SYNAPTIC TRANSMISSION. Figure 222 illustrates several different factors that can alter the

number of impulses transmitted by a synapse in response to a given degree of stimulation of the presynaptic terminals. Note that *hypnotics, anesthetics,* and *acidosis* all depress the transmission of impulses at the synapse, while *alkalosis,* the *mental stimulants* such as *caffeine, benzedrine,* and *strychnine* all greatly facilitate synaptic transmission. Strychnine, when given in sufficiently high dosage, will cause the neurons to discharge spontaneously even in the absence of a presynaptic stimulus. It is in this way that strychnine kills an animal, causing so many neuronal impulses to be transmitted throughout the central nervous system and into the motor system that it causes death by spastic paralysis of the respiratory muscles.

THE INTEGRATIVE FUNCTION OF THE NEURON. To summarize the action of the neuron, one can call it an "integrator," which means a type of calculator that collects information and sums it all together. Signals reach the neuron by way of excitatory and inhibitory presynaptic terminals that in turn are excited by neurons from other parts of the nervous system. If the resultant sum of all the excitatory and inhibitory effects is above the threshold for excitation, the neuron fires. Some of the neurons fatigue rapidly, others slowly. Some of the neurons have high thresholds, others have low thresholds. Some fire at rapid rates, others at slow rates. The varying characteristics of the different neurons and their different connections in the nervous system allow one portion of the nervous system to control one function of the body while another portion controls another function. They allow the sorting of signals to determine their meanings, the performance of special skilled motions, the thinking of specific thoughts, and modification of these thoughts by signals arriving from other parts of the nervous system.

BASIC NEURONAL CIRCUITS

The Neuronal Pool

The central nervous system is divided into many different anatomic parts, in each of which are located many accumulations of neurons called *neuronal pools.* An important feature of the different neuronal pools

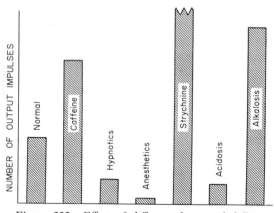

Figure 222. Effect of different drugs and different physiological states on the excitability of a neuron.

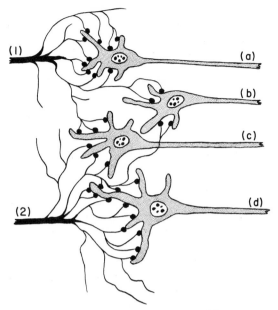

Figure 223. Schematic organization of four neurons in a neuronal pool.

is that each has a different pattern of organization from all the others. That is, the distribution of nerve fibers within the pool, the number of incoming nerve fibers, the number of outgoing fibers, the types of neurons in the pool, and many other features differ from one pool to another. Each pool is organized to perform a specific function. The purpose of the present section is to discuss, first, the general functions of the neuronal pool and, second, the characteristics of special types of pools.

SYNAPTIC CONNECTIONS IN A NEURONAL POOL. A neuronal pool is composed of thousands to millions of neuronal cell bodies. Figure 223 shows four typical neurons in a pool stimulated by two incoming nerve fibers. Note that each incoming fiber branches to supply terminal fibrils to many different neurons in the pool. A fiber may deliver one presynaptic terminal to a given neuron or as many as several hundred. Therefore, when the fiber is stimulated it will not facilitate neurons while exciting others.

AREAS OF DISCHARGE AND FACILITATION IN THE NEURONAL POOL. While still observing Figure 223, let us expand our imaginations until this pool of four neurons becomes a million closely packed neurons with hundreds of millions of terminal nerve fibrils. At the point of entry of a fiber to the pool it usually gives off large numbers of terminal fibrils, but farther away the number of terminal fibrils becomes less and less, as is illustrated in Figure 224. Those neurons that lie in the center of the fiber's "field" are usually supplied with enough presynaptic terminals that they discharge each time the input fiber is stimulated. Therefore, the dark area in Figure 224 is called the *discharge zone.*

In the peripheral portion of the fiber field, the number of terminal fibrils ending on any single neuron is usually too few to cause discharge but nevertheless still enough to cause facilitation. Therefore, this area is called the *facilitated zone.* When a neuron is facilitated, a stimulus from some other source can excite it more easily.

Simple Circuits in Neuronal Pools

The simplest circuit in a neuronal pool is that in which one incoming nerve fiber stimulates one outgoing fiber. This type of circuit does not exist in a precise form, though occasionally it is approximated. For instance, in transmitting certain types of sensory signals from peripheral nerves into the brain, a single incoming impulse into a

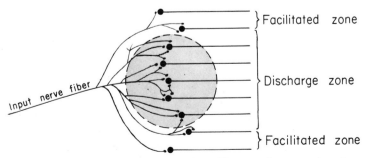

Figure 224. "Discharge" and "facilitated" zones of a neuronal pool.

neuronal pool may cause a single outgoing impulse; more than likely, though, there will be some simple ratio such as 1 to 2, or 2 to 1, or so forth.

Such a circuit obviously acts simply as a relay station, relaying on to additional neurons essentially the same information that enters it.

DIVERGING CIRCUITS. Figure 225 illustrates two types of *diverging circuits.* In the first of these, a single incoming fiber stimulates progressively more and more fibers further and further along the pathway. This can also be called an *amplifying circuit* because an input signal from a single nerve fiber causes an output signal in many different fibers. This type of circuit is exemplified by the system for control of skeletal muscles, because under appropriate conditions stimulation of a single motor cell in the brain sends a signal down to the spinal cord to stimulate perhaps as many as a thousand anterior horn cells, and each of these in turn stimulates approximately 150 muscle fibers. Thus, a single motor neuron of the brain can stimulate as many as 150,000 muscle fibers.

The second diverging circuit in Figure 225 allows signals from an incoming pathway to be transmitted into separate pathways, the same information being relayed in different directions at the same time. This type of circuit is common in the sensory nervous system. For instance, when a limb moves, the same information regarding the movement is transmitted by such a circuit into (a) neuronal pools of the spinal cord, (b) neuronal pools of the cerebellum, (c) neuronal pools of the thalamus, and (d) neuronal pools of the cerebral cortex.

THE CONVERGING CIRCUIT. Figure 226 illustrates two types of *converging circuits,*

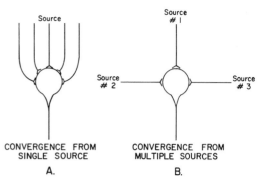

CONVERGENCE FROM SINGLE SOURCE
A.

CONVERGENCE FROM MULTIPLE SOURCES
B.

Figure 226. Converging circuits.

which are opposites of the diverging circuits. In the figure to the left, different nerve fibers from the same source converge on a single output neuron causing especially strong stimulation. To the right is another important type of converging circuit in which input fibers from several different sources converge on the same output neuron. This type of circuit allows signals from many different sources to cause the same effect. Thus, the smell of a pigpen might give a person the thought of a pig, the sight of a pig could do the same, the sound of his grunting could also produce this idea, and touching the pig or eating pork chops might lead to the same thought.

Use of converging circuits to perform integrative functions. Converging circuits can provide integrative functions that are much more complex than the integrative function of single neurons, for in a converging circuit there may be cells that respond to the incoming signal in several different ways at the same time. For instance, one of the neurons might have a very high threshold but when once stimulated might be so powerful that it causes a very intense reaction. In other parts of the circuit there may be neurons that have low thresholds but when stimulated transmit only weak signals. The entire pool, therefore, could transmit very weak output signals usually but very, very powerful signals if a certain degree of input stimulation should be exceeded. Thus, the converging circuit can be organized to perform almost any type of selective function. The individual neurons of the circuit select which signals shall pass, but the arrangement of the fibers and the combinations of cells determine which incoming signals will have the greater effects, and whether the effects will be excitatory or inhibitory.

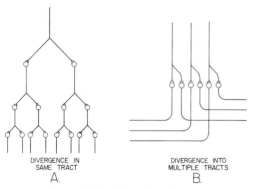

DIVERGENCE IN SAME TRACT
A.

DIVERGENCE INTO MULTIPLE TRACTS
B.

Figure 225. Diverging circuits.

REPETITIVE CIRCUITS. *The reverberating circuit.* Figure 227 shows two different types of circuits which, when stimulated only once, will cause the output cell to transmit a series of impulses. The upper circuit of this figure is a *reverberating circuit* which functions as follows: An incoming signal stimulates the first neuron which then stimulates the second and third cells. However, branches return to the first neuron and restimulate it. As a result, the signal travels once again through the chain of neurons, this process continuing around and around the circuit indefinitely until one or more of the cells fails to fire. The usual cause of failure is fatigue, and until this takes place, the output neuron continues to be stimulated every time the signal goes around the circuit, giving a continuous output signal.

The reverberating circuit is the basis of innumerable central nervous system activities, for it allows a single input signal to elicit a response lasting a few seconds, minutes, or hours. Indeed, the life-long respiratory rhythm is caused by a reverberating circuit that continues to reverberate without fatiguing as long as the person remains alive. It is probable, also, that a special circuit in the brain causes a person to awaken when it reverberates, and allows him to relax into a state of sleep when it stops reverberating.

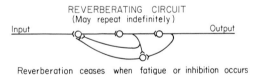

REVERBERATING CIRCUIT
(May repeat indefinitely)

Input   Output

Reverberation ceases when fatigue or inhibition occurs

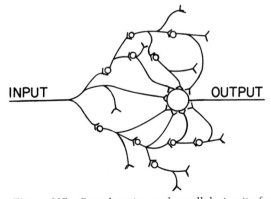

PARALLEL CIRCUIT
(Repeats definite number of times)

INPUT   OUTPUT

Figure 227. Reverberating and parallel circuits for repetitive discharge.

Almost all rhythmic muscular activities, including the rhythmic movements of walking, are probably controlled by reverberating circuits.

A few moments of thought about the reverberating type of circuit will emphasize the extreme number of variable functions it can perform. For instance, the number of neurons in the circuit may be great or small. If the number is great, the length of time required for the signal to go around the circuit will be long; if the number is small, the reverberatory period will be short. As a result, the output signal may repeat itself rapidly or slowly. Furthermore, more than one of the neurons in the circuit can give off output signals. One of the cells in the circuit, for instance, might cause an arm to move upward, then another cell in the circuit a fraction of a second later might cause the arm to move to the right, another cell still later cause it to move downward, and another cell cause it to move to the left. As the reverberatory cycle repeats itself once more, the same motions occur again, causing the arm to move around and around continuously until the cycle stops. When the person wishes to perform this motion, all he has to do is to set off the reverberatory cycle.

The parallel circuit. The lower part of Figure 227 illustrates the *parallel* type of repetitive circuit in which a single input signal stimulates a sequence of neurons that send separate nerve fibers directly to a common output cell. Because a delay of about $1/2000$ second occurs each time an impulse crosses a synapse, the impulse from the first neuron arrives at the output cell $1/2000$ second ahead of the impulse from the second neuron, and impulses continue to arrive at these short intervals until all the neurons have been stimulated. Since there is no feedback mechanism in the circuit, the repetition then ceases entirely.

Parallel circuits in different parts of the central nervous system differ in several ways: First, some circuits are composed of far more neurons than others and give a repetitive output lasting considerably longer periods of time. Second, the parallel circuit can be combined with diverging circuits, so that the signal can be amplified to stimulate the output neuron very strongly during part of the repetitive discharge and weakly during other parts of the discharge. In this way the output signal will have a definite

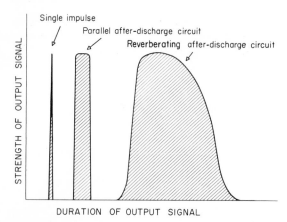

Figure 228. Characteristics of signals caused by (1) a single nerve impulse, (2) discharge of a parallel after-discharge circuit, and (3) discharge of a reverberatory after-discharge circuit.

described, we can discuss briefly the overall responses of typical neuronal pools. Almost never is a neuronal pool simply a relay station; that is, almost never do the same number of impulses leave a pool as enter it. Usually incoming signals diverge, converge, or are changed into repetitive signals whose durations last long after the input signal is over. Also, many incoming stimuli do not excite the neuronal pool at all but instead inhibit it.

Figure 229 illustrates some of the response characteristics of typical neuronal pools. The four different types of responses illustrated here are: (1) low threshold, high amplification response, (2) low threshold, low amplification response, (3) high threshold, high amplification response, and (4) high threshold, low amplification response.

With a little imagination one can readily understand possible functions for each of these types of circuits. For instance, the *low threshold, high amplification* response would be that in which a person reacts vigorously to a very slight sensory input signal, such as stimulation of only a few pain nerve endings. The *high threshold, high amplification* response would be one in which a person subjected to moderately loud noises pays no attention and yet when a very loud noise suddenly occurs, he reacts vigorously. In the same manner, we could easily find typical central nervous system responses representative of the other two response curves in Figure 229.

An important characteristic of neuronal pools is that the response curve to input

amplitude pattern as well as a definite duration.

The parallel type of circuit has certain advantages over the reverberating circuit. For instance, the very variable phenomenon of fatigue is a major factor in determining how long the reverberating circuit will continue to fire, whereas the output duration of the parallel circuit is independent of fatigue. For performing very exact activities such as mathematical calculations the parallel circuit is perhaps very useful, while for rhythmic functions or for greatly prolonged discharges the reverberating circuit is a necessity. Unfortunately, the parallel circuit cannot control functions that last more than a fraction of a second, because one neuron is required in a circuit for each $1/2000$ second. Probably not more than a few dozen neurons are ever organized into parallel circuits, which would limit this type of circuit to activities occurring in less than about $1/50$ second.

Figure 228 illustrates the time duration and amplitude characteristics of the output signals from, first, a one-to-one circuit, second, a parallel after-discharge circuit, and, third, a reverberatory after-discharge circuit.

Relationship of the Output Signal to Input Signal of a Neuronal Pool

Now that the characteristics of some of the important neuronal circuits have been

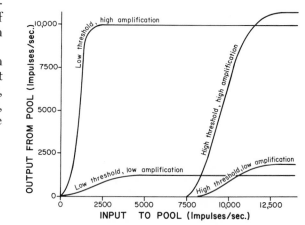

Figure 229. Function curves of different neuronal pools, illustrating the response of different types of circuits to incoming impulses.

signals from one source can be altered by input signals coming from secondary sources. For instance, sensory nerve impulses from the skin into the spinal cord ordinarily cause no reflex skeletal muscle effects. However, strong facilitatory impulses transmitted from the brain down the spinal cord to the neurons that control the muscles can make these neurons so excitable that a very light scratch on the skin elicits a strong contraction of the underlying muscle. Thus, a high threshold, low amplification circuit has been changed into a low threshold, high amplification circuit.

It is now up to the imagination of the student to conceive the many possible ways in which individual types of neurons can be organized into different types of neuronal pools and in which different types of neuronal pools can perform an infinite number of reflex and integrative nervous functions. Unfortunately, the precise circuits in most neuronal pools have not yet been worked out in detail. Nevertheless, the types of circuits that are already known can suggest mechanisms by which all of the functions of the central nervous system could be attained.

REFERENCES

Alpers, B. J.: *Clinical Neurology.* 5th Ed. Philadelphia, F. A. Davis Company, 1963.

Bremer, F.: Central regulatory mechanisms—introduction. *Handbook of Physiology,* Sec. I, Vol. II, p. 1241. Baltimore, The Williams & Wilkins Company, 1960.

Davis, H.: Some principles of sensory receptor action. *Physiol. Rev.* 41:391, 1961.

Eccles, J. C.: Neuron physiology—introduction. *Handbook Physiol.* 1:59, 1959.

Gerard, R. W., and Duyff, J. W. (eds.): Processing of information in the nervous system. *Proc. Intern. Union Physiol. Sci.,* Vol. III, 1962.

Magoun, H. W.: *The Waking Brain.* Springfield, Ill., Charles C Thomas, 1960.

Martin, A. R., and Veale, J. L.: The nervous system at the cellular level. *Ann. Rev. Physiol.* 29:401, 1967.

Sherrington, C. S.: *The Integrative Action of the Nervous System.* New York, Charles Scribner's Sons, 1906.

CHAPTER 23

SOMESTHETIC SENSATIONS AND INTERPRETATION OF SENSATIONS BY THE BRAIN

The term *somesthetic sensation* means sensation from the body. Also physiologists frequently speak of subdivisions of the somesthetic sensory system, including *exteroceptive* sensation, *proprioceptive* sensation, and *visceral* sensation. There is much overlap between these different types of sensations.

Exteroceptive sensations are those normally felt from the skin, such as (1) touch, (2) pressure, (3) heat, (4) cold, and (5) pain.

Proprioceptive sensations are those that apprise the brain of the physical state of the body, including such sensations as (1) tension of the muscles, (2) tension of the tendons, (3) angulation of the joints, and (4) deep pressure from the bottom of the feet. Note that pressure can be considered to be both an exteroceptive sensation and a proprioceptive sensation.

Visceral sensations are those from the internal organs, including such sensations as (1) pain, (2) fullness, and (3) sometimes the sensation of heat. Thus, the visceral sensations are similar to exteroceptive sensations and are functionally the same except that they originate from inside the body.

General Organization of the Somesthetic Sensory System

Figure 230 illustrates the general plan for transmission of somesthetic sensory signals into the brain. The sensations are detected by special nerve endings in the skin, in the muscles, in the tendons, or in the deeper areas of the body, and these emit nerve impulses that are transmitted through nerve trunks into the spinal cord. Upon entering the spinal cord, the sensory nerve fibers branch. Some of the branches end in the spinal cord itself to cause *cord reflexes* which will be discussed in Chapter 26, while the others extend to other areas of the cord and brain. The sensory pathways to the brain terminate in several discrete areas as follows: (1) in sensory areas of the brain stem, including the bulboreticular formation and the central gray area, (2) the cerebellum, (3) the thalamus and other closely allied structures, and (4) the cerebral cortex. The signals transmitted into each of these different areas subserve specific functions, which will become clear later in the chapter. Signals transmitted into the cere-

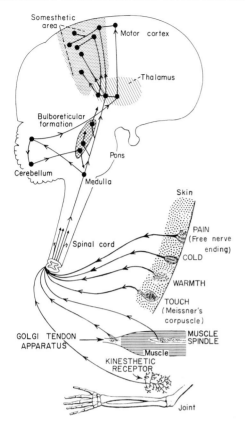

Figure 230. Transmission of sensory signals to the brain, showing the sensory receptors and the nerve pathways for transmitting these sensations into the brain.

bellum occur entirely at a subconscious level and are concerned with subconscious control of motor function. Therefore, the cerebellar component of the somesthetic sensory system will be discussed in connection with motor activities of the body rather than in the present chapter.

FUNCTION OF DIFFERENT NERVOUS SYS-TEM LEVELS OF SOMESTHETIC SENSATION. The sensory nerve fibers that terminate in the spinal cord initiate cord reflexes. These cause immediate and direct motor activities such as contraction of muscles to pull a limb away from a painful object, or perhaps alternate contraction of the limbs to cause walking movements. These reflexes will be discussed in detail in Chapter 26.

The sensory signals that terminate in the lower brain stem cause subconscious motor reactions of much higher and more complex nature than those caused by the cord reflexes. For instance, it is in the lower brain stem that chewing, control of the body trunk, and

control of the muscles that support the body against gravity are all effected.

As the sensory signals travel still farther up the brain and approach the thalamus, they begin to enter the level of consciousness. When sensory signals reach the thalamus their origins are localized crudely in the body, and the types of sensations, called their *modalities of sensation*, begin to be appreciated. Thus, one can then determine whether the sensation is touch, heat, or cold, and so forth. Yet for full appreciation of these qualities, and especially for very discrete localization, the signals must pass on into the cortex. The cortex is a large storehouse of information, of memories of past experiences, and it is this stored information that allows the finer details of interpretation to be achieved.

THE SENSORY RECEPTORS

Some sensory nerve endings are nothing more than small filamentous branches called *free nerve endings,* while others are special *end-organs* that are designed to respond only to special types of stimuli. Figure 230 illustrates some of the more representative sensory receptors, and their functions may be described as follows:

FREE NERVE ENDINGS. Free nerve endings detect the sensations of crude touch, pressure, pain, heat, and cold. These sensations are likely to become somewhat confused with each other because the free endings interconnect extensively. For instance, extreme heat or cold is likely to give one a sensation of pain, or very hard pressure might also be confused with pain.

Despite the fact that free nerve endings transmit only crude sensations, they are by far the most common type of nerve ending. They perform most of the general functions of sensation, while the specific functions, such as discrimination of very slight differences between degrees of touch, warmth, and cold, are left to the more specialized receptors.

SPECIAL EXTEROCEPTIVE RECEPTORS. Also shown in Figure 230 are specialized receptors for detection of cold, warmth, and light touch (Meissner's corpuscles). The reason why each of these end-organs registers only one type of sensation is not known. Presumably the physiologic organization of the end-organ itself allows cold, warmth, or light touch respectively to stimulate the nerve ending by some physical effect on the end-organ.

Still other specialized sensory receptors are present in the skin. For instance, at the base of each hair is a receptor that allows one to feel even the slightest pressure on any hair. In animals some of these hair receptors are so well developed that the associated hairs are called *tactile hairs*. A cat actually helps to guide itself in the darkness by its tactile whiskers. Special end-organs have also been found in the lips and snouts of some rooting animals; these presumably aid in the search for food. Finally, in the sexual organs special endings are possibly responsible for the distinctive qualities of sexual sensations.

THE PROPRIOCEPTOR END-ORGANS. Figure 230 illustrates three different types of proprioceptor end-organs. One of them, the joint kinesthetic receptor, is found in the capsules of joints. These receptors apprise one of the *degree* of angulation of the joints and of the *rate* at which this degree of angulation changes.

Two special end-organs transmit proprioceptive information from the muscles. These are the *muscle spindle* and the *Golgi tendon apparatus*. The muscle spindle is composed of nerve filaments wrapped around specially adapted muscle fibers. The spindle detects the degree of stretch of the muscle. This information is transmitted to the central nervous system to aid in the control of muscular movements. The Golgi tendon apparatus detects the overall tension applied to the tendon, and, therefore, apprises the central nervous system of the effective strength of contraction of the muscle.

Adaptation of the Sensory Receptors

When a stimulus is suddenly applied to a sensory receptor, it usually responds very vigorously at first, but progressively less and less so during the next few seconds or minutes. An example of this is getting into a tub of hot water, which causes an intense burning sensation at first but after a moment's exposure produces a sensation of pleasing warmth. The same is true to varying degrees for all the other sensations. This loss of sensation during prolonged stimulation is called *adaptation* of the sensory receptor.

Adaptation occurs to much greater degrees for some sensations than for others. For instance, the sensations of light touch and of pressure adapt within a few seconds. This allows one to feel an object when he first touches it but to lose the sensation very soon thereafter. Were it not for this rapid adaptation of the light touch receptors, one would feel intense touch sensations, never stopping, from all the body areas in contact with any object such as the seat of a chair, the shoes, and even the clothing. These sensations would continue to bombard the brain to such an extent that one could hardly think of anything else. The value of rapid adaptation of certain of the sensations therefore is obvious.

The sensations of pain and some types of proprioception usually adapt to only a slight extent. Pain sensations are elicited when tissue damage is occurring. As long as the damage continues to occur it is important that the person also continue to be apprised of this fact, so that he will institute appropriate measures to remove the cause of the damage. The long persistence of proprioceptive sensations, also, is desirable because the brain needs to know the physical status of the different parts of the body at all times and not simply immediately after movements have occurred.

Discrimination of Intensity of Sensations—The Weber-Fechner Law

A person can detect the weight of a flea on the tip of his finger or he can detect the weight of a man stepping on the same finger. The difference in weight of these two animals is approximately 70 million times, and yet, from the sensations perceived, a person can estimate the heftiness of the two objects. The reason why one can discriminate such wide differences in intensities is that the number of impulses transmitted by a sensory nerve is roughly proportional to the logarithm of the intensity of sensation rather than to the actual intensity. The logarithmic response of receptors is illustrated in Figure 231, which shows the impulse rate from a

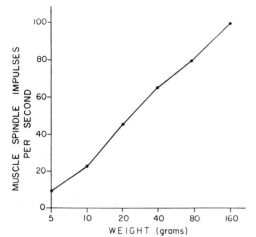

Figure 231. Logarithmic response of the muscle spindle. (Drawn from data from Matthews: *J. Physiol. (Lond.)* 78:1, 1933.)

muscle spindle as different weights are applied to the muscle. Note that the weight scale doubles at each point, and yet the spindle impulses increase linearly. This is a logarithmic type of response; that is, the number of impulses is approximately proportional to the logarithm of the weight rather than to the weight itself.

In the case of the flea and the man, the difference between the logarithms of their weights is only about 8 times even though the difference between the actual weights is one hundred million times. This logarithmic detection of sensation applies not only to somesthetic sensations, but also to the sensations of vision, hearing, taste, and smell; it is called the *Weber-Fechner law.*

Another example of the Weber-Fechner law is the following: If one is holding an object in his hand that weighs 1 ounce and he exchanges this object for another that weighs 1.1 ounce, he will be able barely to discriminate the difference in weight of the two objects. The actual difference in weight is 0.1 ounce. Then he picks up an object that weighs 10 pounds, and exchanges this object for another that weighs 10 pounds and 0.1 ounce. He will be completely unable to discriminate the difference in weight. Instead, the second object must now weigh approximately 11 pounds for him to tell the difference. The increase in weight this time must be 1 pound instead of 0.1 ounce. In each instance, the increase is 10 per cent of the original weight. Therefore, another way of expressing the Weber-Fechner law is:

discrimination of intensity of the different types of sensation is on relative basis rather than on an absolute basis.

Law of Specific Nerve Energies

Even though a different type of nerve receptor is responsible for detecting each type of sensation—pain, touch, pressure, position, and so forth—nevertheless, it is not the receptor itself that determines the type of sensation that a person feels. Instead, it is the point within the brain to which the signal is transmitted. For instance, if a pain nerve fiber is stimulated by crushing it, by heating it, by bending it, or in any other way, the person will still feel pain regardless of how the pain nerve ending has been stimulated. Likewise, a touch nerve fiber can be stimulated by crushing it, burning it, or bending it, and the person will feel nothing but touch even though the nerve fiber itself is being damaged severely, which one generally considers to cause pain.

MODALITY OF SENSATION. The term "modality" of sensation means the specific quality of sensation felt, whether the feeling be one of pain, touch, pressure, position, vision, hearing, equilibrium, smell, or taste, all of which are different modalities of sensation.

It is frequently stated that the thalamus is the primary locus for determining modality of sensation. The reason for stating this is that fiber pathways for different modalities terminate at different points in the thalamus, and destruction of the thalamus makes it difficult to distinguish some types of sensory modalities. However, it has now been learned that some of the pain pathways terminate in even lower areas of the brain stem such as the central gray matter of the mesencephalon and in the hypothalamus, and that these areas are principal sites for detection of pain. Furthermore, the cerebral cortex is known to sharpen one's ability to detect the different modalities, even though the lower areas of the brain can provide crude detection.

Thus, it is becoming apparent that widely scattered regions of the brain operate together to determine the different sensory qualities. As the sensory signals enter the brain from the cord, they are picked apart for their different characteristics. One portion of the brain determines whether or not there is a pain element in the sensation, another whether or not there is a touch element, another the part of the body from which the sensation is coming, and so forth.

DUAL SYSTEM FOR TRANSMISSION OF SOMESTHETIC SENSATIONS

On entering the spinal cord the somesthetic sensory signals may be transmitted up the remainder of the nervous system axis by either one of two pathways called respectively the *dorsal column system* and the *spinothalamic system*. These two pathways, illustrated in Figure 232, are anatomically distinct and also display different characteristics for transmission of signals.

Comparison of the Dorsal Column and Spinothalamic Systems

The dorsal column system is illustrated in Figure 232A. Note that sensory nerve fibers entering this system pass all the way up the same side of the spinal cord until they reach the medulla. Here they terminate on "second order" neurons, which in turn send fibers immediately to the opposite side of the medulla and then up to the posterior part of the thalamus, terminating in an area called the *ventrobasal complex*. Here these neurons synapse with "third order" neurons which send fibers thence to the *somesthetic sensory area* of the cerebral cortex. Note especially that there are three separate neurons in the pathway, the first neuron passing all the way from the receptor on the surface of the body to the medulla oblongata, the second from the medulla to the thalamus,

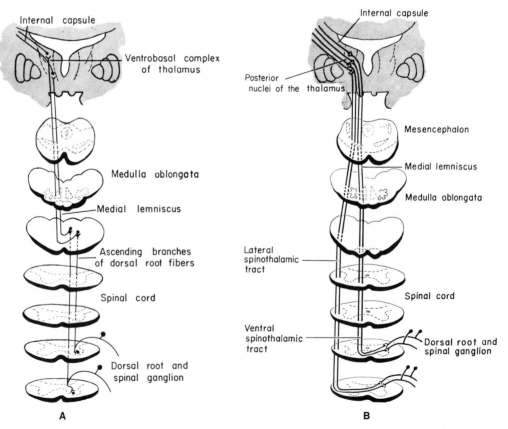

Figure 232. (A) The dorsal column system. (B) The spinothalamic system. (Modified from Ranson and Clark: *Anatomy of the Nervous System.* W. B. Saunders Co.)

and the third from the thalamus to the cortex. Figure 233 shows second order neurons synapsing with third order neurons in the thalamus which then radiate upward to the somesthetic sensory cortex. Note that the leg is represented near the mid-line of the cortex and the face far to the side.

The spinothalamic system is illustrated in Figure 232B. Note that it, too, is composed of three orders of neurons. The first order neurons originate in the receptors in the body and pass to the spinal cord. Here, they terminate almost immediately, synapsing with second order neurons located in the posterior horns of the spinal cord gray matter. The fibers from these neurons in turn cross to the opposite side of the cord and travel up to the brain stem through the lateral and ventral columns of the cord. These neurons terminate at all levels of the brain stem, from the medulla all the way up to the posterior thalamus. In the thalamus, they synapse with third order neurons that transmit signals to the somesthetic cortex.

DISCRETENESS OF SIGNAL TRANSMISSION IN THE DORSAL COLUMN AND SPINOTHALAMIC SYSTEMS. The dorsal column system, as well as the peripheral nerve fibers that connect with it, is composed of very large nerve fibers that transmit signals at velocities of 35 to 100 meters per second. On the other hand, the spinothalamic system and its associated peripheral nerve fibers are composed of very small nerve fibers, some of which are not myelinated at all. These fibers transmit signals at velocities of 1 to 30 meters per second. Since the spinothalamic system conducts signals very slowly, it can be used only for information that the brain can afford to receive after a short delay. On the other hand, the dorsal column system allows transmission of information to the brain within a very small fraction of a second.

Another major difference between the dorsal column and spinothalamic systems is the degree of spatial orientation of the nerve fibers within each of the two tracts. The nerve fibers of the dorsal column system are myelinated so that they are well insulated from each other. Also, there is very little crossover of signals from one part of the tract to other parts where the nerve fibers synapse in the medulla and thalamus. Therefore, when a single receptor is stimulated in the skin or other peripheral area of the body, a very discrete signal is transmitted to a very highly localized point in the cerebral cortex. By contrast, the degree of insulation between the nerve fibers in the spinothalamic system is far less, and there is far greater diffusion of nerve signals sidewise where the spinothalamic pathways synapse in the spinal cord and in the thalamus. As a consequence, stimulation of a single nerve receptor in the periphery causes excitation of a widely dispersed area in the brain. Furthermore, it is often necessary to stimulate many receptors simultaneously to get enough signal strength to cause transmission of the signal into the central regions of the brain at all.

MODALITIES OF SENSATION TRANSMITTED BY THE TWO SYSTEMS. With the foregoing differences in mind, we can now list the types of sensations transmitted in the two systems.

The Dorsal Column System

1. Touch sensations having a high degree of localization of the stimulus and transmitting fine gradations of intensity.

2. Phasic sensations, such as vibratory sensations.

3. Kinesthetic sensations (sensations having to do with body movements).

4. Muscle sensations.

5. Pressure sensations having fine gradations of intensity.

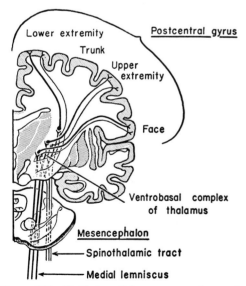

Figure 233. Radiation of the sensory pathways from the thalamus to the somesthetic sensory area in the cortex. (Modified from Brodal: *Neurological Anatomy in Relation to Clinical Medicine.* Oxford University Press.)

The Spinothalamic System
1. Pain.
2. Thermal sensations, including both warm and cold sensations.
3. Crude touch sensations capable of gross localization of the stimulus on the surface of the body.
4. Pressure sensations of a cruder nature than those transmitted by the dorsal column system.
5. Tickle and itch sensations.
6. Sexual sensations.

COLLATERAL SIGNALS FROM THE SPINO-THALAMIC SYSTEM TO THE BASAL REGIONS OF THE BRAIN. Figure 232B illustrates another very important difference between the spinothalamic and dorsal column systems. That is, a large share of the nerve fibers of the spinothalamic system terminate in the medulla, pons, and mesencephalon even before they reach the thalamus, and many of the fibers that do go all the way to the thalamus give off branches into these lower regions of the brain on the way up. On the other hand, the dorsal column system passes through these regions with almost no branches. Thus, the spinothalamic system is much more concerned with sensations that elicit subconscious automatic reactions than is the dorsal column system. On the other hand, the dorsal column system is concerned with very discrete signals that are transmitted primarily into the conscious areas of the brain.

To express this another way, the spinothalamic system is concerned with the older types of sensations, those that occur in even very low forms of animal life, while the dorsal column system is concerned with types of sensation that have appeared recently in the phylogenetic development of the human being.

MECHANISM FOR LOCALIZING SENSATIONS ON SPECIFIC SURFACE AREAS OF THE BODY

Among the most important information that the brain must determine about each somesthetic sensation is its point of origin on the body. Once the location has been determined, it is possible to do something about the condition giving rise to the sensation. To provide this localization ability, the nerve fibers are *spatially oriented* in the nerve trunks, in the spinal cord, in the hindbrain, and in the cerebral cortex. This may be explained as follows.

SPATIAL ORIENTATION OF NERVE FIBERS. The sensory nerve fibers arising in the leg are separated in the spinal cord and brain from the fibers arising in the arm. The fibers arising from adjacent toes are separated from each other, and even the fibers arising from two areas of skin only one centimeter apart are kept separate from each other. In the thalamus, fibers from each area of the body end at a specific point. Therefore, the thalamus is capable of determining, though only roughly, which part of the body is being stimulated. The thalamus is not organized satisfactorily to localize sensations to very discrete areas of the body; instead, this function is performed in the somesthetic area of the cerebral cortex.

THE SOMESTHETIC CORTEX. Figure 234 illustrates the *somesthetic cortex*, which is located immediately posterior to the central sulcus of the brain and extends from the longitudinal fissure at the top of the brain into the fissure of Sylvius at the side. This figure also shows the areas of the somesthetic cortex that receive sensations from each of the different parts of the body. Sensations from the foot, for instance, excite the somesthetic cortex where it dips into the longitudinal fissure. The leg area is approximately at the point where the somesthetic cortex comes out of the longitudinal fissure; then comes the thigh area, the abdomen, thorax, shoulder, arm, hand, fingers, thumb, neck, tongue, palate, and larynx. Thus, the entire body is spatially represented in the somesthetic cortex, each discrete point in

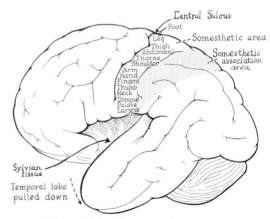

Figure 234.   The somesthetic cortex.

this portion of the brain corresponding to a discrete area only a few millimeters in size on the body surface.

The function of the somesthetic cortex is to localize very exactly the points in the body from which the sensations originate. Though the thalamus is capable of localizing sensations to very general areas, such as to one arm, to a leg, or to the body, it is not capable of localizing sensations to minute places on the body. Instead, the thalamus relays the necessary signals into the somesthetic cortex where a much better spatial representation is available, and there the job of discrete localization is performed.

Not all types of sensations are localized equally well. The sensations of aching pain, crude touch, warmth, and cold are localized to general areas of the body rather than to discrete areas. It seems that the thalamus performs most of the localization function for these modalities of sensations. On the other hand, the sensations of light touch, pressure, and position are very discretely localized by the somesthetic cortex. These are the sensations normally detected by the special end-organs, for instance Meissner's corpuscles for light touch and joint receptors for position sensation.

Function of the somesthetic cortex in interpreting sensory modality. Even when the somesthetic cortex is completely removed, a person still has the ability to detect the type of sensation that is being received, that is, whether it be pain, touch, heat, or cold and so forth, but his appreciation of these sensations is markedly reduced. Therefore, the cerebral cortex is not so much concerned with *determination* of modality of sensation as with its *interpretation.*

The somesthetic association area of the cortex. The area of the cortex a few centimeters behind the somesthetic cortex is called the *somesthetic association area,* because it is in this region that some of the more complex qualities of sensation are appreciated. Signals are transmitted directly into this area from several sources: (1) from the thalamus, (2) from other basal regions of the brain, and (3) backward from the somesthetic cortex. In ways not understood, the somesthetic association area puts all this information together and determines the following characteristics of sensations: (1) shape of an object, (2) relative positions of the parts of the body such as the legs, the hands,

and so forth, (3) texture of a surface, such as whether it is rough, smooth, or undulating, and (4) orientation of one object with respect to another object, in other words the spatial orientation of objects that are felt.

PAIN

The sense of pain deserves special comment because it plays an exceedingly important protective role for our body, apprising us of almost any type of damaging process and causing appropriate muscular reactions to remove the body from contact with the damaging stimuli.

THE STIMULUS THAT CAUSES PAIN. Pain receptors are stimulated when tissues of the body are *being* damaged. For instance, one feels pain while the skin is *being* cut, but shortly after the cut has been made, the pain generally is gone. Indeed, thousands of soldiers on the battlefield in World War II who had been mortally wounded were asked whether or not they felt pain. In most instances, no pain was actually felt except for a few minutes after the damage had been inflicted.

Different types of damaging stimuli that can cause pain are *trauma* to the tissues, *ischemia* of the tissues (lack of blood flow), intense *heat* to the tissues, intense *cold* (especially freezing of the tissues), or *chemical irritation* of the tissues. It is believed that as tissues are damaged they release some substance from the cells that stimulates the pain nerve endings. *Histamine* and *bradykinin* have both been suggested as this substance, though neither has been proved.

Perception of pain. It is frequently said that one person perceives pain more intensely than others. However, experiments in which graded intensities of tissue damage were caused in a large number of different persons showed that all normal persons perceive pain at almost precisely the same degree of tissue damage. For instance, when heat is used to cause tissue damage, almost all persons feel pain between the temperatures of 44 and 46° Centigrade, which is a very, very narrow range.

Reactivity to pain. On the other hand, not all people *react* alike to the same pain, for some react violently to only slight pain

while others can withstand tremendous pain
before reacting at all. This is determined not
by differences in sensitivity of the pain
receptors themselves but, instead, by differences in psychic makeup of the individuals.
Therefore, when a person is said to be extremely "sensitive" to pain, it is meant that
he *reacts* to pain far more than do other
persons and not that he perceives far more
pain.

THE VISCERAL SENSATIONS

The visceral sensations are those that
arise from the internal structures of the body,
such as the organs of the abdomen and chest,
and they also include sensations from inside
the head, from the muscles, the bones, and
other deep structures.

The usual modalities of visceral sensation
are pain, burning sensations, and pressure
(which is often manifested as a sensation of
fullness). Because these modalities are also
exhibited by the exteroceptive sensations,
the visceral sensations are sometimes considered to be part of the exteroceptive
system. Indeed, the pathways of these two
types of sensation are closely related, as will
be noted below.

PATHWAYS FOR VISCERAL SENSATIONS.
Ordinarily one is not at all conscious of his
internal organs, but an inflamed organ can
transmit pain sensations to the central nervous system through two separate pathways,
the *parietal pathway* and the *visceral pathway*, both of which are illustrated for sensations from the appendix in Figure 235. The
parietal pathway is the same as the pathway
for transmission of exteroceptive sensations.
In the case of the appendix, the inflamed
appendix irritates the peritoneum overlying
the appendix, and pain impulses are transmitted from the peritoneum through the
same nerve trunks that carry exteroceptive
sensations from the outside of the abdominal
wall at this point.

The visceral pathway utilizes sensory
nerve fibers in the autonomic nervous system. Figure 236 illustrates sensory fibers
leaving the appendix to enter the sympathetic chain. After traveling upward a few
segments the fibers leave the sympathetic
chain to enter the spinal cord at approximately the T-10 segment in the lower

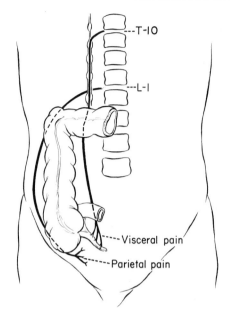

Figure 235. Parietal and visceral pathways for
transmission of pain from the appendix.

thoracic region. Therefore, the visceral
sensations from the appendix enter the spinal
cord at an entirely different point from the
parietal sensations.

REFERRED PAIN. Pain from an internal
organ is often felt on a surface area of the
body rather than being localized in the organ
itself. This is called *referred pain.* It may
be referred to the surface immediately above
the organ, or often to areas considerable
distance away. The mechanism of referred
pain is probably that illustrated in Figure

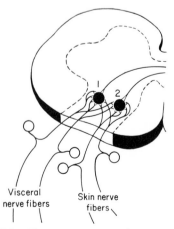

Figure 236. The neurogenic mechanism responsible
for referred pain.

236, which shows two visceral nerve fibers as well as two additional nerve fibers from the skin entering the spinal cord. Both of these types of fibers synapse with the same two neurons in the spinal cord. Therefore, stimulating either the visceral fibers or the skin fibers will send impulses up the spinal cord along the same pathways. Because the person has never had reason to know the location of his internal organs but is very familiar with the location of the different skin areas, the impulses from the visceral fibers are usually interpreted as coming from the skin.

When a visceral sensation is transmitted through a parietal pathway, the referred pain is usually felt on the surface of the body directly over the respective internal organ, but when the sensation is conducted through a visceral pathway, it is usually referred to a skin area quite remote from the organ. The reason for this is that fibers of the visceral pathway usually travel a long distance in the sympathetic chain before entering the spinal cord. Figure 237 illustrates the surface areas to which pain is referred from many of the different organs. For instance, though the appendix lies far to the right and in the lower abdomen, its visceral sensations are referred to an area around the umbilicus. Referred pain from the kidney and ureter occurs near the midline on the anterior abdominal wall, though these organs actually lie in the posterior part of the abdomen.

Visceral pain from the heart is among the most important of the referred sensations. Figure 237 shows the areas to which cardiac pain is often referred, including the upper thorax, the shoulder, and the medial side of the left arm, particularly along the radial artery.

The surface area to which visceral pain is referred usually corresponds to that portion of the body from which the organ originated during embryonic development. For example, the heart originates in the neck of the embryo and so does the arm; therefore, heart pain is frequently referred to the arm. The appendix, as another example, originates from the portion of the primitive gut that develops near the umbilicus, which explains the reference of appendiceal pain to the umbilical region. The other areas of referred pain illustrated in Figure 237 also correspond with the areas of origin of the different organs.

THE STIMULUS FOR VISCERAL PAIN. Cutting through the gut, the heart, the liver, the muscle, or other internal organs with a sharp knife causes almost no pain. Instead, pain from these areas is produced much more easily by (1) stopping the blood flow to the area, (2) application of an irritant chemical over wide areas, (3) stretching the tissues, or (4) spasm of the muscle in the organ.

One of the most important stimuli for visceral sensation is *ischemia*, which means lack of blood flow. The pain of a heart attack, as an example, is caused by poor flow through the coronaries to the heart muscle. Even the skeletal muscles become extremely painful when their blood supply is diminished for a prolonged period of time. The reason for the pain caused by ischemia has never been determined, though it is believed that the lack of blood flow allows metabolic end products, such as acids, to build up in the tissues, and that these in turn produce irritant effects on the pain nerve endings.

Stretching the tissues possibly causes pain by stretching the nerve endings, or this could be caused by producing ischemia, for overstretching the tissues occludes the blood vessels. Likewise, spasm of the muscle in an organ can stretch nerve endings as well as compress the blood vessels; spasm also increases the rate of metabolism so that far more than usual quantities of metabolic end-products are dumped into the tissues. These factors, therefore, could explain why spasm of the gut often causes very intense abdominal pain.

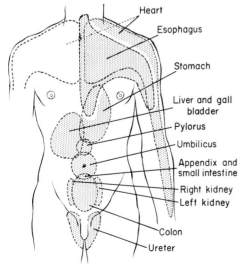

Figure 237. Surface areas of the body to which pain from different organs is referred.

Heart
Esophagus
Stomach
Liver and gall bladder
Pylorus
Umbilicus
Appendix and small intestine
Right kidney
Left kidney
Colon
Ureter

Headache

Headache is another type of referred pain, and it is usually caused by irritation or damage occurring in the tissues inside the head. Figure 238 illustrates the surface areas of the head to which pain from the deep structures is referred. Pain fibers from inside the eye and from the nasal sinuses are transmitted through the first and second divisions of the fifth nerve which also supply the skin areas over the lower forehead and around the eye and nose. Therefore, sinus infections or irritation of the eyes caused by their overuse or by intense light can result in dull aching pain referred diffusely over the frontal and orbital areas of the head.

Irritation occurring inside the skull but above the level of the ears — that is, above the tentorium — causes headache referred to the frontal and temporal surfaces of the head. The reason for this is that the third division of the fifth cranial nerve supplies both the supratentorial areas inside the skull and the surface areas of the skull in the temporal and frontal regions.

Irritative effects beneath the tentorium, in the pocket of the skull beneath the level of the ears, give rise to headache localized over the occipital part of the skull, also shown in Figure 238. Both the subtentorial areas inside the cranial vault and the occipital surface areas of the skull are supplied by the upper cervical spinal nerves, which explains the reference of subtentorial pain to the occipital regions.

MENINGEAL HEADACHE. Headache originating inside the skull is often caused by irritation of the *meninges*, which are the membranes surrounding the brain. Some drastic causes of very severe headache of this type are: (1) meningitis, an infection of the meninges, which causes very intense headache; (2) removal of fluid from the spaces around the brain, allowing the brain to rub freely against the meninges and to irritate them, this too causing very intense headache; and (3) an alcoholic binge, which probably causes headache by irritation of the meninges, for the meninges almost certainly become reddened and inflamed in the same manner as the whites of the eyes the day after an alcoholic bout.

Headaches are believed by many physicians not to result from pain originating in the substance of the brain itself. In fact, a patient who is having a brain operation under local anesthetic — that is, without being asleep — can feel no pain when the surgeon cuts through the brain tissue. Yet when he cuts the meninges, or especially when he cuts one of the major blood vessels supplying the meninges, intense headache is experienced.

THE EVERYDAY HEADACHE. Despite all that we *do* know about the origin of headaches, the modern physician is still completely befuddled by the common everyday headache. It is generally stated that there are two basic types. One, the *migraine* type, supposedly results from spasm of blood vessels supplying some of the intracranial tissues, followed by intense and painful dilatation of these same vessels, lasting for many hours. However, this simple explanation is still quite hypothetical.

The second type is the so-called *tension headache*, which results when a person operates under considerable emotional tension. It is associated with tightening of muscles attached to the base of the skull and perhaps with simultaneous vascular spasm or other effects occurring intracranially. It is possible that the muscles themselves are the source of some of the pain and that the pain is referred to the head, though this theory is extremely hypothetical.

It is also well known that constipation is frequently associated with headache, possibly because of absorption of toxic products from the colon. It is possible that diffuse irritation of the brain by these toxic substances can cause headache, despite the fact that cutting through brain tissue in an awake person does not usually cause much pain.

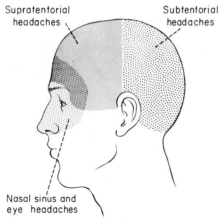

Supratentorial headaches

Subtentorial headaches

Nasal sinus and eye headaches

Figure 238. Surface areas of the head to which different types of headache are referred.

THE INTERPRETATION OF SENSATIONS BY THE BRAIN

Interpretation of Somesthetic Sensations — The Somesthetic Association Area

Figures 236 and 239 illustrate an area of the cerebral cortex, immediately posterior to the somesthetic cortex, called the *somesthetic association area*. Impulses pass into this area from several sources: (1) from the somesthetic cortex, (2) directly from the thalamus, and (3) from various other parts of the cerebral cortex.

The function of the somesthetic association area is to interpret the "meaning" of the somesthetic sensations arriving in the brain. It is here that the brain determines the shapes of objects, their weights, the textures of their surfaces, and their positions in relation to the body. This area also keeps track of the exact position of each part of the body at all times.

Many of the memories of past sensory experiences are stored in the somesthetic association area, and when new sensations similar to the old ones arrive in the brain the parallel nature of the two sensations is immediately discerned. It is in this manner that one associates a new sensation with previous ones and thereby recognizes the nature of the sensation. As more and more sensory experiences accumulate, new sensory experiences can be interpreted on the basis of what is remembered from the past.

Cortical Localization of Other Sensations

The sensations of vision, hearing, taste, and smell, like the somesthetic sensations, are each transmitted from the receptor organs to small circumscribed areas in the cerebral cortex, each of the areas having a different location. These areas are called the *primary cortical areas* for sensations, and from each primary area the impulses go to an *association area*. The primary area for somesthetic sensation, the somesthetic cortex, and the somesthetic association area were discussed above.

CORTICAL AREAS FOR VISION. The *primary visual cortex* is located in the posterior part of the brain on the medial side of each hemisphere, as will be discussed in the following chapter. A small portion of the primary visual cortex extends over the occipital pole, as shown in Figure 239, though most of it is hidden inside the longitudinal fissure. The primary visual area interprets only the more basic meanings of visual sensations such as whether the object is a line, a square, or a star, and also interprets the color of the object.

Accessory visual signals pass in all directions from the primary visual cortex and also from the thalamus into adjacent areas of the cortex called the *visual association area*, which also is shown in Figure 239. This area interprets the deeper meaning of the visual signals. It interprets the interrelations of the different objects and identifies the objects. Finally, the association area interprets the overall meaning of the scene before the eyes.

Interpretation of the written language is one of the most important functions of the visual association area. To accomplish this feat, this area must first discern from the light and dark spots the letters themselves, then from the combination of letters the words, and from the sequence of words the thoughts that they express.

CORTICAL AREAS FOR HEARING. Auditory sensations are transmitted from the ears to a small area called the *primary auditory cortex* in the upper part of the temporal lobe, which will be discussed in Chapter 25. From this area signals pass into the surrounding *auditory association area*, shown in Figure 239. The primary

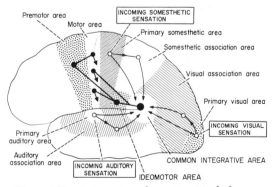

Figure 239. Integration of sensory signals from several different sources into a common thought by the common integrative area of the brain, showing also the primary and association areas for vision, for auditory sensations, and for somesthetic sensations.

auditory cortex interprets the basic characteristics of the sound, such as its pitch and its rhythmicity, while the auditory association area interprets the meaning of the sound. One part of the association area determines whether the sound is noise, music, or speech; then other parts determine the thoughts conveyed by the sound. To interpret the meaning of speech, the auditory association area first combines the various syllables into words, then words into phrases, phrases into sentences, and finally sentences into thoughts.

CORTICAL AREAS FOR SMELL AND TASTE. The *primary cortex for smell* is located on the bottom surface of the brain, possibly in the region of the *uncus*, though it has never been located exactly, and the *primary cortex for taste* is located at the bottommost end of the primary somesthetic area, deep in the fissure of Sylvius. From these primary areas signals pass into surrounding *smell and taste association areas*, and there the sensations are interpreted in the same manner that somesthetic, visual, and auditory sensations are interpreted by their respective association areas.

The Common Integrative Area of the Brain—the Gnostic Area

Figure 239 illustrates the passage of signals from the somesthetic, visual, and auditory association areas into a *common integrative area*, located in the angular gyrus of the brain midway between these three respective association areas. Signals are also transmitted into this area from the taste and smell association areas and directly from the thalamus and other basal areas of the brain. It is here that all the different types of sensations are integrated to determine a common meaning, which is the reason for its name, *common integrative area*. This region of the brain is also called the *gnostic area*, which means the "knowing area." If a person were in a jungle and heard a noise in the brush, saw the leaves moving, and smelled the scent of an animal, he might not be able to tell from any one of these sensations exactly what was happening, but from all of them together he could quite readily assess his danger. It is in the common integrative area that all the thoughts from the different sensory areas are correlated and weighed

against each other for deeper conclusions than can be attained by any one of the association areas alone.

Most of the sensory information arriving in the brain finally is funneled through the common integrative area. For this reason, any damage to this area is likely to leave the person mentally inept; even though the different association areas might still be able to interpret their respective sensations, this information is almost valueless to the brain unless its final meaning can be interpreted. Some signals can be transmitted directly from the association areas to other portions of the brain without going through the common integrative region, but these are so few that the person who loses his entire common integrative area generally becomes an imbecile. This is occasionally the unfortunate result of a brain tumor or a *stroke*. (A stroke is caused by sudden loss of blood supply to an area of the brain because of hemorrhage or thrombosis of a blood vessel.)

DOMINANCE OF ONE SIDE OF THE BRAIN. The common integrative area is located in the angular gyrus of the *left* cerebral hemisphere in at least nine tenths of all people. The angular gyrus of the opposite hemisphere is usually almost totally nonfunctional, though in one tenth of the people it is the common integrative area and the left angular gyrus is non-functional. At birth both angular gyruses probably have almost equal functional abilities, but the right angular gyrus region usually becomes almost totally suppressed as the brain develops, while the left angular gyrus becomes the most important portion of the entire cerebral cortex. Occasionally, though, the relationship between the two is reversed. This phenomenon of *suppression* occurs very commonly in different parts of the brain when signals from two different sources interfere with each other. For instance, when a person is cross-eyed, so that the visual signals from the two eyes do not correspond satisfactorily, the brain automatically suppresses the signals from one of the eyes. As a result, this eye gradually becomes "functionally" blind until almost all vision is lost, while the person develops excellent vision in the other eye.

CONTROL OF MOTOR FUNCTIONS BY THE COMMON INTEGRATIVE AREA. Once the common integrative area has integrated all incoming sensations into a common thought, signals are then sent into other portions of the brain to cause appropriate responses. If the integrated thought indicates that muscular activity is needed, the common integrative area sends impulses into the motor portions of the brain to cause muscular contractions.

The basic functions of the cerebral cortex in motor control will be discussed in much more detail in Chapter 27.

Role of the Thalamus and Other Lower Brain Centers in the Interpretation of Sensations

In the preceding sections of this chapter we have discussed cortical functions as if the cortex were operating almost independently of other parts of the brain. This, however, is entirely untrue. Even when major portions of the sensory cortex are destroyed, the animal is still capable of crude degrees of interpretation of sensation. As has already been pointed out, crude localization and interpretation of many somesthetic sensations can be effected by the thalamus and other related areas. In the case of vision, the cerebral cortex is not needed to interpret the intensity of light—it is needed to interpret the shapes and colors of objects. In the case of hearing, an animal can interpret the direction from which sound is coming without the cerebral cortex, though in general he cannot interpret the finer meanings of sound.

Therefore, whenever it is stated that a particular type of sensation is interpreted in a particular region of the cerebral cortex, it is meant simply that this part of the cortex is responsible for the deeper shades of meaning rather than for total interpretation. Indeed, it would be utterly impossible for the cerebral cortex to operate without preliminary processing of information in the lower regions of the brain.

ACTIVATION OF SPECIFIC PORTIONS OF THE CEREBRAL CORTEX BY THE THALAMUS. During evolutionary development of the brain, the cerebral cortex originated mainly as an outgrowth of the thalamus, for which reason each area of the cerebral cortex is very closely connected to a corresponding discrete area of the thalamus. Thus, the frontal regions of the cerebral cortex have to and fro nerve connections directly with the anterior portions of the thalamus. Likewise,

occipital portions of the cerebral cortex are connected with the posterior thalamus, and temporal portions of the cerebral cortex are connected with lateral areas of the thalamus.

Furthermore, the cortex cannot activate itself but must be activated from lower regions of the brain, an effect that will be discussed in Chapter 29. Especially, signals from specific parts of the thalamus activate specific areas of the cerebral cortex. Thus, in this way, function in the thalamus presumably calls forth information stored in the memory pool of the cerebral cortex. Therefore, the thalamus is actually a type of *control center* for the cerebral cortex.

Therefore, again it must be stressed that even though it is conventional to speak of certain types of sensations being interpreted in specific areas of the cerebral cortex, it is really meant that these cortical areas are the loci of the vast memory information associated with these particular types of sensations. Yet, it is still the lower regions of the brain that are calling the signals, that are controlling which parts of the cerebral cortex will be activated, that are calling forth stored information from the cerebral cortex, and that are responsible for channeling sensory signals into appropriate parts of the cerebral cortex. Furthermore, many basic aspects of sensations are detected in the lower regions of the brain even before the signals reach the cerebral cortex.

REFERENCES

Bishop, P. O.: Central nervous system: Afferent mechanisms and perception. *Ann. Rev. Physiol.* 29:427, 1967.

Finneson, B. E.: *Diagnosis and Management of Pain Syndromes.* Philadelphia, W. B. Saunders Company, 1962.

Goldberg, J. M., and Lavine, R. A.: Nervous system: Afferent mechanisms. *Ann. Rev. Physiol.* 30:319, 1968.

Hardy, J. D.: The nature of pain. *J. Chronic Dis.* 4:22, 1956.

Neff, W. D.: Sensory discrimination. *Handbook of Physiology*, Sec. I, Vol. III, p. 1447. Baltimore, The Williams & Wilkins Company, 1960.

Ostfeld, A. M.: *The Common Headache Syndromes: Biochemistry, Pathophysiology, Therapy.* Springfield, Ill., Charles C Thomas, 1962.

Perl, E. R.: Somatosensory mechanisms. *Ann. Rev. Physiol.* 25:459, 1963.

Weddell, G., and Miller, S.: Cutaneous sensibility. *Ann. Rev. Physiol.* 24:199, 1962.

Wolff, H. G.: *Headache and Other Head Pain.* 2nd Ed. New York, Oxford University Press, 1963.

Zanchetti, A.: Somatic functions of the nervous system. *Ann. Rev. Physiol.* 24:287, 1962.

CHAPTER 24

THE EYE

THE EYE AS A CAMERA

Figure 240 illustrates the general organization of the eye, showing the optical system for focusing the image, and the *retina* on which the image is projected. The optical system is identical to that of a camera, and the retina corresponds to the photographic film. The retina translates the image into nerve impulses and transmits these into the brain through the optic nerve.

The outer envelope of the eye is a very strong bag composed mainly of a thick fibrous structure called the *sclera*. Anteriorly the sclera connects with the *cornea*, which is the clear part through which light enters the eye. The inside of the eye is filled with fluid, and an ovoid clear body called the *crystalline lens* is located approximately 2 mm. behind the cornea. The fluid anterior to the lens is almost pure extracellular fluid and is called *aqueous humor*, while that behind the lens contains a protein matrix that forms a gelatinous but clear fluid called the *vitreous humor*. Light passes first through the clear cornea, then through the aqueous humor, then the lens, and finally the vitreous humor before impinging on the retina.

THE LENS SYSTEM OF THE EYE. The lens system of the eye is composed of the cornea and the crystalline lens. Because the cornea is curved on its outside, light rays passing

from the air into the cornea are *refracted* (bent) in the same way that any optical lens refracts rays. After the rays pass through the cornea and aqueous humor they strike the curved anterior surface of the crystalline lens, where still more bending occurs, and as they pass through the posterior surface of the lens the rays bend again. Thus, light rays are refracted, or bent, at three different interfaces in the eye. This is analogous to the refraction of light at the different surfaces in a compound lens system of a camera, for in the camera the light rays are also refracted at each interface between the lenses and the air.

FUNCTION OF A CONVEX LENS IN FORMING AN IMAGE. Figure 241 shows the focusing of light rays from a distant source and also from two points near the lens. It will be noted that the light rays striking the outer edges of

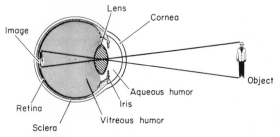

Figure 240. General structure of the eye, showing the function of the eye as a camera.

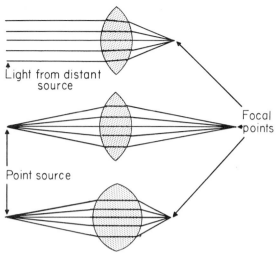

Figure 241. Focusing of parallel light rays and light rays from point sources by convex lenses.

that the rays from the two lights are each focused to focal points on the opposite side of the lens directly in line with the lens center.

Figure 242B illustrates the focusing of light rays from different point sources on a human being's body. Those parts of the body that are very bright are point sources of light, whereas those parts that are dark represent the black spaces between the point sources of light. So far as the eyes are concerned, any object is a mosaic of point sources of light. Light from each source is focused by the lens, and the focal point of each is always directly in line with the center of the lens and the original source. Consequently, an inverted image is formed by the focal points, as shown in the figure.

the convex lens are bent toward the center, while those that strike the lens exactly in the center pass on through without being bent. The reason for this is that the outer rays enter the substance of the lens at an angle, which causes refraction, while those striking the center enter the lens perpendicular to its surface, which does not cause refraction. The light rays from the edges of the lens angle inward to meet those passing through the center, and they all *focus* on a common point called the *focal point*.

Figure 242A shows the focusing of light rays from two different point sources. Note

Abnormalities of the Lens System

The normal eye focuses parallel light rays exactly on the retina. This normal focusing of the eye is called *emmetropia*, and it is illustrated at the top of Figure 243. However, three different abnormalities frequently occur to prevent focusing of light rays precisely on the retina. These are *hypermetropia, myopia, and astigmatism.*

HYPERMETROPIA. Hypermetropia, or *far-sightedness*, is caused by failure of the lens to bend the light rays enough to bring them

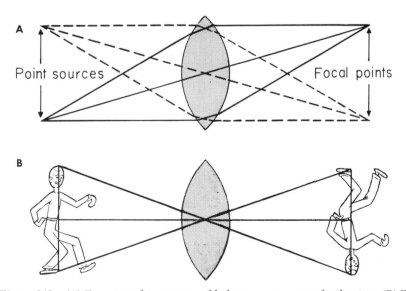

Figure 242. (A) Focusing of two points of light to two separate focal points. (B) Formation of an image by a convex lens focusing light rays from an object.

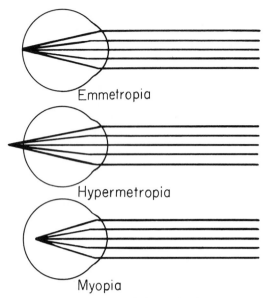

Figure 243. Focusing of light rays by the emmetropic eye, the hypermetropic eye, and the myopic eye.

to a focal point on the retina, as shown in Figure 243. Instead, the light rays are still diffuse when they reach the retina, and thus vision is blurred. Hypermetropia is called far-sightedness because with this type of vision objects can be seen more clearly at a distance than near at hand.

MYOPIA. Myopia, which is also called *near-sightedness,* is caused by too strong a lens system for the distance of the retina

behind the lens. That is, the light rays are focused before they reach the retina, and by the time they do reach the retina they have spread apart again, causing fuzziness of each point in the image.

Myopia is called near-sightedness because the myope can see objects near him with complete clarity, while not being able to focus any objects that are at a far distance. The reason for this is that the closer the object is to the eye, the more the light rays diverge outward so that the lens system cannot bend the rays inward as much as usual. If the object is close enough to the eye, the focal point will fall on the retina rather than in front of it, and the image will be completely clear.

ASTIGMATISM. Astigmatism occurs when one of the components of the lens system becomes egg-shaped rather than spherical. This effect is shown in Figure 244. Either the cornea or the crystalline lens becomes elongated in one direction in comparison with the other direction. Because the radius of curvature is greater in the elongated direction than in the short direction, the light rays entering the lens along this lengthened curvature are focused behind the retina, while those entering along the shortened curvature are focused in front of the retina. In other words, the eye is far-sighted for some of the light rays and near-sighted for the remainder. Therefore, the person with

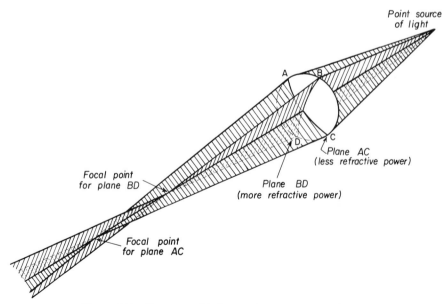

Figure 244. Focusing of light rays by an astigmatic lens.

astigmatic eyes is unable to focus any object clearly, regardless of how far the object is away from the eyes, for when the near-sighted light rays are in focus the far-sighted ones are out of focus and vice versa.

CORRECTION OF THE OPTICAL ABNOR-MALITIES OF THE EYE. Glasses with properly prescribed lenses can be used to correct the abnormalities of the lens system of the eye. Glasses bend the light rays before they enter the eye in an appropriate manner to correct for the excess or deficient refractive power of the eye. Figure 245 shows the correction of myopia and hypermetropia. In the myopic individual the light rays normally focus in front of the retina. To prevent this, a concave lens is placed in front of the eye. This type of lens bends the light rays outward and, therefore, compensates for the excess inward bending of the myopic lens system. By prescribing the appropriate curvature to the concave lens, a myopic person's vision can be made completely normal.

In the hypermetropic eye the lens system normally fails to bend the light rays sufficiently. To correct this abnormality a convex lens is placed in front of the eye so that the light rays will be partially bent even before they reach the eye. With the aid of this preliminary convergence, the lens system of the eye can then bring the rays to a focal point on the retina.

The lens that must be used to correct astigmatism is somewhat more complicated than those used to correct either myopia or hyper-metropia, for it must be fashioned with more curvature in one direction than the other. However, by prescribing a lens with precisely ground curvatures and by placing the curvatures in exactly the right direction in front of the eye, the abnormal refraction of light rays by each portion of the astigmatic eye can be corrected appropriately.

FUNCTION OF THE RETINA

Figure 246 illustrates the anatomy of the retina, showing that it is composed of many different layers of cells. The light rays enter from the bottom of the figure and pass all the way through the retina to the top, finally stiking the *rods* and *cones* and the pigment layer. The rods and cones are nerve receptors that are excited by light. These cells change the light energy into neuronal impulses that are transmitted into the brain.

The pigment layer of the retina contains large quantities of a very black pigment called *melanin*. The function of the melanin is to absorb the light rays after they have passed through the retina and thereby to prevent light reflection throughout the eye. *Albino* persons, who are unable to manufacture melanin in any part of their bodies, have a complete lack of pigment in this layer of the retina. As a result, after passing through the retina the light rays are not absorbed, but instead are reflected so intensely in all

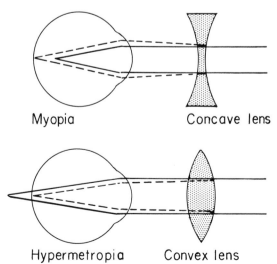

Figure 245. Correction of myopia by a concave lens and of hypermetropia by a convex lens.

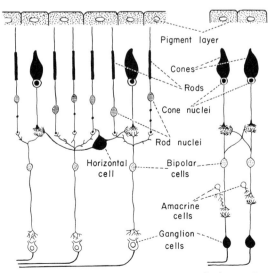

Figure 246. Functional anatomy of the retina.

directions through the eye that they cause all images to become bleached out with light. The albino's vision is usually about two to three times less acute than that of a normal person, and he is so blinded by bright sunlight that he must wear very dark glasses to see at all.

By far the greater number of light receptors in the retina are rods. Light of all colors stimulates the rods, while the cones are stimulated selectively by different colors. Therefore, the cones are responsible for color vision, in contradistinction to the rods, which provide only black and white vision.

The Chemistry of Vision

CHEMISTRY OF ROD EXCITATION. Figure 247 shows the chemical changes that occur in the rods, both when light strikes the retina and during periods in between light stimulations. Vitamin A is the basic chemical utilized by both the rods and cones for synthesizing substances sensitive to light. On being absorbed into a rod, vitamin A is converted into a substance called *retinene*. This then combines with a protein in the rods called *scotopsin* to form a light-sensitive chemical, *rhodopsin*. If the eye is not being exposed to light energy, the concentration of rhodopsin builds up to an extremely high level.

When a rod is exposed to light energy some of the rhodopsin is changed immediately into *lumi-rhodopsin*. However, lumi-rhodopsin is a very unstable compound that can last in the retina only about a tenth of a second or more. It decays almost immediately into another substance called *meta-rhodopsin*, and this compound, which is also unstable, decays very rapidly into retinene and scotopsin.

Thus, in effect, light energy breaks rhodopsin down into the substances from which the rhodopsin itself had been formed, retinene and scotopsin. In the process of splitting rhodopsin, the rods become excited, probably by ionic charges in the intracellular fluid when scotopsin splits away from retinene. These ions last for only a split second. During this slight interval, nerve impulses are generated in the rod and transmitted into the optic nerve and thence into the brain.

After rhodopsin has been decomposed by light energy, its decomposition products, retinene and scotopsin, are recombined slowly by the metabolic processes of the cell to form new rhodopsin. The new rhodopsin in turn can be utilized again to provide still more excitation of the rods. Thus, a continuous cycle occurs. Rhodopsin is being formed continually, and it is broken down by light energy to excite the rods.

CHEMISTRY OF CONE VISION. Almost exactly the same chemical processes occur in the cones as in the rods except that the protein scotopsin of the rods is replaced by three similar proteins called *photopsins* in the three respective types of cones. The chemical differences among the photopsins make the respective types of cones selectively sensitive to different colors, which will be discussed later.

PERSISTENCE OF IMAGES AND FUSION OF FLICKERING IMAGES. Following a sudden flash of light that lasts only one millionth of a second, the eye sees an image of the light that lasts for approximately one-tenth second. The duration of the image is the length of time that the retina remains stimulated following the flash, and this presumably is about as long as the lumi-rhodopsin remains in the rods.

The persistence of images in the retina allows flickering images to *fuse* when one views a moving picture or television screen. The moving picture flashes 16 to 30 pictures per second, and the television screen provides 60 pictures per second. The image on the retina persists from one picture to the

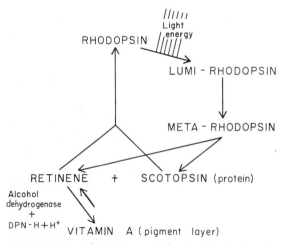

Figure 247. The retinene-rhodopsin chemical cycle responsible for light sensitivity of the rods.

next, which gives one the impression of seeing a continuous picture.

LIGHT AND DARK ADAPTATION. It is common experience to be almost totally blinded when first entering a very bright area from a darkened room, and also to be almost totally blinded when entering a darkened room from a brightly lighted area. The reason for this is that the sensitivity of the retina becomes temporarily out of adjustment with the intensity of the light. To discern the shape, the texture, and other qualities of an object, it is necessary to see both the bright and dark areas of the object at the same time. When the sensitivity of the retina is too great, all areas of the object will be bright enough to stimulate the rods and cones a maximum amount, and, as a result, everything—even dark objects—will seem to be completely bright. This is the experience upon first coming into bright sunlight after seeing a movie. On the other hand, if the sensitivity of the retina is very slight when entering a darkened area, not even the bright portions of the image can stimulate the rods and cones, and as a result, no object can be seen. Fortunately, the retina automatically changes its sensitivity in proprotion to the degree of light energy available. This phenomenon is called light and dark adaptation.

The mechanism of *light adaptation* can be explained by referring once again to Figure 247. When large quantities of light energy continually strike the rods, rhodopsin is broken into retinene and scotopsin almost as rapidly as it is formed, and, because rhodopsin formation is a relatively slow process, the total concentration of rhodopsin in the rods falls to a very low value as the person remains in the bright light. Essentially the same effects occur in the cones. Therefore, the sensitivity of the retina soon becomes greatly depressed in bright light.

The mechanism of *dark adaptation* is essentially opposite to that of light adaptation. When the person enters a darkened room from a lighted area, the quantity of rhodopsin in his rods (and color-sensitive chemicals in his cones) is at first very slight, and, as a result, he cannot see anything. Yet the amount of light energy in the darkened room is also very slight, which means that the rhodopsin being formed in the rods is utilized only very slowly. Therefore, the concentration of rhodopsin builds up during

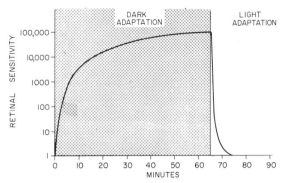

Figure 248.   Dark and light adaptation.

the ensuing minutes until it finally becomes high enough for even a very minute amount of light to stimulate the rods.

During dark adaptation, the sensitivity of the retina can increase as much as one thousand-fold in only a few minutes, and as much as one hundred thousand times in an hour or more. This effect is illustrated in Figure 248, which shows the retinal sensitivity increasing from an arbitrary value of 1 up to a value of 100,000 in one hour after the person has left a very bright area and moved into a completely darkened room. Then, on reentering the bright area, light adaptation occurs, and retinal sensitivity decreases from 100,000 back down to 1 in another 10 minutes, which is a more rapid process than dark adaptation.

Function of the Cones— Color Vision

The cones are different from the rods in several respects. First, they respond selectively to certain colors, some to one color and others to other colors. Second, cones are considerably less sensitive to light than are rods, for which reason they cannot provide vision in very dim light. Third, the cones are connected with the brain in such a way as to provide greater acuity of vision than the rods provide. That is, 10 to 100 rods usually connect with the same optic nerve fiber, which means that impulses transmitted by the rods to the brain do not necessarily originate from one very discrete point on the retina. On the other hand, often only one and never more than a few cones are connected to the same optic nerve fiber.

DETECTION OF DIFFERENT COLORS BY THE CONES. The retina contains three different types of cones, each of which responds to a different spectrum of colors. This is illustrated in Figure 249, which shows light wave lengths at which the three different types of cones—the blue cone, the green cone, and the red cone—respond. Note that the blue cone responds maximally at a wave length of 430 millimicrons, which is a blue color; the green cone at 540 millimicrons, a greenish-yellow color; and the red cone at 575 millimicrons, an orange color. The so-called "red" cone is called the red cone not because its maximal response is in the red range, but because it is the only cone that has any significant response at all above 600 millimicrons, which is red.

DETERMINATION OF THE INTERMEDIATE COLORS BY THE BLUE, GREEN, AND RED CONES. It is quite easy to understand how the blue cones determine that an object is blue, how the green cones determine that an object is green, and how the red cones determine that an object is red, but it is much more difficult to understand how these cones detect the intermediate colors between the three primary ones. This is accomplished by utilizing a combination of cones. For instance, yellow light stimulates the red and green cones approximately equally. When both of these types of cones are stimulated equally, the brain interprets the color as yellow. Also, when the red cones are stimulated about one and one-half times as strongly as the green cones, which occurs when light with a wave length of 580 millimicrons strikes the retina, the brain interprets the color as orange. If both the red and green cones are stimulated, but the green more than the red cones, the color is interpreted as a greenish-yellow. Like-

wise, when both the green and blue cones are stimulated, the color is interpreted as a bluish-green. Thus, by combining the degrees of stimulation of the different cones, not only can the brain distinguish among the three primary colors but also among the colors having intermediate wave lengths.

The intensity of a color is determined by the number of nerve impulses transmitted to the brain by the cones. For instance, if a yellow color stimulates both the green and red cones and the number of impulses transmitted by each cone is only 10 per second, the intensity will be relatively weak, but if the number of impulses transmitted is 100 per second, the intensity will be strong. It is especially important to note that a change in intensity does not change the *ratio* between the degree of stimulation of the two types of cones. The brain interprets color on the basis of this ratio and not on the basis of the actual intensity of stimulation of each cone.

COLOR BLINDNESS. Color blindness can be understood very readily on the basis of Figure 249. Occasionally one of the three primary types of cones is lacking because of failure to inherit the appropriate gene for formation of the cone. The color genes are sex-linked and are found in the female sex chromosome. Since females have two of these chromosomes they almost never have a deficiency of a color gene, but because males have only one female chromosome, one or more of the color genes is absent in about 4 per cent of all males. For this reason, almost all color blind people are males.

If a person has complete lack of red cones, he is able to see green, yellow, orange, and red colors by use of his green cones. However, he is not able to distinguish satisfactorily between these particular colors because he has no red cones to contrast with the green ones. Likewise, if a person has a deficit of green cones, he is able to see all the colors, but he is not able to distinguish between green, yellow, orange, and red colors, because the green cones are not available to contrast with the red. Thus, loss of either the red or the green cones makes it difficult or impossible to distinguish between the colors of the longer wave lengths. This is called *red-green color blindness*. In very rare instances a person lacks blue cones, in which case he has difficulty

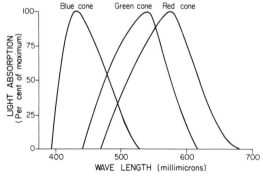

Figure 249. Spectral sensitivity curves for blue, green, and red cones. (Drawn from curves recorded by Marks, Dobelle, and MacNichol, Jr.: *Science* 143:1181, 1964, and by Brown and Wald: *Science* 144:45, 1964.)

distinguishing violet, blue, and green from each other. This type of color blindness is frequently called *blue weakness.*

NEURONAL CONNECTIONS OF THE RETINA WITH THE BRAIN

Figure 250 illustrates the connections of the retina with the brain, showing that the right halves of the two eyes are connected with the right visual cortex and the left halves with the left visual cortex. The *optic nerve* fibers from the nasal half of each retina cross in the *optic chiasm*, located on the bottom surface of the brain, and join the fibers from the temporal half of the opposite retina. Then the combined fibers pass backward through the *optic tract*, synapse in the *lateral geniculate body*, and finally spread through the *optic radiation* into the *visual cortex.*

In addition, fibers pass directly from the optic tract into the *pretectal nuclei;* these fibers carry signals for control of pupillary size in response to light intensity, as will be discussed later in the chapter.

Finally, fibers not shown in Figure 250 pass from the *lateral geniculate body* into the lateral thalamus and also into the *superior colliculus* located in the mid-brain. These areas help to discriminate certain qualities of the visual scene, as will be discussed in the following section.

DISCRIMINATION OF THE VISUAL IMAGE. Even at the level of the retina the visual image begins to be picked apart, and the pattern of stimulation in the visual cortex is markedly different from that in the retina. The retina breaks the visual image into two components. The first component is the *visual pattern*, which is transmitted backward along the optic nerve fibers to the lateral geniculate body. That is, the light areas are stimulated and the dark areas are not stimulated, and appropriate signals are transmitted. The second component is a signal denoting *change* in light intensity. For instance, when a spot of light travels across the retina, each stimulated optic nerve fiber responds very strongly at first, but within a fraction of a second the strength of stimulation dies out to a very low value. The extreme initial strength of stimulation calls one's attention to this moving spot. By this mechanism, the eye can detect movement of a gnat far out in the peripheral vision, though it could not detect even a much larger object if it were standing still.

At the lateral geniculate body, several other aspects of the visual scene seem to be interpreted. One of these is probably *depth perception*, for it is here that signals from the two eyes first come together, and, as we shall see later in the chapter, depth perception depends upon comparing minute differences in shapes of objects as seen by the two separate eyes. The lateral geniculate body is admirably suited for this because visual signals from one eye terminate on one layer of neurons lying immediately above a second layer of neurons stimulated by the opposite eye.

It is also possible that the lateral geniculate body plays a major role in *color vision.* A reason for believing this is that one can look at a red light with the left eye and a green light with the right eye and see a yellow color, thus indicating that combination of colors from the two separate eyes occurs at least to some extent in the brain, perhaps at the lateral geniculate level.

FUNCTION OF THE VISUAL CORTEX IN DISCRIMINATING THE VISUAL IMAGE. Once the visual image reaches the visual cortex, it is immediately changed to an entirely new pattern of stimulation, as illustrated in Figure 251. To the left is the retinal image

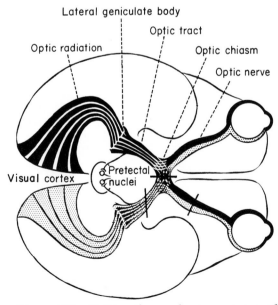

Figure 250. Optic pathways for transmission of visual signals from the retinae of the two eyes to the optic cortex. (From Polyak: *The Retina.* The University of Chicago Press.)

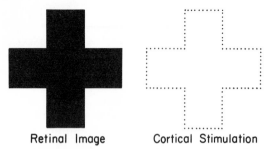

Retinal Image   Cortical Stimulation

Figure 251. Pattern of stimulation in the visual cortex caused by observing a black cross.

of a heavy black cross; to the right we see the pattern of stimulation in the visual cortex. Note that stimulation occurs only at the edges of the cross. This is caused by a mechanism in the visual cortex that allows a neuronal cell to be stimulated if there is contrast between a light area and a dark area. If there is no contrast, the neuron will not be stimulated. Thus, the visual cortex brings out borders and thereby determines the shapes of images. This explains why a simple line drawing of someone's face can be recognized as a picture of some particular individual. That is, the visual cortex itself actually converts the image of a person to a type of line drawing anyway.

Another feature of discrimination by the visual cortex is that it also determines direction of orientation of the lines and borders in the image. Thus, if the cross shown in Figure 251 were leaning slightly to the right, an entirely different set of neurons would be stimulated, thus denoting this leaning.

Signals are transmitted into the visual association areas, located on all sides of the visual cortex, both directly from the visual cortex and from the lateral geniculate body. In these areas the finer meanings of the visual signals are interpreted. For instance, the picture of a letter is interpreted here as letter A, B, C, or so forth, and a combination of letters is interpreted as a word. Still further away from the primary visual cortex, the combination of words is interpreted as a thought.

DETERMINATION OF LUMINOSITY BY BASAL REGIONS OF THE BRAIN. A monkey that has lost its cerebral cortex can still determine light and dark even though it cannot discriminate shapes of objects. This fact indicates that overall light intensity of the visual scene is interpreted by basal levels of the brain rather than by the cortex.

Perhaps the lateral thalamus and superior colliculus play roles in this effect. At any rate, after this interpretation is made, these lower centers presumably send this information to the visual association areas where it can be used for overall interpretation of the visual image.

Fields of Vision

A means used to determine the extent of a person's normal vision, and also to determine abnormalities of vision, is to plot his *fields of vision.* The person is asked to close one eye and to look straight forward with the other eye. A small spot of light is moved above his central point of vision as far as he can see, then below as far as he can see, then to the right, to the left, and in all directions. In this way his ability to see in all areas away from his central-most point of vision is plotted.

Figure 252 shows the normal field of vision of the right eye. Far out to the right one can actually see objects at right angles (90 degrees) from the direction in which the eye is looking. To the left the nose is in the way, and the person can see objects only 50 degrees away from the central point of vision. Likewise, in the upward direction the orbital ridge is in the way, and in the downward direction the cheek bone is in the way. Were it not for these structures around the eye, the field of vision would be considerably greater.

In the field of vision is a *blind spot* caused

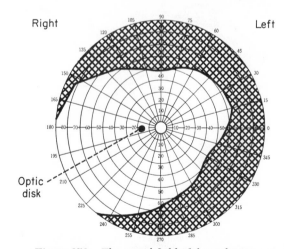

Figure 252. The visual field of the right eye.

by the *optic disk*, which is the point where the optic nerve enters the eyeball. At this point no rods or cones are present. The blind spot of each eye is located approximately 15 degrees to the temporal side of the central point of vision. However, the blind spots of the two eyes are on opposite sides in the respective fields of vision. The blind spots normally are not noticeable in one's vision.

Plotting the fields of vision provides a means for determining and locating damage in the retina or in the neuronal tracts from the retina to the brain. Referring back to Figure 250, it will be noted that three dark lines have been placed respectively across the right optic nerve, the optic chiasm, and the right optic tract. If the right optic nerve has been sectioned, the field of vision of the right eye will be zero, or, in other words, the eye will be completely blind. If the optic chiasm has been sectioned, the nasal half of each retina will be blind. This means that the *lateral* half of each visual field will be blind, because the lateral half of the field is registered by the nasal half of the retina. Finally, if the right optic tract is destroyed, the left halves of the visual fields of both eyes will be blind. Obviously, also, damage in any portion of the visual cortex or at any other point in the visual transmission system will cause loss of vision in respective areas in the fields of vision. The point in the eye or in the brain at which the damage has occurred can often be discerned from the pattern of visual loss.

Visual Acuity

Visual acuity means the degree of detail that the eye can discern in an image. The usual method for expressing visual acuity is to compare the person being tested with the normal person. If at a distance of 20 feet he can barely read letters of exactly the size that the normal person can barely read, he is said to have 20/20 vision, which is normal. If the letters must be as large as those that the normal person can read at a distance of 40 feet, his vision is said to be 20/40, or, in other words, his vision is one-half normal. If he can barely read letters at 20 feet distance that the normal person can read at 100 feet, his vision is 20/100, or, in other words, his vision is one-fifth normal. An occasional person has better than normal vision, so that he can read at 20 feet what the normal person can read only at 15 feet, in which case his vision is said to be 20/15.

RELATIONSHIP OF THE CONES TO VISUAL ACUITY. A person normally has very acute vision only in the central area of his visual field. The reason for this is that a small central area of the retina, only a millimeter or so in diameter, called the *fovea*, is especially adapted for acute vision. No rods are present in the fovea, and the cones there are considerably smaller in diameter than those in the peripheral portions of the retina. Also, the nerve fibers and blood vessels are all pulled to one side, so that light can pass with ease directly to the deep layers of the retina where the cones are located. Finally, and especially important, each of the cones of this region connects through an almost direct pathway with the brain so that the impulses from each cone do not become confused with impulses from other cones.

EFFECT OF THE OPTICS OF THE EYE ON VISUAL ACUITY. Figure 253 illustrates some of the optical factors that also influence the acuity of vision. It shows the images of point sources of light focused on the retina. If sources are so close together that both of the images focus on the same cone, then they will be interpreted as a single light rather than as two distinct lights. Likewise, if two lights are focused on two adjacent cones, they still will be interpreted as a single light, because a single light very frequently does stimulate two adjacent cones. Therefore, for the retina to interpret two points of light as separate lights, the images

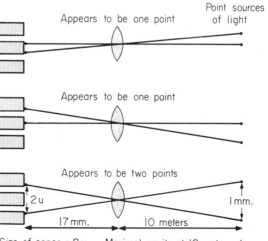

Figure 253. Focusing of points of light on cones in the fovea, illustrating the limits of maximal visual acuity.

must impinge on cones separated by an un-excited cone between them. And because the diameter of each foveal cone is about 2 to 3 microns, the images of point sources of light must be about 2 to 3 microns apart before they can be seen as two separate points. Thus, in the lower part of Figure 253, it is shown that two points of light 10 meters away from the eye and 1 mm. apart can barely be discerned by the normal eye as two separate lights rather than as a single one. To express this another way, the maximal acuity of vision at 10 meters is 1 mm.

The normal lens system of the eye is sufficiently effective so that visual acuity is determined mainly by the diameters of the cones in the fovea. When the optics of the eye are abnormal, visual acuity is determined instead by the ability of the lens system to focus the image on the retina. Consequently, a person with severe myopia, hypermetropia, or astigmatism may have a visual acuity of 20/40, 20/100, 20/200, or so forth, depending on the degree of optical abnormality.

Depth Perception

The eyes determine the distance of an object by two principal means. These are, first, by the *size of the image* on the retina, and second, by the phenomenon called *parallax*. To determine the distance of an object from the eyes by the image size, the person must have had previous experience with the object and know its actual size. For example, our previous experience with other persons is that their average height is somewhere between 5 and 6 feet. Therefore, even when using one eye we can determine relatively accurately the distance of a person by the size of his image on the retina. If he is nearby, his image is very large, but, if he is far away, his image is very small.

The second means for determining the distance of an object, parallax, depends on slight differences between the shapes and positions of the images on the retinae of the two eyes. This effect, which is shown in Figure 254, occurs because the two eyes are set several inches apart. In this figure the two eyes are observing a ball near the eyes and a block at a distance. In the left eye the image of the block lies to the right of the image of the ball, but in the right eye, the image of the block lies to the left of the image of the ball.

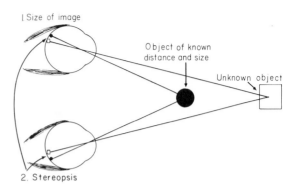

Figure 254. Mechanisms of depth perception.

In other words, the images of these two objects are actually reversed on the retinae because of the angles from which the two eyes observe them, and the brain interprets their relative distances by the degree to which they are reversed. Obviously, this is an extreme example of parallax, but even when looking at a single object the images in the two eyes are slightly different. It must be noted particularly that this mechanism interprets only the *relative distances* of objects and not the actual distances. However, if the distance of one of the objects is known, that of the second object is also known. For practical considerations, a person always has his hands or other parts of his body to use as objects of known distance, which can be used as reference points for determining the distances of unknown objects.

NEUROGENIC CONTROL OF EYE MOVEMENTS

If the eyes are to function satisfactorily as a camera, their line of sight must be appropriately directed so that the portion of the image that is to be seen will fall exactly on the fovea. In addition to this, the lens system of the eye must be focused for the distance of the object, and the pupil of the eye must be enlarged or contracted in proportion to the amount of light available, thus aiding in the light and dark adaptation of the eyes. It can be seen, then, that several different sets of eye muscles must be controlled very exactly to attain visual efficiency.

Positioning of the Eyes

Each eye is positioned by three different pairs of muscles, which are illustrated in

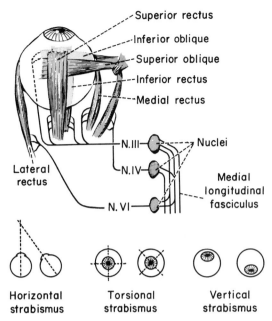

Figure 255.   The extraocular muscles.

Figure 255. One pair of these muscles is attached superiorly and inferiorly to the eyeball to move it up or down; another pair is attached horizontally to move the eye from side to side; and another pair is attached around the eyeball respectively on the bottom and top so that it can be rotated in either direction.

To control the eye movements, the visual association areas of the optic cortex must determine first whether or not the eyes are pointing toward the object; if not, impulses are transmitted through *oculomotor centers* in the brain stem to move the eyes in the appropriate direction. The visual association areas also determine whether or not the two eyes are receiving the same image on the corresponding portions of the two retinae — that is, whether or not the images are *fused* with each other. If not, one or both of the eyes are adjusted up or down, to one side or the other, or rotated so that the images will fuse. These minute adjustments of the eye positions are so important to vision that almost as much of the visual association areas in the brain is concerned with eye movements as with the interpretation of the meaning of visual signals.

MOVEMENTS OF THE EYES FOR FOLLOWING MOVING OBJECTS.   Special mechanisms are also available in the visual association areas and cerebellum for allowing the eyes to follow moving objects. To achieve this,

the eyes must move slowly in the same direction that the object is moving, neither falling behind nor getting ahead of the object. If the image does begin to fall behind or to move ahead, the visual association areas immediately send corrective signals to the eye muscles.

The cerebellum is involved in this mechanism because the object's course of movement must be predicted ahead of time; the predictive ability of the cerebellum allows the eyes to move along with the object rather than having to wait for the object to get to a new position before bringing the eyes along too.

Focusing of the Eyes

The eyes are kept focused on the visual field by changing the curvature of the crystalline lens as the field's distance changes. This is accomplished by the following neuromuscular mechanism:

Figure 256 shows the lens of the eye suspended from the sides of the eyeball by the suspensory ligaments, of which there are about seventy. The lens is actually an elastic ovoid structure having a clear envelope and clear viscous fluid in its center. When the ligaments are not pulling on the lens it assumes a round shape, but when the ligaments are tightened it flattens out. Normally, the ligaments are tight and the lens is flattened.

A smooth muscle called the *ciliary muscle* attaches to the suspensory ligaments of the lens where they connect with the eyeball.

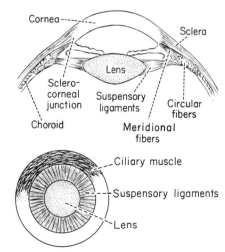

Figure 256.   The focusing (accommodation) mechanism of the eye.

This muscle is composed of *meridional fibers* and *circular fibers*. The meridional fibers extend from the ends of the suspensory ligaments anteriorly to the sclerocorneal junction; when they contract they pull the ends of the suspensory ligaments forward and loosen them. This allows the elastic lens to become thickened and the curvature of the lens to become greater, increasing the focusing power. The circular fibers extend all the way around the eye. When they contract they act as a sphincter, tightening progressively into a smaller and smaller circle, thus again loosening the suspensory ligaments so that the lens becomes still more convex and develops more focusing power.

Focusing, like positioning of the eyes, is accomplished by signals initiated in the visual association areas. When the image on the retina is out of focus, the fuzziness of the image initiates appropriate reactions to cause a change in the tension in the ciliary muscle. If the focus then becomes progressively worse, the visual association area immediately recognizes this and causes the tension in the ciliary muscle to begin changing in the other direction. As the focus becomes better and better it finally reaches a point at which the image has its greatest acuity. If the tension in the muscle keeps on changing, the eye goes out of focus in the opposite direction. Again the visual association area changes the direction of contraction of the ciliary muscle. Thus, the focusing mechanism keeps "searching" all the time for the image of maximal acuity, and even when the distance of the object from the eye changes tremendously, this searching mechanism can still find the appropriate focus within about one-fifth second.

Control of the Pupil

The pupil of the eye is the round opening of the iris through which light passes to the interior. The iris can constrict until the pupillary diameter is no greater than 1.5 mm., or it can relax until the diameter becomes 8 to 9 mm. Constriction of the pupil is caused by contraction of the *pupillary sphincter*, a circular muscle around the pupillary opening that is controlled by the parasympathetic nerves to the eye. Dilation of the pupil is caused by relaxation of the sphincter or by contraction of *radial muscle fibers* that extend from the edge of the pupil to the outer

border of the iris; this muscle is controlled by the sympathetic nerves to the eye.

Because the amount of light that enters the eye is proportional to the area of the pupil and not to the diameter, the amount of light entering the eye can be varied approximately forty-fold. This provides a mechanism for light and dark adaptation in addition to the retinal mechanism for light and dark adaptation. Changes in pupillary size can occur in less than one second, in contrast with retinal adaptation which requires several minutes.

The size of the pupil is regulated by a *pupillary light reflex*. When the retina is stimulated by light, some of the resulting impulses, instead of passing all the way to the visual cortex, leave the optic tract as shown in Figure 250 and pass to the pretectal nuclei of the brain stem. From here signals are transmitted to the brain stem centers that control the muscles of the eye and finally back to the iris. When the light intensity in the eyes is great, the size of the pupillary opening diminishes. On the other hand, when the light intensity becomes slight, the opening of the pupil increases.

The pupillary light reflex is frequently absent in a person with syphilis of the central nervous system, for this disease has a special predilection for destroying the pretectal nuclei.

REFERENCES

Alpern, M.: Distal mechanisms of vertebrate color vision. *Ann. Rev. Physiol.* 30:279, 1968.

Brindley, G. S.: *Physiology of the Retina and Visual Pathway.* Baltimore, The Williams & Wilkins Company, 1960.

Brown, P. K., and Wald, G.: Visual pigments in single rods and cones of the human retina. *Science* 144: 45, 1964.

Davson, H. (ed.): *The Eye.* 4 volumes. New York, Academic Press, 1962.

Davson, H.: *Physiology of the Eye.* 2nd Ed. Boston, Little, Brown & Co., 1962.

Hubel, D. H.: The visual cortex of the brain. *Sci. Amer.* 209(5):54, 1963.

Linksz, A.: *An Essay on Color Vision and Clinical Color Vision Tests.* New York, Grune & Stratton, 1964.

Marks, W. B., Dobelle, W. H., and MacNichol, E. E., Jr.: Visual pigments of single primate cones. *Science* 143:1181, 1964.

Neisser, U.: Visual search. *Sci. Amer.* 210(2):94, 1964.

Smelser, G. K. (ed.): *The Structure of the Eye.* New York, Academic Press, 1961.

Wolbarsht, M. L., and Yeandle, S. S.: Visual processes in the Limulus eye. *Ann. Rev. Physiol.* 29:513, 1967.

CHAPTER 25

HEARING, TASTE, AND SMELL

The function of the ear is to convert sound energy into nerve impulses. Figure 257 illustrates the general organization of the ear, showing the *external ear*, the *auditory canal*, the *tympanic membrane*, and the *ossicular system*, composed of the *malleus, incus*, and *stapes*, that transmits sound into the *cochlea*. The cochlea is also called the inner ear, and it is here that sound is converted into nerve impulses.

TRANSMISSION OF SOUND TO THE INNER EAR

CHARACTERISTICS OF SOUND AND SOUND WAVES. Sound is caused by compression waves traveling through the air. The transmitter of the sound, whether it be another person's voice, a radio speaker, or some noise-making device, creates the sound by alternately compressing the air and then relaxing the compression. For instance, a vibrating violin string creates sound by moving back and forth. When the string moves forward it compresses the air, and when it moves backward it decreases the amount of compression even below normal. This alternate compression and evacuation of the air produces sound.

Sound waves travel through the air in very much the same way that waves travel over the surface of water. Thus, compression of the air adjacent to a violin string builds up extra pressure in this region, and this in turn causes the air a little further away to become compressed. The pressure in this second region then compresses the air still further away, and this process continues on and on until the wave finally reaches the ear.

Function of the Tympanic Membrane (Tympanum) and Ossicles

When sound waves strike the tympanic membrane, the alternate compression and

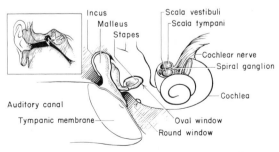

Figure 257. General organization of the ear, showing the external ear, the ossicular system, and the cochlea.

evacuation of the air adjacent to the membrane cause it to move backward and forward. The center of this membrane is connected to the handle of the *malleus;* this in turn is connected to the *incus,* and the incus to the *stapes.* These bones are called the *ossicles.* They are suspended in the *middle ear* by ligaments, so that they can rock back and forth. Movement of the handle of the malleus, therefore, causes the stapes also to move back and forth against the *oval window* of the cochlea, thus transmitting the sound into the cochlear fluid.

TRANSFORMATION OF THE PRESSURE OF THE SOUND WAVES BY THE OSSICULAR SYSTEM. If sound waves were applied directly to the oval window, they would not have enough pressure to move the fluid in the cochlea backward and forward to produce adequate hearing, because fluid has many times as much inertia as air, and a correspondingly greater amount of pressure is required to cause movement of fluid. The tympanic membrane and the ossicular system transform the pressure of the sound waves into a usable form by the following means:

The sound waves are collected by a very large membrane, the tympanic membrane, the area of which is approximately 70 square millimeters, or 22 times that of the oval window which has an area of only 3.2 square millimeters. Therefore, 22 times as much sound energy is collected as could be collected by the oval window alone, and all of this is transmitted through the ossicles to the oval window. Thus, the pressure of movement of the foot of the stapes is increased to about 22 times that which could be effected by applying sound waves directly to the oval window. This pressure is now sufficient to move the fluid of the cochlea backward and forward.

Transmission of Sound into the Cochlea by the Bones of the Head

The cochlea lies in a bony chamber inside the temporal bone of the skull. Therefore, any vibration of the skull also causes vibration of the fluid in the cochlea. Ordinarily, sound waves in the air cause almost no vibration in the skull bones, but clicking of the teeth or holding vibrating devices such as a tuning fork or special sound vibrators against the skull can cause bone vibrations. In this way, instead of the fluid vibrating inside the cochlea, the cochlea vibrates around the fluid, allowing the cochlea to react to the sound as well as if it had entered via the tympanic membrane and ossicles.

FUNCTION OF THE COCHLEA

PHYSIOLOGIC ANATOMY OF THE COCHLEA. The cochlea is a membranous device formed of coiled tubes. The outside appearance of the cochlea is illustrated in Figure 257, and a cross-section of its structure is shown in Figure 258. If the coiled tubes were stretched out in the manner illustrated in Figure 260, their total length would be about 3.5 cm.

The cross-sectional diagram of Figure 258 shows that the cochlea is actually composed of three separate tubes, lying side by side, called the *scala vestibuli,* the *scala tympani,* and the *scala media.* All of these tubes are filled with fluid, and they are separated from each other by membranes. The membrane between the scala vestibuli and scala media is so thin that it never obstructs the passage of sound waves. Its function is simply to separate the fluid of the scala media from

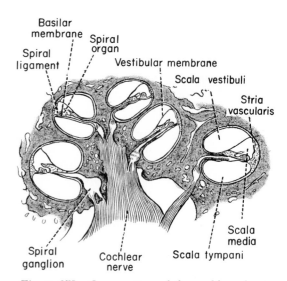

Figure 258. Cross-section of the cochlea, showing its coiled tubes composed of the scala vestibuli, the scala tympani, and the scala media. The cochlear nerve leading from the basilar membrane is also illustrated. (From Goss: *Gray's Anatomy of the Human Body.* Lea & Febiger.)

that of the scala vestibuli. The two fluids have separate origins, and the difference between them is believed to be important for proper operation of the sound receptor cells. On the other hand, the membrane separating the scala media from the scala tympani, called the *basilar membrane*, is a very strong structure that does impede the sound waves. It is supported by about 25,000 reed-like, stiff, thin spines that project into the membrane from one side. These spines, called *basilar fibers*, stretch most of the distance across the membrane. Located on the surface of the basilar membrane are the sound receptor cells called the *hair cells*.

Figure 259 depicts the cochlea diagrammatically after it has been uncoiled. In this figure the membrane between the scala vestibuli and the scala media has been eliminated because it does not affect sound transmission in the cochlea in any way. So far as the transmission of sound is concerned, the cochlea is composed of two separate tubes rather than three, and the two tubes are separated by the basilar membrane. The basilar fibers near the oval and round windows are short, but they become progressively longer, like the reeds of a harmonica, farther and farther up the cochlea, until at the tip of the cochlea the fibers are about two and one-half times as long as those near the base.

Resonance of Sound in the Cochlea

CONDUCTION OF SOUND IN THE COCHLEA. As each sound vibration enters the cochlea, the oval window at first moves inward and pushes the fluid of the scala vestibuli up into the cochlea, as indicated by the arrows in Figure 259. The sudden pressure in the scala

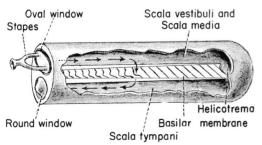

Figure 259. Movement of fluid in the cochlea when the stapes is thrust forward.

vestibuli bulges the basilar membrane into the scala tympani; this pushes fluid in this chamber toward the round window and causes it to bulge outward. Then, as the sound vibration causes the stapes to move backward, the procedure is reversed, with the fluid moving in the opposite direction along the same pathway and the basilar membrane bulging now into the scala vestibuli.

RESONANCE IN THE COCHLEA. A phenomenon called *resonance* occurs in the cochlea to allow each sound frequency to vibrate a different section of the basilar membrane. These vibrations are similar to those that occur in many musical instruments and can be explained as follows: When the string of a violin is pulled to one side it becomes stretched a little more than usual, and this stretch makes the string then move back in the other direction. However, as it moves it builds up momentum and, therefore, does not stop moving when it reaches its normal straight position. The momentum causes the string to become stretched once again but this time in the opposite direction, and the string then moves back in the first direction. The cycle continues over and over again so that once the string starts vibrating it continues.

Two factors determine the frequency at which a string will vibrate. First, the greater the *tension* developed by the string, the more rapidly it turns around and moves in the opposite direction, and the higher is the frequency of vibration. The second factor is the *mass* of the string. The greater its mass, the greater is the momentum developed during vibration, and the longer it will take for the string to change direction of movement; therefore, the lower will be the sound frequency. These same two factors, mass and elastic tension, apply to resonance in the cochlea. The vibrating mass in the cochlea is the fluid that vibrates back and forth between the oval and round windows, while the elastic tension is mainly the tension developed by the basilar fibers.

When *high frequency* sound enters the oval window, the sound wave travels along the basilar membrane only a short distance, as shown in Figure 260A, before a point of resonance is reached. The basilar membrane at this point vibrates "in tune" or "in resonance" with the frequency of the sound. As a result, the membrane moves back and forth forcibly at this point, while the move-

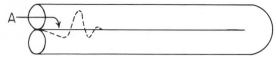

High frequency

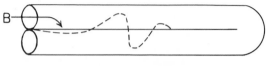

Medium frequency

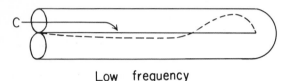

Low frequency

Figure 260. Diagrammatic representation of sound waves traveling along the basilar membrane and resonating at different points for high, medium, and low frequency sounds.

ment is very slight. When a *medium frequency* sound enters the oval window, the wave travels much further along the basilar membrane before the area of resonance is reached, as illustrated in Figure 260B. And, finally, a *low-frequency* sound wave travels almost all the way to the end of the membrane before it reaches its resonance point, as illustrated in Figure 260C.

High frequency sound waves resonate near the base of the cochlea for two reasons: First, the mass of fluid from the oval and round windows to the basal regions of the basilar membrane is very slight. Second, the basilar fibers, which provide the elastic tension for the vibrating system, have far greater rigidity at the basal end of the membrane than toward its tip. The combined effect of these two factors, the low mass and the greater rigidity, causes the basilar membrane to resonate at very high frequencies near its base.

On the other hand, when a sound wave travels all the way to the tip of the basilar membrane, the mass of fluid that must move is very great. Also, the basilar fibers are longer and less rigid near the tip of the cochlea. The combined effect of these two factors, the great mass and the low rigidity of the elastic component, makes the basilar

membrane near this end resonate at a very low frequency. Similarly, intermediate frequency sounds resonate at intermediate points along the membrane. Figure 261A illustrates the amplitude of vibration of different parts of the basilar membrane for a medium frequency sound wave, and Figure 261B represents the degree of vibration of the membrane for several different sound frequencies from very low to very high.

DETERMINATION OF PITCH BY THE COCHLEA. Located on the surface of all the basilar fibers are special nerve receptors, called *hair cells,* that are stimulated when the basilar fibers vibrate back and forth. These are shown in Figure 262. When the hair cells near the base of the cochlea are stimulated, the brain interprets the pitch of the sound as that of a very high frequency. When the hair cells in mid-cochlea are stimulated, the brain interprets the sound as one of intermediate pitch, and stimulation of those at the tip of the cochlea is interpreted as low pitch. It can be seen, therefore, that the pitch of a sound is determined by the point in the cochlea at which the basilar membrane vibrates.

DETERMINATION OF LOUDNESS OF A SOUND. The loudness of a sound is determined by the intensity of movement of the basilar fibers. The greater the displace-

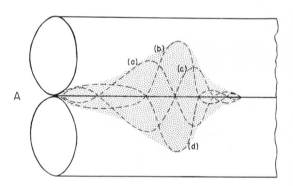

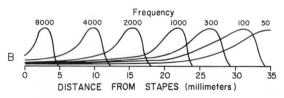

Figure 261. *(A)* Amplitude pattern of vibration of the basilar membrane for a medium frequency sound. *(B)* Amplitude patterns for sounds of frequencies between 50 and 8000 per second, showing resonance at different points for the different frequencies.

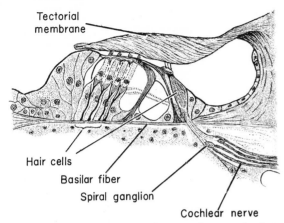

Tectorial membrane

Hair cells

Basilar fiber

Spiral ganglion

Cochlear nerve

Figure 262. The organ of Corti, showing the hair cells and the tectorial membrane pressing against the projecting hairs. (Modified from Bloom and Fawcett: *A Textbook of Histology.* 8th Ed.)

ment back and forth, the more intensely are the hair cells stimulated, and the greater are the number of impulses transmitted into the brain to indicate the degree of loudness. For instance, if a hair cell near the base of the cochlea transmits only one impulse per second, the frequency of the sound will still be interpreted as one of very high pitch, but the loudness of the sound will be almost zero. If the same hair cell is stimulated 1000 times per second, the pitch of the sound will remain the same, but the loudness will be extreme.

Conversion of Sound Energy into Nerve Impulses

The method by which vibration of the basilar fibers converts sound energy into nerve impulses is not completely understood, but experiments have shown that vibration of the basilar fibers causes the hair cells located on top of them to generate electrical potentials. Movement of the basilar membrane in one direction causes the potential above the membrane to become positive and below negative. Then, when the membrane vibrates in the opposite direction, the potential shifts polarity. It is believed that this electrical potential elicits impulses in the many filamentous nerve fibers that connect with the hair cells, as illustrated in Figure 262, and that from these nerve fibers the signals pass through the cochlear nerve into the brain.

The hair cells of the organ of Corti are almost identical with the hair cells of the vestibular apparatus for maintaining equilibrium, which will be described in Chapter 26. It is known that the hair cells in the latter apparatus are stimulated by bending the hairy projections that extend from the ends of the cells. Therefore, it can be presumed that the hair cells of the cochlea are also stimulated by bending their hairs when sound vibrates the basilar membrane. A suggested mechanism by which this could occur involves the *tectorial membrane,* which is shown in Figure 262. Vibration of the basilar membrane causes the hair cells lying on its upper surface to rock back and forth; this is believed to brush the projecting hairs against the tectorial membrane or to cause fluid to flow back and forth over the hairs in the narrow channel between the hair cells and the tectorial membrane. Movement of the hairs excites the cells and generates impulses in the small filamentous cochlear nerve endings that entwine the hair cells.

TRANSMISSION OF SOUND INTO THE CENTRAL NERVOUS SYSTEM

Figure 263 shows the pathways for transmission of sound impulses from the cochlea into the central nervous system. After passing through the cochlear nerve, the

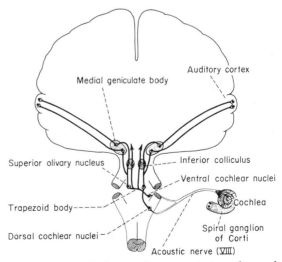

Auditory cortex

Medial geniculate body

Superior olivary nucleus

Inferior colliculus

Ventral cochlear nuclei

Trapezoid body

Cochlea

Dorsal cochlear nuclei

Spiral ganglion of Corti

Acoustic nerve (VIII)

Figure 263. Pathways for transmission of sound impulses from the cochlea into the central nervous system.

impulses are transmitted through at least five different structures, which may be listed as follows: (1) the *dorsal* and *ventral cochlear nuclei,* (2) the *trapezoid body* and *superior olivary nucleus,* (3) the *inferior colliculus,* (4) the *medial geniculate body,* and (5) the *auditory cortex.*

FUNCTION OF LOWER AUDITORY CENTERS. The function of the lower auditory centers is not well understood, but some of them are probably responsible for localization of the direction from which sound is coming. For instance, if sound comes from the left, it reaches the left ear before it reaches the right. If the sound is coming from directly in front, it reaches both ears simultaneously, but as its direction changes from directly ahead to one side, the time lag becomes progressively greater. The time lag is presumably interpreted in some of the lower auditory centers, to determine the direction from which the sound is coming.

Another function of the auditory centers in the brain stem is reflex production of rapid motions of the head, of the eyes, or even of the entire body in response to auditory signals. Some of these signals pass into the cerebellum and bulboreticular formation to alter a person's equilibrium, to make his head jerk to one side, or to make his eyes roll in one direction. Some of these rapid responses can occur even though the cerebral portions of the auditory system have been destroyed.

Figure 263 shows that the auditory signals from each ear are transmitted approximately equally into the auditory pathways on both sides of the brain stem and cerebral cortex. Therefore, destruction of one of the pathways will not greatly affect one's hearing ability in either ear.

FUNCTION OF THE AUDITORY CORTEX. The function of the auditory cortex has already been discussed in Chapter 23. In brief, the *primary auditory cortex* is located in the middle of the superior gyrus of the temporal lobe. This area receives the sound signals and interprets them as different sounds. However, the signals must also be transmitted into surrounding *auditory association areas* before their meanings become clear. Finally, the signals are transmitted into the *common integrative center* of the cortex, where the overall meaning of all combined auditory, visual, and other types of sensation is determined.

Destruction of one auditory cortex reduces hardly at all one's ability to hear, but destruction of both cortices greatly depresses the hearing. It is especially interesting, though, that even with both auditory cortices destroyed, an animal is still capable of hearing sounds of extreme intensity. It seems that the medial geniculate bodies of the thalamus are the regions that receive these sounds, for destruction of these areas causes total deafness. This illustrates again that many sensory functions of the brain can be performed by the thalamus independently of the cerebral cortex.

DEAFNESS AND HEARING TESTS

Any damage to the ossicular system, the cochlea, the cochlear nerve, or to the pathways for transmission of sound to the auditory cortex can cause partial or total deafness. The types of deafness are generally separated into two categories, *conduction deafness* and *nerve deafness.* The term conduction deafness means deafness caused by failure of sound waves to be conducted from the tympanic membrane through the ossicular system into the cochlea. Nerve deafness, on the other hand, means failure of the sound impulses to reach the auditory cortex because of damage to the cochlea itself or to the neurogenic transmission system for sound.

One of the most common causes of *conduction deafness* is repeated blockage of the *auditory tube.* This tube connects the middle ear with the nose. Its function is to keep the pressure inside the middle ear equal to the pressure of the air, so that no pressure difference will exist between the two sides of the tympanic membrane. If the auditory tube becomes plugged because of a cold, because of allergic swelling of the nasal membranes, or for some other cause, then the air in the middle ear becomes absorbed, and in its place a serous fluid collects. Also, the tympanic membrane is pulled inward because of lowered pressure in the middle ear. Fibroblasts tend to grow into the serous fluid and cause fibrous tissue between the ossicles and the walls of the middle ear. If this process continues long enough, the ossicles finally become so firmly bound to the walls of the middle ear that sound conduction into the cochlea becomes almost nil.

Nerve deafness is characteristic of old age; almost all older people normally develop at least some degree of this type of deafness, especially for very high frequency sounds. This is probably caused by the aging process in the cochlea itself, though the auditory pathways in the central nervous system possibly degenerate to some extent also.

Hearing Tests

Almost any type of sound instrument can be used to check a person's hearing; the most common of these for many years has been the *tuning fork*. After striking a tuning fork the sound can usually be heard for about 30 seconds if the fork is held near the normal ear. If the person has conduction deafness, the ear is unable to hear the sound, but placing the butt of the fork against the skull will allow transmission of vibrations to the cochlea via the bones of the skull. If the cochlea is still functioning properly, the sound is then heard even though it could not be heard by air conduction. On the other hand, if the person has nerve deafness, he is unable to hear the tuning fork either by air or bone conduction.

THE AUDIOMETER. In recent years a special sound emitter, called an audiometer, has been used to measure the degree of deafness. This is an electronic apparatus capable of generating sound of all frequencies in an earphone or in a vibrator placed against the bone of the skull. The apparatus is calibrated so that the zero mark corresponds to the intensity of sound required for a normal person barely to hear it. If the person is deaf for sound of a particular frequency, the sound intensity must be increased far above the zero level before he can hear it, and the amount of extra sound energy that must be added is said to be the *hearing loss* for that particular frequency.

Figure 264 shows an *audiogram* from a person who has conduction deafness. Approximately 40 to 60 decibels of extra sound energy had to be transmitted into the earphone at each frequency for the person to hear the sound. However, when the skull vibrator to cause bone conduction was used instead of the earphone, no extra energy was required; indeed, at the high frequencies the hearing was even better than normal for

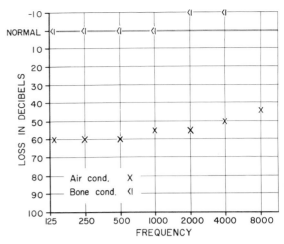

Figure 264. An audiogram from the ear of a person with conduction deafness.

bone conduction. This illustrates that bone conduction was normal, which means that the cochlea and auditory pathways were also normal. However, conduction of sound through the ossicular system was greatly impaired.

The decibel system for expressing sound energy is a logarithmic scale rather than a linear one. Thus a change of 10 decibels is a 10-fold change in energy, 20 decibels is a 100-fold change, 30 decibels is 1000-fold, and 60 decibels is 1,000,000-fold. For example, in the audiogram of Figure 264, the hearing loss at most frequencies is about 60 decibels, which means that this deaf person actually had a hearing loss of about 1,000,000-fold.

Sounds that the normal ear hears frequently vary in intensity a hundred million-fold or more. For instance, the intensity of sound in a very noisy factory is about one million times as great as that of a quiet whisper. Therefore, a person with 60 decibels hearing loss—a one million-fold loss—can still hear sounds of very strong intensity.

TASTE

The senses of taste and smell are called the *chemical senses* because their receptors are excited by chemical stimuli. The taste receptors are excited by chemical substances in the food that we eat, while the smell receptors are excited by chemical substances in the air.

The Taste Bud

The sensory receptor for taste is the *taste bud*, shown in Figure 265. It is composed of epithelioid *taste receptor cells* arranged around a central pore in the mucous membrane of the mouth. Several very thin hairlike projections called *microvilli* protrude from the surface of each taste cell and thence through the pore into the mouth. These microvilli provide the receptor surface for taste.

Interweaving among the taste cells is a branching network of two or three taste nerve fibers that is stimulated by the taste cells.

Before a substance can be tasted it must be dissolved in the fluid of the mouth, and it must diffuse into the taste pore around the microvilli. Therefore, highly diffusible substances such as salts or other small molecular compounds generally cause greater degrees of taste than do less diffusible substances such as proteins or others of very large molecular size.

THE FOUR PRIMARY SENSATIONS OF TASTE. Psychologically, we can detect four major types of taste called the *primary* taste sensations. They are: (1) salty, (2) sweet, (3) bitter, and (4) sour.

Until recent years it was believed that four enitrely different types of taste buds existed, each type detecting one particular primary taste sensation. It has now been learned that every taste bud has some degree of sensitivity for all the primary taste sensations. However, each bud usually has a greater degree of sensitivity to one or two of the taste sensations than to the others. The brain detects the type of taste by the ratio of stimulation of the different taste buds. That is, if a bud that detects mainly saltiness is stimulated to a higher intensity than buds that respond more to the other tastes, the brain interprets this as a sensation of saltiness even though the other buds are stimulated to a lesser extent.

Almost any electrolyte will stimulate the salty buds. These buds determine the proportion of salts and other electrolytic substances present in food. The most familiar salt that stimulates these buds is sodium chloride, common table salt. Therefore, we normally associate the function of these buds almost entirely with the content of table salt in food.

The sweet taste buds detect the amount of sugars in food. This is one of the means by which animals determine whether or not fruits are ripe and whether or not unknown foods are nutritious. Essentially all wild foods that are sweet are safe to eat and contain a considerable amount of nutrition.

The bitter taste buds provide a protective function, for they detect principally the poisons in wild plants. The alkaloidal compounds of wild herbs, for instance, are among the poisons detected by the bitter taste buds. These same alkaloids, when given in appropriate quantities, are frequently very valuable drugs. Many drugs such as quinine, though of very bitter taste, have a beneficial effect when used properly, and yet can be lethal when used improperly.

The sour taste buds detect the degree of acidity of food—that is, they detect the concentration of hydrogen ion in the mouth. If the degree of acidity is slight, such as that of dilute vinegar, the food is usually very palatable; if the acidity is extremely strong, the taste is very unpleasant, and the food is rejected.

Regulation of the Diet by the Taste Sensations

The taste sensations obviously help to regulate the diet. For example, the sweet taste is usually pleasant, causing an animal to choose foods that are sweet. On the other hand, the bitter sensation is always unpleasant, causing bitter foods, that are often poisonous, to be rejected. The sour taste is sometimes pleasant and sometimes un-

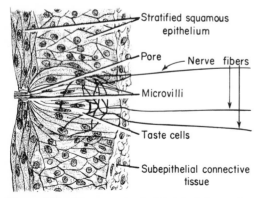

Stratified squamous epithelium

Pore

Nerve fibers

Microvilli

Taste cells

Subepithelial connective tissue

Figure 265. The taste bud. (Modified from Bloom and Fawcett: *A Textbook of Histology.* 8th Ed.)

pleasant, and, likewise, the salty taste is sometimes pleasant and sometimes unpleasant. The pleasantness of these types of taste is often determined by the momentary state of nutrition of the body. If a person has been long without salt, then, for reasons which are not yet understood, the salty sensation becomes extremely pleasant. If a person has been eating an excess of salt, the salty taste is very unpleasant. The same is true for the sour taste and to a less extent for the sweet taste. In this way the quality of the diet is automatically varied in accord with the needs of the body. That is, lack of a particular type of nutrient often intensifies one or more of the taste sensations and causes the person to choose foods having a taste characteristic of the deficient nutrient.

IMPORTANCE OF SMELL TO THE SENSATION OF TASTE. Much of what we call taste is actually smell, because foods entering the mouth give off odors that spread into the nose. Often, a person who has a cold states that he has lost his sense of taste, but on testing for the four primary sensations of taste they are all still present.

The smell sensations, which are discussed in following sections of the chapter, function along with taste sensations to help control the appetite and the intake of food.

Transmission of Taste Signals to the Central Nervous System

Figure 266 shows the pathways for transmission of taste signals into the brain stem

and then into the cerebral cortex. The signals pass from the taste buds in the mouth to the *tractus solitarius* located in the medulla. Then signals are transmitted to the *thalamus,* and from there to the *primary taste cortex* of the opercularinsular region, as well as into surrounding taste association areas and into the common integrative region, which integrates all sensations.

TASTE REFLEXES. One of the functions of the taste apparatus is to provide reflexes to the salivary glands of the mouth. To do this, impulses are transmitted from the tractus solitarius into nearby nuclei that control secretion by the parotid, submaxillary, and other salivary glands. When food is eaten, the quality of the taste sensations, operating through these reflexes, determines whether the output of saliva will be great or slight.

SMELL

The sense of smell is vested in two small areas, called the *olfactory epithelium,* located respectively on each side in the upper reaches of the nasal cavity. The location of these areas is illustrated in Figure 267, which shows to the right a cross-section of the air passages of the nose, and also shows the olfactory epithelium connected with the nervous system.

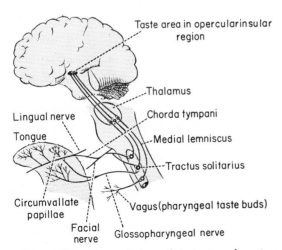

Figure 266. Transmission of taste impulses into the central nervous system.

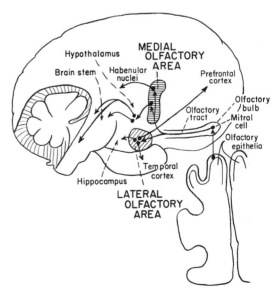

Figure 267. Organization of the olfactory system, and pathways for transmission of olfactory impulses into the central nervous system.

The Olfactory Receptors

The olfactory epithelium contains numerous nerve receptors, called *olfactory cells,* which are shown in Figure 268. These are a special type of nerve cell that projects small root-like structures called *olfactory hairs* outward from the nasal epithelium into the overlying mucus. It is the olfactory hairs that detect the different odors.

The means by which odors excite the olfactory hairs are not well understood. However, those odors most easily smelled are, first, the very highly volatile substances, and, second, substances that are highly soluble in fats. The necessity for volatility is easily understood, for the only means by which an odor can reach the high spaces of the nose is by air transport to this region. The reason for the fat solubility seems to be that the olfactory hairs themselves are outcroppings of the cell membrane of the olfactory cell, and all cell membranes are composed mainly of fatty substances. Presumably, then, the odoriferous substance becomes dissolved in the olfactory hairs, and this produces nerve impulses in the olfactory cells.

THE PRIMARY SENSATIONS OF SMELL. It has been very difficult to study individual olfactory cells, for which reason we are not yet sure what primary chemical stimuli excite the various types of olfactory cells. Yet on the basis of crude experiments, the following primary sensations have been postulated:

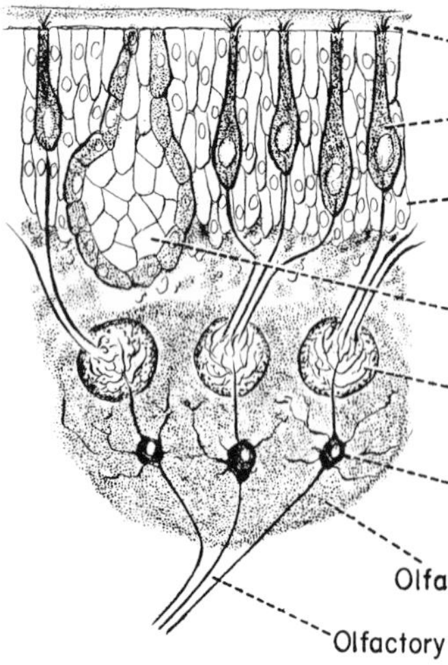

1. Camphoraceous
2. Musky
3. Floral
4. Pepperminty
5. Ethereal
6. Pungent
7. Putrid

Thus, smell, like all the other sensations, is probably subserved by a few discrete types of cells that give rise to specific primary olfactory sensations. However, the preceding listing is mainly conjecture and probably is in error.

ADAPTATION OF SMELL. Smell, like vision, can adapt tremendously. On first exposure to a very strong odor, the smell may be very strong, but after a minute or more the odor will hardly be noticeable. The olfactory receptors apprise the person of the presence of an odor, but then do not keep belaboring him with its presence. This is especially valuable when one must work in pungent surroundings.

MASKING OF ODORS. Unlike the eye's ability to see a number of different colors at the same time, the olfactory system detects the sensation of only a single odor at a time. However, the odor may be a combination of many different odors. If both a putrid odor and a sweet odor are present, the one that dominates the other is the one that has the greater intensity, or, if both are of about equal intensity, the sensation of smell is between that of sweetness and putridness. The ability of a high intensity odor to dominate is called *masking.* This effect is used in hospitals, toilets, and other areas to make pungent surroundings pleasant; incense may be burned or some odoriferous but pleasant smelling substance may be evaporated into the air to mask the less desirable odors.

Transmission of Smell Signals into the Nervous System

Because smell is a subjective phenomenon that can be studied satisfactorily only in human beings, very little is known about transmission of smell signals into the brain. Smell pathways terminate in two major areas of the brain called the *medial olfactory area* and the *lateral olfactory area* respectively, both illustrated in Figure 268. The medial olfactory area lies in the very middle of the brain, while the lateral olfactory area lies laterally on the under surface of the brain covered over by the temporal lobe.

Olfactory hairs
Olfactory cell
Sustentacular cells
Bowman's gland
Glomerulus
Mitral cell
Olfactory bulb
Olfactory tract

Figure 268. The olfactory membrane, showing especially the olfactory cells with their cilia protruding into the overlying mucus of the nose. (Modified from Bloom and Fawcett: *A Textbook of Histology.* 8th Ed.)

The medial olfactory area is responsible primarily for the primitive functions of the olfactory system, such as stimulation of salivation in response to smell, licking the lips, and causing an animal to stalk a juicy meal. On the other hand, the lateral olfactory area is very closely associated with higher functions of the nervous system; direct pathways pass from the lateral olfactory area into the temporal cortex, the hippocampus, and the prefrontal cortex, all important regions of cortical function. The lateral olfactory area is concerned with complicated responses that depend on olfactory stimuli. Thus, recognition of a certain type of smell as belonging to a particular animal is a function believed to be performed by this area. And recognition of various delectable or detestable foods presumably is also a function of this area. In human beings, tumors located in this region frequently cause the person to perceive very abnormal smells which may be of any type, pleasant or unpleasant, and continuing sometimes for months on end.

REFERENCES

Davis, H.: Biophysics and physiology of the inner ear. *Physiol. Rev.* 37:1, 1957.

Katsuki, Y.: Comparative neurophysiology of hearing. *Physiol. Rev.* 45(2):380, 1965.

Moulton, D. G., and Beidler, L. M.: Structure and function in the peripheral olfactory system. *Physiol. Rev.* 47(1):1, 1967.

Oakley, B., and Benjamin, R. M.: Neural mechanisms of taste. *Physiol. Rev.* 46(2):173, 1966.

Pfaffmann, C.: Taste, its sensory and motivating properties. *Amer. Sci.* 52:187, 1964.

Schwartzkopff, J.: Hearing. *Ann. Rev. Physiol.* 29:485, 1967.

Wever, E. G., and Lawrence, M. *Physiological Acoustics.* Princeton, Princeton University Press, 1954.

Zotterman, Y.: *Olfaction and Taste.* New York, The Macmillan Company, 1963.

CHAPTER 26

MOTOR FUNCTIONS OF THE SPINAL CORD AND LOWER BRAIN STEM

In the past few chapters we have considered primarily the sensory functions of the central nervous system. To make use of sensory information, the nervous system employs *motor* mechanisms to control the muscles and some of the glands of the body, which are collectively called the *effectors*. The purpose of the next few chapters will be to discuss these motor mechanisms, beginning in the present chapter with a discussion of the motor function of the spinal cord and lower brain stem.

Though most people have an inherent intuition that it is only the conscious elements of the brain that cause muscle movements and other motor activities, this is the farthest from the truth, for perhaps the greater proportion of our motor activities are actually controlled by lower regions of the central nervous system, specifically the spinal cord and lower brain stem, which operate primarily at a subconscious level.

Physiologic Anatomy of the Spinal Cord

Figure 269A illustrates a cross-section of the spinal cord, showing that it is composed of two major portions, the *white* and the *gray matter*. The gray matter, which lies deep inside the cord, has the appearance of double horns protruding anteriorly and posteriorly. The cell bodies of the neurons of the cord are located in the gray matter. The white matter, which comprises all other portions of the cord, is composed of fiber tracts. Several long *descending tracts* originate in the brain and pass down the cord to terminate on the neurons in the gray matter, and several other long tracts, the *ascending tracts*, originate in the cord and then pass upward to the brain.

In addition to the long descending and ascending tracts, many fibers called *propriospinal fibers* pass from one region of the cord to another, as shown in Figure 269B. This figure also illustrates the general pattern of organization of each segment of the spinal cord. Sensory nerve fibers enter the posterior horns of the gray matter on each side; some fibers synapse with neurons here while others pass directly up the cord. The neurons in turn send fibers in all directions through the cord. Some of these fibers enter the propriospinal tracts and pass upward or downward to adjacent segments, others pass to the internuncial cells, and a few pass directly to

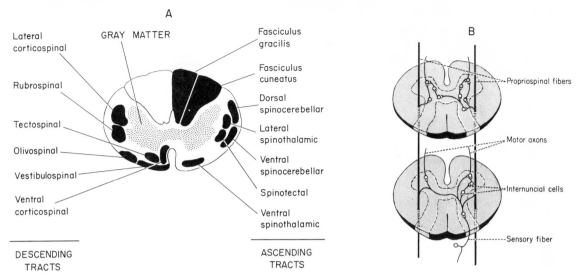

Figure 269. (A) Fiber tracts and gray matter of the spinal cord. (B) Two segments of the cord, illustrating the internuncial mechanisms that integrate cord reflexes and the propriospinal fibers that interconnect the cord segments.

the anterior motor neurons. The anterior motor neurons then send fibers from the cord into the spinal nerves.

THE SIMPLE CORD REFLEXES

A neurogenic reflex is the transmission through nervous pathways of a signal initiated somewhere in the body to cause a reaction somewhere else in the body. A very large portion of the nervous system, especially the cord and the basal regions of the brain, functions as one great mass of reflexes. For example, the passage of somesthetic sensations from the skin to the cord or brain to cause some muscular reaction is a reflex.

Two essentials must always be present for a reflex to occur: a *receptor* organ and an *effector* organ. All of the sensory nerve receptors are receptor organs for reflexes, and all the muscles of the body, whether they be smooth muscles or skeletal muscles, are effector organs. Glandular cells that can be stimulated by nerve impulses are also effector organs.

The Axon Reflex

The simplest neuronal reflex is the axon reflex shown to the left in Figure 270; this

reflex involves only part of a single neuron. Each sensory nerve fiber in a spinal nerve normally has a hundred or more branches, some of which terminate in the skin while others terminate on blood vessels or on other structures beneath the skin. It will be recalled that regardless of what point of a neuron is stimulated, the nerve impulse travels over the entire membrane of the fiber. Therefore, stimulating a single sensory receptor in the skin causes an impulse to travel backward into all the branches of the fiber as well as upward to the cord. When the impulse reaches the terminals near blood vessels, a hormone, possibly histamine, acetylcholine, or some other similar vasodila-

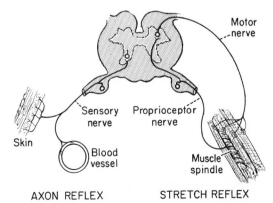

Figure 270. The neuronal mechanisms of the axon and stretch reflexes.

tory substance, is secreted, and this causes the blood vessels to dilate.

The axon reflex occurs principally when the skin is damaged. For example, scratching the skin with a pin elicits an axon reflex, which in turn causes blood to flow more rapidly than usual. The skin becomes reddened for several millimeters on each side of the scratch mark. In this way the axon reflex provides increased blood flow to damaged tissues and aids in their repair.

The Proprioceptor Reflexes

The spinal cord is the neurogenic locus of many different reflexes that originate in the proprioceptor receptors of the muscles, tendons, and other deep tissues. Signals from the receptors pass into the gray matter of the cord and cause motor impulses to be transmitted back to the muscles to cause appropriate muscular responses. Some of the proprioceptor reflexes are the following:

THE STRETCH REFLEX. Sudden stretch of a muscle causes a reflex that makes the muscle contract. This reflex, the nervous pathway of which is illustrated to the right in Figure 270, is one of the simplest of all cord reflexes, for it involves only a single sensory neuron and single motor neuron. Sudden stretch of the muscle stimulates muscle spindles in the muscle, and impulses from these excite the anterior motor neurons that control the same muscle. These, in turn, transmit reflex signals back to the muscle to cause it to contract.

The knee jerk and other muscle reflexes. One of the best known examples of the stretch reflex is the *knee jerk,* shown in Figure 271. Almost every doctor elicits this reflex when he performs a physical examination. A small hammer is used to strike the patellar tendon immediately below the kneecap. The sudden jolt to the tendon stretches the quadriceps muscle, which in turn stretches muscle spindles and sends a barrage of impulses into the spinal cord. Then a reflex signal passes back to the quadriceps muscle, causing a sudden contraction that makes the lower leg jerk forward.

A muscle jerk similar to the knee jerk can be elicited in any muscle of the body by suddenly striking its tendon, or even by striking the muscle itself. The only essential for

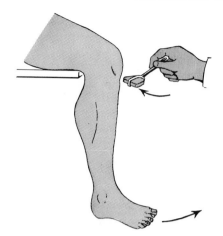

Figure 271. Method for eliciting the knee jerk.

eliciting such a reflex is to stretch the muscle suddenly. Muscle reflexes of this type are elicited by physicians for two major purposes: First, if the reflex can be demonstrated, it is certain that both the sensory and motor nerve connections are intact between the muscle and the spinal cord. Second, muscle reflexes can help determine the degree of excitability of the spinal cord. When a large number of facilitatory impulses are being transmitted from the brain to the cord the muscle reflexes will be so active at times that simply tapping the patellar tendon with the tip of one's finger might make the leg jump a foot or more. On the other hand, the cord may be intensely inhibited by other impulses from the brain, in which case almost no degree of pounding on the muscles or tendons can elicit a response.

Relationship of the stretch reflex to muscle tone. Muscle tone is ordinarily maintained by impulses that arise continuously in the muscle spindles. These pass into the spinal cord to cause reflex contraction of the muscles, which keeps the muscles in a slightly taut state all of the time. When this basic mechanism for maintaining muscle tone is absent, a signal from the brain must be far stronger than normal to elicit a response.

Servo-positioning function of the muscle spindle. Figure 272 illustrates the relation of the muscle spindle to the surrounding skeletal muscle fibers. The muscle spindle is composed of a receptor area in its center called the *annulospiral receptor,* and a *muscular area* composed of *intrafusal muscle fibers* at both ends. The intrafusal muscle

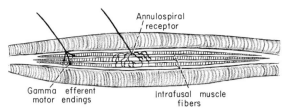

Figure 272. Arrangement of the muscle spindle with respect to the surrounding muscle fibers.

fibers are controlled by a special set of nerve fibers from the spinal cord called *gamma efferent fibers.*

The muscle spindle operates to cause the muscle to contract to a particular length as follows: Impulses transmitted from the spinal cord through the gamma efferent nerves shorten the intrafusal muscle fibers. If the skeletal muscle fibers do not shorten simultaneously, the central portion of the muscle spindle becomes stretched and excites the annulospiral receptors. Impulses are then transmitted into the spinal cord, causing reflex contraction of the skeletal muscle fibers. Thus, the original contraction of the intrafusal fibers causes automatic reflex contraction of the skeletal muscle fibers. In other words, the muscle *follows* the degree of contraction of the intrafusal fibers, which is called *servo-positioning* of the muscle. This type of contraction is believed to be used in the performance of postural movements or in positioning limbs to exact positions with respect to the rest of the body.

Damping function of the stretch reflex. If a limb or some other part of the body is suddenly knocked out of position by some outside force, the muscles are stretched and an immediate stretch reflex is elicited which tends to oppose movement of the limb, thereby attempting to maintain the status quo. This is called the *damping function* of the stretch reflex. This function also provides smooth motions rather than jerky motions in response to signals from the brain, because most signals from the brain arrive in the cord in sudden bursts rather than continuously.

THE TENDON PROTECTIVE REFLEX. A small, branching receptor that looks like a bush and called the *Golgi tendon apparatus* is found in all muscle tendons. It detects the degree of stretch of the tendons, and its function is to protect the tendon and muscle from excessive stretch. The reflex that performs this function is the following: Ordinarily, light and moderate degrees of stretch cause no reflex whatsoever; however, if the degree of stretch on the tendon exceeds a certain critical value, an immediate reflex occurs to *inhibit* the anterior motor neurons that innervate the muscle. The muscle immediately relaxes, and the excessive stretch is removed from the muscle. Obviously, this is an important protective reflex to prevent damage to the muscle or tendon. It is a more complicated reflex than the stretch reflex, requiring several intermediate neurons in the cord gray matter between the input sensory nerve and the anterior motor neuron.

THE EXTENSOR THRUST REFLEX. An even more complex proprioceptor reflex that helps support the body against gravity is the *extensor thrust reflex.* Pressure on the pads of the feet causes automatic tightening of the extensor muscles of the legs. This reflex is initiated by pressure receptors in the bottom of the foot. The signal passes to the *internuncial cells* (intermediate cells between sensory input and motor output) in the cord. Here it is amplified and diverged into an appropriate pattern of impulses to tighten the extensor muscles, allowing the animal to keep his leg stiffened automatically when standing.

THE MAGNET REACTION. A proprioceptor reflex closely allied to the extensor thrust reflex, but still more complicated, is the so-called *magnet reaction.* In an animal that has had its spinal cord cut so that impulses from the brain will not interfere, one can place the tip of his finger on the pad of the animal's foot, then move his finger in all directions, and the foot will follow the finger. Moving the finger to one side causes appropriate proprioceptor reflexes to make the limb move in the direction of the force. This reaction is an aid to equilibrium of the animal, for excess pressure on one side of the foot indicates that he is falling in that direction, and automatic stiffening of the leg in that direction helps to prevent the fall.

The Flexor Reflex

Pain causes automatic withdrawal of any pained portion of the body from the object causing the pain. This is called a *flexor reflex* or sometimes a *pain reflex,* a *nociceptive reflex,* or a *withdrawal reflex.*

The neuronal mechanism of the flexor reflex is shown to the left in Figure 273. On entering the gray matter of the cord, the pain signal stimulates internuncial cells that transmit a pattern of impulses through appropriate anterior motor neurons to cause withdrawal of the area of the body that is being pained. In Figure 273 the pain stimulus is initiated in the right hand, and the biceps muscle becomes excited and pulls the hand away.

The flexor reflex normally lasts for only a split second after the painful stimulus has been applied, but this short period of time is usually sufficient to pull away from the stimulus.

The Crossed Extensor Reflex

When a flexor reflex occurs in one limb, impulses also pass to the opposite side of the cord, where they stimulate the internuncial cells controlling the extensor muscles of the opposite limb, thus causing it to extend. This is called a *crossed extensor* reflex. For example, a painful stimulus applied to the right hand causes the left arm to extend, as shown in Figure 273. Such a reflex pushes the person away from the painful object. That is, at the same time that he withdraws his pained hand on the right, he pushes his entire body away on the left.

Withdrawal reflexes are present in all

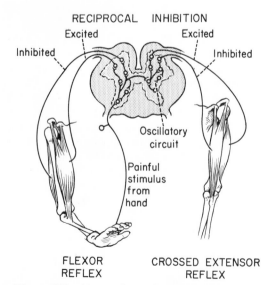

RECIPROCAL INHIBITION

Excited   Excited

Inhibited   Inhibited

Oscillatory
circuit

Painful
stimulus
from
hand

FLEXOR   CROSSED EXTENSOR
REFLEX   REFLEX

Figure 273. Neuronal mechanisms of the flexor reflex, of the crossed extensor reflex, and of reciprocal inhibition.

parts of the body, though organized somewhat differently in different areas. For instance, if a needle pricks the small of a person's back, he automatically arches forward, which causes withdrawal. Though this reflex is slightly different from the flexor and crossed extensor reflexes of the limbs, its result is the same.

Reciprocal Inhibition

A very important feature of most reflexes, especially illustrated by the flexor and crossed extensor reflexes, is the phenomenon called *reciprocal inhibition.* That is, when a reflex excites a muscle, it ordinarily inhibits the antagonistic muscle at the same time. For example, when the flexor reflex of Figure 273 excites the biceps muscle, it simultaneously inhibits the opposing triceps muscle. Also, the crossed extensor reflex excites the triceps but inhibits the biceps. In all parts of the body where opposing muscles exist, a corresponding reciprocal inhibition circuit is present in the spinal cord. This obviously allows greater ease in the performance of desired activities.

RHYTHMIC REFLEXES OF THE CORD

The reflexes of the spinal cord are not limited to single reactions; frequently rhythmic reflexes also occur. That is, "patterns of activity" in the cord can make a portion of the body move back and forth in a rhythmic motion. Two especially important types of rhythmic reflexes controlled entirely by the spinal cord are the scratch reflex and the walking reflexes.

The Scratch Reflex

The scratch reflex, though not of importance in the human being, is one of the most important of all protective mechanisms in lower animals. A dog can actually scratch away a flea or other irritating object even after his spinal cord has been completely transected.

The scratch reflex depends on two separate integrative abilities of the spinal cord:

First, the cord must provide the rhythmic to and fro motion of the leg. This is accomplished by the reverberating circuits in the internuncial cells that send impulses during part of the cycle to one group of muscles, then to the opposing muscles, contracting alternately both sets of muscles to cause the rhythmic motion. The second cord mechanism utilized in the scratching act pinpoints the area of the body that needs scratching. For instance, a flea moving across the back of a sleeping dog is followed exactly by the scratching paw. When the flea crawls across the midpoint of the animal's belly this paw stops scratching, but the paw on the opposite side of the body immediately finds the flea and begins to scratch. Thus, the scratch reflex is one of the most complicated of all cord reflexes for it requires, first, localization of the irritation, second, coordinate movement of the paw into position, and third, reverberating motion of the limb.

Walking Reflexes

The spinal cord also is capable of providing the rhythmic to and fro walking movements of the legs. These movements, like the scratching act, are controlled by reverberating circuits in the internuncial cells that contract alternately the antagonistic pairs of muscles. Reciprocal inhibition also plays a part, for this keeps the muscles on the two sides of the body operating at opposite phase to each other, allowing one leg to move forward while the opposite one moves backward. In lower animals, fiber pathways also pass from the hindlimb region of the cord to the forelimb region, to cause appropriate phasing of the hind and forelimb movements. This cord control of walking movements with appropriate phasing of the limbs is illustrated by Figure 274, which shows a dog with his spinal cord transected at the neck. He is nevertheless still exhibiting continuous, rhythmic, to and fro movements of his limbs.

AUTONOMIC REFLEXES OF THE CORD

In addition to the cord reflexes for controlling skeletal muscular action, the spinal cord also harbors reflex circuits that help to control the visceral functions of the body. These reflexes are initiated by sensory receptors in the viscera, and the signals are transmitted through sensory nerves to the internuncial cells of the cord gray matter, where appropriate patterns of reflex responses are determined. Then the signals pass to *autonomic motor neurons* located in the cord gray matter about midway between the anterior and posterior horns. These cells send impulses into the sympathetic chain and back to the viscera to cause autonomic stimulation.

The Peritoneal Reflex

One of the most important of the autonomic reflexes is the peritoneal reflex. Whenever any portion of the peritoneum is damaged, this reflex slows up or stops all motor activity in the nearby viscera. For example, in appendicitis, movement of food through the gastrointestinal tract stops almost completely, preventing further irritation of the inflamed appendix and allowing the reparative processes of the body to function at optimum efficiency. Without this reflex many would die of appendicitis before aid could be given.

Walking movement

Figure 274. Walking movements in a dog whose spinal cord has been transected at the neck.

Vascular Control by the Cord

The blood pressure is normally controlled by vasomotor impulses transmitted from the brain, but after the spinal cord has been transected these impulses can no longer reach the blood vessels. Yet cord reflexes are still capable of modifying local vascular blood flow in response to such factors as pain, heat, and cold. For instance, when one of the viscera is pained tremendously by overdistention, such as overdistention of the urinary bladder, the blood pressure may rise to twice normal. Though these effects of cord reflexes on the blood pressure are not important normally, they do illustrate the capability of the cord to help control many of the involuntary reactions of the autonomic nervous system.

The Bladder and Rectal Reflexes

Among the most important of the autonomic reflexes of the cord are the bladder and rectal reflexes that cause automatic emptying of the urinary bladder and the rectum when they become filled. When either the bladder or the rectum becomes excessively full, sensory signals are transmitted into the internuncial cells of the lower end of the cord, as shown in Figure 275. Appropriate signals are then transmitted through parasympathetic nerves back to the bladder or colon. Here they excite the main body of the respective organ, but at the same time inhibit the internal sphincter of the urethra or anus, thereby causing emptying.

In persons whose spinal cords have been cut, these automatic reflexes are sometimes

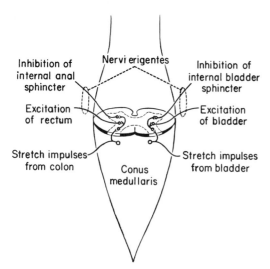

Figure 275. Neuronal circuits of the bladder- and rectal-emptying reflexes.

Labels in figure:
Nervi erigentes
Inhibition of internal anal sphincter
Inhibition of internal bladder sphincter
Excitation of rectum
Excitation of bladder
Stretch impulses from colon
Stretch impulses from bladder
Conus medullaris

effective enough to evacuate the bladder every half hour or more and the colon one or more times each day. In the normal person, however, these reflexes are inhibited by impulses from the brain until an opportune time arises for evacuation. Unfortunately, this inhibition frequently leads to constipation or painful distention of the bladder.

MOTOR FUNCTIONS OF THE LOWER BRAIN STEM

The lower brain stem is composed of the medulla, the pons, and the mesencephalon. It is in these areas that many of the centers for blood pressure control, respiration, and gastrointestinal regulation are located, all of which are discussed in detail in other chapters. The purpose of the present section is not to give an overall discussion of the lower brain stem, but to present its motor functions and their relation to the cord and the cerebrum. The motor functions of the lower brain stem can be divided into two major parts: first, its function in helping the animal support his body against gravity, and, second, its function in the maintenance of equilibrium.

THE BULBORETICULAR FORMATION OF THE BRAIN STEM. Figure 276 shows the *bulboreticular formation*, comprising most of the lower brain stem, and some of its connections with other parts of the body. Stimulation of this formation transmits impulses down the *reticulospinal, tectospinal, vestibulospinal,* and *rubrospinal tracts* into the cord to control the muscles throughout the body.

The bulboreticular area receives incoming fibers from many sources: fibers direct from all areas of the body through the spinal cord, fibers from the cerebellum, from the equilibrium apparatus of the ear, from the motor cortex, and from the basal ganglia. Thus, the bulboreticular formation is an integrative area for combining and coordinating (1) sensory information from the body, (2) motor information from the motor cortex, (3) equilibrium information from the vestibular apparatuses, and (4) proprioceptor information from the cerebellum. With this information available it controls many of the involuntary muscular activities.

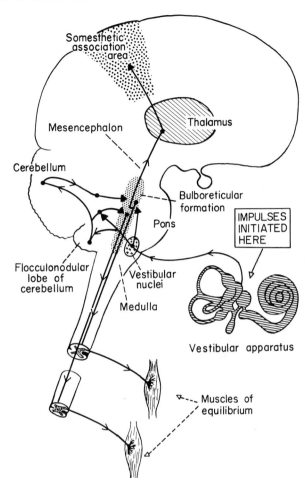

Figure 276. Nervous mechanisms of equilibrium.

Support of the Body Against Gravity

Widespread stimulation of the upper bulboreticular formation excites mainly the extensor muscles of the body, causing the trunk and the limbs to become stiffened, and making it possible to stand erect. Without this stiffening, the body would immediately crumple because of the pull of gravity.

SUPPRESSION OF THE BULBORETICULAR AREA AND DECEREBRATE RIGIDITY. When one wants to sit rather than to stand, the excitation of the muscles by the bulboreticular formation must be inhibited. This is accomplished mainly by suppressor impulses from the basal ganglia. When the basal ganglia are damaged, or when all the fiber tracts from the cerebrum are cut, the bulboreticular formation automatically becomes so active that the animal becomes rigid all over. This phenomenon, called *decerebrate*

rigidity, is illustrated in the dog of Figure 277. It shows the necessity of the basal ganglia to inhibit the bulboreticular formation when motor functions other than the support of the body are to be performed.

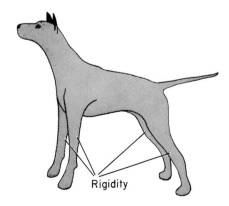

Figure 277. Rigidity in a dog whose brain has been transected above the bulboreticular formation.

COORDINATION OF CONTRACTION OF ANTAGONISTIC POSTURAL MUSCLES. Another function of the basal ganglia in relation to the bulboreticular formation is to help coordinate the contraction between antagonistic postural muscles. The degree of contraction of antagonistic muscles must be controlled appropriately at all times to keep one of the muscles from overbalancing the other. Signals transmitted from the bulboreticular formation up to the basal ganglia and then back again to the bulboreticular formation are probably responsible for this coordination. Without this neuronal circuit, the antagonistic muscles reverberate back and forth, trying to establish appropriate degrees of contraction but never being able to do so. This is probably the cause of the tremor that occurs in Parkinson's disease, which will be described in the following chapter.

Equilibrium Function of the Lower Brain Stem

In addition to providing the neurogenic mechanisms for support of the body against gravity, the bulboreticular formation is also capable of maintaining equilibrium by varying the degree of tone in the different extensor muscles. Most of the equilibrium reflexes are initiated by the vestibular apparatuses located on each side of the head, adjacent to each internal ear.

FUNCTION OF THE VESTIBULAR APPARATUS. The vestibular apparatus is shown in Figure 278, and its connections with the equilibrium areas of the central nervous system are shown in Figure 276. This apparatus detects the position of the head in space—that is, it determines whether the head is upright with respect to the gravitational pull of the earth or whether it is lying back, upside down, or in another position. It also detects sudden changes in movement. To perform these functions the vestibular apparatus is divided into two separate physiologic sections, the *utricle* and the *semicircular canals.*

The utricle. On the wall of the utricle is a structure called the *macula*, which is illustrated in Figures 278A and B. Nerve cells in the base of the macula project "hairs" upward into a gelatinous mass in which are located many minute bone-like calcified

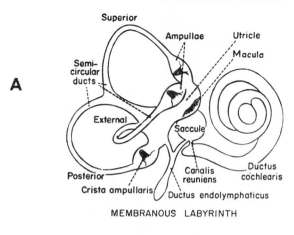

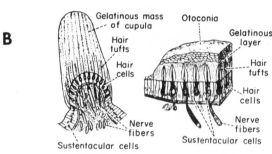

Figure 278. Functional structures of the vestibular apparatus.

granules, like very small granules of sand, called *otoconia*. When the head is bent to one side, the weight of the otoconia pushes the hairs to that side, thereby stimulating the nerves. In this way the utricle supplies the equilibrium regions of the central nervous system with the information needed to maintain balance.

The utricle also helps the person maintain his balance when he suddenly begins to move forward, to one side, or in any other linear direction. When he begins to move forward, the inertia of the otoconia causes them to lag behind the movement of the rest of the body and, therefore, to bend the hairs backward. This gives the sensation of falling off balance in the backward direction. As a result, the person leans forward to correct this imbalance, which explains why an athlete leans forward automatically when he first starts to run.

On the other hand, when one wishes to decelerate, he must lean backward. Again it is the otoconia of the utricle that initiate automatically this backward leaning, for when one brakes himself the momentum of the otoconia keeps them moving forward while the rest of the body slows. This bends the hairs of the macula in the forward direction, making the person feel as if he were falling head first toward the ground. As a result, the equilibrium mechanism leans the body backward automatically.

The semicircular canals. The *semicircular canals* are small circular tubes that contain fluid, as shown in Figure 278A. The vestibular apparatus on each side of the head has three separate canals located respectively in the three planes of space, one in the horizontal and the other two in the two vertical planes. If one suddenly turns his head in any direction, the fluid in one or more of the semicircular canals lags behind the movement of the head because of the inertia of the fluid. This is the same effect as that observed when one suddenly rotates a glass of water; the glass rotates, but the water remains still. As the fluid moves in the semicircular canal it flows against the *crista ampullaris,* shown in Figures 278A and B, which is a valve-like leaflet located at one end of each canal. This structure contains hair tufts from hair cells like those in the macula, and bending these tufts to one side or the other gives a person the sensation that his head is beginning to turn.

The information transmitted from the semicircular canals apprises the nervous system of sudden changes in direction of movement. With this information available, the bulboreticular formation can correct for any imbalance that is likely to occur *even before it does occur.* This is particularly important when one is changing direction of movement rapidly, as when playing a fast game of almost any type.

FUNCTION OF THE CEREBELLUM IN EQUILIBRIUM. In addition to transmitting signals into the bulboreticular formation, the semicircular canals and utricle also send information to the *flocculonodular lobes* of the cerebellum. Since one of the major functions of the cerebellum is to predict future position of the body in space, which will be discussed in the following chapter, the function of the flocculonodular lobes is probably to predict ahead of time when a state of imbalance is about to occur. This allows appropriate corrective signals to the bulboreticular formation even before the person falls off balance, and prevents imbalance from ever occurring rather than necessitating attempts to correct it after it has already occurred. Persons who lose their cerebellum lose this ability to predict ahead of time, and, as a result, must perform all movements slowly or else fall.

THE NECK PROPRIOCEPTOR RECEPTORS AND THEIR RELATIONSHIP TO THE VESTIBULAR MECHANISMS. The vestibular apparatuses detect only the position of the *head,* not of the *body,* in relation to the pull of gravity. To translate this information from the head to the whole body, the relationship of the head to the body must be known. This knowledge is provided by proprioceptor receptors in the neck. For instance, when the head is bent backward, the vestibular apparatuses send information that the position of the head is changing, but at the same time the neck proprioceptors send information that the head is angulating backward in relation to the rest of the body. The two sets of impulses negate each other and the tautness of the postural muscles remains exactly the same. In other words, the proprioceptor reflexes from the neck are as necessary for regulating equilibrium as the complicated reflexes initiated by the vestibular apparatuses.

PERIPHERAL PROPRIOCEPTOR AND VISUAL MECHANISMS OF EQUILIBRIUM. After the vestibular apparatuses have been destroyed, a person can still maintain his equilibrium provided he moves slowly. This is accomplished mainly by means of proprioceptor information from the limbs and surfaces of the body and visual information from the eyes. If he begins to fall forward, the pressure on the anterior parts of his feet increases, stimulating the pressure receptors. This information transmitted to his brain helps to correct the imbalance. At the same time, his eyes also detect the lack of equilibrium, and this information too helps to correct the situation.

Unfortunately, the visual and proprioceptor systems for maintaining equilibrium are not organized for rapid action, which explains why a person without his vestibular apparatuses must move slowly.

OVERALL CONTROL OF LOCOMOTION

From the foregoing discussions of the motor functions of the spinal cord, lower brain stem, and cerebrum, it is now possible to construct the overall pattern of locomotion. First, the person must support himself against gravity. This is accomplished partly by the extensor thrust mechanism of the spinal cord that allows stiffening of the limbs when pressure is applied to the pads of the feet; in addition to this, the bulbo-reticular formation transmits impulses continuously into the extensor muscles to keep the body and limbs stiff. Also, the degree of stiffening of the different parts of the body is varied by the equilibrium apparatus so that when one tends to fall over, appropriate muscles are contracted to bring the body back into the upright position.

Once the body is supported against gravity and maintained in a state of equilibrium, locomotion then depends on rhythmic motion of the limbs. Rhythmic circuits in the spinal cord are capable of providing the to and fro movement of the limbs, and the movements of the opposing limbs are kept in opposite phase with each other by the reciprocal inhibition mechanisms of the cord. Thus, most of the functions of locomotion can be provided by the cord and brain stem, but the cerebral cortex must control these functions in accord with the desires of the individual. When he wishes to move forward, to stop, or to turn to one side, his motor cortex and basal ganglia simply initi-ate the action, stop it, or change it by sending *command signals*. The cord provides the stereotyped actions required to perform the actual movements. In this way, the energy of the portion of the brain operating at a conscious level is conserved to perform other mental feats.

REFERENCES

Creed, R. S., Denny-Brown, D., Eccles, J. C., Liddell, E. G. T., and Sherrington, C. S.: *Reflex Activity of The Spinal Cord.* New York, Oxford University Press, 1932.

Denny-Brown, D.: *The Basal Ganglia: And Their Relation to Disorders of Movement.* New York, Oxford University Press, 1962.

Eldred, E.: Posture and locomotion. *Handbook of Physiology.* Sec. I, Vol. II, p. 1067. Baltimore, The Williams & Wilkins Company, 1960.

Eldred, E., and Buchwald, J.: Central nervous system. *Ann. Rev. Physiol.* 29:573, 1967.

Granit, R.: Muscular tone. *J. Sport Med.* 2:46, 1962.

Hunt, C. C., and Perl, E. R.: Sinal reflex mechanisms concerned with skeletal muscle. *Physiol. Rev.* 40: 538, 1960.

Jasper, H. H., et al. (eds.): *The Reticular Formation of the Brain.* Boston, Little, Brown and Company, 1958.

Liddell, E. G. T.: *The Discovery of Reflexes.* New York, Oxford University Press, 1960.

Lloyd, D. P. C.: Spinal mechanisms involved in somatic activities. Handbook of Physiology, Sec. I, Vol. II, p. 929. Baltimore, The Williams & Wilkins Company, 1960.

CHAPTER 27

FUNCTION OF THE CEREBRAL CORTEX, BASAL GANGLIA, AND CEREBELLUM FOR CONTROL OF MUSCLE MOVEMENT

Though a major share of the motor functions of the body can be performed without the cerebral cortex, one of the distinguishing features of the human being is his ability to carry out extremely complex voluntary types of muscle activity. He can perform the intricate tasks of talking, writing, construction of delicate instruments, and performance of specialized patterns of movement required for dance routines, basketball, or football. All these activities involve a high degree of control by the cerebral cortex.

The basal ganglia are large masses of neurons located deep in the cerebrum but having close connections with the cerebral cortex and thalamus. These areas control many stereotyped movements that are much more complex than the involuntary postural types of movement controlled by the lower brain stem but yet much less complex than the intricate voluntary movements controlled by the cerebral cortex.

The cerebellum never functions alone to control motor movements but instead functions always in association with the other motor portions of the brain as follows: In the absence of the cerebellum, movements are usually performed in a jerky, tremulous manner and without the fine degree of control necessary for accuracy, as will be explained later in the chapter. The cerebellum provides special damping functions for smoothing out the tremors and jerkiness that would otherwise be present in muscular activity.

FUNCTION OF THE PRIMARY AND SECONDARY MOTOR AREAS OF THE CORTEX

THE PRIMARY MOTOR CORTEX. The *primary motor cortex* is located in front of the central sulcus of the brain immediately an-

321

terior to the somesthetic cortex, as illustrated in Figure 279. This figure also shows several other areas that are associated with motor function, such as areas in the posterior part of the brain that help to choose the words for phonation and help to control eye fixation. Anterior to the primary motor cortex are areas that control automatic head rotation, contralateral eye movements, and the actual process of word formation. These areas are part of the *secondary motor cortex*, which is described below.

The primary function of the primary motor cortex is to control discrete movements of the different skeletal muscles in the body. Stimulation of a point area in the primary motor cortex causes a specific muscle on the opposite side of the body to contract. However, not all muscles are controlled to the same degree of precision by the motor cortex; for instance, stimulation of a tiny area of the motor cortex may cause half of the trunk muscles on the opposite side of the body to contract all at the same time, while stimulation of another small area equally large may cause only one of the small thumb muscles to contract. Thus the "degree of representation" in the primary motor cortex is more than 100 times as great for the thumb as for the trunk muscles. Figure 280 illustrates schematically the degree of representation of the different muscles, showing that the hand and mouth muscles constitute about two thirds of the entire representation in the primary motor cortex. Note also that the very top of the primary motor cortex controls the toe, ankle, and foot muscles on the op-

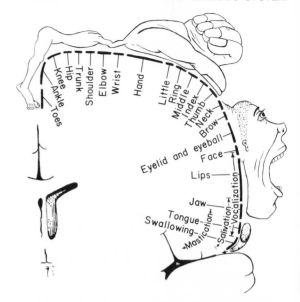

Figure 280. Degree of representation of the different body muscles in the primary motor cortex. (From Penfield and Rasmussen: *The Cerebral Cortex of Man.* The Macmillan Company.)

posite side of the body, while the lower part of the primary motor cortex controls the face muscles on the opposite side.

THE SECONDARY MOTOR CORTEX. The *secondary motor cortex*, also known as the *premotor cortex*, lies anterior to the primary motor cortex, between it and the prefrontal lobes, as illustrated in Figure 279. This area controls more coordinate movements than does the primary motor cortex, often exciting many muscles at a time. It sends impulses either (a) to lower centers of the brain, which in turn transmit signals to the muscles or (b) to the primary motor cortex which in turn excites the individual muscles.

The secondary motor cortex seems to control "patterns" of movement. If one will stop to think for a moment, most of the motions he performs with his hands or other portions of his body fall into no more than a few hundred patterns of motion; these are used over and over again in different sequences. Each time a particular pattern of movement is used, a neuronal circuit in the secondary motor cortex seems to become progressively more facilitated until eventually the movement can be performed rapidly and precisely without fumbling. This is theoretically the learning process for control of muscular activities. The reason for believing that the secondary motor area is such a locus for these circuits is that removal of this area

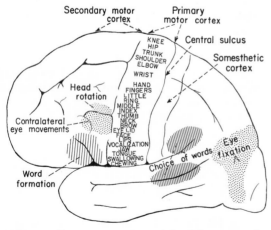

Figure 279. Representation of the different muscles of the body in the motor cortex, and location of other cortical areas responsible for certain types of motor movements.

causes loss of coordinate movements but does not cause loss of discrete movements of muscles. The greater the number of patterns of movements stored in the secondary motor area and the more precise they become, the more skilled the person becomes in the performance of physical tasks.

OVERALL CONTROL OF MOTOR ACTIVITY. Most of what we know about the overall control of motor activity is based on studies in human beings who have had destruction of certain parts of the brain. These studies have shown the following:

Removal of a small area of *primary motor cortex* causes loss of discrete movements in a given area of the body but not loss of many coordinate movements in the same area.

Removal of a small area of *secondary motor cortex* causes loss of some coordinate movements and *simple* patterns of movement in a specific area of the body but not loss of discrete movements in the same area.

Removal of the *somesthetic cortex* and *somesthetic association areas* makes it impossible for a person to perform *complex learned patterns* of movement but does not prevent his *attempting* to perform purposeful acts; nor does it prevent his performing *simple* patterns of movement.

Serious damage in the brain stem, particularly in the *reticular substance of the mesencephalon*, can cause serious disruption of almost any type of motor activity.

On the basis of these studies, the following theory of motor control has been formulated: The thought that initiates motor activity supposedly begins in the *common integrative area* of the cortex operating in association with the *thalamus* and *mesencephalon*. These areas do not control the actual muscle movements. Instead, they "command" the motor and sensory regions of the brain to perform a particular activity that has been performed before and the memory of which is stored in *the sensory portion of the brain*, including especially *the somesthetic cortex and somesthetic association area*. The motor cortex, both the primary and secondary areas, sends impulses to the muscles which begin the movement, and, simultaneously, sensory signals from the moving parts of the body are transmitted back to the sensory cortex, which determines whether the movement is going in the proper direction. If it is not, *corrective signals* are sent from the sensory

cortex to the motor cortex to make the muscles move properly. Thus, the pattern in the sensory region controls the motor cortex which acts as a *servo system* to make the muscles "follow" the sensory pattern.

However, some activities must be performed so rapidly that feedback impulses cannot return all the way from the peripheral parts of the body to the sensory cortex in time to effect proper control. In these instances, less complicated patterns of movement are believed to be stored in the secondary motor cortex, and these, so the theory goes, can be called forth without invoking the feedback control system of the sensory cortex.

Functions of lower centers. Figure 281 illustrates diagrammatically the interrelationships of the different parts of the brain in the control of motor functions. This figure shows that lower centers as well as the cortex occupy a central position in motor control. The different parts of the cortex communicate with each other mainly through the thalamus and mesencephalon, so that the sensory cortex probably initiates most motor activities by stimulating lower centers of the brain which in turn excite the motor cortex. Also, many of the gross body movements and primitive movements, such as movements of the eyes, rotation of the head, and movements of equilibrium, can be controlled directly by lower centers in the mesencephalon, pons, and medulla without participation of the cortex at all.

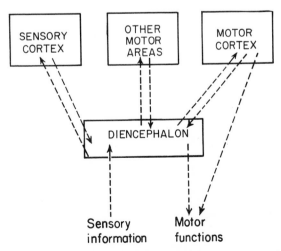

Figure 281. Principal pathways for performance of motor functions.

TRANSMISSION OF MOTOR SIGNALS TO THE SPINAL CORD AND MUSCLES

The Corticospinal Tract

Figure 282 illustrates the *corticospinal* or *pyramidal tract*, which transmits impulses from the motor cortex downward through the basal regions of the brain, into the medulla, and finally into the spinal cord. Almost all the fibers in the corticospinal tract cross to the opposite side in the medulla or in the upper segments of the spinal cord. As a result, stimulation of the left motor cortex contracts the muscles on the right side of the body; and, vice versa, stimulation of the right motor cortex contracts the muscles on the left side of the body.

The corticospinal tract is made up of two different types of nerve fibers. One type comes from large neurons in the primary motor cortex called *Betz cells* or *pyramidal cells*. Each primary motor cortex contains approximately 30,000 of these cells, and each one of the cells transmits signals to a discrete muscle, causing it to contract. The second group of fibers in the corticospinal tract is composed of very small fibers that originate in many small neurons of the entire motor cortex and some even from the somesthetic sensory cortex. These have much less ability to stimulate the muscles but do have the ability to facilitate the neurons in the spinal cord that control the muscles. This facilitation makes these cells more active than usual, causing an increase in the basic tone of the muscles and also making them excessively responsive to signals transmitted by the Betz cells.

The Internuncial Cells of the Cord

Figure 283 illustrates a cross-section of the spinal cord, showing several of the fiber tracts that pass downward from the brain. The *corticospinal tract* is in the lateral and posterior portion of the cord. Nerve fibers from this tract terminate on *internuncial cells* located in the cord *gray matter*. Motor signals are usually transmitted through several of the internuncial cells in rapid sequence before finally reaching the anterior motor neurons that control the muscles. The internuncial cells may be arranged in reverberating or parallel circuits for causing pro-

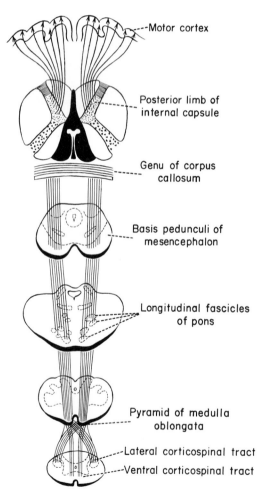

Figure 282. The corticospinal tract (also called the pyramidal tract) through which motor signals are transmitted from the motor cortex to the spinal cord. (Modified from Ranson and Clark: *The Anatomy of the Nervous System.* 10th Ed.)

Labels in figure:
- Motor cortex
- Posterior limb of internal capsule
- Genu of corpus callosum
- Basis pedunculi of mesencephalon
- Longitudinal fascicles of pons
- Pyramid of medulla oblongata
- Lateral corticospinal tract
- Ventral corticospinal tract

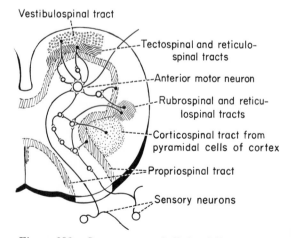

Figure 283. Convergence of all the different motor pathways on the anterior motor neuron.

Labels in figure:
- Vestibulospinal tract
- Tectospinal and reticulospinal tracts
- Anterior motor neuron
- Rubrospinal and reticulospinal tracts
- Corticospinal tract from pyramidal cells of cortex
- Propriospinal tract
- Sensory neurons

longed discharges, divergent circuits for amplifying the signal, or convergent circuits for integrating signals from several sources.

A number of other fiber pathways besides those of the corticospinal tract terminate on the internuncial cells. These include *propriospinal fibers* that originate in other portions of the spinal cord, *sensory fibers* from the surface of the body, and fibers from the brain transmitted through the *rubrospinal tract*, the *reticulospinal tract*, and the *tectospinal tract*. Thus, the internuncial cells are not controlled entirely by the corticospinal tract, but instead by at least six or eight different types of fibers arriving from different sources in the nervous system and also from the peripheral nerves. These other fibers are mainly responsible for the automatic and subconscious movements of the muscles, in contrast to the conscious movements controlled by the corticospinal tract.

Some signals reaching the internuncial cells are inhibitory rather than excitatory. These include especially signals from some parts of the reticulospinal tracts and from the vestibulospinal tracts.

The function of the internuncial cells is to integrate the signals from all possible sources and then to determine the type of motor signal to be directed to the anterior motor neurons. For instance, if a strong excitatory signal enters the internuncial cells from a sensory nerve at the same time that a signal enters from the corticospinal tract, the resultant stimulation of the anterior motor neuron will be much greater than will occur when a signal comes only from the corticospinal tract.

The Anterior Motor Neurons

The final pathway for transmission of motor signals from the brain to the muscles is through the *anterior motor neurons* which are very large neurons located in the anterior horns of the cord gray matter. On receiving impulses from the internuncial cells, these cells subtract the inhibitory signals from the excitatory signals, and the result determines the rate at which impulses will be transmitted into the motor nerve.

Fibers from the anterior motor neurons pass through the spinal nerves to all the skeletal muscles. Each fiber divides an average of approximately 150 times, and

each of the terminal fibrils innervates one neuromyal junction on a muscle fiber. Therefore, each anterior motor neuron controls approximately 150 muscle fibers which constitute the motor unit described in Chapter 7.

Summary of Motor Transmission Through the Corticospinal Tract

Discrete signals for muscular control originate in the Betz cells of the primary motor cortex, and many less discrete signals originate in other parts of both the motor and sensory cortex. These signals are transmitted through the corticospinal tract into the spinal cord where they excite the internuncial cells of the cord gray matter. Here they are integrated with signals from other regions of the brain, and from peripheral nerves, to determine the overall degree of activity. If signals are arriving simultaneously from the corticospinal tract and other tracts, such as from the rubrospinal, tectospinal, and vestibulospinal tracts, the degree of activity will be much greater than if signals are arriving only through the corticospinal tract. Likewise, signals entering from the spinal nerves help to determine the degree of activity. The integrated signals from the internuncial cells then pass to the anterior motor neurons and from these to the muscles.

FUNCTION OF THE BASAL GANGLIA

The basal ganglia are composed of large masses of neurons located deep in the substance of the cerebrum and in the upper part of the mesencephalon. These are shown in Figure 284. They include the *caudate nucleus*, the *putamen*, the *globus pallidus*, the *subthalamic nucleus*, the *substantia nigra*, and several other less important nuclei.

CONTROL OF SUBCONSCIOUS MOVEMENTS BY THE BASAL GANGLIA. Before attempting to discuss the function of the basal ganglia in man, we should speak briefly of their better known functions in lower animals. In birds, for instance, the cerebral cortex is very poorly developed, while the basal ganglia are highly developed. These ganglia perform essentially all the motor functions, even controlling the voluntary movements in much the same manner that the motor

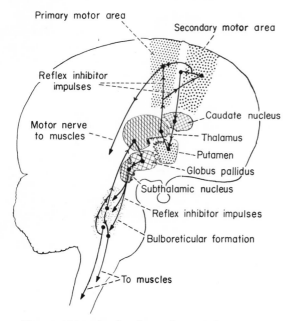

Figure 284. The basal ganglia and their connections with the cerebral cortex and other structures in the brain stem.

cortex of the human being controls voluntary movements. In the cat, and to a less extent in the dog, removal of the cerebral cortex prevents discrete types of motor functions but does not interfere with the cat's ability to walk, to eat perfectly well, to fight, to develop rage, to have periodic sleep and wakefulness, and even to participate naturally in sexual activities. However, if a major portion of the basal ganglia is destroyed, only gross stereotype movements remain, which are controlled by the very primitive lower areas in the brain stem.

In the human being, many of the potential functions of the basal ganglia are suppressed by the cerebral cortex, but if the cerebral cortex becomes destroyed in a very young human being, many of the voluntary motor functions do develop. The person will never be able to develop very discrete movements, particularly of the hands, but he can learn to walk, to control his equilibrium, to eat, to rotate his head, to perform almost any type of postural movement, and to carry out other subconscious movements. On the other hand, destruction of a major portion of the caudate nucleus almost totally paralyzes the opposite side of the body except for a few stereotyped reflex movements integrated in the cord or lower brain stem.

Unfortunately, little is known about the function of most of the individual basal ganglia other than the fact that they operate together in a loosely knit unit to perform the subconscious movements. Yet the following are a few of the discrete bits of information that are known about their individual functions.

(1) The *caudate nucleus* controls gross intentional movements of the body, these occurring mainly subconsciously but aiding in the overall control of body movements.

(2) The *putamen* operates in conjunction with the caudate nucleus to control gross intentional movements. Both of these nuclei also function in cooperation with the motor cortex to control many of the patterns of movement.

(3) The *globus pallidus* probably controls the "background" positioning of the gross parts of the body when a person begins to perform a complex movement pattern. That is, if a person wishes to perform a very exact function with one of his hands, he first positions his body appropriately and then tenses the muscles of the upper arm. These functions are believed to be initiated by the globus pallidus.

(4) The *subthalamic nucleus* controls walking movements and perhaps other types of gross rhythmic body motions.

Transmission of Signals from the Basal Ganglia Through the Extrapyramidal System

In previous sections of the chapter we noted that signals from the cerebral cortex to the spinal cord that cause voluntary motor activities are transmitted through the pyramidal tract (called also the corticospinal tract). The basal ganglia, on the other hand, do not transmit their motor signals through this tract but instead over short pathways into neuronal centers of the lower brain stem. From there the signals are relayed down the cord through the (1) reticulospinal tracts, (2) vestibulospinal tracts, (3) rubrospinal tracts, and (4) propriospinal tracts.

The basal ganglia receive large numbers of fibers from the secondary motor areas which lie in front of the motor cortex and also some directly from the motor cortex and sensory cortex.

This entire system for transmitting motor

signals down the axis of the nervous system is called the *extrapyramidal system* because it does not utilize the pyramidal tract (corticospinal tract) as does the system for direct control of voluntary muscle movement, as explained earlier. In general, one can say that the extrapyramidal system controls the various postural movements and the background tone of the different muscles in contrast to control of discrete voluntary movements by the corticospinal system.

Abnormalities Associated with Damage of the Basal Ganglia

Even though we do not know all the precise functions of the basal ganglia, we do know many abnormalities that develop when portions of the basal ganglia are destroyed, as follows:

CHOREA. Chorea is random, uncontrolled sequences of motor movement occurring one after the other. Normal progression of movements cannot occur. Instead, the person may perform a normal sequence of movements for a few seconds and then suddenly jump to a new sequence, this jumping to a new sequence occurring again and again without stopping. This is caused by widespread damage in the *caudate nucleus* and *putamen*.

ATHETOSIS. Athetosis is characterized by slow, writhing movements of peripheral parts of the body. For instance, the hand or arm may undergo worm-like movements, such as twisting of the arm to one side, then to the other side, then backward, then forward, repeating the same activity over and over again. The damage is always in the *globus pallidus*.

HEMIBALLISMUS. Hemiballismus is an uncontrollable succession of violent movements of large areas of the body. For instance, a leg may suddenly kick forward, and this may be repeated once every few seconds, or sometimes only once in many minutes. The damage that causes this is in the *subthalamus*.

PARKINSON'S DISEASE. This disease is characterized by *tremor* and *rigidity* of the musculature in either widespread or isolated areas of the body. The typical person with full-blown Parkinson's disease walks in a crouch like an ape, except that his muscles are obviously tense, his face is mask-like and he jerks all over with a violent tremor approximately 6 to 8 times per second. Yet when he attempts to perform voluntary movements, the tremor stops temporarily. This disease is caused by destruction of the *substantia nigra*, one of the less conspicuous basal ganglia that extends downward into the mesencephalon.

COORDINATION OF MOTOR MOVEMENTS BY THE CEREBELLUM

Movements of parts of the body are affected greatly by their inertia and momentum. That is, a limb requires a certain force to start it moving, but once started it keeps on moving until an opposing force stops the motion. The cerebral cortex is not organized to take these physical factors into consideration. Instead, the *cerebellum* makes the automatic adjustments that keep these factors from distorting the patterns of activity.

The cerebellum, whose incoming and outgoing fiber pathways are shown in Figure 285, is a large structure located posterior to the brain stem. It receives signals from the proprioceptive receptors located in all joints, in all muscles, in the pressure areas of the body, and anywhere else that signals informing of the physical state of the body can be obtained. Signals are transmitted into the cerebellum, too, from the equilibrium apparatus of the ear, and even from the eyes to depict the visual relationship of the body to its surroundings. Finally, the cerebellum receives information directly from the motor cortex of all motor signals that are being sent to the muscles. In summary, the cerebellum is a collecting house for all possible information on the instantaneous physical status of the body.

Figure 285A illustrates the fiber tracts that transmit the various types of information into the cerebellum. These include (1) the *spinocerebellar tracts* that carry proprioceptor information from the body, (2) the *cortico-ponto-cerebellar tract* that carries motor information from the cerebrum to the cerebellum, and (3) the *vestibulocerebellar tract* that carries impulses from the equilibrium apparatus of the ear.

Once the cerebellum has collated its information on the physical status of the body, it transmits its analysis to other areas of the brain through the fiber tracts shown in Figure 285B, including especially (1) the *fastigio-*

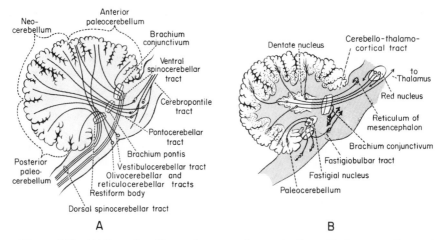

Figure 285. (A) Incoming fiber pathways to the cerebellum. (B) Outgoing fiber pathways from the cerebellum.

bulbar tracts that go from the cerebellum into many of the structures of the brain stem, and (2) the *cerebello-thalamo-cortical tracts* that pass to the thalamus and then to the motor regions of the cerebral cortex.

FEEDBACK CONTROL OF MOTOR FUNC-

TION. The cerebellum acts as a feedback mechanism to help control the functions of the motor cortex; this is illustrated in Figure 286. It receives information from the cortex of the muscular movements that it *intends to perform,* while simultaneously receiving

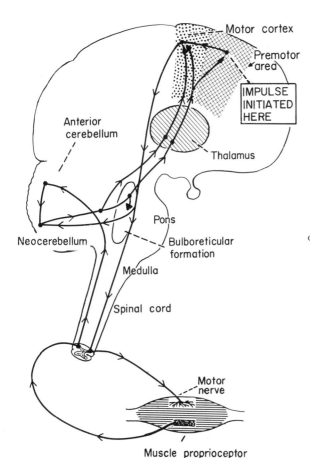

Figure 286. Feedback circuits of the cerebellum for damping motor movements.

proprioceptive information directly from the body apprising it of the movements *actually performed*. After comparing the intended performance with the actual performance, "corrective" signals are sent back to the motor cortex to bring the actual performance in line with the intended one. Some of the specific functions of the cerebellum are the following:

DAMPING FUNCTION OF THE CEREBELLUM. When one moves his hand rapidly to a new position he can normally stop it exactly at the desired point, but without the cerebellum this cannot be accomplished. If the cerebellum is removed, the momentum of the arm carries the hand beyond the projected point until some other part of the brain can detect the fact that the hand has gone too far. The eyes and the proprioceptor mechanisms of the somesthetic cortex finally do detect the overshoot, and they set into play opposing muscular forces to bring the hand back to the projected point. Again momentum is built up during this return motion, and the hand moves too far in the opposite direction. Once more the cerebral mechanisms detect the overshoot and bring the hand back to approach the point. Gradually the successive overshoots become less and less until the hand at last reaches the desired point.

To prevent this effect of momentum, the cerebellum collects information from the moving parts of the body while they are actually moving and determines how much momentum is affecting the movements. Then, even before each part reaches its destination, the cerebellum sends "feedback" signals to the motor cortex to initiate appropriate "braking" contractions of opposing muscles to slow up and stop the movement at the proper point. In this way the finger can be brought to rest at a desired position without the annoying overshoot. This overall mechanism to prevent overshoot is called the *damping* function of the cerebellum.

Ataxia. Lack of the damping function of the cerebellum causes the condition called *ataxia*, which means incoordinate contraction of the different muscles. For example, if a person who has lost his cerebellum tries to run, his feet will overshoot the necessary points on the ground for maintenance of equilibrium and he will fall immediately.

Even when walking, his gait will be very severely affected, for placement of the feet can never be precise; he falls first to one side, then over-corrects to the other side, and must correct again and again, giving him a broken gait. Ataxia can also occur in the hands and may be so severe that it becomes impossible for the person to write, to nail a nail, or to perform any other precise movements without overshooting.

PREDICTIVE FUNCTION OF THE CEREBELLUM. Another function of the cerebellum closely allied to the damping function is its ability to predict the position of the different parts of the body ahead of time. When the leg is moving very rapidly forward while a person is running, the cerebellum, operating in conjunction with the somesthetic cortex, predicts where the leg will be at each instant during the next few hundredths of a second. It is because of this prediction that the person can send appropriate signals to the leg muscles, directing the exact point on the ground where the foot is to be placed to keep him from falling to one side or the other.

The predictive function also applies to the relationship of the body to surrounding objects, for without a cerebellum a person running toward a wall cannot predict when he will likely reach the wall. Monkeys that have had portions of the cerebellum destroyed have been known to run so rapidly toward walls, being unable to predict how rapidly they are approaching, that they bash their brains.

EQUILIBRIUM FUNCTION OF THE CEREBELLUM. A certain portion of the cerebellum, called the *flocculonodular lobes*, is concerned particularly with equilibrium of the body—that is, maintenance of the body in the upright position despite the pull of gravity. When the person tends to fall to one side, information to this effect enters the brain stem and cerebellum from the equilibrium mechanism of the ear, from the visual sensations of the eyes, and from many of the proprioceptors of the body. This information is all summed together, and the necessary signals to readjust the equilibrium are sent to the postural muscles. This equilibrium mechanism was considered in detail in the previous chapter, which deals with cord and lower brain stem functions.

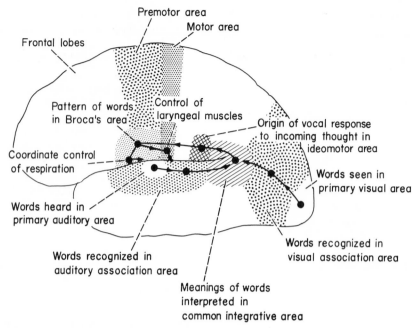

Figure 287. Pathways in the brain for communication, showing reception of thoughts by the auditory and visual pathways and control of speech by motor pathways.

THE CONTROL OF SPEECH

The major characteristic of human beings that sets them apart from other animals is their ability to communicate one with the other. This depends on two highly developed functions of the brain: first, the ability to interpret speech and, second, the ability to translate thought into speech. These communicative functions require the highest degree of operative perfection in almost all parts of the brain. Therefore, a description of the act of communication provides an overall review of the integrative functions of the central nervous system.

INTERPRETATION OF COMMUNICATED IDEAS. Ideas are generally communicated from one person to another either by sounds or written words. In the case of sounds, the information enters the primary auditory cortex in the superior part of the temporal lobe, as shown in Figure 287. The sounds then are interpreted as words and the words as sentences in the auditory association areas. The sentences are interpreted as thoughts in the common integrative region. Similarly, combinations of letters seen by the eyes are interpreted as words and the words as sentences by the visual association

areas, and the sentences become thoughts in the common integrative region.

MOTOR FUNCTIONS OF SPEECH. The common integrative region of the brain develops thoughts that it wishes to communicate to someone else. To do this it operates in association with the facial region of the somesthetic sensory cortex to initiate a sequence of impulses, each impulse representing perhaps a syllable or a whole word, and transmits these into the secondary motor area that controls the larynx and mouth. This area is called *Broca's area*, or it is sometimes called simply the *speech center*. It is actually no more a speech center than the common integrative area, but it is here that the muscular patterns for forming different sounds by the larynx and mouth are controlled. Impulses arriving from the common integrative area set off a sequence of patterns in the speech center, and these in turn form the words. In addition to controlling the larynx and mouth, Broca's area sends impulses into an allied region of the secondary motor cortex that controls respiration. At the same time that the laryngeal and mouth movements occur, the respiratory muscles are contracted to provide appropriate air flow for the speech process.

REFERENCES

Fadiga, E., and Pupilli, G. C.: Teleceptive components of the cerebellar function. *Physiol. Rev.* 44:432, 1964.

Jung, R., and Hassler, R.: The extrapyramidal motor system. *Handbook of Physiology*, Sec. I, Vol. II, p. 863. Baltimore, The Williams & Wilkins Company, 1960.

Laursen, A. M.: Higher functions of the central nervous system. *Ann. Rev. Physiol.* 29:543, 1967.

Marchiafava, P. L.: Activities of the central nervous system: motor. *Ann. Rev. Physiol.* 30:359, 1968.

Oscarsson, O.: Functional organization of the spino- and cuneocerebellar tracts. *Physiol. Rev.* 45(3): 495, 1965.

Paillard, J.: The patterning of skilled movements. *Handbook of Physiology*, Sec. I, Vol. II, p. 1679. Baltimore, The Williams & Wilkins Company, 1960.

Patton, H. D., and Amassian, V. E.: The pyramidal tract: its excitation and functions. *Handbook of Physiology*, Sec. I, Vol. II, p. 837. Baltimore, The Williams & Wilkins Company, 1960.

Penfield, W., and Rasmussen, T.: *The Cerebral Cortex of Man.* New York, The Macmillan Company, 1950.

CHAPTER 28

THE AUTONOMIC NERVOUS SYSTEM AND HYPOTHALAMUS

The autonomic nervous system operates involuntarily to control many of the internal functions of the body. Various aspects of these have already been described in relation to many functions of the body such as the bladder and rectal reflexes, peritoneal reflexes, the control of arterial pressure, and the control of many internal organs. The purpose of the present chapter is to present the overall function of the autonomic nervous system, pointing out its special relationships to the remainder of the nervous system.

SYMPATHETIC AND PARASYMPATHETIC DIVISIONS OF THE AUTONOMIC NERVOUS SYSTEM

The autonomic nervous system is considered to have two separate divisions, the *sympathetic* and the *parasympathetic*. The reasons for this division are: First, the anatomic distributions of the nerve fibers in the two divisions are distinct from each other. Second, the effects of the two divisions on the organs are often antagonistic to each other. Third, the types of hormones secreted at the nerve endings are usually different in the two systems.

ANATOMY OF THE SYMPATHETIC NERVOUS SYSTEM. Figure 288 illustrates the anatomy of the sympathetic nervous system, showing the sympathetic chain that lies to each side of the spinal cord, and the chain's connections with the cord and peripheral organs. The sympathetic chain is connected with the spinal cord in a peculiar manner, shown in Figure 289. Sympathetic fiber tracts leave the cord through the anterior roots of the spinal nerve. Then, after traveling less than a centimeter, they pass through a small whitish nerve called the *white ramus* into the *sympathetic chain*. From here, fibers travel in two directions. Some pass into visceral *sympathetic nerves* that innervate the internal organs of the body, while the others return through another small nerve called the *gray ramus* back into the spinal nerve. These latter fibers then travel all through the body along the spinal nerves to supply the blood vessels, the sweat glands, and even the pilo-erector muscles that cause the hairs to stand on end.

Sympathetic fibers enter the sympathetic chain through the white rami only in the thoracic and upper lumbar regions, and none enter the chain in the neck, lower lumbar, or sacral regions. To supply the head with sympathetic innervation, sympathetic fibers from the thoracic chain extend upward into the neck and then to all the structures of the head. Also, sympathetic fibers pass downward from the chain into the lower abdomen and legs.

Visceral sensory fibers in the sympathetic nerves. Sensory fibers pass through the sympathetic nerves along with the sympathetic fibers. These arise in the internal organs, then enter the sympathetic nerves, and finally travel by way of the white rami into the spinal nerves. From here they enter the posterior horns of the cord gray matter and either cause autonomic cord reflexes or transmit sensations to the brain in the same manner that sensations are transmitted

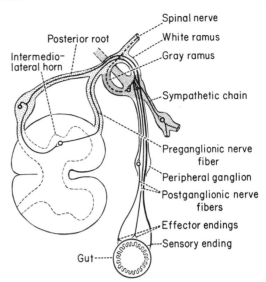

Figure 289. Connections between the spinal cord, a spinal nerve, and the sympathetic chain.

from the surface of the body. These sensations are the visceral sensations that were considered in detail in Chapter 23.

Preganglionic and postganglionic neurons of the sympathetic nervous system. Figure 288 shows bulbous enlargements located periodically along the sympathetic chain and at different points among the sympathetic nerves after they leave the chain. These enlargements are called *ganglia,* and they contain neuronal cell bodies.

Sympathetic signals are transmitted from the spinal cord to the periphery through two successive neurons. The cell body of the first neuron is located in the spinal cord in the lateral gray matter. The fiber from this neuron passes into the sympathetic system as illustrated in Figure 289. It synapses with a second neuron in either a ganglion of the sympathetic chain or a more peripheral ganglion. The fiber from the second neuron passes directly to the organ to be controlled. The first neuron, from the cord to the ganglion, is called the *preganglionic neuron;* the second, from the ganglion to the organ, is called the *postganglionic neuron.* Thus, the sympathetic motor system is different from the skeletal motor system, for skeletal muscles are stimulated through a single neuron rather than through two.

ANATOMY OF THE PARASYMPATHETIC NERVOUS SYSTEM. Figure 290 shows the parasympathetic nervous system. The fibers

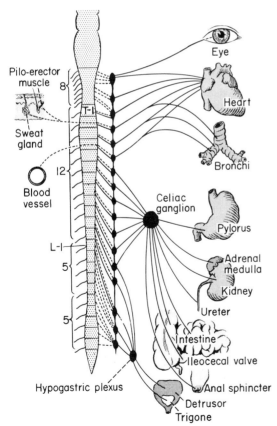

Figure 288. Anatomy of the sympathetic nervous system. (The dashed lines represent the gray rami.)

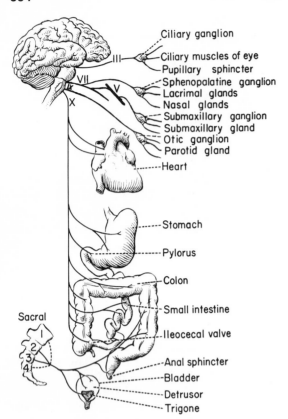

Ciliary ganglion
Ciliary muscles of eye
Pupillary sphincter
Sphenopalatine ganglion
Lacrimal glands
Nasal glands
Submaxillary ganglion
Submaxillary gland
Otic ganglion
Parotid gland
Heart
Stomach
Pylorus
Colon
Small intestine
Ileocecal valve
Anal sphincter
Bladder
Detrusor
Trigone

Sacral

Figure 290. Anatomy of the parasympathetic nervous system.

of this system originate mainly in the tenth cranial nerve, called the *vagus nerve*. However, a few fibers originate in the third, seventh, and ninth cranial nerves, and also in several of the sacral segments of the spinal cord. The vagus nerve supplies parasympathetic fibers to the heart, the lungs, and almost all of the organs of the abdomen. The other cranial nerves supply the head, and the sacral fibers supply the urinary bladder and the lower parts of the colon and the rectum. However, since approximately 90 per cent of all the parasympathetic fibers of the body pass through the vagus nerve, most physiologists, in thinking about the parasympathetic system, think almost automatically of the vagus nerve itself.

Preganglionic and postganglionic neurons of the parasympathetic system. The parasympathetic system is like the sympathetics in that impulses travel through preganglionic and postganglionic neurons. The cell bodies of the preganglionic neurons are in the brain stem or sacral cord, and their fibers usually pass all the way to the wall of the organ to be stimulated, where the postganglionic

neurons are located. The postganglionic fibers then travel only a few millimeters, or at most a few centimeters, before they reach their final destination on muscle fibers or glandular cells. This is different from the sympathetics, whose postganglionic cells are located in the sympathetic ganglia long distances from the respective organs.

ADRENERGIC VERSUS CHOLINERGIC NERVES IN THE AUTONOMIC NERVOUS SYSTEM. One of the major differences between parasympathetic and sympathetic nerves is that the postganglionic neurons of the two systems usually secrete different hormones. The postganglionic neurons of the parasympathetic system secrete *acetylcholine,* for which reason these neurons are said to be *cholinergic.* Those of the sympathetic system secrete mainly norepinephrine; another name for norepinephrine is noradrenaline, for which reason most postganglionic neurons of the sympathetic system are said to be *adrenergic.* However, a few are cholinergic like the parasympathetic neurons, as is noted below.

Actions of the Parasympathetic and Sympathetic Nervous Systems on Different Organs

EFFECT ON THE EYE. Table 8 lists the effects on different organs caused by parasympathetic or sympathetic stimulation. This table shows that the sympathetic system dilates the pupil of the eye, allowing increased quantities of light to enter, and the parasympathetics constrict the pupil, thus decreasing the amount of light. The parasympathetics also control the ciliary muscle that focuses the lens for far and near vision.

SECRETION OF THE DIGESTIVE JUICES. The secretion of digestive juices by some of the glands of the gastrointestinal tract is controlled by the parasympathetics, while the sympathetics have very little effect on most of the glands. The salivary glands of the mouth and the fundic glands of the stomach normally are almost entirely controlled by the parasympathetics. On the other hand, the glands in the intestines are controlled to a slight extent by the parasympathetics, but mainly by local effects in the intestines themselves.

EFFECTS ON THE SWEAT GLANDS. The

TABLE 8. *Autonomic Effects on Various Organs of the Body*

ORGAN	EFFECT OF SYMPATHETIC STIMULATION	EFFECT OF PARASYMPATHETIC STIMULATION
Eye: Pupil	Dilated	Contracted
Ciliary muscle	None	Excited
Gastrointestinal glands	Vasoconstriction	Stimulation of thin, copious secretion containing many enzymes
Sweat glands	Copious sweating (cholinergic)	None
Heart: Muscle	Increased activity	Decreased activity
Coronaries	Vasodilated	Constricted
Systemic blood vessels:		
Abdominal	Constricted	None
Muscle	Dilated (cholinergic)	None
Skin	Constricted or dilated (cholinergic)	None
Lungs: Bronchi	Dilated	Constricted
Blood vessels	Mildly constricted	None
Gut: Lumen	Decreased peristalsis and tone	Increased peristalsis and tone
Sphincters	Increased tone	Decreased tone
Liver	Glucose released	None
Kidney	Decreased output	None
Bladder: Body	Inhibited	Excited
Sphincter	Excited	Inhibited
Male sexual act	Ejaculation	Erection
Blood glucose	Increased	None
Basal metabolism	Increased up to 50%	None
Adrenal cortical secretion	Increased	None
Mental activity	Increased	None

sweat glands are stimulated by fibers from the sympathetic nervous system. However, these fibers are different from the usual sympathetic fibers, for they are mainly cholinergic rather than adrenergic. Also, they are stimulated by the nervous centers in the brain that control the parasympathetics, rather than by the centers that control the sympathetics. Therefore, despite the fact that the fibers supplying the sweat glands are anatomically sympathetic, they can be considered physiologically to be parasympathetic fibers.

EFFECTS ON THE HEART. Stimulation of the sympathetic nervous system increases the heart's activity and also dilates the coronaries so that increased nutrition will be available. On the other hand, parasympathetic stimulation decreases the activity of the heart while constricting the coronaries to conserve blood flow. These effects were discussed in Chapter 11.

CONTROL OF THE BLOOD VESSELS. Perhaps the most important function of the sympathetic nervous system is to control the blood vessels in the body. Most vessels are constricted by sympathetic stimulation, though a few, the coronaries and skeletal muscle vessels for instance, are dilated. By controlling the peripheral blood vessels, the sympathetic nervous system is capable of regulating both the cardiac output and arterial pressure; constriction of the veins and venous reservoirs increases the cardiac output, and constriction of the arterioles increases the peripheral resistance, which elevates the arterial pressure.

The parasympathetics, when they affect

the blood vessels at all, usually dilate them, but this effect is so slight and occurs in such a few areas of the body, the brain for instance, that it can be almost totally ignored.

EFFECT ON THE LUNGS. The bronchi are dilated by sympathetic stimulation. However, the sympathetic system has only a slight vasoconstricting effect on the blood vessels of the lungs. This is different from the very strong effect on the blood vessels of the remainder of the body.

CONTROL OF GASTROINTESTINAL MOVEMENTS. About 75 per cent of all the parasympathetic nerve fibers are distributed to the gastrointestinal tract, which indicates that by far the most important function of this entire system is regulation of gastrointestinal activities. Parasympathetic stimulation increases peristalsis and at the same time decreases the tone of the sphincters. Peristalsis propels the food forward while the open sphincters between the different segments of the gastrointestinal tract allow the food to move forward with ease. During extreme parasympathetic stimulation, food can actually pass all the way from the mouth to the anus in approximately 30 minutes, though the normal transmission time is about 24 hours.

Sympathetic stimulation, on the other hand, inhibits peristalsis and tightens the sphincters. This slows down the movement of food through the gastrointestinal tract.

RELEASE OF GLUCOSE FROM THE LIVER. Sympathetic stimulation causes rapid breakdown of glycogen into glucose in the liver and then release of the glucose into the blood. This increased level of glucose in the blood provides a quick supply of nutrition for the tissue cells, an effect especially valuable during exercise.

EFFECT ON THE KIDNEYS. Sympathetic stimulation causes intense vasoconstriction of the renal blood vessels and greatly decreases the output of urine. This is a very important mechanism for the regulation of blood volume and arterial pressure, for when need be, sympathetic stimulation can cause fluid to be retained in the circulatory system, increasing the blood volume and the venous return to the heart. These effects over a period of hours or days increase the cardiac output and raise the blood pressure to the desired level.

EMPTYING OF THE BLADDER. Emptying of the bladder is caused mainly by the para-

sympathetic system, which excites the muscular wall of the bladder while inhibiting the sphincters that normally hold back the flow of urine. On the other hand, sympathetic stimulation prevents emptying of the bladder. Though the sympathetic effect ordinarily is not important, occasionally when a person has severe peritoneal inflammation in the region of the bladder, a resulting reflex excites the sympathetics so greatly that the person becomes unable to urinate.

CONTROL OF SEXUAL FUNCTIONS. The autonomic nervous system also enters into the control of the sexual acts of both the male and the female. In the male the parasympathetics cause erection, and the sympathetics cause ejaculation. In the female the parasympathetics cause erection of the erectile tissue around the vaginal opening, which causes tightening, and they also cause the female to secrete large quantities of mucus which facilitates the sexual act. The effect of the sympathetics on the female sexual act is not well understood, but it is believed that these nerves might initiate reverse uterine peristalsis during the female climax.

METABOLIC EFFECTS. Generalized sympathetic stimulation increases the metabolism of all cells of the body. The sympathetic nerve fibers are so widespread in all tissues that at least some norepinephrine secreted by their nerve endings seems to reach every functioning cell. The norepinephrine increases the rates of the chemical reactions in all cells, thereby increasing the overall rate of metabolism of the body. In this way the sympathetics can keep a person warm when he tends to become too cold, and during exercise or other states of activity they can make the body perform greater quantities of work than would be possible otherwise. These metabolic effects can be brought about in only a few seconds, and they can be stopped in another few seconds when the need for increased metabolism is over.

STIMULATION OF THE ADRENAL CORTEX. When the body is subjected to physical stress of sufficient intensity to damage tissues, the sympathetics usually become stimulated, and this initiates the secretion of adrenocortical hormones in the following manner: Stimulation of the sympathetic nerve endings in the anterior pituitary gland causes this gland to secrete a hormone called *adrenocorticotropic hormone*, which in turn flows by way of the blood stream to the adrenal cortex where it promotes the output of adrenocortical hormones. These hormones then help the tissue cells repair damage that has resulted from the stress.

STIMULATION OF MENTAL ACTIVITIES. A final important effect of sympathetic stimulation is an increase in the rate of mental activity. This probably results from the increased rate of metabolism in the neuronal cells.

Overall Function of the Sympathetic Nervous System

In general, when the sympathetic centers of the brain become excited they stimulate almost all of the sympathetic nerves at once. As a result, the blood pressure rises, the rate of metabolism increases, the degree of mental activity is enhanced, and the glucose in the blood increases. Considering all of these effects together, it can be seen that mass discharge of the sympathetic nervous system prepares the body for activity. When a person's sympathetic nerves have been destroyed he cannot "generate steam"; in other words, he simply cannot reach the high levels of vigor often necessary to perform rapid, forceful, and excited activities.

SPECIAL FUNCTIONS OF THE SYMPATHE-TIC NERVOUS SYSTEM. In addition to the mass discharge mechanism of the sympathetic system, certain physiological conditions can stimulate localized portions of the system independently of the remainder. Two of these are the following:

Heat reflexes. Heat applied to the skin causes a reflex through the spinal cord and then back again to the skin to dilate the blood vessels in the heated area. Also, heating the blood flowing through the heat control centers of the hypothalamus increases the degree of vasodilatation of all the skin blood vessels without significantly affecting the vessels deep in the tissues.

Shift of blood flow to the muscles during exercise. During exercise two separate portions of the sympathetic nervous system are stimulated, the vasodilator fibers to the muscles and the vasoconstrictor fibers to most other parts of the body. Nerve impulses pass directly from the motor cortex to the sympathetic regions of the brain stem to initiate these effects. Vasodilatation in the muscles allows blood to flow through them with great ease, while vasoconstriction elsewhere, except in the heart and brain, decreases almost all other bodily blood flow. Thus, in exercise the sympathetic nervous system causes a massive shift of blood flow to the active muscles.

FUNCTION OF THE ADRENAL MEDULLA. The adrenal medulla is the central portion of the adrenal gland. The medulla is surrounded by the adrenal cortex, which was previously mentioned. These two portions of the adrenal gland are actually two separate glands, for they secrete entirely different types of hormones, and regulation of each is entirely different from the other.

The glandular cells of the medulla are modified postganglionic sympathetic neurons that secrete norepinephrine and epinephrine directly into the blood when stimulated by sympathetic impulses. These hormones then flow throughout the body, causing sympathetic effects everywhere. Therefore, every time a mass discharge occurs in the sympathetic nervous system, in addition to direct stimulation of the organs by the sympathetic nerves, large quantities of norepinephrine and epinephrine also flow to all points in the body. These hormones then exert almost exactly the same effects on the different organs as direct sympathetic stimulation by the nerves. For instance, they increase the activity of the heart, inhibit peristalsis in the gut, increase the rate of metabolism of all cells, and so forth. In effect, then, the adrenal medullae provide a second means for causing all the sympathetic activities. They also allow those parts of the body that do not have sympathetic nerve fibers to respond to sympathetic stimulation.

The quantity of norepinephrine and epinephrine secreted by the adrenal medullae is great enough so that perhaps one fourth to one half of all the functions of the sympathetic nervous system are mediated by this means rather than by direct stimulation of the different organs. In fact, when all the sympathetic nerves except those to the adrenal medullae are destroyed, the sympathetic system will still function almost normally because the adrenals compensate by secreting extra quantities of hormones. Conversely, loss of adrenal medullary secretion is hardly discernible because of compensation by the nerves. To stop entirely the action of the sympathetic nervous system, the function of both the adrenal medullae and all the sympathetic nerve fibers that go directly to the organs must be blocked.

Parasympathetic and Sympathetic Tone

Impulses normally are transmitted continuously, though at a relatively slow rate, through all the fibers of both the parasympathetic and sympathetic systems. This allows continuous stimulation of the internal structures, which is called *tone* or the *tonic effect* of the system. Tone allows each system to exert both positive and negative effects on a structure—that is, by increasing the number of impulses above the normal value, the effect can be increased; on the other hand, by decreasing the number of impulses below the normal the effect can be decreased. As an example, if the sympathetic vasoconstrictor fibers to the blood vessels were normally dormant, it would be possible for sympathetic regulation only to constrict the vessels and never to dilate them; however, because of the normally persistent tone, the blood vessels are always partially constricted so that the sympathetics can either further constrict the vessels by increasing their stimulation or they can dilate the vessels by decreasing their stimulation. These principles apply throughout the parasympathetic and sympathetic systems and are responsible for a higher degree of effectiveness than would be possible otherwise.

CONTROL OF THE AUTONOMIC NERVOUS SYSTEM BY NERVE CENTERS IN THE BRAIN

Not much is known about the exact centers in the brain that regulate the discrete autonomic functions. However, stimulation of widespread areas in the reticular substance of the medulla, pons, and mesencephalon can cause all of the effects of the sympathetic and parasympathetic systems.

The hypothalamus can regulate many, though not all, of the functions of the autonomic system. Finally, stimulation of localized areas of the cerebral cortex can cause general or sometimes discrete autonomic effects.

IMPORTANT AUTONOMIC CENTERS OF THE BRAIN STEM. Figure 291 shows a tentative

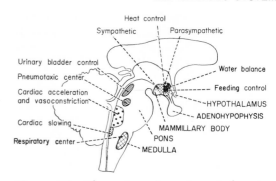

Figure 291. The major autonomic control centers of the brain stem.

outline of the centers in the hindbrain and hypothalamus that regulate different autonomic functions of the internal organs. It will be noted from this figure that the vascular system and respiration are controlled mainly by areas in the reticular substance of the medulla and pons. The urinary bladder is controlled by centers in the mesencephalon, but these centers are usually suppressed by areas in the cerebral cortex that provide conscious regulation of bladder emptying.

AUTONOMIC CENTERS OF THE HYPOTHALAMUS. Figure 292 gives much more detailed locations of different control centers in the hypothalamus. A major share of the control functions of the hypothalamus are transmitted through hormonal systems. Some of the specific functions of the hypothalamus are the following:

Cardiovascular regulation. In general, stimulation of the posterior hypothalamus increases the arterial pressure and heart rate, while stimulation of the preoptic area in the anterior portion of the hypothalamus has exactly opposite effects, causing marked decrease in both heart rate and arterial pressure. These effects are transmitted through the cardiovascular control centers of the lower brain stem and thence through the autonomic nervous system.

Regulation of body temperature. Centers in the anterior hypothalamus are directly responsive to blood temperature, becoming more active with increasing blood temperature and less active with decreasing temperature. These areas in turn control other parts of the brain stem to regulate the body temperature, the details of which will be described in Chapter 33. Basically this mechanism (a) controls the amount of blood flow through the skin thereby controlling the rate of heat loss from the skin, (b) con-

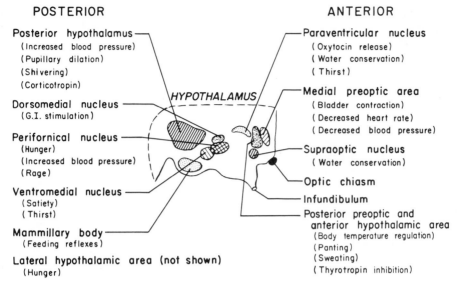

Figure 292.   Functional anatomy of the hypothalamus.

trols sweating, (c) controls shivering which greatly increases heat production, (d) controls secretion of adrenal medullary hormones which also stimulate heat production, and (e) controls thyroid hormone production which has a direct effect on heat production in all cells of the body.

Regulation of body water. A center located in the supraoptic nucleus controls antidiuretic hormone secretion which in turn controls rate of water loss through the kidneys. This mechanism is discussed in detail in Chapter 18 in relation to kidney function and in Chapter 34 in relation to function of the hypophyseal hormones.

A closely allied region of the hypothalamus, the paraventricular area, controls drinking. It does this by transmitting signals into other parts of the brain to create the conscious feeling of thirst, thereby making the person seek water.

Regulation of feeding. Stimulation of *lateral hypothalamic areas* causes an animal to have a voracious appetite and an intense desire to search for food. This area is frequently called the *hunger center.* Located more medially, in the ventromedial nuclei, are areas which when stimulated block all feeling of hunger. Therefore, these areas are called the *satiety center.*

Control of excitement and rage. In the very middle of the hypothalamus is a nucleus called the *perifornical nucleus* which, when stimulated, causes an animal to become greatly excited along with develop-

ment of elevated blood pressure, dilated pupils, and symptoms of rage such as hissing, arching his back, and assuming a stance ready to attack. Thus, the hypothalamus plays a major role in overall behavior of an animal.

Hypothalamic control of endocrine functions. Some areas of the hypothalamus also secrete so-called *neurosecretory substances.* These are then transported through special veins from the hypothalamus down to the anterior pituitary gland (adenohypophysis) where they promote formation of different anterior pituitary hormones. It is in this way that the hypothalamus controls the secretion of most anterior pituitary hormones and thereby controls many of the metabolic functions of the body. These neurosecretory substances and their more specific functions will be discussed in Chapter 34.

Summary. A number of discrete areas of the hypothalamus have now been found that control specific vegetative functions of the body. However, these areas overlap so much that the foregoing separation of different areas for different hypothalamic functions is partially artificial. Most important of all, the hypothalamus plays perhaps the key role in setting the basic tenor of bodily function.

IMPORTANT CEREBRAL CENTERS. Stimulation of the autonomic nervous system by the cerebral cortex occurs principally during emotional states. Many discrete centers, especially in the prefrontal lobes and temporal regions of the cortex, can increase or decrease the degree of excitation of the hypothalamus centers. Also, the thalamus and closely related structures deep in the cerebrum help to regulate the hypothalamus. Therefore, both conscious and subconscious portions of the cerebrum can cause autonomic effects. Sudden psychic shock, which originates in the cerebral cortex, can cause fainting. This usually results from powerful stimulation of the vasodilator fibers and the cardiac inhibitory fibers (vagus nerves) which cause the arterial pressure to fall precipitously. On the other hand, extreme degrees of psychic excitement, which can result from either conscious or subconscious stimulation in the cerebrum, can stimulate the vasomotor system to increase the blood pressure and cardiac output.

Unfortunately, our present knowledge of the relation of the cerebrum to the autonomic nervous system is so scanty that it can be only descriptive rather than explanatory of the actual functional relationships.

REFERENCES

Bleier, R.: *The Hypothalamus of the Cat.* Baltimore, The Johns Hopkins Press, 1961.

Burn, J. H.: *The Autonomic Nervous System: for Students of Physiology and of Pharmacology.* Philadelphia, F. A. Davis Company, 1963.

Burnstock, G., and Holman, M. E.: Smooth muscle: autonomic nerve transmission. *Ann Rev. Physiol.* 25:61, 1963.

Ciba Foundation Symposium: *Adrenergic Mechanisms.* Boston, Little, Brown & Company, 1961.

Cross, B. A.: The hypothalamus in mammalian homeostasis. Symposium. *Soc. Exp. Biol.* 18:157, 1964.

Hess, W. R.: *The Functional Organization of the Diencephalon.* New York, Grune & Stratton, 1958.

Soderberg, U.: Neurophysiological aspects of homeostasis. *Ann. Rev. Physiol.* 26:271, 1964.

von Euler, U. S.: *Noradrenaline.* Springfield, Ill., Charles C Thomas, 1956.

von Euler, U. S.: Neurotransmission in the adrenergic nervous system. *Harvey Lect.* 55:43, 1961.

CHAPTER 29

INTELLECTUAL PROCESSES; SLEEP AND WAKEFULNESS; BEHAVIORAL PATTERNS; AND PSYCHOSOMATIC EFFECTS

Another title for this chapter could be "How do the different parts of the brain function together?" That is, what makes us think? What makes our levels of consciousness increase and decrease during sleep and wakefulness? What influences our inner feelings? And, how do our intellectual processes express themselves in the form of bodily function?

Unfortunately, it is these integrative aspects of the brain that are least well understood, for which reason much of what we believe is based on inference from psychological tests rather than from direct experiments in the brain itself.

ROLE OF THE CEREBRAL CORTEX IN THE THINKING PROCESS

At several points we have discussed various aspects of cerebral function and its relation to some phases of the thinking process. To review briefly, the cerebral cortex is in one sense a large outgrowth of the thalamus and related basal areas of the brain, and there are direct communications back and forth between respective portions of the cerebral cortex and thalamus. Whenever a particular part of the thalamus is stimulated a corresponding part of the cerebral cortex also becomes stimulated. Therefore, we believe that the thalamus has the ability to call forth activity in specific portions of the cerebral cortex. On the other hand, it is in the cerebral cortex that most of the neurons of the central nervous system are located, 9 billion out of a total of only 12 billion in the entire brain. Therefore, we must ask ourselves the question: what is the purpose of this great mass of neurons in the cerebral cortex? The answer to this seems to be that the cerebral cortex is a vast storehouse for memories, a place where data can be collected and held for days, months, or years until needed at a later time.

INTERPRETATION OF INCOMING SIGNALS. The cerebral cortex is a large sheet of thin folded tissue lying on the surface of the brain and composed of six separate neuronal cell layers as illustrated in Figure 293, each layer performing a different function. Some of these layers send nerve fibers into deeper areas of the brain; others send fibers to adjacent regions of the cortex. It is believed, also, that some of the layers act as *comparators* to compare new incoming signals with memories stored from the past. That is, if the incoming information to one layer fits exactly with past memory information stored in an overlying layer, one could immediately identify the new information as something that had been experienced before. Furthermore, if the previous experience had been associated with some specific quality of sensation such as pain or happiness, displeasure or pleasure, then this too would be remembered.

NATURE OF THOUGHTS. We all know intuitively what a thought is, but, even so, a thought is very difficult to define neurophysiologically. To attempt this, let us first describe what happens in the brain when a person sees with his eyes for the first time a new exciting visual scene. Upon opening his eyes to the scene, nerve signals are transmitted by way of the optic nerve first

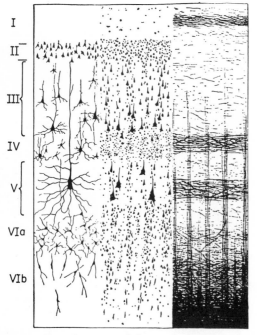

Figure 293. Structure of the cerebral cortex, showing to the left the layers of cell bodies and to the right the layers of connecting fibers. (From Ranson and Clark (after Brodmann): *Anatomy of the Nervous System.*)

into many areas at the base of the brain, including the *lateral geniculate body,* the *superior colliculus,* the *lateral thalamus,* and *midregions of the mesencephalon.* From the lateral geniculate body secondary impulses are transmitted almost immediately to the visual cortex located in the occipital lobe, as described in Chapter 24. These first signals to the cortex terminate in the *primary visual cortex,* but a few milliseconds later signals reach the visual *association cortex,* and soon thereafter they reach the *angular gyrus,* the *hippocampus,* the *prefrontal* cortex, and some other parts of the brain. It is the totality of all these signals that gives one the thought expressed by the visual scene. Therefore, a thought can be described as a specific pattern of signal transmission throughout the brain.

Every thought has specific qualities. It may be pleasant or it may be unpleasant. It may have an element of luminosity, or it may have an element of sound. It may have a sensation of tactile feeling, or it may have a sensation of taste. It may have a characteristic of repetitiveness or of uninterrupted continuity. There is much reason to believe that specific areas in the brain, many of these areas located in the basal regions of the brain, interpret these individual qualities. For instance, the characteristic of pleasure or displeasure seems to be interpreted by several closely associated but antagonistic areas located in the hypothalamus, in the mid-portion of the mesencephalon, and in other surrounding areas. When a visual scene or other type of sensory input to the brain is pleasant, the pleasure areas of these basal regions receive a signal. If they are unpleasant, the displeasure areas receive a signal. Likewise, there appear to be basal regions that can detect luminosity signals telling one that the thought is a visual thought, and other areas that can tell one that the signal has the characteristic of sound.

Finally, every thought has its own specific details, such as contrasts between one segment of the incoming signal and other segments—for instance, contrasts between light and dark areas in the visual scene, or contrasts between different frequencies of sound or different intensities of tactile sensations from one second to the next. It is memories of these details that seem to be stored in the cerebral cortex; therefore, it is presumably the cerebral cortex that is responsible for identification of the fine details of thoughts.

MEMORIES. Memory is the ability to recall thoughts that were originally initiated by incoming sensory signals. Probably most of the memory process occurs in the cerebral cortex, primarily because three-quarters of the neurons of the brain are located here. Yet, we know that essentially every area of the central nervous system can participate in the phenomenon of memory. Indeed, experiments have shown that even the spinal cord can hold crude memories for at least as long as a few minutes to perhaps a few hours.

But what is the nature of a memory? And what causes a memory to persist sometimes for a short time and sometimes for a very long time? There appear to be at least two different types of memory, which can be called short-term memory and long-term memory.

Short-term memory. Short-term memory may be defined as persistence of an incoming thought for a few seconds or a few minutes without causing any permanent imprint on the brain. Some short-term memories are probably caused by continued reverberation of signals within the brain for a short time after the initiating sensation is gone. That is, the incoming thought stimulates neuronal cells connected in reverberating circuits. These cells stimulate secondary cells which then stimulate tertiary cells, and finally the signal gets back to the original cells. Thus, the signal goes around and around the circuit for seconds or minutes after the incoming sensation is over, and as long as these reverberations persist the person still retains the thought in his mind. In support of this concept is the fact that such reverberating signals can be demonstrated for as long as an hour back and forth between the cerebral cortex and the thalamus after a strong sensory signal barrages the brain. Furthermore, these reverberating signals are localized to certain areas of the cerebral cortex depending upon the type of sensation that is experienced.

Long-term memory. We all know that some memories last for years even though reverberating signals in the brain cannot possibly persist longer than a few hours at most. These long-term memories almost certainly result from the following mechanism: When a signal passes through a particular set of neuronal synapses, these synapses become *facilitated* for passage of similar signals at a later date. Therefore, when a thought enters the brain, it facilitates those synapses that are used for that particular thought, and this makes it easier for one to recall that same thought at some later date.

Yet, passage of a signal through a synapse only one time usually will not cause sufficient facilitation for the thought to be remembered. Therefore, we must ask the question: How is it that a single sensory experience lasting for only a few seconds can sometimes be remembered for years thereafter? The answer to this seems to be that the reverberating signals of the short-term memory mechanism send this same thought through the same synapses many thousands of times during the course of an hour or more after the initial sensory experience is over. If an animal is struck on the head immediately after experiencing a very strong sensory experience, the blow to the head can block the reverberating signals and thereby stop the short-term memory. In such an instance, the long-term memory also fails to develop. Therefore, a persisting short-term memory for a period of an hour or more seems to be essential to develop the long-term memory "imprint" or "engram." However, once this engram has been established, almost any stray signal in the brain can at some later date set off a sequence of signals exactly like those originally initiated by the incoming sensation, whereupon the person experiences the same original thought.

DETERMINATION OF WHICH MEMORIES TO REMEMBER—ROLE OF THE HIPPOCAMPUS AND OF PAIN OR PLEASURE. Determination of whether or not a memory will be stored for years or is almost immediately forgotten seems to be made by some of the basal regions of the brain and not by the cerebral cortex. For instance, the hippocampus, a very old portion of the cerebral cortex located on each side of the brain stem is essential for storage of the many if not most memories. If one has experienced the same thought many times before and therefore has become habituated to it, the hippocampus fails to be stimulated. On the other hand, if the thought evinces either pain or pleasure or some other very strong quality, then the hippocampus does become stimulated. In ways not yet understood, this elicitation of signals in the hippocampus is believed to cooperate with other basal regions of the brain to cause the reverberating signals between the cortex and thalamus that lead to permanent storage of the memory.

KNOWLEDGE. Many persons will be surprised to know that the human being is born already having certain types of knowledge. For instance, a newborn baby knows to suck on the breast and even to search for the breast. He knows to cry when pained and to smile, even without being trained to do so, when pleased. Some of the lower animals are born with still other types of knowledge such as the ability to stand and walk and the ability to search out food. Indeed, a major share of the useful knowledge of some lower animals is inherited.

Yet, the human mind is born with much less knowledge than that of many lower animals. Instead, knowledge accumulated in the mind of the human being is generally of an adaptive type based on previous experience, made up almost entirely of memories rather than based on inherited neuronal connections. It is this difference between the lower animal mind and the human mind that gives the human being his great breadth of abilities, one person becoming a mathematical genius, another a vast storehouse of linguistic data, another a depository of jokes, and so forth.

But we also know that the process of forgetfulness allows the character of knowledge in the mind to change with time so that a person's mind may be a great depository of book learning in his school days and yet in later years be filled with practical experience that takes the place of forgotten book learning. Psychological tests show that the quantity of knowledge in a person's mind generally increases during the first 39 years of his life, reaching a peak at this time. Beyond that age the total amount of stored knowledge gradually declines. This does not mean, though, that the amount of stored knowledge of a specific type might not continue to increase on into very old age.

Function of the Angular Gyrus Area (The Common Integrative Area) of the Cerebral Cortex

It was pointed out in Chapter 23 that most of the sensory input signals to the brain eventually funnel information into the common integrative area of the angular gyrus region in the brain's dominant hemisphere, usually the left hemisphere. This area is located at the juncture of the parietal lobe of the brain, the occipital lobe, and the temporal lobe. The primary and association areas for somesthetic and taste sensations are located in the parietal lobes, those for vision are located in the occipital lobes, and those for sound in the temporal lobes. On the under surface of the brain, also closely related to the angular gyrus, are areas which represent smell. Therefore, the angular gyrus is singularly located for confluence of information from these many different sources.

Destruction of the dominant angular gyrus of an adult almost completely destroys his intellect, destroys his ability to recognize his surroundings, and also destroys his ability to perform useful functions. For this reason this portion of the cortex is perhaps the one most important area of all. In a newborn child, destruction of this area is not nearly so serious because the angular gyrus in the opposite hemisphere of the brain can then develop the same functions, but in the adult this opposite hemisphere has become suppressed so that its functions will not interfere with those of the dominant hemisphere. It is very difficult to develop this angular gyrus once the process of suppression has taken place.

The fact that the angular gyrus area is very important to intellectual function of the brain does not mean that it is in this area that all important memories are stored. It merely means that this is one of those key points in the brain where, if the neuronal connections are destroyed, the other parts of the brain cannot function satisfactorily. Other aspects of angular gyrus function were discussed in Chapter 23.

Function of the Prefrontal Areas of the Cerebral Cortex

The prefrontal areas of the cerebral cortex are the most anterior portions of the brain, lying in the front 5 to 7 centimeters of the frontal lobes, anterior to the primary and secondary motor control areas discussed in Chapter 27. These areas are frequently called "silent" areas because no grossly observable effects occur when the prefrontal areas are destroyed. Instead, the person simply loses a major share of what we generally call his intellectual ability, especially his ability for abstract thought.

FUNCTION OF THE PREFRONTAL AREAS IN ABSTRACT THINKING. An animal that has lost his prefrontal areas loses the ability to keep small bits of information in his mind for longer than a few seconds at a time. For instance, if food is placed on one side of the cage and his attention is drawn to the other side, he forgets the locus of the food when he is no longer looking at it. It is believed that the prefrontal areas in the human being perform this same function, holding many small bits of information for short periods of time and yet to be forgotten a moment later. In performing mathematical problems it is necessary to store information from each stage of the logical process into a small corner of the mind and then to come back to this information a few seconds or a few minutes later. Thus, after storing and processing several hundred bits of information, the solution to the problem can be solved. Essentially the same mechanisms are used in abstract thinking related to legal processes, or diagnosis of diseases by the physician, or analysis of complex business problems.

THE PREFRONTAL AREAS AS THE LOCUS OF AMBITION, CONSCIENCE, PLANNING, AND WORRYING. Obviously, without the ability to think through the consequences of one's actions a person might act in too great haste and perform activities which he would regret in the future. Thus, it is well demonstrated that persons who have lost their prefrontal areas appear also to have lost much of their conscience.

Likewise, the qualities of planning for the future, of worrying about one's activities, or developing ambition are dependent upon his ability to think through interrelationships of all his actions. Here again, the person who loses his prefrontal areas likewise loses all these qualities. On the other hand, he very fortunately is often exempt from worry for the remainder of his life.

EFFECT ON BEHAVIOR CAUSED BY LOSS OF THE PREFRONTAL AREAS. Occasionally the prefrontal areas are completely destroyed by disease or trauma, or sometimes they are destroyed surgically because of harmful patterns of thought that have developed and cannot be stopped in other ways. When the prefrontal areas are gone, the common integrative region in the angular gyrus and surrounding regions of the parietal and temporal lobes are left mainly in charge of the thought patterns of the brain. The person is likely to exhibit extreme reactions, some of which are very happy in nature while others have the characteristics of extreme temper. Often the responses are rapid and are likely to lead to disastrous results. As long as the person is unprovoked he has very much the same personality as a giddy teenager, responding emotionally to everything but without much thought. When provoked, he is likely to fall into a state of rage which cannot be easily quelled at the moment, but which will be totally forgotten a few minutes later.

Control of Motor Function by the Thought Processes

Once incoming sensory signals have been interpreted, the brain then determines appropriate responses to the sensations. This determination begins in the angular gyrus region of the brain, the same area where the sensory signals are interpreted, the area called the common integrative area. Indeed, removal of the common integrative area makes one unable to determine almost any complex course of action. Even though he can still perform all discrete patterns of activity, he simply cannot put these patterns of activity together to make them meaningful. For instance, he can speak individual words or sometimes even phrases, but rarely can he speak an intelligent sentence and almost never put the sentences together into a complex thought.

Once the course of action has been decided in the common integrative area, a *sensory* program of action is then activated which the motor parts of the brain are required to follow. This effect was discussed in Chapter 27 but may be re-explained briefly as follows: Almost every motor function that one performs is a learned function, and the memory of this is stored in the sensory regions of the brain. This memory is called a *sensory engram*. If he wishes to reperform the same act, command signals are supposedly sent to make the primary and secondary motor cortices reduplicate the original sensory engram. In other words, the motor portion of the brain acts as a servo system simply to follow directions from the sensory cortex. For instance, in the process of talking, the thought to be expressed occurs first in the common integrative region of the brain; then the motor speech area in the frontal cortex is activated to express each succeeding phase of the thought as it is decreed in the common integrative area. It is for this reason that destruction of the common integrative and surrounding areas of the brain is so vastly damaging to all intellectual functions.

SLEEP AND WAKEFULNESS

One of the most important mysteries of brain function is its diurnal cycle of sleep and wakefulness. Even when a person remains in total darkness or in total light he still maintains approximately the same sleep-wakefulness cycle, with a periodicity of once every 24 hours. Ordinarily the nervous system shows signs of fatigue shortly before sleep ensues, and it shows signs of having become considerably rested at the termination of sleep. It seems, then, that neuronal fatigue plays an important part in causing sleep, and that sleep in turn relieves the fatigue.

Electrical studies of the brain indicate that while a person is awake many nerve impulses pass continuously through the nervous system, never ceasing. However, when the person is asleep, much fewer impulses are present. Thus, the state of wakefulness seems to be caused by a high degree of activity in the cerebrum, while the state of sleep is caused by a low degree of activity. Therefore, any theory to explain sleep and wakefulness must also explain these changing degrees of cerebral activity during the two states.

THE RETICULAR ACTIVATING SYSTEM. Stimulation of portions of the mesencephalon (midbrain) and thalamus greatly increases the activity of the cortex, an effect shown diagrammatically in Figure 294. Stimulation anywhere along the arrows will cause impulses to spread upward and eventually to excite the cortex. This system is called the *reticular activating system*, and it is divided into two separate parts, the *mesencephalic part* and the *thalamic part*.

The mesencephalic part of the reticular activating system is composed mainly of the reticular substance in the mesencephalon and the reticular substance of the upper pons. Stimulation of this region causes very diffuse flow of impulses upward through widespread areas of the thalamus and thence to widespread areas of the cortex, causing generalized increase in cerebral activity.

The thalamic portion of the reticular activating system differs from the mesencephalic part in that stimulation here activates *localized* regions of the cerebral cortex. Stimulation posteriorly in the thalamus activates posterior parts of the cerebral cortex while stimulation anteriorly activates anterior parts of the cortex. Thus, signals from specific parts of the thalamus can call forth activity in specific parts of the cortex rather than activating the entire cortex.

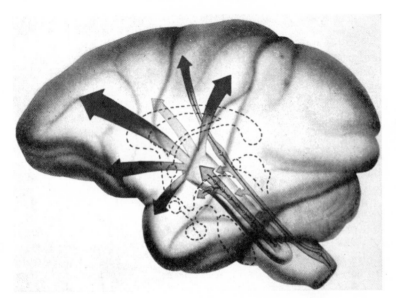

Figure 294. The reticular activating system, showing by the arrows passage of impulses from the reticular substance of the mesencephalon upward through the thalamus to all parts of the cortex. (From Lindsley: *Reticular Formation of the Brain.* Little, Brown and Co.)

It is likely that the function of the thalamic portion of the reticular activating system is to direct one's attention to memories or thoughts stored in specific parts of the cerebral cortex and thereby to allow his consciousness to consider individual thoughts at a time. On the other hand, *activation of the mesencephalic portion of the reticular activating system causes generalized wakefulness.*

The arousal reaction. The reticular activating system itself must be stimulated to action by input signals from other sources. When an animal is asleep, the reticular activating system is in an almost totally dormant state; yet almost any type of sensory signal can immediately activate the system. For instance, pain stimuli from any part of the body, or proprioceptor signals from the vestibular apparatuses, from the joints, and so forth can all cause immediate activation of the reticular activating system. This is called the *arousal reaction*, and it is the means by which sensory stimuli awaken us from deep sleep.

Cerebral stimulation of the reticular activating system. Signals from the cerebral cortex can also stimulate the reticular activating system and thereby increase its activity. Fiber pathways to both the mesencephalic and thalamic portions of this system are particularly abundant from the *somesthetic cortex*, the *motor cortex*, the *frontal cortex*, and parts of the cortex that deal mainly with emotions, the *cingulate gyrus* and the *hippocampus*. Whenever any of these cortical regions becomes excited, impulses are transmitted into the reticular activating system, thereby increasing the degree of activity of the reticular activating system.

The Feedback Theory of Wakefulness and Sleep

From the above discussion we can see that activation of the reticular activating system intensifies the degree of activity of the cerebral cortex, and in turn increased activity in the cerebral cortex increases the degree of activity of the reticular activating system. Therefore, it is fairly evident that a so-called "feedback loop" could develop whereby the reticular activating system excites the cortex and the cortex in turn re-excites the reticular activating system, thus setting up a cycle that causes continued intense excitation of both of these regions. Such a cycle is shown in Figure 295.

The reticular activating system is also involved in another feedback loop as follows: It sends impulses down the spinal cord to activate the muscles of the body. In turn, the muscle activation excites muscle proprioceptors that send sensory impulses upward to re-excite the reticular activating system. Thus, here again, activity of the reticular activating system sets off another cycle that in turn further increases the activity in the reticular activating system. This cycle is also shown in Figure 295.

One can readily see, therefore, that once either or both of these feedback loops involving the reticular activating system should become excited, the resulting increased activity in the reticular activating system would stimulate all other parts of the brain, thereby creating a state of wakefulness.

On the other hand, one might wonder how *sleep* could possibly occur once the different feedback loops should become activated. A probable answer to this is that

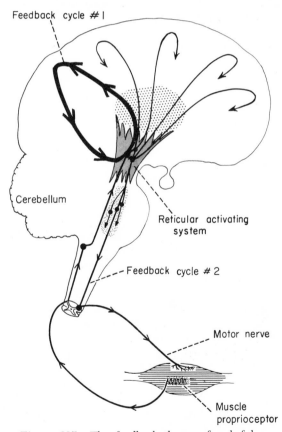

Figure 295. The feedback theory of wakefulness, showing two feedback cycles passing through the reticular activating system: one to the cerebrum and back (#1), and another to the peripheral muscles and back (#2).

all synapses of the central nervous system eventually fatigue, so that they either stop or nearly stop transmitting impulses. Therefore, after a prolonged period of wakefulness, one would expect the synapses in the feedback loops to become fatigued and therefore the feedback loops to stop functioning, thus allowing the reticular activating system to become dormant, which is the state of *sleep.*

We also know that a person can exist in all levels of wakefulness and sleep. The feedback theory of wakefulness postulates that the degree of wakefulness depends on the number of feedback loops activated at any one time. Figure 295 illustrates a large number of activating pathways passing from the mesencephalic activating system upward through the thalamus to the cortex. If only one of these pathways becomes activated, then obviously the degree of wakefulness would be very slight. If large numbers of the pathways should become activated simultaneously, then the degree of wakefulness would be greatly enhanced. One would expect that after a long period of wakefulness, progressively more and more of the feedback loops would become inactive and a person would fade into a state of sleep. Then upon arousal all of the feedback loops might be activated simultaneously, which explains why one could go from a state of sleep to complete wakefulness in a few seconds.

Effects of Wakefulness

Wakefulness is associated with three major effects: (1) increase in the degree of activity of the cerebrum, (2) transmission of signals directly from the reticular activating system to the muscles, and (3) excitation of the sympathetic nervous system.

The increase in cerebral function is caused by the great number of impulses transmitted from the wakefulness center upward through the thalamus to the cerebral cortex. These impulses continually impinge on the cerebral neurons, facilitating them to activity.

Stimulation of the reticular areas of the brain stem increases the degree of tone in all the muscles of the body. This makes the muscles more excitable than would otherwise be true and prepares them for immediate activity, which explains why a person who is wide awake has the feeling of muscular readiness.

The sympathetic stimulation caused by wakefulness increases the blood pressure slightly; it increases the rate of metabolism in all the tissues of the body; and, in general, it simply makes the body ready to perform increased amounts of work.

Effects of Sleep

Sometimes it is hard to understand why a person needs to sleep at all. Certain parts of the body such as the heart never rest and still are capable of functioning throughout life. One might reason that sleep is a measure to conserve the energies of most parts of the body when they are not needed. Some animals, as a matter of fact, have carried this principle of conservation so far that they pass into a state of very prolonged and often deep sleep called *hibernation,* which lasts throughout the entire winter.

A special value of sleep seems to be to reestablish appropriate balance of excitability among the various portions of the nervous system. As a person becomes progressively fatigued, some parts of his central nervous system lose excitability more than others, so that one part may overbalance the others. In fact, extreme fatigue can even precipitate severe psychotic disturbances. Yet after prolonged sleep, all parts of the nervous system usually will have returned once again to appropriate degrees of excitability and to a state of serenity.

Lack of sleep does not *directly* affect the intrinsic functions of the different organs. However, lack of sleep often causes severe autonomic disturbances, and these in turn *indirectly* lead to gastrointestinal upsets, loss of appetite, and other detrimental effects. In this way loss of sleep can affect the whole body as well as the nervous system itself.

Brain Waves

Electrical impulses called *brain waves* can be recorded from all active parts of the brain and even from the outside surface of the head. The character of these waves is closely related to the degree of sleep and wakefulness. When a person is awake but not thinking hard, continuous waves at a rate of approximately 10 to 12 per second can be recorded from almost all parts of the cerebral cortex. These are called *alpha waves.* The impulses that cause them probably originate in the reticular activating system and then spread into the cerebral cortex. These are believed to be the signals that keep the cortex facilitated during wakefulness. The alpha waves are illustrated by the recording at the top of Figure 296.

When any part of the brain becomes very active, for example the motor region initiating muscular activities, additional waves having a frequency sometimes as high as 50 cycles per second, and intensities usually greater than those of the alpha waves, take the place of the normal alpha waves. These, called *beta waves,* are shown by the second recording of Figure 296.

During sleep the alpha and beta waves are replaced by a few straggling waves occurring approximately once every one to two seconds. These are the "sleep waves" or *delta waves,* as shown by the third recording of Figure 296.

ABNORMAL BRAIN WAVE PATTERNS. Various abnormalities of the brain can cause strange brain wave patterns. Two of these, caused by different types of *epilepsy,* are shown in the fourth and fifth records of Figure 296. The fourth record shows a "spike and dome" picture which occurs in *petit mal* epilepsy. In this disease the person suddenly becomes unconscious for 3 to 10

seconds at a time. Such episodes may occur every few minutes, every few hours, or only once in many months, and when they do occur the person usually continues, while unconscious, whatever physical activity he is already doing even though he might be walking across a crowded street. Petit mal epilepsy seems to result from some abnormality of the wakefulness regions of the brain. The transmission of the normal alpha waves to the cerebral cortex is temporarily stopped. Instead, the spike and dome pattern is transmitted, and the person falls asleep for a few seconds until the alpha wave pattern picks up again.

The bottom record of Figure 296 shows the brain waves in *grand mal* epilepsy. In this condition the cerebral cortex becomes extremely excited, and many very strong signals spread over the brain at the same time. When these reach the motor cortex they cause rhythmic movements, called *clonic convulsions,* throughout the body. Grand mal epilepsy probably results from abnormal reverberating cycles developing in the reticular activating system. That is, one portion of the system stimulates another portion, this stimulates a third portion, and this in turn restimulates the first portion, causing a cycle that continues for two to three minutes until the neurons of the system fatigue so greatly that the reverberation ceases. At the beginning of a grand mal attack a person may experience very violent, abnormal, hallucinatory thoughts; once most of the brain is involved, he no longer has any conscious thoughts at all, because signals are then being transmitted in all directions rather than through discrete thought circuits. Therefore, even though his brain is violently active, he becomes unconscious of his surroundings. Following the attack his brain is so fatigued that he sleeps at least a few minutes and sometimes as long as several days.

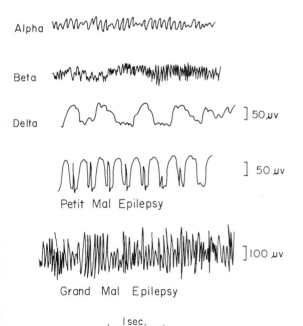

Alpha

Beta

Delta]50 µv

]50 µv

Petit Mal Epilepsy

]100 µv

Grand Mal Epilepsy

| 1 sec. |

Figure 296. Brain wave patterns in the normal person and in two persons with different types of epilepsy.

BEHAVIORAL FUNCTIONS OF THE BRAIN: THE LIMBIC SYSTEM

Behavior is a function of the entire nervous system, not of any particular portion. However, most involuntary aspects of behavior are controlled by the so-called *limbic system,* which is illustrated in block diagram form in Figure 297. This figure shows most

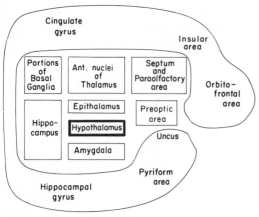

Figure 297.   The limbic system.

of the central structures of the brain with a surrounding ring of cerebral cortex called the *limbic cortex*. The great mass of cerebral cortex lies still beyond this ring. This ring of limbic cortex consists of (1) the *uncus*, the *pyriform area*, and the *hippocampal gyrus* on the very bottom of the brain, (2) the *cingulate gyrus* lying deep in the longitudinal fissure of the brain, and (3) the *insular* and *orbital frontal areas* lying on the bottom anterior portion of the brain. All of these areas of the cerebral cortex are phylogenetically old—that is, they were among the earliest portions of the cerebral cortex to evolve.

Perhaps the most important part of the limbic system, from the point of view of behavior, is the hypothalamus, the autonomic functions of which were discussed in Chapter 28. Many of the surrounding portions of the limbic system, including especially the hippocampus, the amygdala, and the thalamus, transmit major portions of their signals through the hypothalamus to cause varied effects in the body, such as to stimulate the autonomic nervous system or to participate in causing such feelings as pain, pleasure, or sensations related to feeding, sex, anger, and so forth.

Some aspects of limbic control are transmitted through the endocrine system, for it will be remembered that the hypothalamus, in addition to controlling the autonomic nervous system, also controls secretion of many of the hypophyseal hormones. This will be discussed in detail in Chapter 34, but for the time being it should be noted that the limbic system operating through the hypothalamus and hypophysis can control (1) the rates of secretion of all three sex hormones, which together control the various sexual drives of the person, and (2) the rates of secretion of thyroid hormone, growth hormone, and various adrenocortical hormones, which together control most of the person's day by day cellular metabolic functions.

Therefore, the limbic system can be said to control the inner being of the person.

With this background in mind, now let us discuss some of the specific mechanisms for control of behavior by the limbic system.

Pleasure and Pain; Reward and Punishment

One of the most important recent discoveries in the field of behavior is the so-called "pleasure and pain" or "reward and punishment" system of the brain. Certain areas in the thalamus, hypothalamus, and mesencephalon, when stimulated, make the animal feel intense punishment as if he were being severely pained. Yet stimulation of other closely related areas cause exactly the opposite effect, making the animal appear to be experiencing extreme pleasure. One of the experimental methods for studying the reward and punishment centers is to implant an electrode in one or the other of these two areas and then to allow the animal itself to press a lever to control the stimulus, as illustrated in Figure 298. If the electrode is implanted in the reward area, the animal will

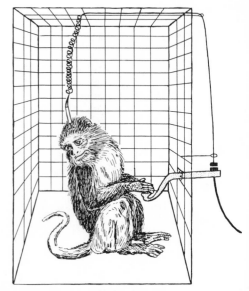

Figure 298.   Technique for localizing reward and punishment centers in the brain of a monkey.

press the lever continually. Indeed, it would rather stimulate its reward center than eat, even though it might be starving. On the other hand, if the electrode is placed in the punishment area, it will avoid stimulation by all means possible.

RELATIONSHIP OF REWARD AND PUNISHMENT TO LEARNING: FUNCTION OF THE HIPPOCAMPUS. Experiments on learning and memory have shown that an animal remembers sensory stimuli that cause either reward or punishment but fails to remember sensory stimuli that fail to excite either the reward or punishment area. For instance, a food that is very pleasant to the taste is remembered, or, likewise, a food that is exceedingly unpleasant is remembered. On the other hand, food that causes neither pleasure nor displeasure is rapidly forgotten. Similarly, a very painful stimulus, such as touching a hot iron, is remembered well, whereas simply touching a book or stick of wood is forgotten in a few seconds.

Therefore, in the process of memory, two different components must be present for a sensory experience to be remembered. The first component is the sensory experience itself, and the second component is an experience of either reward or punishment — that is, an experience of pleasure or pain.

Function of the Limbic Cortex

Even though the limbic cortex is the oldest part of the cerebral cortex, its precise function is least understood of almost all portions of the brain. Electrical stimulation in different parts of the limbic cortex can cause such effects as excitement, depression, increased movement, decreased movement, on rare occasions the phenomenon of rage, on other occasions intense degrees of docility, and so forth. However, these effects cannot be elicited repeatedly from specific points in the limbic cortex. Therefore, it is very difficult to state precise functions for the limbic cortex.

The probable function of the limbic cortex is to act as an association area for control of most of the behavioral functions of the body. It presumably stores information about past experiences such as pain, pleasure, appetite, various smells, sexual experiences, and so forth. This store of information is then combined with other information channeled into the limbic areas from surrounding regions of the cerebral cortex, such as from the prefrontal areas and from the sensory areas of the posterior part of the brain. This association of information then presumably provides stimuli for initiating appropriate behavioral responses for each respective occasion, whether this behavior be rage, docility, excitement, or lethargy, and so forth. Thus, we believe this to be the part of the cerebral cortex that plays the greatest role in controlling the emotions and other patterns of behavior.

The Defensive Pattern — Rage

Stimulation of the *perifornical nuclei of the hypothalamus*, located in the very middle of the hypothalamus, gives an animal the most intense sensation of punishment and simultaneously causes him to (1) develop a defense posture, (2) extend his claws, (3) lift his tail, (4) hiss, (5) spit, (6) growl, and (7) develop piloerection, wide-open eyes, and dilated pupils. Furthermore, even the slightest provocation causes an immediate savage attack. This is the pattern of behavior that has been called simply *rage*. It can occur in decorticated animals, illustrating that the basic behavioral patterns for defense and rage are controlled from the lower regions of the brain.

Exactly opposite emotional behavioral patterns occur when the reward centers are stimulated, namely, docility and tameness. During such stimulation the animal becomes completely amenable to almost any type of treatment.

Functions of the Amygdala

The amygdala is a complex group of nuclei located immediately beneath the surface of the cerebral cortex in the anterior pole of each temporal lobe. In lower animals, the amygdala is concerned primarily with olfactory stimuli, but in human beings it is much larger and operates in very close association with the hypothalamus to control many behavioral patterns. It is believed that the normal function of the various amygdaloid nuclei is to help control the overall pattern of behavior demanded for each social occasion.

Stimulation of various parts of the amygdala can transmit signals through the hypothalamus to cause (1) increase or decrease in arterial pressure, (2) increase or decrease in heart rate, (3) increase or decrease in gastrointestinal activity, (4) defecation or urination, (5) pupillary dilatation or constriction, (6) piloerection, which means hair standing on end, or (7) secretion of various adenohypophyseal hormones.

In addition, the amygdala can transmit signals to areas of the lower brain stem to cause (1) changes in the degree of muscle tone throughout the body, (2) postural movements, such as raising the head or bending the body, (3) circling movements, (4) rhythmic movements, or (5) movements associated with eating, such as licking, chewing, and swallowing.

Finally, excitation of other portions of the amygdala can cause sexual excitement, including erection, copulatory movements, ejaculation, ovulation, uterine activity, and premature labor.

From this foregoing list of functions, one can well understand how the amygdala can play a major role in controlling the body's overall pattern of behavior.

PSYCHOSOMATIC EFFECTS

A psychosomatic effect is a bodily, or *somatic*, effect produced by psychologic stimulation. The brain can produce such effects in three general ways: (1) by transmission of signals through the autonomic nervous system, (2) by transmission of signals to the muscles through the bulboreticular area, and (3) by control of certain of the endocrine glands. Figure 299 illustrates the general neuronal mechanisms believed to be responsible for psychosomatic effects. It shows conscious signals beginning either in the frontal cortex or in the thalamus, then being transmitted to the hypothalamus, and finally going through the centers of the hindbrain to the somatic regions.

PSYCHOSOMATIC EFFECTS TRANSMITTED THROUGH THE AUTONOMIC NERVOUS SYSTEM. Almost any emotion can affect the autonomic nervous system. For example, very intense agitation will increase the excitability of the reticular activating system, because this system in turn excites sympathetic activity throughout the body. Therefore, generalized sympathetic stimulation of the organs is one of the most common of all

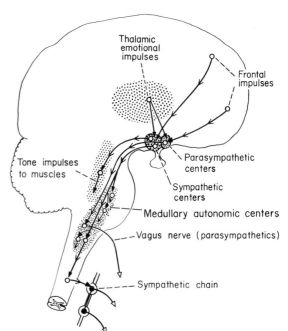

Figure 299. Neuronal circuits responsible for psychosomatic effects.

psychosomatic effects. Many emotions such as excitement, anxiety, or rage often discharge the sympathetics *en masse*, causing marked increase in arterial pressure, palpitation of the heart, and cold chills over the skin.

Psychologic effects also often stimulate the parasympathetic centers of the hypothalamus. The emotions of worry, depression, and lethargy, all of which have effects opposite to those that excite the sympathetic system, often stimulate the parasympathetics. On occasion, however, both of the systems may be stimulated simultaneously. Fear, for instance, can cause extreme sympathetic stimulation resulting in elevation of arterial pressure, while at the same time stimulating the parasympathetics to elicit such intense gastrointestinal activity that the person has diarrhea.

TRANSMISSION OF PSYCHOSOMATIC EFFECTS THROUGH THE SKELETAL MUSCLES. The reticular activating system sends signals downward through the spinal cord and thence directly to the muscles. Therefore, the same emotions that excite the sympathetics and reticular activating system usually increase the tone of the muscles throughout the body. Sometimes the tone becomes so intense that it causes muscle tremor, which explains why certain emotions can culminate in actual shaking.

On the other hand, the emotions that normally stimulate the parasympathetics usually decrease the activity of the bulboreticular formation. As a result, the muscular tone decreases to a very low level, which explains the muscular *asthenia* (muscular weakness) that is characteristic of some psychic states.

TRANSMISSION OF PSYCHOSOMATIC EFFECTS THROUGH GLANDS. The nervous system controls several of the endocrine glands either completely or partially. For instance, the sympathetic nervous system controls the adrenal medulla, and the hypothalamus controls almost all the activities of the adenohypophysis. The hypothalamus effects its control by secreting *neurosecretory substances* within the substance of the hypothalamus; these are carried through minute veins from the hypothalamus to the adenohypophysis where they cause secretion of several major hormones, *growth hormone, corticotropin, thyrotropin,* and three *gonadotropins.* These hormones in turn control growth rate, protein metabolism, overall rate of metabolism, and most sexual functions. Also, the hypothalamus controls the secretion of *antidiuretic hormone* by neurons located in the supraoptic nuclei; this hormone in turn controls the degree of retention of water by the kidneys.

Obviously, therefore, many psychosomatic effects can be mediated through the endocrine glands. For instance, psychic effects that overly stimulate the hypothalamus can cause hyperthyroidism, the thyroid gland secreting excess thyroid hormone and thereby increasing the rate of metabolism of all cells of the body. Likewise, psychic signals can affect the output of sex hormones and thereby cause failure of ovulation, excess menses, diminished menses, infertility, and other sexual abnormalities.

Psychosomatic Diseases

Occasionally abnormal cerebral function can cause physical disorders. Perhaps the most common psychosomatic disease is extreme tension throughout the body, resulting from excessive stimulation of the sympathetics and from increased tone of the peripheral muscles. This is often described simply as a "nervous breakdown."

Psychosomatic effects can also cause abnormal function of individual organs. For instance, stimulation of the sympathetics can so decrease gastrointestinal activity that constipation results. On the other hand, excessive stimulation of the parasympathetics can increase the degree of gastrointestinal activity so greatly that severe diarrhea results. A very common psychosomatic disorder is palpitation of the heart caused by excitement, anxiety, or other emotional states.

Occasionally the dysfunction caused by a psychosomatic disorder is so great that tissues are actually destroyed. For instance, stimulation of the parasympathetics to the stomach can cause so much secretion of gastric juices that they eat a hole into the wall of the stomach or upper intestine. This causes the condition called *peptic ulcer.* Such patients can be treated by operative removal of portions of the stomach, by neutralizing the gastric juices with special drugs, or in some instances by psychiatric treatment to alleviate the emotional condition that is initiating the excessive secretion of gastric juices.

PSYCHOSOMATIC PAIN. Many psychosomatic disorders can lead to pain which is called simply psychosomatic pain. For instance, a stomach ulcer causes intense burning in the pit of the stomach. Spasm of the gut caused by excess parasympathetic stimulation causes cramps in the abdomen. And, occasionally, overstimulation of the heart can even cause cardiac pain. If the psychic condition that causes the functional abnormality can be corrected, then the pain likewise will be corrected.

MYTHS ABOUT PSYCHOSOMATIC DISEASE. Despite the many different ways in which psychosomatic disease can come about, this subject has been greatly overemphasized by persons not familiar with it. It is a common myth that a person can worry so much about the function of one of his organs that he thereby creates disease in that particular organ. Except in a few instances this is not true. Most psychosomatic diseases exhibit regular patterns such as general states of tension, constipation or diarrhea, ulcer, and a few others of a similar nature.

REFERENCES

Brazier, M. A. B.: The analysis of brain waves. *Sci. Amer.* 206(6):142, 1962.

Broadbent, D. E.: Attention and the perception of speech. *Sci. Amer.* 206(4):143, 1962.

Brutkowski, S.: Functions of prefrontal cortex in animals. *Physiol. Rev.* 45(4):721, 1965.

Galambos, R., and Morgan, C. T.: The neural basis of learning. *Handbook of Physiology*, Sec. I, Vol. III, 1471. Baltimore, The Williams & Wilkins Company, 1960.

Hernandez-Peon, R.: *The Physiological Basis of Mental Activity.* New York, American Elsevier Publishing Co., 1963.

Jouvet, M.: Neurophysiology of the states of sleep. *Physiol. Rev.* 47(2):117, 1967.

Kandel, E. R., and Spencer, W. A.: Cellular neurophysiological approaches in the study of learning. *Physiol. Rev.* 48(1):65, 1968.

Kleitman, N.: *Sleep and Wakefulness.* Chicago, University of Chicago Press, 1963.

Magoun, H. W.: *The Waking Brain.* 2nd Ed. Springfield, Ill., Charles C Thomas, 1962.

Olds, J.: Hypothalamic substrates of reward. *Physiol. Rev.* 42:554, 1962.

Penfield, W.: *The Excitable Cortex in Conscious Man.* Springfield, Ill., Charles C Thomas, 1958.

THE GASTROINTESTINAL AND METABOLIC SYSTEMS

CHAPTER 30

GASTROINTESTINAL MOVEMENTS AND SECRETION, AND THEIR REGULATION

The function of the gastrointestinal system, illustrated in Figure 300, is to provide nutrients for the body. Food, after entering the mouth, is propelled through the esophagus into the stomach, and then through the small and large intestines before finally emptying out the anus. While the food passes through the gastrointestinal tract, digestive enzymes secreted by the gastrointestinal glands act on the food, breaking it into simple chemical substances that can be absorbed through the intestinal wall into the circulating body fluids. The general functions of the gastrointestinal tract, therefore, can be divided into (1) propulsion and mixing of the gastrointestinal contents, (2) secretion of digestive juices, (3) digestion of food, and (4) absorption of food. The first two of these functions are discussed in the present chapter, and the remaining two in the following chapter.

Physiologic Anatomy of the Gastrointestinal Tract

The gastrointestinal tract is essentially a long muscular tube with an inner lining that secretes digestive juices and absorbs nutrients. Figure 301 illustrates a typical cross-section of the gut, showing that most of the outer part is *smooth muscle* arranged in two layers, a *longitudinal layer* and a *circular layer*. Contraction of the longitudinal muscle shortens the gut, and contraction of the circular layer constricts it. The inner lining of the gut is called the *mucosa*, and it is covered by an *epithelium*. Small glands called *mucosal glands* penetrate into the deeper layers of the lining. These glands secrete digestive juices. Protruding from the lining of parts of the gut are many small *villi* which are responsible for most of the absorption of nutrients.

THE INTRAMURAL NERVE PLEXUS. One of the primary controllers of gastrointestinal function is a plexus of nerves called the *intramural nerve plexus* which is divided into two subdivisions, the *myenteric plexus* which lies mainly between the two muscle layers of the gut and *Meissner's plexus* which extends into the inner lining. The intramural plexus is present in the wall of the gut all the way from the esophagus to the anus, forming an intertwining web of nerve

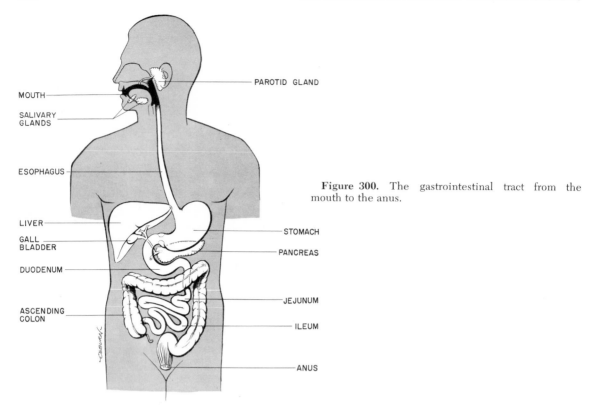

MOUTH

SALIVARY
GLANDS

ESOPHAGUS

LIVER

GALL
BLADDER

DUODENUM

ASCENDING
COLON

PAROTID GLAND

STOMACH

PANCREAS

JEJUNUM

ILEUM

ANUS

Figure 300. The gastrointestinal tract from the mouth to the anus.

fibers and nerve cell bodies. It controls muscular contraction in the gut and controls secretion by many of the glands.

Parasympathetic nerve fibers from the brain to the gut, carried mainly in the vagus nerve, terminate in the intramural plexus and, when stimulated, increase the degree of activity of the nerve network. *Sympathetic nerve fibers* from the spinal cord to the gut also terminate in the intramural plexus

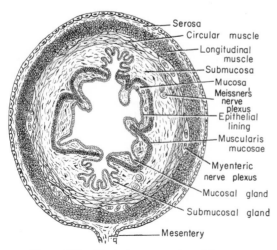

Serosa
Circular muscle
Longitudinal muscle
Submucosa
Mucosa
Meissner's nerve plexus
Epithelial lining
Muscularis mucosae
Myenteric nerve plexus
Mucosal gland
Submucosal gland
Mesentery

Figure 301. Typical cross-section of the gut.

or directly in the gut wall itself and, when stimulated, have exactly the opposite effect on activity of the plexus, decreasing its level of activity.

GASTROINTESTINAL MOVEMENTS

Two basic types of movement occur in the gastrointestinal tract, *propulsive movements* and *mixing movements*, one of which keeps the food moving along the gut and the other of which keeps it mixed. The characteristics of the movements in different parts of the gastrointestinal tract exhibit some differences that need to be described separately for certain portions of the gut. However, let us first consider the general characteristics of the movements.

Propulsive Movements of the Gastrointestinal Tract—Peristalsis

Food is moved along the gastrointestinal tract by *peristalsis* which is caused by slow advancement of a circular constriction, as

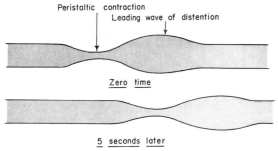

Peristaltic contraction
Leading wave of distention

Zero time

5 seconds later

Figure 302. Peristalsis.

illustrated in Figure 302. This has very much the same effect as encircling one's fingers tightly around a thin tube full of paste and then pulling the fingers along the tube. Any material in front of the fingers will be squeezed forward.

MECHANISM OF PERISTALSIS. Peristalsis is caused by nerve impulses that move along the myenteric nerve plexus. Stimulation of any single point of the plexus causes impulses to travel around the gut and lengthwise in both directions. The impulses traveling around the gut constrict it, and those traveling lengthwise cause the constriction to move forward. The usual rate of movement of this constriction is a few centimeters per second.

The usual stimulus that initiates peristalsis is *distention* of the gut, for this excites the local myenteric plexus, causing a circular constriction to advance along the gut.

LAW OF THE GUT. Even though peristalsis can move in both directions along the gut, it most frequently moves toward the anus. The probable reason for this is that the myenteric plexus itself is "polarized" in this direction. When a portion of the gut becomes distended, it causes contraction of the gut on the headward side of the distention and relaxation on the anal side. The contraction pushes the food forward and the relaxation allows easy forward movement; the movement, therefore, is normally analward rather than backward. This is called the *law of the gut.*

Mixing Movements in the Gastrointestinal Tract

The mixing movements consist of two basic types, (1) *weak peristaltic movements* that fail to move the food forward but nevertheless do succeed in mixing the intestinal

contents adjacent to the wall of the gut, and (2) *segmental movements,* which are isolated constrictions that occur at many points along the gut at the same time. The segmental movements occur rapidly, several times each minute, and each time "chop" the food into new segments. The precise characteristics of the mixing movements are quite different in the different parts of the gut, for which reason they will be described specifically for each part of the gastrointestinal tract.

Let us now begin at the upper end of the gastrointestinal tract and describe both the propulsive and mixing movements as the food passes analward.

Swallowing

Swallowing is initiated by a bolus of food pushed backward on the tongue into the pharynx. The bolus stimulates *swallowing receptor areas* located all around the opening of the pharynx, and impulses pass to the brain stem to initiate a series of automatic muscular contractions as follows:

(1) The soft palate is pulled upward to close the posterior part of the nose from the mouth.

(2) The vocal cords and larynx close strongly, and the epiglottis swings backward to prevent food from going into the trachea.

(3) The muscular sphincter that normally keeps the esophagus closed becomes relaxed, followed immediately by upward movement of the larynx that pulls the esophagus open.

(4) The pharyngeal muscles then constrict to force the bolus of food from the pharynx downward into the esophagus.

Nervous control of swallowing. Figure 303 illustrates the nervous pathways involved in the swallowing mechanism, showing that the swallowing receptors transmit impulses from the posterior mouth and throat mainly through the *trigeminal nerve* into the reticular substance of the *medulla oblongata* where the swallowing center is located. Once this center has been activated, the sequence of muscular reactions listed above occurs automatically and usually cannot be stopped. Nerve impulses go by way of the *glossopharyngeal* and *vagus nerves* to the pharyngeal and laryngeal regions to move the bolus of food into the upper esophagus. Then impulses from the vagi activate the proximal portion of the esophagus to push the food on toward the stomach.

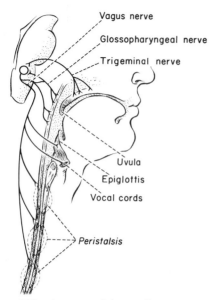

Figure 303.　Anatomy of the swallowing mechanism.

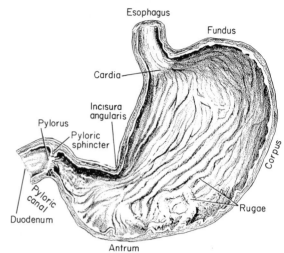

Figure 304.　Physiological anatomy of the stomach.

ESOPHAGEAL STAGE OF SWALLOWING. The musculature of the pharynx and of the upper one third of the esophagus is different from that of the rest of the gastrointestinal tract, for this muscle is skeletal muscle, controlled directly by nerves from the brain, in contrast to the remainder of the esophagus and gastrointestinal tract that is composed of smooth muscle and is only indirectly controlled by the central nervous system through the effects of the autonomic nervous system on the intramural plexus. Therefore, all contractions of the pharynx and upper one third of the esophagus are initiated directly by vagal and glossopharyngeal nerve impulses, and without these nerves the act of swallowing becomes paralyzed.

However, once food has reached the middle third of the esophagus, the distention elicits a typical peristaltic wave called the *secondary peristaltic wave* of the esophagus that pushes the food the rest of the way into the stomach. The entire time required for food to pass from the pharynx to the stomach is about five to ten seconds.

FUNCTION OF THE CARDIAC CONSTRICTOR. A centimeter or more of the esophagus proximal to the stomach is normally maintained in a state of tonic contraction. This area of the esophagus is called the *cardiac constrictor* or simply the *cardia;* it is illustrated in Figure 304. When swallowed food reaches the lower end of the esophagus, its progress is momentarily slowed by the cardia, but the approach of the esophageal peri-

staltic wave causes relaxation of the cardia which allows the food to move on into the stomach.

On the other hand, contractions in the stomach ordinarily will not push food backward through the cardia because there normally is no backward relaxation effect. Fortunately, therefore, stomach contractions normally propel food only forward rather than backward.

Motor Functions of the Stomach

The motor functions of the stomach are threefold: (1) storage of large quantities of food immediately after a meal, (2) mixing of this food with gastric secretions, and (3) emptying of the food from the stomach into the small intestine. The basic functional parts of the stomach are illustrated in Figure 304. Physiologically the stomach can be divided into two major parts, the *corpus* and the *antrum.*

STORAGE FUNCTION OF THE STOMACH. Food, on emptying from the esophagus into the stomach, first enters the corpus which is an extremely elastic bag that can store very large quantities of food. Furthermore, the tone of the corpus is normally very slight, so that even extreme amounts of food do not increase the pressure in the stomach greatly. Thus, the corpus is mainly a receptive organ for holding food until it can be utilized by the remainder of the gastrointestinal tract.

MIXING IN THE STOMACH—CHYME. The gastric glands, which cover most of the mucosa of the corpus, secrete large quantities of digestive juices which come into contact immediately with the stored food. Weak, rippling peristaltic waves, called *tonus waves* or *mixing waves*, pass along the stomach wall approximately once every 20 seconds. These begin in any part of the corpus and spread for variable distances toward the antrum, and they become more intense when food is present in the stomach than when it is not present.

The mixing waves mix the gastric secretions with the outermost layer of food and gradually move the mixture toward the antral part of the stomach. On entering the antrum, the waves become stronger, and the food and gastric secretions become mixed to a greater and greater degree of fluidity. As the food becomes thoroughly mixed with the gastric secretions, the mixture takes on a milky white appearance and is then called *chyme.*

PROPULSION OF CHYME THROUGH THE STOMACH AND EMPTYING OF THE STOMACH. The tonus or mixing waves are rarely strong enough to push chyme through the pylorus into the duodenum. However, occasional very powerful peristaltic contractions also occur, beginning either in the corpus or antrum, and these generate as much as 50 mm. Hg pressure in the prepyloric portion of the antrum. This is enough pressure to push open the pyloric sphincter and propel the chyme on into the duodenum.

Regulation of stomach emptying. Emptying of the stomach is controlled mainly by the intensity of the strong peristaltic waves. The pylorus itself usually remains slightly constricted all the time. Therefore, a certain degree of pressure is required before chyme can pass through. A weak peristaltic wave fails to move the chyme but a strong peristaltic wave succeeds. Among the different factors that determine whether or not the peristaltic wave will succeed are the following:

Degree of fluidity of the chyme. Obviously, the better the food has become mixed with gastric secretions the more easily it can flow through the narrow passageway of the pylorus. Therefore, ordinarily, food will not pass out of the stomach until it has been thoroughly mixed.

Quantity of chyme already present in the small intestine. When a large amount of chyme has already emptied into the small intestine, particularly when a large portion of this is still present in the duodenum, a reflex called an *enterogastric reflex* spreads backward through the myenteric plexus from the duodenum to the stomach to inhibit peristalsis. In this way, the duodenum keeps itself from becoming overfilled.

Presence of acids and irritants in the small intestine. The gastric secretions, as will be discussed later in the chapter, are highly acidic, but the acid in the chyme entering the duodenum is ordinarily neutralized by pancreatic secretions that also empty into the duodenum. Until the acid becomes neutralized by pancreatic juice, irritation of the wall of the duodenum elicits an enterogastric reflex similar to that elicited by distention; this too inhibits the peristaltic waves in the stomach, thus stopping gastric emptying. In this way, the duodenum protects itself from too much acidity. Likewise, any other irritant also causes an enterogastric reflex that will do the same thing.

Presence of fats in the small intestine — enterogastrone. When fats enter the small intestine from the stomach, they extract from the mucosa of the duodenum and jejunum a hormone called *enterogastrone,* which is immediately absorbed into the blood and carried to the stomach. Here it inhibits stomach peristalsis and slows up stomach emptying. This mechanism allows adequate time for fat digestion to occur in the small intestine. Proteins and carbohydrates, on the other hand, have very little inhibitory effect on stomach emptying. Fortunately, both of these foods are digested much more easily in the intestine than is fat.

In summary, gastric emptying is determined mainly by fluidity of the contents in the stomach and the state of the duodenum. If the duodenum is already filled, has irritant substances in it, or has fat in it, emptying will proceed very slowly, but if the duodenum is empty and the contents of the stomach are very fluid, emptying will proceed rapidly.

HUNGER CONTRACTIONS. In addition to the tonus and peristaltic contractions of the stomach, a third type of very intense contraction, called *hunger contractions,* occur when the stomach has been empty for 8 to 20 hours. These are normally very powerful rhythmic peristaltic contractions but occasionally become a strong tetanic contraction lasting for as long as 2 to 10 minutes. They cause a tight sensation in the pit of the stomach called *hunger pangs,* and they intensify the desire for food.

Movements of the Small Intestine

PROPULSIVE MOVEMENTS. It is in the small intestine that the most typical peristalsis occurs, for distention of any portion of the small intestine with chyme initiates a peristaltic wave. Peristalsis is far more intense when the parasympathetic nerves are stimulated, and sympathetic stimulation can inhibit greatly or totally block peristalsis.

MIXING CONTRACTIONS – SEGMENTATION. The presence of chyme in the small intestine initiates a type of contraction called *segmentation,* which is illustrated in Figure 305. When the small intestine becomes distended, many constrictions occur either regularly or irregularly along the distended area. As shown in Figure 305, the intestine becomes "chopped" into small sausage-like vesicles. The constrictions then relax, but others occur at different points a few seconds later. Thus, repetitive "chopping" of the chyme keeps it mixed continually while it is in the small intestine.

Both the segmentation and peristaltic movements of the small intestine are controlled by the myenteric plexus.

EMPTYING OF INTESTINAL CONTENTS AT THE ILEOCECAL VALVE. Figure 306 shows the *ileocecal valve* where the small intestine empties into the large intestine. Note that the small intestine actually protrudes forward into the colon. This projection of the ileocecal valve prevents the contents of the colon from regurgitating into the small intestine; instead the lips of the ileocecal valve simply close on themselves when pressure builds up in the colon.

Emptying of the small intestine at the ileocecal valve occurs in very much the same way that the stomach empties; that is, peristaltic waves in the small intestine build

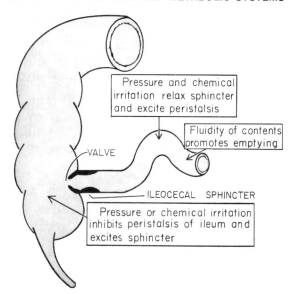

Figure 306.   Emptying at the ileocecal valve.

up pressure behind the valve and push chyme forward into the colon, but if the colon has become too full, myenteric reflexes can inhibit peristalsis and thereby slow up or stop the emptying.

Movements of the Colon

The functions of the colon are (1) absorption of water and electrolytes from chyme and (2) storage of fecal matter until it can be expelled. The first half of the colon, illustrated in Figure 300, is concerned mainly with absorption and the distal half with storage. And, except when the bowels are to be emptied, the movements of the colon are usually very sluggish.

MIXING MOVEMENTS. The mixing movements of the colon are similar to the segmentation movements of the small intestine. Concentric contractions of the colon break it up into large pockets called *haustrations,* which are illustrated in Figures 300, 306, and 307. The circular constrictions last for about 30 seconds, and after another few minutes occur in nearby but not the same areas. Thus, fecal material is slowly "dug" into and rolled over in much the same manner that one spades the earth.

Ordinarily, 500 milliliters of chyme is emptied into the colon each day, and about 400 milliliters of this, mainly the water and electrolytes, is reabsorbed before defecation takes place, leaving an average volume of feces of 100 milliliters each day.

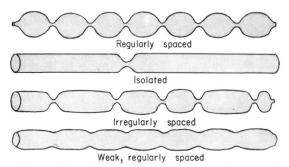

Regularly spaced

Isolated

Irregularly spaced

Weak, regularly spaced

Figure 305.   Segmentation movements of the small intestine.

PROPULSIVE MOVEMENTS. The typical peristaltic movements that occur in the small intestine do not occur in the colon, for the colon has no peristaltic movements at all 99 per cent of the time. Yet when the colon becomes overfilled, several strong peristaltic movements, called *mass movements,* occur one after another. These propel the fecal material long distances, sometimes all the way from the ascending colon to the descending colon. After a few minutes the mass movements cease, to reappear again many hours later when some part of the colon becomes overfilled again.

DEFECATION. When the mass movements of the colon have succeeded in moving fecal material into the rectum, a special reflex called the *defecation reflex* occurs, which causes emptying of the rectum and lower parts of the colon. Figure 307 shows that filling the rectum excites nerve endings that send impulses into the lower part of the spinal cord. These cause reflex impulses to be transmitted through the sacral parasympathetic nerves to the descending colon, sigmoid flexure, rectum, and *internal anal sphincter,* causing relaxation of the sphincter and contraction of the gut wall. This reflex thus causes emptying of the bowels if the *external anal sphincter* is also relaxed. However, the external anal sphincter is a skeletal muscle that guards the outer opening of the anus, controlled by voluntary skeletal nerves that can be relaxed or tightened according to the will of the person. If the time is not propitious for emptying the bowels, tightening of the external anal sphincter prevents defecation despite the defecation reflex. On the other hand, relaxation allows defecation to take place.

If a person prevents defecation when the defecation reflex occurs, the reflex usually dies out after a few minutes but returns a few hours later. Also, a person can frequently initiate a defecation reflex at will by tightening his abdominal muscles, which compresses the rectal wall and elicits the typical reflex. Unfortunately, this elicited reflex is usually much weaker than the natural reflex, so that defecation is less efficacious under these conditions than when the reflex has been elicited naturally.

Special Types of Gastrointestinal Movements

ANTIPERISTALSIS AND VOMITING. Occasionally, some intensely irritating substance enters the gastrointestinal tract. An immediate effect of this is an increase in the rate of local secretion of mucus, which helps to protect the interior of the gut. Simultaneously, the local gut wall contracts intensely. For reasons not too well understood, these intense local contractions elicit *antiperistalsis,* meaning peristalsis backward toward the mouth, rather than forward peristalsis. Food cannot move backward from the colon into the small intestine because of the ileocecal valve, but it can move all the way from the tip of the small intestine back into the stomach.

On reaching the stomach, the irritative material is then rapidly expelled by the vomiting process as follows: The intense irritation of the gut causes impulses to be transmitted through the visceral sensory nerves into the brain. These cause a sensation of *nausea,* and, if the impulses are strong enough, they will also cause an automatic reflex integrated in the medulla oblongata called the *vomiting reflex.* This reflex causes closure of the airway into the trachea, relaxation of the cardia of the stomach, and very tight contraction of both the diaphragm and the abdominal muscles. The squeezing action of the diaphragm and abdominal muscles on the stomach pushes food out of the stomach upward through the esophagus and mouth. This is the vomiting process.

GASTROCOLIC AND DUODENOCOLIC REFLEXES. Almost everyone is familiar with the natural desire to defecate following

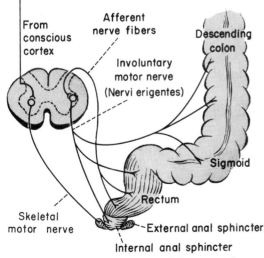

Figure 307.   Anatomy of the defecation reflex.

either a heavy meal or the first meal of the day. The cause of this is the *gastrocolic* and *duodenocolic reflexes*, mainly the latter. These reflexes are elicited by increased filling of the stomach and duodenum, which in turn transmit impulses downward along the myenteric plexus to the colon to cause increased excitability of the entire colon, initiating both mass movements and defecation reflexes.

THE PERITONEAL REFLEX. Irritation of the peritoneum in any way, whether caused by cutting the peritoneum during an abdominal operation, by infection of the peritoneum, or even by a severe blow to the abdomen that causes trauma of the peritoneum, will elicit a *peritoneal reflex* that strongly excites the sympathetic nerves to the gut. These nerves in turn *inhibit* gastrointestinal activity and thereby stop or slow movement of chyme along the intestinal tract. Obviously, lack of movement aids in the repair of the peritoneal damage.

MUCOSAL REFLEXES. Irritation inside the gut or distention of the gut ordinarily excites the intramural plexus rather than inhibiting it. Reflexes occur mainly locally, causing increased local secretion and increased local motor activity. The secretion dilutes the irritating factor, and the motor activity moves the irritant on through the gut. When the irritation is great, it causes *diarrhea*. And if the degree of irritation increases still more, it can cause diarrhea from the lower part of the intestinal tract while causing vomiting from the upper intestinal tract. Obviously, these reflexes are protective in nature and prevent irritating substances or infectious processes inside the intestinal tract from causing severe permanent damage.

GASTROINTESTINAL SECRETIONS

Glands are present throughout the gastrointestinal tract to secrete chemicals that mix with the food and digest it. These secretions are of two types: first, mucus which protects the wall of the gastrointestinal tract, and, second, enzymes and allied substances that break the large chemical compounds of the food into simple compounds.

MUCUS. Mucus is secreted by every portion of the gastrointestinal tract. It contains a large amount of mucoprotein that is resistant to almost all digestive juices. Mucus lubricates the passage of food along the mucosa, and it forms a thin film everywhere to prevent the food from excoriating the mucosa. Also, it is *amphoteric*, which means it is capable of neutralizing either acids or bases. All these properties of mucus make it an excellent substance to protect the mucosa from physical damage, and to prevent the digestion of the wall of the gut by the digestive juices.

Salivary Secretion

Saliva is secreted by the *parotid, submaxillary, sublingual,* and smaller glands in the mouth. Saliva is about half *mucus* and half a solution of the enzyme *ptyalin*. The function of mucus is to provide lubrication for swallowing. Without mucus one can hardly swallow. If one mixes food with water to take the place of mucus, approximately 10 times as much water as mucus is necessary to provide the same degree of lubrication. The function of the ptyalin in the saliva is to begin the digestion of starches and other carbohydrates in the food. Ordinarily the food is not exposed in the mouth to saliva long enough for more than 5 to 10 per cent of the starches to become digested. However, the mixed saliva and food is normally stored in the corpus of the stomach for 30 minutes to several hours, during which time the saliva may digest as much as 50 per cent of the starches.

REGULATION OF SALIVARY SECRETION. The *superior and inferior salivatory nuclei* located in the brain stem, as shown in Figure 308, control secretion by the salivary glands. These nuclei in turn are controlled mainly by taste impulses and other sensory impulses from the mouth. Foods that have a pleasant taste ordinarily cause the secretion of large quantities of saliva, while some unpleasant foods may decrease salivary secretion so greatly that swallowing is made very difficult. Also, the sensation of smooth-textured foods inside the mouth increases salivation, while the sensation of roughness decreases salivation. This effect presumably allows those foods that will not abrade the mucosa to be swallowed with ease, and causes the rejection of abrasive foods.

THE PHASES OF SALIVARY SECRETION. In addition to the salivation that occurs while food is actually in the mouth, salivation fre-

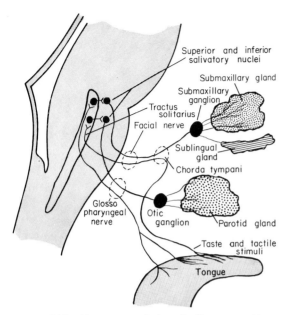

Figure 308. Nervous regulation of salivary secretion.

quently occurs even before food enters the mouth—that is, when a person is thinking about or smelling pleasant food—and it continues to occur even after the food has been swallowed. Therefore, salivary secretion can be divided into three phases, the *psychic phase*, the *gustatory phase*, and the *gastrointestinal phase*. The psychic phase presumably makes the mouth ready for food and aids in the secretion of saliva as the food is presented to the mouth. The gustatory phase supplies the saliva that mixes with the food while one is chewing, and the gastrointestinal phase continues the secretion of saliva even after the food has passed for storage into the stomach. Secretion during the gastrointestinal phase is especially likely to be abundant when one has swallowed irritant foods. The saliva, on being swallowed, helps to neutralize the irritant substance, thereby relieving any irritation of the stomach that might be occurring.

Esophageal Secretions

The esophagus secretes only *mucus*. Normally, food passes from the mouth through the esophagus and into the stomach in about 7 seconds. This food has not been subjected to the mixing movements of the gastrointestinal tract and, therefore, is in its most abrasive state. Fortunately the esophagus is supplied with a great abundance of mucous glands that secrete mucus to protect the mucosa from excoriation.

Gastric Secretions

MUCUS. The primary function of the gastric secretions is to begin the digestion of proteins. Unfortunately, the wall of the stomach is itself composed of protein substances. Therefore, the surface of the stomach must be exceptionally well protected at all times against digestion. Perhaps the most abundant secretion of mucus in any part of the gastrointestinal tract occurs in the stomach. The entire surface of the stomach is covered by very small *mucous cells*, the outer half of which are composed almost entirely of mucus; these prevent gastric secretions from ever touching the deeper layers of the stomach wall. In the antral region of the stomach, where the powerful peristaltic movements occur and where excoriation of the stomach wall is particularly likely to occur, mucus is secreted not only by the mucous cells on the surface of the mucosa but also by large mucous glands that extend deep into the mucosa. In the absence of mucus secretion, a hole is eaten in the wall of the stomach in only a few hours. This hole is called a *stomach ulcer.*

DIGESTIVE SUBSTANCES. The major digestive substances secreted by the stomach are *hydrochloric acid* and *pepsin.* Hydrochloric acid activates the pepsin, which is an enzyme that begins the digestion of proteins.

Less abundant enzymes secreted by the stomach are *gastric lipase* for beginning the digestion of fats, and *rennin* for aiding in the digestion of *casein,* one of the proteins in milk. These enzymes are secreted in such minor quantities that they are of almost no importance.

The total quantity of stomach secretion each day is about 2000 ml.

REGULATION OF GASTRIC SECRETION. *Neurogenic mechanisms.* Stomach secretion is regulated by both neurogenic and hormonal mechanisms, as shown in Figure 309. Some of the neurogenic mechanisms are quite similar to those regulating salivary secretion. For instance, food in the stomach can cause local nervous reflexes that occur entirely in the wall of the stomach itself to cause local secretion. Also, signals from the stomach mucosa to the medulla of the brain can cause reflexes back to the stomach

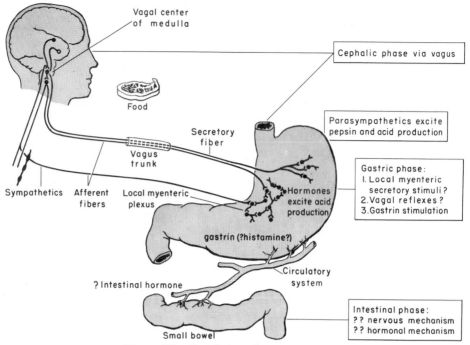

Figure 309.   Regulation of gastric secretion.

through the vagus nerves to cause secretion. In addition, secretory signals from the medulla to the stomach can be excited by impulses originating in various other areas of the brain, particularly the cerebral cortex.

The gastrin mechanism. Gastric secretion is also regulated to some extent by a hormone called *gastrin*. When meats and certain other protein foods reach the antral portion of the stomach, they cause the hormone *gastrin* to be extracted from the antral mucosa and to be absorbed into the blood stream. This hormone then passes by way of the blood to the fundic glands of the stomach, and causes them to secrete a strongly acidic gastric juice. The acid, in turn, greatly aids in the digestion of the meats that first initiated the gastric mechanism. In this way the stomach helps to tailor-make the secretion to fit the particular type of food that is eaten.

The phases of gastric secretion. Large amounts of stomach juices are often secreted when pleasant food is simply thought of or particularly when it is smelled. This is called the *psychic phase* of stomach secretion. It prepares the stomach for food that is to be eaten. The second phase of gastric secretion is the *gastric phase*, which is the secretion that occurs while the food is in the stomach

itself. This is caused mainly by reflexes initiated by food in the stomach and by the gastrin mechanism. Finally, even after the food has left the stomach, gastric secretions continue for several hours. This is called the *intestinal phase* of gastric secretion. It is probably caused by hormones that pass to the stomach in the blood after being extracted from the intestinal mucosa by the food. Ordinarily, the amount of secretion during the intestinal phase is only about 10 per cent of the total during the other two phases.

Pancreatic Secretions

The pancreas, shown in Figure 310, is a large gland located immediately beneath the stomach. It empties about 1200 ml. of secretions each day into the upper portion of the small intestine 1½ inches beyond the pylorus. These secretions contain large quantities of *amylase* for digesting carbohydrates, *trypsin* and *chymotrypsin* for digesting proteins, *pancreatic lipase* for digesting fats, and other less important enzymes. It is obvious from this list that the pancreatic secretions are as important for digesting the

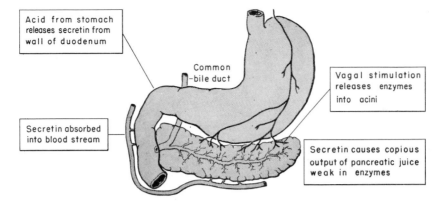

Acid from stomach releases secretin from wall of duodenum

Common -bile duct

Vagal stimulation releases enzymes into acini

Secretin absorbed into blood stream

Secretin causes copious output of pancreatic juice weak in enzymes

Pancreozymin is absorbed like secretin and produces large quantities of enzymes.

Figure 310.   Regulation of pancreatic secretion.

food as any others of the entire gastrointestinal tract.

In addition to the digestive enzymes, pancreatic secretions contain large amounts of *sodium bicarbonate,* which react with the hydrochloric acid from the stomach to form sodium chloride and carbonic acid. The carbonic acid then is absorbed into the blood, becomes water and carbon dioxide, and the carbon dioxide is expired through the lungs. The net result is an increase in the quantity of sodium chloride, a neutral salt, in the intestine. Thus, pancreatic secretions neutralize the acidity of the chyme coming from the stomach. This is one of the most important functions of the pancreatic secretion.

REGULATION OF PANCREATIC SECRETION.
The secretin mechanism and neutralization of chyme. When chyme enters the upper small intestine it causes a polypeptide called *secretin* to be released from the intestinal mucosa; the quantity of secretin released is especially abundant when the chyme is highly acidic. The secretin in turn is absorbed into the blood and carried to the glandular cells of the pancreas. There it causes the cells to secrete large quantities of fluids containing extra large amounts of sodium bicarbonate. The bicarbonate then reacts with the acid of the chyme to neutralize it. Thus, the secretin mechanism is an automatic process to prevent excess acid in the upper small intestine.

When satisfactory neutralization does not occur, the acidic chyme, containing also large quantities of the protein-digesting enzyme pepsin, is likely to erode through the wall of the duodenum, the uppermost part of the small intestine, causing an ulcer. In fact, ulcer of the duodenum is about four times as common as ulcer of the stomach, because this area is not nearly as well protected by mucous glands as is the stomach.

The pancreozymin mechanism. At the same time that secretin is extracted from the intestinal mucosa, another hormone, *pancreozymin,* also is extracted in response mainly to proteins in the chyme, but to a less extent in response to fats and carbohydrates. Pancreozymin, like secretin, passes by way of the blood to the pancreas, but, unlike secretin, it causes the secretory cells to secrete large quantities of digestive enzymes instead of sodium bicarbonate. These enzymes, on entering the duodenum, begin digesting the foods.

Vagal regulation. Stimulation of the vagus nerve also causes the secretory cells of the pancreas to secrete highly concentrated enzymes. The quantity, however, is usually so small that the enzymes remain in the ducts of the pancreas and later are floated into the intestinal tract by the copious secretion of fluid that follows secretin stimulation.

Vagal stimulation of pancreatic secretion seems to be a by-product of the vagal reflexes to the stomach. That is, some of the reflex impulses initiated by food in the stomach return to the pancreas rather than to the stomach. This allows preliminary formation of pancreatic enzymes even before the food enters the intestine. However, vagal stimulation of pancreatic secretion is probably unimportant in comparison with the hormonal stimulation by secretin and pancreozymin.

Liver Secretion

The liver, shown in Figure 311, secretes a solution called *bile* that contains a large quantity of *bile salts,* a moderate quantity of *cholesterol,* a small quantity of the green pigment *bilirubin* which is a waste product of red blood cell destruction, and a number of other less important substances. The only substance in bile that is of importance to the digestive functions of the gastrointestinal tract is the bile salts. The remainder of the contents is actually waste products being *excreted* from the body fluids by this route.

The bile salts are not enzymes for digesting foods, but they act as a powerful *detergent* (a substance that lowers the surface tension at the surface between water and fats). This helps the mixing movements of the intestine break the large fat globules of the food into small globules, thus allowing the lipases of the intestinal tract, which are water soluble, to attack larger surface areas of the fat, and to digest it. Without this action of bile almost none of the fats in the food would be digested.

REGULATION OF BILE SECRETION. Liver secretion, unlike secretion by other gastrointestinal glands, does not increase and decrease significantly in response to food in the intestine. The only hormone known to affect it is secretin, which can increase the output of bile some 10 to 20 per cent but does not have the powerful effect on the liver that it has on the pancreas.

Storage of bile in the gallbladder. Even though bile secretion is a continuous process, the flow of bile into the gastrointestinal tract is not continuous. As illustrated in Figure 311, a circular muscle around the outlet where the common bile duct empties into the duodenum, called the *sphincter of Oddi,* normally blocks the flow of bile into the gut. Instead, the bile flows into the *gallbladder,* which is attached to the side of the common bile duct. Much of the fluid and electrolytes of the bile is then reabsorbed into the blood by the gallbladder mucosa. This concentrates by as much as twelve-fold the bile acids, cholesterol, and bilirubin, which cannot be reabsorbed, and allows the gallbladder, even though it has a maximum volume of only 50 ml., to accommodate the active components (bile acids) of an entire day's liver secretion of bile (600 ml.).

Emptying of the gallbladder. When food enters the small intestine, two mechanisms simultaneously cause the gallbladder to empty into the small intestine. First, the fatty substances in the food extract from the intestinal mucosa a hormone called *cholecystokinin,* which then passes through the blood to the gallbladder and causes the muscular wall to contract. Second, the presence of food in the duodenum causes duodenal peristalsis, and the peristaltic waves send periodic nerve signals to the sphincter of Oddi through the myenteric plexus to open it. This combination of gallbladder contraction and opening of the sphincter of Oddi allows the stored bile to empty into the intestine, and the bile salts immediately begin their emulsifying action on the fats.

Gallstones. Gallstones are caused mainly by a fatty waste product, *cholesterol,* which is normally excreted in the bile. Cholesterol

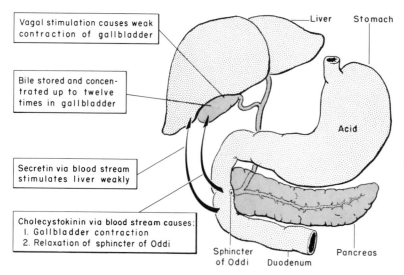

Figure 311. Bile secretion, bile storage in the gallbladder, and the cholecystokinin mechanism for promoting gallbladder emptying.

Vagal stimulation causes weak contraction of gallbladder

Bile stored and concentrated up to twelve times in gallbladder

Secretin via blood stream stimulates liver weakly

Cholecystokinin via blood stream causes:
1. Gallbladder contraction
2. Relaxation of sphincter of Oddi

Liver   Stomach

Acid

Sphincter of Oddi   Duodenum   Pancreas

is relatively insoluble, but is normally held in solution in the bile by chemical attraction to the bile salts. Often, though, when bile becomes concentrated in the gallbladder, too much water is removed from the bile, and the cholesterol becomes too concentrated to remain in solution. Small crystals of cholesterol then begin to precipitate, these sometimes growing to fill the entire bladder.

A means for preventing the formation of gallstones is to eat a diet low in fat, for cholesterol is formed in great abundance in response to a high fat diet. Once gallstones have been formed, however, the only treatment is removal of the stones, or, preferably, removal of the gallbladder itself along with the stones. Absence of the gallbladder does not greatly affect the digestion of fats, because bile continues to be excreted into the intestine, though now it enters the gut almost all the time rather than periodically.

Secretion in the Small Intestine

The small intestine secretes the enzymes *sucrase*, *maltase*, and *lactase* for splitting disaccharides into monosaccharides, the final digestion products of carbohydrates. Also secreted are large quantities of *peptidases* for performing the final steps in protein digestion, and small quantities of *lipases* for splitting fats.

However, secretion in the small intestine does not occur in the usual manner. Instead, the digestive enzymes are formed in the epithelial cells lining the intestinal wall, and much of the digestive process occurs either inside these cells or in close proximity to them. Also, some of the cells slough off into the intestine and then discharge small amounts of the enzymes that act directly on the food of the chyme.

MUCUS SECRETION. The small intestine secretes along its entire surface large quantities of mucus, which provide the same protective function in this part of the gastrointestinal tract as in the stomach, the esophagus, and elsewhere. In the duodenum, an especially abundant amount of mucus is secreted by large mucous glands, called *Brunner's glands*, lying deep in the mucosa. The function of this secretion is to protect this portion of the intestinal tract from the powerful digestive action of pepsin and hydrochloric acid in the chyme newly arrived from the stomach. Once the chyme has been neutralized by pancreatic juice, however, it no longer has such a strong tendency to digest the wall of the intestine, which explains why Brunner's glands are needed only in this uppermost region of the intestinal tract.

QUANTITY OF SECRETION. The total amount of secretion of the small intestine is about 3 liters per day, which compares with about 500 ml. of saliva, 2000 ml. of gastric juice, 1200 ml. of pancreatic juice, and 600 ml. of bile. In other words, the total quantity of intestinal secretion is almost as much as that of all the remaining gastrointestinal glands put together.

ABSORPTION FROM THE SMALL INTESTINE. Almost all the secretions that enter the small intestine, including saliva, gastric secretion, pancreatic secretion, bile, and intestinal secretion, are absorbed before entering the large intestine. Any fluid that is ingested is absorbed as well. That is, 8 liters or more of fluid is absorbed by the mucosa of the small intestine each day. A remaining 500 ml. passes along with the chyme into the large intestine, and most of it is absorbed there before the feces are expelled. Thus, fluid circulation occurs continually between the body fluids and the gastrointestinal tract. Because the fluid utilized in forming the gastrointestinal secretions is mainly extracellular fluid, any loss of fluid from the gastrointestinal tract, such as by vomiting or diarrhea, is actually a loss of extracellular fluid and can cause one to become extremely dehydrated.

REGULATION OF SECRETION OF THE SMALL INTESTINE. Most secretion in the small intestine is probably regulated by local nervous reflexes. That is, food distending the intestinal tract or irritating the intestinal mucosa initiates reflexes in the intramural plexus to simulate secretion by the intestinal mucosa.

However, there is also a hormonal mechanism for regulating intestinal secretion. Though not as important quantitatively as the nervous reflex mechanism, the hormonal system perhaps helps to determine the types of enzymes secreted. Food in the small intestine extracts a mixture of hormones called *enterocrinin* from the mucosa, and these hormones are said to stimulate the intestinal glands to secrete appropriate enzymes for digesting the types of food present. For instance, it is claimed that a large quantity of proteins in the intestine causes the concentration of peptidases in the intestinal juices to be greater than the concentration of the other enzymes. Likewise, carbohydrates promote the secretion of sucrase, maltase, and lactase, and fats promote the secretion of lipases.

Secretions of the Large Intestine

The large intestine, like the esophagus, performs no digestive functions. Therefore, its only significant secretion is mucus. The entire mucosa is coated with mucous cells that provide lubrication for the passage of feces from the ileocecal valve to the anus and also protect the large intestine from digestion by the enzymes emptied from the small intestine. The portions of the large intestine near the ileocecal valve are usually protected against the digestive hormones better than the distal portions. Consequently, during severe diarrhea the rapid flow of digestive enzymes from the small intestine into the distal colon is very likely to cause extreme irritation. Prolonged and severe diarrhea sometimes initiates a condition called *ulcerative colitis*, which occasionally leads to holes in the colon that cause death. The mucus secreted in the large intestine normally protects against this.

REFERENCES

Bockus, H. L.: *Gastroenterology.* 2nd Ed. 3 volumes. Philadelphia, W. B. Saunders Company, 1963-1965.

Brooks, F. P.: The control of the secretion of pancreatic juice and bile. *Ann. Intern. Med.* 55:528, 1961.

Davenport, H. W.: *Physiology of the Digestive Tract.* Chicago, Year Book Medical Publishers, 1961.

Farrar, G. E., Jr., and Bower, R. J.: Gastric juice and secretion: Physiology and variations in disease. *Ann. Rev. Physiol.* 29:141, 1967.

Gregory, R. A.: *Secretory Mechanisms of the Gastrointestinal Tract.* Baltimore, The Williams & Wilkins Company, 1962.

Grossman, M. I.: The digestive system. *Ann. Rev. Physiol.* 25:165, 1963.

Jenkins, G. N.: *The Physiology of the Mouth.* 3rd Ed. Philadelphia, F. A. Davis Company, 1965.

Kerr, A. C.: *The Physiological Regulation of Salivary Secretion in Man.* New York, The Macmillan Company, 1960.

Magee, D. F.: *Gastrointestinal Physiology.* Springfield, Ill., Charles C Thomas, 1962.

Popper, H., and Schaffner, F,: *Liver: Structure and Function.* New York, McGraw-Hill Book Company, 1957.

Taylor, W. H.: Proteinases of the stomach in health and disease. *Physiol. Rev.* 42:519, 1962.

Truelove, S. C.: Movements of the large intestine. *Physiol. Rev.* 46(3):457, 1966.

CHAPTER 31

DIGESTION AND ASSIMILATION OF CARBOHYDRATES, FATS, AND PROTEINS

The term *digestion* means the splitting of large chemical compounds in the foods into simpler substances that can be used by the body. The term *assimilation* includes several functions that may be listed: (1) absorption of the digestive end-products into the body fluids, (2) transport of these to the cells where they will be used, and (3) chemical change of some of them into other substances that are specially needed for various purposes. The function of the digestive and assimilative processes is to provide nutrients for the chemical reactions of metabolism.

DIGESTION, ABSORPTION, AND DISTRIBUTION OF CARBOHYDRATES

Carbohydrates are composed of carbon, hydrogen, and oxygen. The basic unit of a carbohydrate is a *monosaccharide*, the most common of which in the food is *glucose*, and

which has the following chemical formula:

$$
\begin{array}{c}
CHO \\
| \\
H-C-OH \\
| \\
HO-C-H \\
| \\
H-C-OH \\
| \\
H-C-OH \\
| \\
CH_2OH
\end{array}
$$

Two other monosaccharides very frequently present in food are *fructose* and *galactose*, the formulas for which are the same as that of glucose except that some of the "H" and "OH" radicals are transposed.

Glucose and other monosaccharides are usually *polymerized* (combined together) into larger chemical compounds such as *starches*, *glycogens*, *pectins*, and *dextrins*. By far the most common carbohydrate of the diet is starch, which is a polymer of glucose. The glucose molecules in starches are joined together in the following manner:

It will be noted from this formula that the successive molecules of glucose are combined with each other by a *condensation* process, which means that one glucose molecule loses a hydrogen ion and the next loses a hydroxyl ion. The hydrogen and hydroxyl ions combine to form water, and the two glucose molecules connect together at the points where the ions were removed.

In addition to starches, another common source of carbohydrates is the disaccharides, which are combinations of only two molecules of monosaccharides. The common disaccharides in the diet are maltose, sucrose, and lactose. *Maltose* is a combination of two glucose molecules, and it is derived mainly by splitting starches into their disaccharide components, all of which are maltose. *Sucrose* is a combination of one molecule of glucose and one molecule of fructose. It is the same as cane sugar. *Lactose* is a combination of one molecule of glucose and one molecule of galactose. It is the sugar present in milk.

SCHEME OF DIGESTION OF CARBOHYDRATES. The schema below shows the digestion of the three most common carbohydrates in the diet, the starches, lactose, and sucrose. Starches and other large carbohydrates are digested principally by *ptyalin* in the saliva and *amylase* in the pancreatic juice, but perhaps to a slight extent also by *hydrochloric acid* in the stomach and *intestinal amylase* in the small intestine. The resulting product of these reactions is the disaccharide maltose.

The intestinal secretions contain the enzymes *maltase, lactase,* and *sucrase,* which split maltose, lactose, and sucrose into their respective monosaccharides. The resulting products of carbohydrate digestion are glucose, galactose, and fructose as shown by the schema. Because all the monosaccharides derived from maltose are glucose and half of those derived from the other two carbohydrates are glucose, it is evident that this substance is by far the most abundant end product of carbohydrate digestion. On the

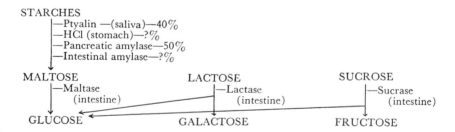

Carbohydrate digestion breaks the starch or other carbohydrate polymers into their component monosaccharides. To do this, one molecule of water must be added to the compound at each point where two successive monosaccharides are joined. This is a process of *hydrolysis,* which is opposite to the condensation process by which the successive monosaccharides are combined with each other. The secretions of the digestive tract contain enzymes that catalyze this hydrolysis process.

average, about 80 per cent of the monosaccharides formed by digestion is glucose, 10 per cent galactose, and 10 per cent fructose.

Absorption of Monosaccharides

ROUTE OF ABSORPTION. The monosaccharides are absorbed into the blood capillaries of many million small *villi* which are shown lining the intestinal wall in Figure

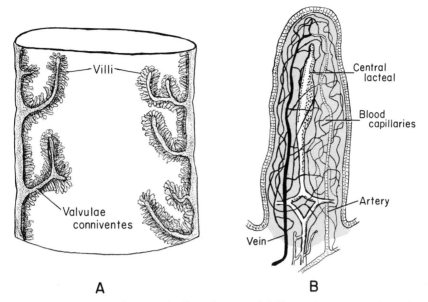

Figure 312. (A) Distribution of villi and mucosal folds on the inner surface of the small intestine. (B) Structure of a villus, showing the blood vessel system and the central lacteal.

312A, and the structure of which is shown in Figure 312B. The small vein from the villus empties into the portal venous system that flows into the liver, so that after absorption the monosaccharides must pass through the liver before entering the general circulation.

MECHANISM OF ABSORPTION. Monosaccharides are absorbed from the gastrointestinal tract by *active absorption*. The mechanism of this process is essentially the same as that for absorption from the tubules of the kidney, which was discussed in detail in Chapter 17. The monosaccharides combine with a carrier substance in the epithelial cells, and are transmitted in this combined form from the intestinal lumen to the opposite side of the cells, where they are then released from the carrier into the capillary blood. For this process to occur, energy must be expended by the epithelial cells, which is the reason why the process is called "active" absorption. Active absorption is very important because it allows absorption to occur even when monosaccharides are present in the intestine in extremely small concentrations, concentrations even smaller than those in the blood itself.

Fate of the Monosaccharides in the Body

GLUCOSE IN THE BLOOD AND EXTRACELLULAR FLUID—CONVERSION OF FRUCTOSE AND GALACTOSE TO GLUCOSE. Immediately after the monosaccharides are absorbed

from the gut, the glucose is rapidly transported throughout the body fluids in its present form. However, most of the fructose and galactose molecules are absorbed by the liver cells, converted into glucose, and then returned in this form to the blood, which means that these, too, are transported throughout the body in the form of glucose. Therefore, glucose is the basis of essentially all of the chemical reactions of carbohydrates in the body.

The concentration of glucose in the blood and extracellular fluid is approximately 90 mg. in each 100 ml., while the concentrations of fructose and galactose are usually very slight because of their rapid conversion to glucose.

TRANSPORT OF GLUCOSE THROUGH THE CELL MEMBRANE—EFFECT OF INSULIN. Before glucose can be used by the cells it must be transported through the cell membrane. Unfortunately, the pores of the cell membrane are too small to allow glucose to enter by the process of diffusion. Here again it must be transported by an active process, the general principles of which are shown in Figure 313. Glucose first combines with a carrier, believed to be a polypeptide, in the cell membrane. Then it is transported to the inside of the cell where it breaks away from the carrier.

In some way that has not yet been completely explained, the hormone *insulin* greatly enhances the transport of glucose through the cell membrane. Some possible

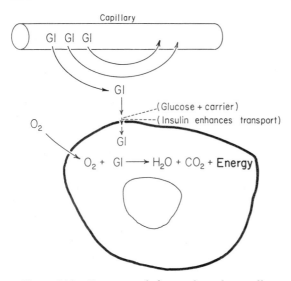

Figure 313. Transport of glucose from the capillary into the extracellular fluid and thence from the extracellular fluid into the cell to be utilized for energy.

ways in which insulin might do this are: (1) by catalyzing the reaction between glucose and the carrier, (2) by removing glucose from the carrier on the inside of the cell, or (3) by forming one of the components of the carrier mechanism itself. Regardless of which of these might be correct, the rate at which glucose can be transported through the cell membrane is determined mainly by the amount of insulin available. When the pancreas fails to secrete insulin, 5 to 10 times less glucose enters than the cell needs. When insulin is secreted in excessive abundance, glucose enters so rapidly that glucose metabolism becomes much greater than normal. It is obvious, therefore, that the rate of carbohydrate metabolism is regulated in accordance with the rate of insulin secretion by the pancreas.

REGULATION OF BLOOD GLUCOSE LEVEL. *Buffer effect of the liver.* After a meal, large quantities of monosaccharides are absorbed into the blood, and the glucose in the portal blood coming from the intestines rises from the normal level of 90 mg. per 100 ml. to as high as double this. However, this blood flows immediately into the liver, which removes about two thirds of the excess glucose before it can reach the general circulation. In this way the liver keeps the general blood concentration of glucose from ever rising above 120 to 130 mg. per 100 ml., despite very rapid absorption from the intestines.

The mechanism by which the liver removes the glucose from the portal blood is the following: Glucose is first absorbed through the cellular membranes into the liver cells under the influence of insulin. Then it is converted to *glycogen*, a polymer of glucose, and stored until a later time. When the blood glucose level falls to lower values several hours after a meal, the glycogen is split back into glucose, which is transferred out of the liver into the blood.

In essence, then, the liver is a buffer organ for blood glucose regulation, for it keeps the blood glucose level from rising too high and from falling too low.

Effect of insulin production by the pancreas. After a person eats a large meal, the rise in blood glucose concentration stimulates the pancreas to produce large quantities of insulin. The insulin in turn promotes rapid transport of glucose into the cells, thus decreasing the blood glucose level back toward normal. Therefore, in addition to the liver buffer mechanism, this pancreatic production of increased quantities of insulin also aids in preventing excessive rises in blood glucose concentration.

Effect of epinephrine, sympathetic stimulation, and glucagon. A low blood glucose level stimulates the sympathetic centers of the brain, causing secretion of norepinephrine and epinephrine by the adrenal glands and excitation of all the sympathetic nerves throughout the body. Also, the low glucose concentration directly stimulates the pancreas to secrete the hormone glucagon. This glucagon, the norepinephrine, epinephrine, and sympathetic stimulation all cause liver glycogen to split into glucose which is then emptied back into the blood. This returns the blood glucose concentration back toward normal, acting as a protective mechanism against low blood glucose levels.

GLUCONEOGENESIS. Another effect that occurs when the blood glucose level falls too low is the formation of glucose from proteins and to a much less extent from fats. This phenomenon is called *gluconeogenesis*. The importance of gluconeogenesis is that it provides glucose to the blood even during periods of starvation. Glucose, unfortunately, is not stored to a major extent in the body, for only 300 grams at most is stored in the form of glycogen in both the liver and all the remainder of the body, mainly the muscles. Ordinarily this amount is not sufficient by itself to maintain the blood glucose concentration at normal values for more than 24 hours. However, as the blood glucose level falls below normal, gluconeogenesis begins and continues until an adequate supply of glucose is available again.

Later in this chapter it will be noted that most of the cells of the body can utilize fats for energy when glucose is not available. However, the neurons of the brain are unable to utilize fats, and without an adequate supply of glucose these cells begin to die. This is the major reason why it is very important that the blood glucose concentration remain essentially normal even during long periods of starvation.

Energy from Glucose

Almost the only function of glucose in the body is to provide energy, though rarely is the glucose molecule used as a building stone for synthesis of other needed compounds. Energy is derived from glucose by two means: first, by splitting the molecule of glucose into smaller compounds, which liberates a small amount of energy, and, second, by splitting hydrogen ions away from the molecule and oxidizing these to form water, which liberates an extremely large amount of energy. These mechanisms of energy release are discussed in the following chapter.

DIGESTION, ABSORPTION, AND DISTRIBUTION OF NEUTRAL FATS

Neutral fats, like carbohydrates, are composed of carbon, hydrogen, and oxygen, though the relative abundance of oxygen in fats is considerably less than in carbohydrates. A representative molecule of neutral fat and its digestive end-products are the following:

It is evident from these formulas that a fat molecule contains two major components: first, a glycerol nucleus and, second, three fatty acid radicals. Each fatty acid radical is combined with the glycerol by a condensation process, which was noted previously as the means by which monosaccharides also combine with each other to form complicated carbohydrates.

The differences among various fats lie in the composition of the fatty acids in the molecule. Most fats in the human body have fatty acids with 16 or 18 carbon atoms in their chains. The fats containing the longer chain fatty acids are a little more solid than those containing the shorter fatty acids. Other than this, the chemical and physical properties of most fats do not vary significantly from one to the other.

Some of the fatty acids in the body and in the diet are *unsaturated*, which means that at various points in the carbon chain the atoms are bonded together by double bonds rather than single bonds, and that there is a corresponding lack of two hydrogen atoms. The unsaturated fats are needed to form a few special structures of the cells, but otherwise even these perform the same functions as the saturated fats in providing energy for the metabolic processes.

Digestion of Fats

The digestion of fats, like that of carbohydrates, is a *hydrolysis* process. It is catalyzed by enzymes called *lipases*, secreted in the stomach, pancreatic, and intestinal juices. The diagram below shows the complete schema of fat digestion.

$$\text{Fat} \xrightarrow{\text{(Gastric lipase + HCl)}} \begin{cases} \text{Fatty acids} \\ \text{Glycerol} \\ \text{Glycerides} \end{cases} \begin{array}{l} \text{not over} \\ \text{a few} \\ \text{per cent} \end{array}$$

$$\text{Fat} \xrightarrow{\text{(Bile + Agitation)}} \text{Emulsified fat}$$

$$\text{Emulsified Fat} \xrightarrow[\text{(Intestinal lipase)}]{\text{(Pancreatic lipase)}} \begin{cases} \text{Fatty acids} \\ \text{Glycerol} \\ \text{Glycerides} \end{cases} \begin{array}{l} 40\% \ (?) \\ \\ 55\% \ (?) \end{array}$$

Tristearin   Stearic acid   Glycerol

Though a minute quantity of fat is digested in the stomach, most of it is digested by the pancreatic and intestinal lipases after it enters the small intestine. The *bile salts* from the liver aid greatly in this digestion by acting as a detergent which allows the mixing movements of the intestines to break the fatty globules of the food into small emulsified globules, thereby providing far more surface area on which the water soluble lipases can act. In the absence of bile salts, most of the fat passes on into the feces in an undigested state.

The end products of fat digestion are *fatty acids, glycerol,* and *glycerides.* Glycerides are composed of a glycerol nucleus with one or two of the fatty acid chains still attached. Though the end result of complete fat digestion is the splitting of the fats entirely into fatty acids and glycerol, the process goes to completion in only about 40 per cent of the fat molecules, leaving many glycerides still among the digestive products.

Because glycerides pass through the intestinal membrane with almost the same ease as glycerol and fatty acids, the digestive process is quite adequate for absorption to occur. In fact, a very small amount of finely emulsified neutral fat that has not been digested at all can even be absorbed through the membrane.

Absorption of the End Products of Fat Digestion

The end products of fat digestion, like the monosaccharides, are also absorbed by the villi of the intestinal mucosa; but, unlike the monosaccharides, they are absorbed into the *central lacteal,* a lymphatic vessel, instead of into the blood. This is shown in Figure 312. As the glycerol, fatty acid, and glyceride molecules pass through the wall of the villus, they are reconverted into neutral fat. Thus, the digestion of fat into small chemical compounds allows it to *diffuse* through the intestinal wall, but, once through, the fat again becomes neutral fat as in the food.

Bile salts, in addition to their effect in helping to emulsify fat globules in the intestines, also aid the absorption of fats by performing a so-called *hydrotropic* action. This makes fatty acids and glycerides more soluble, allowing increased effectiveness of absorption.

Fat absorption, like glucose absorption, also occurs by an active process but in a different manner. It will be recalled that glucose is transported actively through the cell membrane and thence into the blood. Fatty acids and glycerides, on the other hand, because they are soluble in the cell membrane, simply diffuse into the mucosal cell. Then inside the cell, the fatty substances are actively absorbed into the endoplasmic reticulum of the cell, whence it is expelled into the interstitial fluid and thence into the central lacteals of the villi.

TRANSPORT OF FAT THROUGH THE LYMPHATICS. Lymph is "milked" from the central lacteals into the abdominal lymphatics by rhythmic contraction of the villi. This contraction is caused by a hormone, *villikinin,* which is released from the intestinal mucosa when fats are in the chyme. After leaving the central lacteals, the neutral fat is transported upward through the *thoracic duct,* the major lymphatic channel of the body, to empty into the blood circulation at the juncture of the internal jugular and subclavian veins.

Chylomicrons. The fat absorbed into the central lacteal aggregates into small fatty globules of about one micron in diameter called *chylomicrons.* These immediately adsorb proteins to their surface, which keeps them suspended in the lymph and keeps them from sticking to each other or to the walls of the lymphatics or blood vessels. It is in this form that fats are transported from the intestine, through the lymphatics, and into the blood.

After a fatty meal the level of chylomicrons in the circulating blood reaches a maximum in approximately two to four hours, sometimes becoming as much as 1 to 2 per cent of the blood, but within 2 to 3 hours almost all of them will have been deposited in the fat tissue of the body or in the liver.

Fat Tissue

Fat tissue is a special type of connective tissue that has been modified to allow the storage of neutral fat. It is found beneath the skin, between the muscles, between the various organs, and in almost all spaces not filled by other portions of the body. The cytoplasm of fat cells sometimes contains as much as 95 per cent neutral fat. These cells do not use the fat for any purpose other than to store it until it is needed elsewhere in the body.

Fat tissue provides a *buffer* function for fat in the circulating fluids. After a fatty meal the high concentration of fats in the blood is very soon lowered by deposition of the extra fat in the fat tissue. Then when the body needs fat for energy or other purposes it can be mobilized from the fat tissues and returned to the circulating blood. One of the means by which the body causes this release of fat is by secreting adrenocortical hormones, for these hormones increase the permeability of the fat cells and allow much of the fat to escape into the circulation. Because of this buffer function of the fat tissues, the fat in the fat cells is in a constant state of flux. Ordinarily half of it is removed every eight days, and new fat is deposited in its place.

Transport of Fats in the Body Fluids

NONESTERIFIED FATTY ACIDS. Most fat is transported in the blood from one part of the body to another in the form of *nonesterified fatty acids*, which are a combination of free fatty acids with albumin, one of the plasma proteins. Every fat depository of the body is well supplied with an enzyme called *lipoprotein lipase* which is capable of splitting the neutral fat of the tissues into glycerol and fatty acids. The fatty acids immediately combine with albumin in the blood and are then transported to other tissues of the body where they are released from the albumin. Some combine with glycerol in new fat depot areas to form new neutral fat. Others enter tissue cells where they are split into smaller molecules that are used to supply energy, as is explained below.

Concentration of nonesterified fatty acids. Even though almost all fat transport in the body is in the form of nonesterified fatty acids, these are normally present in the blood in a concentration of only about 10 milligrams per cent, or one-tenth the concentration of glucose. However, the fatty acids are transported to their destination within a few minutes and can therefore account for tremendous amounts of available energy to the cells. When fats are being used in great quantities by the cells, the blood concentration of fatty acids increases as much as four-fold or more.

LIPOPROTEINS. Lipoproteins are minute fatty particles covered by a layer of adsorbed protein. These are suspended in a colloid form in the plasma and to a less extent in other extracellular fluids. The *chylomicrons* are a type of lipoprotein because they are composed of lipid substances (neutral fat, phospholipids, and cholesterol) and a layer of adsorbed protein. However, in addition to the chylomicrons, large numbers of much smaller lipoprotein particles are also present in the blood. These are formed almost entirely in the liver, and their function is probably to transport neutral fat, phospholipids, and cholesterol from the liver to the different cells of the body.

Synthesis of Fat from Glucose and Proteins

Much of the fat in the body is not derived directly from the diet but instead is synthesized in the body. The fat cells themselves are capable of synthesizing a small amount of fat, but most of it is synthesized in the liver and then transported to the fat cells. Both glucose and amino acids derived from proteins can be converted into fat, but by far the most important source is glucose.

When an excess of glucose is in the diet and sufficient insulin is secreted by the pancreas to cause all the glucose to enter the cells, almost all the extra glucose not used immediately for energy passes into the liver and is converted into fat. This is then transported by the blood to the fat tissues and deposited. Thus, the fat tissues provide a means for storing energy derived from carbohydrates as well as from fats. This conversion of other foods to fats explains why eating any type of food, whether it be fat, carbohydrate, or protein, can increase the amount of fat tissue.

Functions of the Liver in the Utilization of Fat

The liver is undoubtedly the most important organ of the body for controlling fat utilization by the body. The liver, in addition to converting excess glucose into fat, converts fat into substances that can be used elsewhere in the body for special purposes. For example, before the tissue cells can make full use of the fats, some of the fats must be desaturated; some must be converted into the fatty substances, *cholesterol* and *phospholipids*, needed for cellular structures; and others are broken into smaller molecules that can be used easily by the cells for energy. The liver performs all these functions.

When the body is depending mainly on fats instead of glucose for energy, the quantity of neutral fat in the liver gradually increases. This is enhanced by adrenocortical hormones, for they cause the fat cells to mobilize their fat.

Energy from Fats

Fatty acids can be utilized for energy by almost all cells of the body with the exception of the neuronal cells of the brain. However, about 40 per cent of the fatty acids used for energy are first split in the liver into acetoacetic acid and then transported to the cells for use as explained in the following chapter.

The first stage in the utilization of fats for energy is to split the neutral fat into glycerol and fatty acids. The glycerol, being very similar to some of the breakdown products of glucose, can then be used for energy in very much the same manner as glucose. However, by far the major amount of energy in the fat molecule is in the fatty acid chains, and before these can be used for energy they must be split into smaller chemical compounds. Ordinarily, this is accomplished by a chemical process called *alternate oxidation* of the carbon chain, which is illustrated by the following reaction:

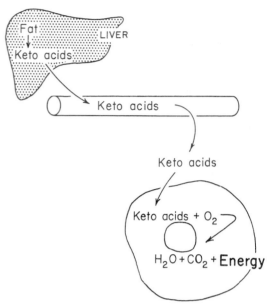

Figure 314. Formation of keto acids in the liver, their transport to the cells, and their utilization for energy.

The acetoacetic acid is called a *keto acid*, and it in turn can change into several other closely related froms of keto acids, or even into acetone.

The keto acids are highly diffusible through cellular membranes. Therefore, as shown in Figure 314, the keto acids formed in the liver diffuse immediately into the blood and are transported to all cells throughout the body. There the keto acids are oxidized in the same manner as acetic acid and

$$CH_3-\overset{\overset{O_2}{\downarrow}}{CH_2}-CH_2-\overset{\overset{O_2}{\downarrow}}{CH_2}-CH_2-\overset{\overset{O_2}{\downarrow}}{CH_2}-CH_2-\overset{\overset{O_2}{\downarrow}}{CH_2}-CH_2-\overset{\overset{O_2}{\downarrow}}{CH_2}-CH_2-\overset{O}{\overset{\|}{C}}-OH \longrightarrow$$

$$6CH_3-\overset{O}{\overset{\|}{C}}-OH$$

The net result is the formation of many molecules of *acetic acid*. This is then oxidized in the cells in a manner almost identical with the oxidation of glucose, giving tremendous amounts of energy to the cells, as will be explained in the following chapter.

In the liver, most of the acetic acid molecules formed from the breakdown of fatty acids condense two at a time to form *acetoacetic acid* in the following manner:

glucose to provide energy for cellular functions.

FAT-SPARING EFFECT OF CARBOHYDRATES. As long as sufficient glucose is available to supply the energy needs of the cells, it is burned in preference to fatty acids and the keto acids, and, if more than enough glucose is available, the excess is converted into fat. Thus, when glucose is available, the burning of fat stops; for this reason glucose is said to be a *fat sparer*.

$$CH_3-\overset{O}{\overset{\|}{C}}-\boxed{OH+H}CH_2-\overset{O}{\overset{\|}{C}}-OH \longrightarrow CH_3-\overset{O}{\overset{\|}{C}}-CH_2-\overset{O}{\overset{\|}{C}}-OH+H_2O$$

Phospholipids and Cholesterol

Two additional substances, *phospholipids* and *cholesterol*, which have some physical properties similar to those of neutral fats, are present in large quantities throughout the body. The phospholipids are composed of glycerol, fatty acid, and a phosphate side chain. Cholesterol is composed mainly of a sterol nucleus that is synthesized from acetic acid. The chemical structures of these substances are the following:

$$H_2C-O-\overset{\displaystyle O}{\overset{\|}{C}}-(CH_2)_7-CH=CH-(CH_2)_7-CH_3$$

$$HC-O-\overset{\displaystyle O}{\overset{\|}{C}}-(CH_2)_{16}-CH_3$$

$$H_2C-O-\overset{\displaystyle O}{\overset{\|}{P}}-O-CH_2-CH_2-NH_2$$
$$OH$$

A phospholipid

Cholesterol

Both phospholipids and cholesterol are fat soluble and are slightly water soluble. They are synthesized in all cells of the body, though to a much greater extent in the liver cells than in other cells. Both are transported in lipoproteins from the liver to other parts of the body.

Both phospholipids and cholesterol are major constituents of cell membranes and membranes of intracellular structures such as the nuclear membrane, the membranes of the endoplasmic reticulum, the membranes of mitochondria, of lysosomes, and so forth.

The precise functions of the phospholipids and cholesterol in the membranes is yet unknown, though they possibly play roles in the transport of substances through cellular membranes.

DIGESTION, ABSORPTION, AND DISTRIBUTION OF PROTEINS

Proteins are large molecules made up of many *amino acids* joined together. Amino acids, in turn, are small organic compounds that have an amino radical, $-NH_2$, and an acidic radical, $-COOH$, both on the same molecule. Twenty-three different amino acids are known to be present in the body proteins; the formulas of these are on page 381.

Thirteen of the amino acids can be synthesized in the body from other amino acids, but ten of them cannot. These ten are called *essential amino acids*, for they must be in the diet in order for the human body to form the proteins necessary for life.

Amino acids combine with each other to form proteins by means of *peptide linkages*, an example of which is illustrated by the equation that is shown at the bottom of this page. It will be noted that the product of the two combined acids, which is called a *peptide*, still has an amino radical and an acid radical, both of which can provide reactive points for combinations with additional amino acids. Most proteins contain several hundred to several thousand amino acids combined in this manner. The nature of the protein is determined by the types of amino acids in the protein and also by the pattern in which they are joined.

Digestion of Proteins

Referring to the preceding equation, it is evident that a peptide linkage is another example of condensation, which is the same means by which the component parts of fats and carbohydrates are combined. Therefore, the digestion of protein, like that of carbohydrates and fats, is accomplished by a process of hydrolysis.

The schema on page 380 shows the sequence of protein digestion, which begins with *pepsin* action in the stomach. Pepsin is

$$\underset{O}{\overset{NH_2}{R-CH-\overset{\|}{C}}}\overset{H}{-(OH+H)-\underset{R}{N}-CH-COOH}\longrightarrow \underset{O}{\overset{NH_2}{R-CH-\overset{\|}{C}}}\overset{H}{-\underset{R}{N}-CH-COOH}+H_2O$$

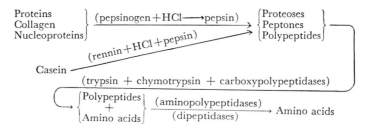

secreted into the stomach in the form of *pepsinogen,* a substance that has no digestive properties, but once it comes in contact with the hydrochloric acid of the stomach it is activated to form pepsin. The hydrochloric acid also provides an appropriate reactive medium for pepsin, for it can split proteins only in acid surroundings. Another enzyme called *gastricsin* is secreted along with pepsin and is responsible for a small part of the stomach's digestive function. However, since gastricsin is very similar to pepsin, it is considered here to be part of the pepsin system.

Protein is digested in the stomach into *proteoses, peptones,* and *polypeptides,* all of which are smaller combinations of amino acids than proteins — the proteoses are nearly as large as proteins, the peptones are intermediate in size, and the polypeptides are combinations of only a few amino acids. After entering the small intestine, these substances are further split by *trypsin* and *chymotrypsin* of the pancreatic juice into small polypeptides. Then the small polypeptides are finally split by *peptidases* of the pancreatic and intestinal juices into amino acids. Thus, the final products of protein digestion are the basic components of proteins, the amino acids.

Absorption of Amino Acids

Amino acids are absorbed from the gastrointestinal tract in almost exactly the same manner as monosaccharides, that is, by active processes. The precise nature of the active processes is not known, but it has been shown that poisons that prevent metabolism in the intestinal mucosa block almost all absorption of amino acids. This is true also of absorption of monosaccharides.

After absorption through the intestinal mucosa, the amino acids pass into the capillaries of the villi and thence into the portal blood, flowing through the liver before entering the general circulation.

Amino Acids in the Blood

All of the different amino acids are in the blood and extracellular fluid in small quantities. However, their total concentration is only about 30 mg. in each 100 ml. of fluid; or, to express this another way, the total concentration of all the 23 different amino acids together is only about one third that of glucose. The reason for this small concentration is that the amino acids, on coming in contact with cells, are absorbed very rapidly.

BUFFER ACTION OF THE LIVER AND TISSUE CELLS FOR REGULATING BLOOD AMINO ACID CONCENTRATION. The liver acts as a buffer for amino acids in the same manner that it acts as a buffer for glucose. When the blood concentration of amino acids rises very high, a large proportion of them is absorbed into the liver cells, where they can be stored, probably combined together to form small protein molecules. When the amino acid concentration in the blood falls below normal, the stored amino acids pass back out of the liver cells into the blood to be used as needed elsewhere in the body.

Most other cells of the body also have this ability to store amino acids to at least some extent and to release these into the blood when its content of amino acids falls. As a result, amino acids are in a state of continual flux from one part of the body to another. If the amount of amino acids in the cells of one area falls too low, then amino acids will enter these cells from the blood and will be replaced by amino acids released from other cells. This continual flux of the amino acids among the various cells is illustrated by the diagram of Figure 315.

Effect of adrenocortical hormones on amino acid flux. Recent research has shown that adrenocortical hormones aid in the mobilization of amino acids from one area of the body to another. Though the precise nature of this mobilization function of adrenocortical hormones is yet unknown, it

AMINO ACIDS

Glycine

$$H-\underset{\underset{NH_2}{|}}{\overset{\overset{H}{|}}{C}}-COOH$$

Alanine

$$H-\underset{\underset{H}{|}}{\overset{\overset{H}{|}}{C}}-\underset{\underset{NH_2}{|}}{\overset{\overset{H}{|}}{C}}-COOH$$

Serine

$$H-\underset{\underset{OH}{|}}{\overset{\overset{H}{|}}{C}}-\underset{\underset{NH_2}{|}}{\overset{\overset{H}{|}}{C}}-COOH$$

Cysteine

$$H-\underset{\underset{SH}{|}}{\overset{\overset{H}{|}}{C}}-\underset{\underset{NH_2}{|}}{\overset{\overset{H}{|}}{C}}-COOH$$

Aspartic Acid

$$\begin{array}{c} COOH \\ H-C-NH_2 \\ H-C-H \\ COOH \end{array}$$

Glutamic Acid

$$\begin{array}{c} COOH \\ H-C-NH_2 \\ H-C-H \\ H-C-H \\ COOH \end{array}$$

Hydroxylysine

$$H_2N-\underset{\underset{H}{|}}{\overset{\overset{H}{|}}{C}}-\underset{\underset{H}{|}}{\overset{\overset{OH}{|}}{C}}-\underset{\underset{H}{|}}{\overset{\overset{H}{|}}{C}}-\underset{\underset{NH_2}{|}}{\overset{\overset{H}{|}}{C}}-COOH$$

Cystine

$$\begin{array}{c} H-\overset{H}{C}-\overset{H}{C}-COOH \\ S \quad NH_2 \\ S \\ H-\overset{H}{C}-\overset{H}{C}-COOH \\ H \quad NH_2 \end{array}$$

Tyrosine

$$HO-\langle\rangle-\underset{\underset{H}{|}}{\overset{\overset{H}{|}}{C}}-\underset{\underset{NH_2}{|}}{\overset{\overset{H}{|}}{C}}-COOH$$

Diiodotyrosine

$$HO-\langle\rangle-\underset{\underset{H}{|}}{\overset{\overset{H}{|}}{C}}-\underset{\underset{NH_2}{|}}{\overset{\overset{H}{|}}{C}}-COOH$$

Thyroxine

$$HO-\langle\rangle-O-\langle\rangle-\underset{\underset{H}{|}}{\overset{\overset{H}{|}}{C}}-\underset{\underset{HN_2}{|}}{\overset{\overset{H}{|}}{C}}-COOH$$

Proline

$$\begin{array}{c} H_2C-CH_2 \\ H_2C \quad C-COOH \\ N \quad H \\ H \end{array}$$

Hydroxyproline

$$\begin{array}{c} H \\ HO-C-CH_2 \\ H_2C \quad C-COOH \\ N \\ H \end{array}$$

ESSENTIAL AMINO ACIDS

THREONINE

$$H-\underset{\underset{H}{|}}{\overset{\overset{H}{|}}{C}}-\underset{\underset{OH}{|}}{\overset{\overset{H}{|}}{C}}-\underset{\underset{H}{|}}{\overset{\overset{NH_2}{|}}{C}}-COOH$$

METHIONINE

$$CH_3-S-\underset{\underset{H}{|}}{\overset{\overset{H}{|}}{C}}-\underset{\underset{H}{|}}{\overset{\overset{H}{|}}{C}}-\underset{\underset{NH_2}{|}}{\overset{\overset{H}{|}}{C}}-COOH$$

VALINE

$$\begin{array}{c} H \\ H-C \\ H \\ \quad \overset{H}{\underset{NH_2}{C}}-COOH \\ H \\ H-C \\ H \end{array}$$

LEUCINE

$$\begin{array}{c} H \\ H-C \\ H \\ \quad \overset{H\;H}{\underset{NH_2}{C}}-COOH \\ H \\ H-C \\ H \end{array}$$

ISOLEUCINE

$$H-\underset{\underset{H}{|}}{\overset{\overset{H}{|}}{C}}-\underset{\underset{H}{|}}{\overset{\overset{H}{|}}{C}}-\underset{\underset{CH_3}{|}}{\overset{\overset{H}{|}}{C}}-\underset{\underset{NH_2}{|}}{\overset{\overset{H}{|}}{C}}-COOH$$

LYSINE

$$H-\underset{\underset{NH_2}{|}}{\overset{\overset{H}{|}}{C}}-\underset{\underset{H}{|}}{\overset{\overset{H}{|}}{C}}-\underset{\underset{H}{|}}{\overset{\overset{H}{|}}{C}}-\underset{\underset{H}{|}}{\overset{\overset{H}{|}}{C}}-\underset{\underset{NH_2}{|}}{\overset{\overset{H}{|}}{C}}-COOH$$

ARGININE

$$H_2N-\overset{NH}{C}-\underset{\underset{H}{|}}{\overset{\overset{H}{|}}{N}}-\underset{\underset{H}{|}}{\overset{\overset{H}{|}}{C}}-\underset{\underset{H}{|}}{\overset{\overset{H}{|}}{C}}-\underset{\underset{NH_2}{|}}{\overset{\overset{H}{|}}{C}}-COOH$$

PHENYLALANINE

$$\langle\rangle-\underset{\underset{H}{|}}{\overset{\overset{H}{|}}{C}}-\underset{\underset{NH_2}{|}}{\overset{\overset{H}{|}}{C}}-COOH$$

TRYPTOPHAN

$$\begin{array}{c} -\overset{H}{\underset{CH}{C}}-\overset{H}{\underset{H}{C}}-\overset{}{\underset{NH_2}{C}}-COOH \\ N \\ H \end{array}$$

HISTIDINE

$$\begin{array}{c} HC-N \\ \| \quad CH \\ C-N-H \\ H-C-H \\ H-C-NH_2 \\ COOH \end{array}$$

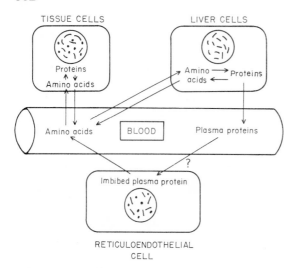

Figure 315. Flux of amino acids from one part of the body to another, and reversible equilibrium between the proteins of the different areas of the body.

is believed that they increase the rate of active transport of amino acids through the cellular membranes and thereby promote their rapid transfer from tissue to tissue. When an area of the body is damaged and is in need of amino acids for repair of its cells, the rate of adrenocortical secretion becomes greatly increased, and the resultant mobilization of amino acids supplies the needed materials.

The Tissue Proteins and Their Synthesis

The proteins of the cells perform two major functions. First, they provide most of the structural elements of the cells, and, second, they are the enzymes that control the different cellular chemical reactions. Therefore, the types of proteins in each cell determine its functions. Each cell is capable of synthesizing its own proteins, and this synthesis is controlled by the genes of the cell's nucleus in the manner described in Chapter 4. Basically, this process is the following:

REGULATION OF PROTEIN SYNTHESIS BY THE GENES. The nucleus of each cell of the human body contains 46 chromosomes arranged in 23 pairs. Each of these chromosomes contains many hundred *deoxyribose nucleic acid* molecules. It is believed that each of these molecules is a separate *gene*, and that its function is determined by its intrinsic chemical structure and also by its position in the chromosome thread.

Each gene of the nucleus controls the formation of a corresponding type of *ribose nucleic acid* that is transported to the cytoplasm of the cell. This has a slightly different chemical composition from the deoxyribose nucleic acid of the nuclear gene, and it in turn acts as a "template" to control the formation of a protein by the ribosomes in the cytoplasm. Since it is the proteins that perform the structural and enzymatic functions of the cell, the nuclear genes regulate in a roundabout way the entire function of the cell.

FORMATION OF PLASMA PROTEINS. The proteins in the plasma are of three different types: *albumin*, which provides the colloid osmotic pressure in the plasma; *globulins*, which provide the antibodies; and *fibrinogen*, which is used in the process of blood clotting. Almost all these are formed in the liver and then released into the blood, though a small portion, the globulins in particular, are formed by reticuloendothelial cells, plasma cells, and large lymphocytes. The plasma proteins are the same as many of the intracellular proteins except that they have been extruded into the circulating blood. The method of extrusion from the liver cells and plasma cells is unknown, but in the case of the large lymphocytes, it has been observed that the cellular membrane actually ruptures and empties its cytoplasmic contents into the lymph. Presumably some similar method is utilized by the liver and plasma cells.

Whenever the concentration of proteins in the plasma falls too low to maintain normal colloid osmotic pressure, the production of plasma proteins by the liver increases markedly. Though the means by which this is effected is unknown, it obviously is of great value in maintaining normal circulatory dynamics, for, if the colloid osmotic pressure of the blood should vary either above or below normal, the transfer of fluids through the capillary membranes into and out of the tissue spaces would become abnormal.

CONVERSION OF PROTEINS INTO AMINO ACIDS. All cells synthesize far more proteins than are absolutely necessary to maintain life of the cells. Therefore, if amino acids are needed elsewhere in the body, some of the cellular proteins can be reconverted into amino acids and then transported in this form. The reconversion process is catalyzed by enzymes, called *kathepsins*, that are in all cells, probably mainly in the lysosomes. The quantity of proteins in a cell is determined by a balance between their rate of synthesis and their rate of destruction. Even the plasma proteins circulating in the blood are subject to reconversion into amino acids, for they can be imbibed by reticuloendothelial cells and perhaps other cells and then split into amino acids by the intracellular enzymes.

The constant balance between amino acids and proteins in the cells, and between amino acids and the plasma proteins, is shown in Figure 315. By this constant interchange of amino acids, the proteins in all parts of the body are maintained in reasonable equilibrium with each other. If one tissue suffers loss of proteins or if the blood suffers loss of plasma proteins, many of the proteins in the remainder of the body will soon be converted into amino acids, which are transported to the appropriate point to form new protein. For example, in widespread cancer that is using extreme quantities of amino acids for formation of new cancer cells, the amino acids are derived continually from the tissue proteins, leading to serious debility. Also, when large quantities of blood are lost, the plasma proteins are replenished back to normal within approximately 5 to 7 days by the transfer of amino acids from the tissue proteins to the liver where new plasma proteins are formed.

Use of Amino Acids to Synthesize Needed Chemical Substances

Most metabolic reactions in the cells require special chemical substances to keep them operating, and most of these chemicals are synthesized from amino acids. For instance, the muscles require large quantities of *adenine* and *creatine* to cause muscular contraction, the blood cells require large amounts of *heme* to form hemoglobin, and the kidneys require large quantities of *glutamine* to be used in forming ammonia.

Also, many of the hormones secreted by the endocrine glands are synthesized from amino acids. These include epinephrine and thyroxine which are synthesized from tyrosine, histamine synthesized from histidine, several pituitary hormones that are themselves small proteins, parathyroid hormone which is a small protein, and insulin which is a very large polypeptide.

Derivation of Energy from Amino Acids

In addition to the use of amino acids for synthesizing new proteins or other chemical substances, some of them are also used for energy, as shown in Figure 316. The first step in using proteins for energy is to remove the amino radical. This process is called *deamination*, and it *occurs in the liver*. The removed amino radical is converted into ammonia, which in turn combines with carbon dioxide to form urea. This is excreted by the kidneys into the urine. Once again the liver is extremely important for one of the metabolic processes.

Referring back to the formulas of the amino acids, it will be evident that removal of the amino radical from certain of these acids still leaves very complicated chemical compounds. Some of these cannot then be utilized by the body because of their nature, and, therefore, are excreted directly into the gastrointestinal tract and lost in the feces. At least 60 per cent of the deaminated amino acids, however, have sufficiently simple chemical structures so that they can enter into the same cellular reactions as glucose and keto acids. These are often directly oxidized to form water and carbon dioxide, liberating energy in the process, or, if energy is not needed at the moment, they can be converted into fat or carbohydrate and later utilized for energy in the form of keto acids or glucose.

CONVERSION OF PROTEINS TO FATS OR CARBOHYDRATES. Ordinarily the body's cells must synthesize about 45 gm. of new proteins each day to replace the proteins being destroyed by the natural processes of wear and tear. If extra quantities of amino acids above the amount needed for this pur-

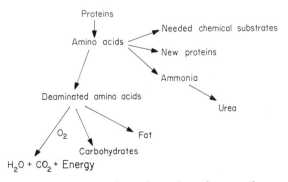

Figure 316. Complete schema for utilization of proteins in the body.

pose are eaten, these normally are deaminated and converted to fats or carbohydrates, or are used for energy. In other words, even normally, excess proteins are not stored in the body in the form of proteins.

PROTEIN-SPARING EFFECT OF CARBOHYDRATES AND FATS DURING STARVATION. When an insufficient quantity of protein is eaten, the major portion of the energy needed for the chemical processes of the cells, is derived from carbohydrates and fats as long as these are available. This is called the *protein-sparing* effect of these substances. However, when the stores of these have been greatly depleted, amino acids begin to be mobilized and deaminated to be used for energy. One can live for another few days on this energy derived from the proteins, but this process rapidly depletes the cells of their functional elements and soon leads to death.

ABSORPTION OF ELECTROLYTES AND WATER

ABSORPTION OF ELECTROLYTES. Electrolytes are absorbed from the gastrointestinal tract in almost exactly the same manner as from the tubules in the kidneys as described in Chapter 17. Sodium, for instance, is *actively absorbed;* that is, it combines with a carrier in the epithelial cells and is transported through the intestinal membrane in this form to be released on the opposite side into the blood.

Though less definitive experiments are available for absorption of other electrolytes from the gastrointestinal tract, it is believed that potassium, calcium, magnesium, phosphates, and iron are all also actively absorbed in a similar manner.

ABSORPTION OF WATER. Water is absorbed by *diffusion.* This means simply that random motion of the water molecules eventually carries them through the epithelial pores into the extracellular fluid.

Water absorption is controlled almost entirely by crystalloidal osmotic forces that operate as follows: When the monosaccharides, amino acids, and electrolytes are absorbed from the small intestine by active absorption, the crystalloidal osmotic pressure of the intestinal fluids becomes very slight because of the loss of the crystalloids. On the other hand, the crystalloidal osmotic pressure of the interstitial fluid on the opposite side of the epithelial membrane becomes increased. As a result an osmotic pressure gradient develops which causes water to be absorbed by osmosis from the intestinal lumen into the extracellular fluids. It is in this way that eight or more liters of gastrointestinal fluid are absorbed from the gastrointestinal tract each day.

REFERENCES

Awapara, J., and Simpson, J. W.: *Comparative Physiology: Metabolism.* Ann. Rev. Physiol. 29:87, 1967.

Booth, C. C.: Absorption from the small intestine. *Scient. Basis Med. Ann. Rev.* 171, 1963.

Boyer, P. D., Lardy, H., and Myrback, K. (eds.): *The Enzymes.* 2nd Ed. 8 volumes. New York, Academic Press, 1959–1963.

Jacobson, E. D.: The gastrointestinal circulation. *Ann. Rev. Physiol.* 30:133, 1968.

Kleiber, M.: *The Fire of Life: An Introduction to Animal Energetics.* New York, John Wiley & Sons, 1961.

Olson, R. E., and Vester, J. W.: Nutrition-endocrine interrelationships in the control of fat transport in man. *Physiol. Rev.* 40:677, 1960.

Stacey, M., and Barker, S. A.: *Carbohydrates of Living Tissues.* Princeton, N. J., Van Nostrand Company, 1961.

Tepperman, J.: *Metabolic and Endocrine Physiology.* Chicago, Year Book Medical Publishers, 1962.

Wilson, T. H.: *Intestinal Absorption.* Philadelphia, W. B. Saunders Company, 1962.

CHAPTER 32

RELEASE OF ENERGY FROM FOODS, AND NUTRITION

The major function of all the digestive and metabolic processes of the body is to provide energy for performing the various bodily functions. Energy is required to lift an arm, to move a leg, or to do any activity employing muscular contraction. It is needed for the secretion of digestive juices, the development of membrane potentials in nerves and other cells, the synthesis of new chemical compounds, and active absorption of substances from the gastrointestinal tract or kidney tubules. In short, almost all functions performed by the body require energy that in turn must be supplied by the ingested food. The digestion, absorption, and intermediary steps in preparing the food for energy release were discussed in the previous chapter. The final steps for release of energy from the foods—that is, the end stages of metabolism—are described in the present chapter.

ADENOSINE TRIPHOSPHATE AS THE COMMON PATHWAY OF ALMOST ALL ENERGY

The cells do not use the actual foods for their immediate supply of energy. Instead, they use almost entirely a chemical compound called *adenosine triphosphate* (ATP) for this energy. The foods, in turn, are then used to synthesize more adenosine triphosphate after it has been used. The importance of this compound to the function of the cell was pointed out in Chapter 3. In the present chapter is explained the role of adenosine triphosphate in the overall utilization of energy by the body.

The formula for adenosine triphosphate is shown at the bottom of the page. Extremely large amounts of energy are stored in this molecule at the *bonds* where the last two phosphate radicals join with the remainder of the molecule. These bonds ($\sim$) are called *high energy phosphate bonds.* Every time a cell needs energy a phosphate radical is broken away from adenosine triphosphate at the high energy bond, and this liberates the needed energy. Each mole of adenosine triphosphate releases 8000 calories of energy for each of the high energy bonds that is broken. In short, there is a storehouse of adenosine triphosphate in each cell that provides the necessary energy for muscular contraction, development of membrane potentials, active absorption, active secretion, synthesis of chemical compounds, and other functions performed by the cells, but this adenosine triphosphate must be replenished continually.

Formation of
Adenosine Triphosphate

USE OF ENERGY FROM CARBOHYDRATES TO FORM ADENOSINE TRIPHOSPHATE. In the preceding chapter it was noted that carbohydrates are digested to form glucose, or are changed into glucose after absorption. Then the glucose is used by the cells for energy. Part of the energy is released from glucose by a process called *glycolysis* that does not require oxygen, but by far the major amount of energy is released when the glucose is oxidized.

Glycolysis. In glycolysis, the glucose molecule, which has six carbon atoms, is split by a series of cellular enzymes into two smaller molecules having only three carbon atoms. Then the three-carbon-chain molecules are modified to form *pyruvic acid,* which has the following formula:

$$CH_3—\overset{\overset{\textstyle O}{\|}}{C}—COOH$$

During glycolysis a small amount of energy is released from the glucose molecule, and this energy is used to form adenosine triphosphate. By splitting glucose to form two pyruvic acid molecules, energy is liberated without the expenditure of any oxygen. This is illustrated by the first stage of the reaction in Figure 317.

Oxidative release of energy from carbohydrates. After the glucose has been split into pyruvic acid molecules, these can then be metabolized with oxygen to form carbon dioxide and water. This reaction is shown

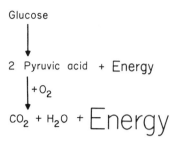

Figure 317. Derivation of energy from glucose by glycolysis and by oxidation.

by the second stage in Figure 317. The oxidative metabolism of pyruvic acid provides about eighteen times as much energy as the glycolytic breakdown of glucose to form pyruvic acid. Therefore, by far the major amount of energy liberated from carbohydrates for the performance of cellular function is derived from *oxidative metabolism.*

The precise chemical reactions by which pyruvic acid is split into smaller molecules and oxidized have been worked out in great detail, the general principles of which are shown in Figure 318. The reactions of the first stage, called the *tricarboxylic acid cycle* or *Krebs cycle* split the pyruvic acid molecule into carbon dioxide and hydrogen; the carbon dioxide is removed by enzymes called *decarboxylases*, and hydrogen atoms are removed by *dehydrogenases.* In the second stage, called *oxidation*, the hydrogen reacts with oxygen to form water.

When hydrogen atoms are split away from pyruvic acid by the dehydrogenases, they

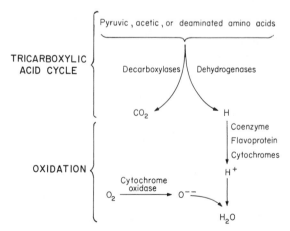

Figure 318. Splitting of pyruvic acid, acetic acid, or deaminated amino acids into carbon dioxide and hydrogen in the tricarboxylic acid cycle, and oxidation of the released hydrogen atoms by the cellular oxidative enzymes.

immediately combine with a substance called *coenzyme*. Then, under the influence of other enzymes, hydrogen atoms are passed to *flavoprotein molecules* and finally to *cytochrome molecules*. At this point the hydrogen atoms are released into the surrounding fluid as hydrogen ions. Simultaneously, dissolved oxygen that has been carried to the tissues by hemoglobin is changed into oxygen ions by *cytochrome oxidase*. Obviously, the presence of ionic hydrogen and ionic oxygen in the same solution provides two highly reactive substances that immediately form water molecules. Thus, the hydrogen atoms removed from the pyruvic acid become oxidized with oxygen to form water. During the various stages for oxidizing hydrogen, energy is released to form still more adenosine triphosphate molecules.

One might ask why it is necessary for the hydrogen and oxygen to go through the complicated stages of the above reactions, for it is well known that hydrogen and oxygen can combine with each other very rapidly simply by being burned together. The answer to this question is that the indirect procedure is required to channel the released energy in the proper direction to form new adenosine triphosphate.

The quantity of adenosine triphosphate formed in the different stages of glucose metabolism is the following: For each molecule of glucose metabolized, 2 molecules of adenosine triphosphate are formed during glycolysis, 2 are formed in the tricarboxylic cycle, and 34 are formed during oxidation of hydrogen, making a total of 38 molecules of adenosine triphosphate for each molecule of glucose metabolized.

The total amount of energy in each mole of glucose is 686,000 calories. Of this amount, 266,000 become stored in the form of ATP. The remainder is lost as heat caused by the chemical reactions. Thus, the overall *efficiency* of energy transfer from glucose to ATP is 39 per cent, the remaining 61 per cent of the energy becoming heat, which represents wasted energy.

USE OF ENERGY FROM FATS AND PROTEINS TO FORM ADENOSINE TRIPHOSPHATE. It was pointed out in the preceding chapter that fatty acids are split into acetic acid, and that amino acids derived from proteins are deaminated to form deaminated amino acids. Then the same decarboxylases and dehydrogenases that remove carbon dioxide and hydrogen from pyruvic acid do the same for the acetic or deaminated amino acids, and the hydrogen atoms are oxidized as explained above for the carbohydrates. Large amounts of energy are released, especially during the oxidation of the hydrogen atoms, to synthesize adenosine triphosphate.

The quantity of energy derived in this manner from fats and proteins represents about 80 per cent of the total energy derived by the cells in contrast to only 20 per cent from oxidation of glucose. The reason for this is that half or more of the carbohydrate eaten by a person is first stored in the body as fat and then later used for energy in the form of fatty acids.

REGULATION OF CELLULAR OXIDATION BY ADENOSINE DIPHOSPHATE (ADP). The oxidation reactions that result in the formation of adenosine triphosphate cannot occur unless adenosine diphosphate is available from which the adenosine triphosphate can be formed. Therefore, the oxidative breakdown of foodstuffs is controlled by the presence or absence of adenosine diphosphate. Every time adenosine triphosphate is used by the cells for energy it becomes adenosine diphosphate, and the newly formed adenosine diphosphate immediately initiates reactions with the foodstuffs to cause release of new energy that is used to convert the adenosine diphosphate back into adenosine triphosphate. These interrelationships are shown in Figure 319. Then, when all the adenosine diphosphate has been resynthesized into adenosine triphosphate, the synthetic processes cease.

Interaction of Adenosine Triphosphate with Phosphocreatine

Another substance that contains high energy phosphate bonds, phosphocreatine, is also present in the cells, in quantities several times as great as those of adenosine triphosphate. When adequate amounts of adenosine triphosphate are available, much of it is used to form phosphocreatine in accordance with the reactions shown in Figure 319, and as rapidly as it is used more adenosine triphosphate is formed. This results in the build-up of large quantities of both adenosine triphosphate and phosphocreatine. When the cell demands energy rapidly

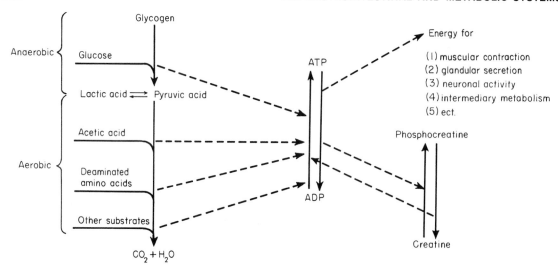

Figure 319. Overall schema for transfer of energy from the food to the functional elements of the cells.

and in large amounts, energy is released from adenosine triphosphate directly to the functional elements of the cells. Then the energy stored in the phosphocreatine is used immediately to reform new adenosine triphosphate. This reaction occurs much more rapidly than the oxidative reactions, and provides a very rapid source of extra energy that can be used to keep the cell functioning at a high rate of metabolism for a short period of time, even though the oxidative methods for reconstituting ATP are slow to respond.

AN OVERALL SCHEMA FOR THE ENERGY MECHANISMS OF THE CELL. Figure 319 illustrates an overall schema for the chemical reactions that provide energy for cellular functions. This schema shows that the breakdown of adenosine triphosphate to adenosine diphosphate releases the needed energy for muscular contraction, glandular secretion, neuronal activity, intermediary metabolism, and other energy functions of the cell. Within a few seconds, phosphocreatine also breaks down, providing energy for resynthesis of much of the adenosine triphosphate. During the ensuing minutes, the adenosine triphosphate and phosphocreatine are resynthesized by energy removed from the foodstuffs, part of it by glycolysis and other anaerobic procedures, but 90 per cent or more by the oxidation of pyruvic acid, acetic acid, deaminated amino acids, and a few other substances such as alcohol, glycerol, and lactic acid.

THE OXYGEN DEBT. For a few seconds at a time one can perform very strenuous feats of exercise requiring energy release many times that which can be sustained over a long period of time. This immediate burst of energy is provided mainly by the adenosine triphosphate and phosphocreatine stored in the cells. During the next few minutes, while the adenosine triphosphate and phosphocreatine are being resynthesized by the oxidative processes of metabolism, an extra quantity of oxygen must be utilized. Also, some of the stores of oxygen in the hemoglobin of the blood, in the myoglobin of muscle, and in the dissolved state in the body fluids will have been used during the rapid burst of energy, and this, too, must be replenished after the exercise is over. The extra oxygen that must be used to restore completely normal conditions after exercise is over is called the *oxygen debt*.

In essence, the ability to develop an oxygen debt explains why an athlete continues to breathe very hard for many minutes after running a race.

NUTRITION

The term *nutrition* means the supplying of foods that are required to keep one alive and healthy. These foods include carbohydrates and fats, which supply most of the body's energy, and proteins, vitamins, and minerals, which are required for synthesis of special structures and special chemical compounds needed by the body.

Foods Supplying Energy

Peoples in various parts of the world have widely differing diets, and some even obtain different proportions of their energy from the various types of foodstuffs in comparison with others. Approximately 45 per cent of the energy in the food of the average American diet is in the form of carbohydrate, 40 per cent fat, and 15 per cent protein. In other less prosperous parts of the world, the abundance of fat and protein in the diets is often less than half these values, while the energy derived from carbohydrate sometimes rises to as much as 80 per cent of the total.

THE CALORIE AS A MEASURE OF ENERGY. The energy of a food is measured in terms of the amount of heat liberated by complete breakdown of the food into end products, and this is expressed in *Calories*, a unit for measuring heat. One Calorie is the amount of heat required to raise the temperature of 1 kilogram of water 1 degree Centigrade.

The Calorie is a very good measure of food energy, for most of the energy released in the body eventually becomes heat anyway. For example, the chemical reactions for extracting energy from the foods are so inefficient that 61 per cent of the energy becomes heat as ATP is formed, and by the time the ATP is used to perform cellular functions 75 to 90 per cent of the energy becomes heat. The remaining 10 to 25 per cent is converted into muscular action or other functional activities of the body. Then after these functions are performed almost all of this remaining energy still becomes heat. As an example, a large amount of energy is used to pump blood around the circulatory system. Then, as the blood flows through the circulatory system, the energy that has been imparted to the blood is converted into heat because of friction between the blood and the walls of the vessels. In this way, all the energy expended by the heart eventually becomes heat. Likewise, almost all the energy expended by the skeletal muscles of the body eventually becomes heat, because most of it is used to overcome the friction of the joints and the viscosity of the tissues, and these two effects in turn convert the energy into heat.

ENERGY CONTENT OF THE DIFFERENT TYPES OF FOODS. The amount of energy released to the body by 1 gm. of each of the three types of foods respectively is the following:

	Calories
Carbohydrate	4.1
Fat	9.3
Protein	4.1

From the figures shown here it is evident that 1 gm. of fat supplies more than twice as much energy as 1 gm. of either carbohydrate or protein. For this reason, fats in the diet are very deceptive. Often one thinks that he is eating a relatively nonfat diet, and yet may be obtaining as much energy from fat as from carbohydrates. A second reason why fat is deceptive is that it often occurs in a pure form in foods, while carbohydrates and proteins are generally diluted several-fold with water. When one eats potatoes with butter, the fat of the butter generally contains almost as much energy as the entire potato, for two reasons: First, the butter supplies $2\frac{1}{4}$ times as much energy per gram as the starch of the potato, and, second, the starch comprises only about one-sixth the bulk of the potato, because most of the potato is water.

Protein Requirements of the Body

However low the concentration of amino acids in the blood, the liver still deaminates some of them all the time, and these either are excreted or are used for energy. Because of this continual loss of amino acids, the proteins of the body diminish constantly unless they are replenished by proteins in the diet. Normally, approximately 45 gm. of protein are required each day to replace this continual loss.

PARTIAL PROTEINS. Certain types of proteins do not contain all the essential amino acids in the proportion in which they are needed in the body. When one eats such a protein his requirement will be considerably above 45 gm. per day. For instance, the proteins from vegetables or grains are composed of ratios of the various amino acids different from those found in the human body. On the other hand, the proteins of animal origin, in general, have almost exactly the same amino acid compositions as those of the human being. Therefore, the person who requires 45 gm. of animal protein in his diet each day might require as much as 65 gm. of vegetable protein, the exact amount depending on the type of vegetable eaten. The proteins that cannot supply the right proportions of the different amino acids are called *partial proteins,* because they supply only part of all the needed amino acids.

Special Need for Unsaturated Fat in the Diet

A small amount of highly unsaturated fat is essential for nutrition of animals, and the body cannot desaturate fat sufficiently by its own metabolic processes to supply it. The types of unsaturated fatty acids usually available in the diet are *arachidonic, linoleic,* and *linolenic* acids. These are believed to be needed by the cells to form cellular structures. Without them animals develop skin sores, mental changes, and other evidences of general cellular debility. Whether or not these same effects would occur in the human being is not known, because such a small amount of these substances is required in the diet that no human being has ever been known to suffer from a deficiency of them.

Daily Energy Requirements

An average man of 70 kilograms who lies in bed all day long and does nothing else except eat and exist usually requires about 1850 Calories of energy each day. If he simply sits in a chair, another 200 or more Calories are required. Therefore, about 2000 Calories per day is the normal basal amount of energy required simply for living. In addition, any type of exercise requires still more energy, which is shown by the values in Table 9 for different types of activity. From this table it is evident that walking upstairs requires approximately 17 times as much energy as lying in bed asleep. However, this tremendous rate of energy utilization can be continued only for short periods of time. Over long periods, a well conditioned worker can average as much as 6000 to 8000 Calories of energy expenditure each 24 hours, or, in other words, as much as four times the basal rate.

Vitamin Requirements of the Body

The vitamins are chemical compounds needed in only minute quantities by the

TABLE 9. *Energy Expenditure per Hour During Various Types of Activity for a 70 Kg. Man (M. S. Rose)*

FORM OF ACTIVITY	CALORIES PER HOUR
Sleeping	65
Awake lying still	77
Sitting at rest	100
Standing relaxed	105
Dressing and undressing	118
Tailoring	135
Typewriting rapidly	140
"Light exercise"	170
Walking slowly (2.6 miles per hour)	200
Carpentry, metal working, industrial painting	240
"Active exercise"	290
"Severe exercise"	450
Sawing wood	480
Swimming	500
Running (5.3 miles per hour)	570
"Very severe exercise"	600
Walking very fast (5.3 miles per hour)	650
Walking up stairs	1100

body to perform special functions. The daily requirements of each of the vitamins is given in Table 10, and the vitamin content of different foods is given in Table 11. In general, eating a balanced diet will provide an adequate quantity of all the different vitamins. Occasionally, though, an abnormality makes it impossible to utilize one of the vitamins, in which case a vitamin deficiency disease can occur even in the presence of a normally satisfactory diet.

The precise physiologic functions of most

TABLE 10. *Daily Requirements of the Vitamins*

VITAMIN	DAILY REQUIREMENT
A	5000 units
Thiamin	1.8 mgm.
Riboflavin	1.8 mgm.
Niacin	18.0 mgm.
Ascorbic acid	75.0 mgm.
D (children and during pregnancy)	400 units
E	unknown
K	none
P	unknown
Folic acid	0.5 mgm. (?)
B_{12}	2.0 μgm. (?)
Inositol	unknown
Pyridoxine	2.0 mgm. (?)
Pantothenic acid	unknown
Biotin	unknown
Para-aminobenzoic acid	unknown

TABLE 11. *Vitamin Content of Various Foods (Modified from W. H. Eddy and G. Dalldorf)*

VITAMIN A UNITS/100 GM.	THIAMINE MGM./100 GM.	RIBOFLAVIN MGM./100 GM.	NIACIN MGM./100 GM.	VITAMIN C MGM./100 GM.
Apricots 5000	Barley 0.450	Dried beans 0.30	Asparagus 1.2	Asparagus 45
Broccoli 4000	Dried beans 0.540	Almonds 0.50	Barley 4.7	Brussel sprouts 95
Butter 3500	Buckwheat 0.450	Beef 0.20	Soybeans 4.0	Cabbage 65
Carrots 6000	Dried cowpeas 0.900	Cheese 0.55	Dried beans 3.0	Butter 70
Collards 8000	Egg yolk 0.320	Chicken 0.30	Beef 5.4	Cauliflower 82
Chard 9000	Whole wheat 0.585	Collards 0.25	Chicken 8.5	Lemon juice 53
Endive 15,000	Pork meat 1.0	Eggs 0.40	Collards 2.3	Mustard greens 115
Kale 20,000	Lamb meat 0.330	Lamb 0.24	Corn 1.3	Orange juice 45
Liver 70,000	Beef meat 0.120	Liver 2.60	Lamb 8.0	Fish 20
Mustard greens 10,000	Liver 0.400	Milk 0.18	Liver 17.0	Potatoes 10
Sweet potatoes 3000	Millet 0.700	Peanuts 0.45	Mackerel 2.1	Lean meats 5
Pumpkin 7000	Brown rice 0.370	Pork 0.24	Pork 7.0	Liver 25
Spinach 25,000	Wheat germ 2.0	Grains 0.10	Salmon 6.2	
			Wheat 4.3	

	PANTOTHENIC ACID MGM./100 GM.	PYRIDOXINE MGM./100 GM.	INOSITOL MGM./100 GM.	BIOTIN MGM./100 GM.	FOLIC ACID MGM./100 GM.
Meats	0.8	0.1	60	0.007	0.15
Eggs	1.4	0.02	33	0.009	0.90
Milk	0.3	0.13	18	0.005	0.005
Cereals	0.4	0.05	65	0.005	0.10
Fruits	0.2	0.05	95	0.003	0.05
Vegetables	0.5	0.15	70	0.007	0.08

of the vitamins in the body are not known, but from the physical effects caused by lack of them in the diet, their functions can at least be speculated upon as follows.

VITAMIN A. Vitamin A is used by the eyes to synthesize the light-sensitive retinal pigments used by the rods and cones for vision. This was discussed in detail in Chapter 24. Also, lack of vitamin A causes the epithelial structures, such as the skin, the intestinal mucosa, and the germinal epithelium of the ovaries and testes, to become highly *keratinized*, or horny. This leads to scaliness of the skin, failure of growth of young animals, failure of reproduction, and even hardening of the cornea, causing corneal opacity and blindness.

THIAMINE. Thiamine forms a compound in the cells called *thiamine pyrophosphate*, which is part of a decarboxylase that removes carbon dioxide from pyruvic acid and other substances. Without thiamine the oxidative processes for releasing energy from foodstuffs become deficient, and this can cause almost any type of abnormality in the body. It especially affects the functions of the nervous system, the heart, and the gastrointestinal system.

Lack of thiamine causes pathologic changes in neurons and in the myelin sheaths of nerve fibers, often resulting in actual destruction of the cells or irritation or degeneration of peripheral nerves. Frequently the fibers of peripheral nerves become so irritable that a condition called *polyneuritis* results, which manifests itself by excruciating pain along the course of the nerves.

In the heart, thiamine deficiency decreases the strength of the muscle. The heart becomes greatly dilated and pumps with little force, resulting in *congestive heart failure.*

In the gastrointestinal tract, thiamine deficiency causes weakness of the intestinal muscle, poor secretion of digestive juices, and poor maintenance of the intestinal mucosa. Severe indigestion, constipation, lack of appetite, and other symptoms very often develop.

A thiamine deficient person who has neuritis, enlargement of the heart, and gastrointestinal symptoms all at the same time is said to have *beriberi.*

NIACIN AND RIBOFLAVIN. Niacin and riboflavin are utilized in the body to form respectively coenzyme and flavoprotein. Earlier in the chapter it was pointed out that these two substances are required during the oxidation of hydrogen to combine with hydrogen atoms after they are removed from pyruvic acid and other substrates. Without niacin and riboflavin, therefore, the oxidation of foods becomes deficient, and the cell fails to receive adequate quantities of energy.

Niacin deficiency leads especially to a discoloration of the skin, which becomes darkened on exposure to sunlight. Along with this, severe muscular weakness, diarrhea, and one or more psychoses often occur. This is the clinical picture called *pellagra.*

Deficiency of riboflavin is not as likely to lead to such severe difficulties as deficiency of niacin, because other types of flavoproteins that can perform almost the same functions as the protein derived from riboflavin are present in the body. Nevertheless, riboflavin deficiency, when occurring along with deficiencies of niacin or thiamine, can greatly intensify the symptoms observed in pellagra or beriberi. A common result of riboflavin deficiency alone is cracking at the angles of the mouth, called *cheilosis.*

VITAMIN B₁₂ AND FOLIC ACID. These two substances are needed by the bone marrow to form red blood cells and are also needed in all the other tissues of the body for adequate growth. When they are lacking, the red cells released into the blood are few in number, and those that are released are considerably larger than normal, poorly formed, and very fragile. On studying the bone marrow one finds that the new red cells being formed have abnormal structures, for which reason it is believed that vitamin B_{12} and folic acid are necessary for the formation of the structural elements of the cells. However, lack of these vitamins does not affect the formation of hemoglobin.

Red cells are affected more than other cells of the body by lack of these two substances, probably because these cells are produced much more rapidly than most other cells. This is supported by the fact that cancer tissues are affected in very much the same way as red cells, for cancer cells also are produced much more rapidly than normal cells. In fact, purposeful deficiency of these two vitamins can actually be used as a means for slowing up the growth of some cancers.

LESS IMPORTANT VITAMIN B COMPOUNDS. For approximately the first 20 years after beriberi was discovered to be a nutritional disease, it was believed that the vitamin extract used to treat this disease contained only one single vitamin, and this was called *vitamin B.* Later, many different vitamins were discovered in the extract, and these have come to be known as the *vitamin B complex.* The vitamins included in this complex are thiamine (vitamin B₁), riboflavin (vitamin B₂), niacin, vitamin B₁₂, folic acid, *pyridoxine, pantothenic acid, biotin, inositol, choline,* and *para-aminobenzoic acid.* The functions of the latter six of these vitamins in the human being are not well known, but some of the functions of pyridoxine and pantothenic acid, the two most important of them, are the following:

Pyridoxine. Pyridoxine is needed for synthesis of some of the amino acids and perhaps other compounds. Dietary lack of this vitamin in lower animals can cause *dermatitis, decreased rate of growth,* development of a *fatty liver, anemia,* and different types of *mental symptoms.* The effects of lack in the human being are not too well known, because pyridoxine deficiency almost never occurs without simultaneous deficiency of other vitamins of the B complex.

Pantothenic acid. Pantothenic acid is used in the body to form a special chemical called *coenzyme A,* which catalyzes the acetylation of many substances in the cells. For example, pyruvic acid must be acetylated before it can be oxidized, and acetylcholine is formed from choline by the process of acetylation. Pantothenic acid deficiency has never been known to develop in the human being, because this substance is widespread in almost all foods of the diet. In animals, deficiency can be created artificially, and the results are retarded growth, failure of reproduction, graying of the hair, dermatitis, fatty liver, and many other effects. All these effects testify to the importance of pantothenic acid to the metabolism of the body.

ASCORBIC ACID (VITAMIN C). The major function of ascorbic acid is to maintain normal intercellular substances throughout the body, though the mechanisms by which this occurs are unknown. These include the connective tissue fibers that hold the cells together, the intercellular cement substance between the cells, the matrix of bone, dentin of the teeth, and other substances excreted by the cells into the intercellular spaces.

Deficiency of ascorbic acid in the diet causes failure of wounds to heal because of failure to deposit new fibers and new cement substance. Also, it causes bone growth to cease and blood vessels to become so fragile that they bleed on the slightest provocation. In short, lack of vitamin C leads to loss of integrity of many of the tissues of the body, and the resulting picture, characterized especially by bleeding of the gums, splotchy hemorrhages beneath the skin, and many other internal abnormalities, is the disease called *scurvy*.

VITAMIN D. Vitamin D promotes calcium and phosphate absorption from the gastrointestinal tract. Without adequate quantities of vitamin D the bones become deficient in both of these substances, and in very severe vitamin D deficiency the level of ionic calcium in the blood may fall so low that muscular tetany develops. All these effects of vitamin D deficiency will be discussed in relation to calcium metabolism and parathyroid hormone in Chapter 36.

VITAMIN E. In lower animals, lack of vitamin E can cause the male germinal epithelium in the testes to degenerate, causing sterility; in the female it can cause reabsorption of a fetus even after conception. For these reasons vitamin E is sometimes called the *antisterility* vitamin. However, deficiency of this vitamin in the human being has never been proved, and the precise chemical means by which vitamin E functions also has not been determined.

VITAMIN K. Vitamin K usually is not in the diet in large quantities. Yet the normal person does not experience vitamin K deficiency because a large amount of this substance is synthesized in the colon by bacteria and is then absorbed. If the bacteria of the colon are destroyed by administration of antibiotic drugs, vitamin K deficiency usually develops within the next few days.

Vitamin K is required for the formation of prothrombin by the liver. Therefore, deficiency of vitamin K depresses blood coagulation so that there is excessive bleeding. This subject was discussed in greater detail in Chapter 10.

Mineral Requirements of the Body

Table 12 gives the amounts of different minerals and some other substances in man, and Table 13 gives the daily requirements of minerals. Some of these minerals have already been discussed in relation to other phases of the physiology of man. For instance, sodium, chloride, and calcium are major constituents of the extracellular fluid, and potassium, phosphate, and magnesium are major constituents of the intracellular

TABLE 12. *Content in Grams of a 70 Kg. Adult Man*

Water	41,400	Mg	21
Fat	12,600	Cl	85
Protein	12,600	P	670
Carbohydrate	300	S	112
Na	63	Fe	3
K	150	I	0.014
Ca	1,160		

fluid. These minerals are responsible for development of electrical potentials at the cell membrane and for the maintenance of proper osmotic equilibria between the extracellular and intracellular fluids. In addition, calcium and phosphate are major constituents of bone, and phosphate forms a great number of different chemicals used for a myriad of functions inside all cells, some of which were discussed earlier in this chapter. The remaining minerals that need special comment are iron, iodine, cobalt, copper, zinc, and fluorine.

IRON. About two thirds of the iron of the body is in the hemoglobin in the blood; most of the remainder is stored in the liver in the form of *ferritin*. The iron can be mobilized when needed, and carried in the blood to the bone marrow, where it is used to form hemoglobin.

Iron is also present in some of the enzymes of the cells—especially in the cytochromes. Therefore, a second function of iron is to aid in the oxidation of foods in the tissue cells.

IODINE. Iodine is used by the thyroid gland in the synthesis of thyroxine, a hormone that increases the rate of metabolism of the body. The functions of thyroxine and its relation to iodine metabolism will be discussed in Chapter 34.

COPPER AND COBALT. Copper and cobalt both affect the formation of red blood cells. Copper, in some way yet unknown, helps to catalyze the formation of hemoglobin. Very rarely does copper deficiency exist,

TABLE 13. *Daily Requirements of Minerals*

Na	3.0 gm.	I	250.0 μgm.
K	2.5 gm.	Mg	unknown
Cl	2.5 gm.	Co	trace
Ca	1.0 gm.	Cu	trace
PO₄	1.5 gm.	Zn	trace
Fe	12.0 mgm.	F	trace

however, so that this is almost never a cause of hemoglobin deficiency. Cobalt is an essential element in vitamin B_{12}, and in this way is essential for maturation of red blood cells. When cobalt in other forms besides vitamin B_{12} is in the diet in excess, the bone marrow, for reasons not yet understood, produces far too many red blood cells, causing polycythemia.

ZINC. Zinc forms part of the structure of the enzyme *carbonic anhydrase,* which is present in many parts of the body, especially in the red blood cells and in the epithelium of the kidney tubules. This substance catalyzes the reaction of carbon dioxide and water to form carbonic acid and also catalyzes the same reaction in the reverse direction. It causes carbon dioxide to combine with water about 250 times as rapidly as it would otherwise. This allows the blood to transport carbon dioxide far more easily than could be possible in the absence of this enzyme.

In the kidney tubules, carbonic anhydrase catalyzes the reactions that cause the secretion of hydrogen ions into the tubular fluid, thereby helping to regulate the acid-base balance of the body fluids.

In addition to the utilization of zinc in carbonic anhydrase, insulin is stored in the pancreas in the form of a zinc compound.

FLUORINE. Fluorine in the diet protects against *carious* teeth. It does not make the teeth any stronger, but is believed to inactivate bacterial secretions that can cause tooth decay. Only a small trace of this substance in the drinking water of the growing child usually provides a major degree of protection against tooth decay throughout life, which is the reason why city water supplies are now usually fluorinated.

REGULATION OF FOOD INTAKE

Food intake is regulated by the sensations of *hunger* and *appetite.* Hunger means craving for food, and the term appetite is often used in the same sense as hunger except that it usually does not imply actual discomfort as is frequently the case with hunger. Also, appetite often means a desire for specific foods instead of for food in general.

The term *satiety* means the opposite of hunger, a feeling of complete fulfillment in the quest for food. Satiety results from a filling meal.

NEURAL CENTERS FOR REGULATION OF FOOD INTAKE. Stimulation of the *lateral hypothalamus* causes an animal to eat voraciously, while stimulation of the *medial nuclei of the hypothalamus* causes complete satiety even in the presence of highly appetizing food. Therefore, we can call the lateral hypothalamus the *hunger center* or the *feeding center,* while we can call the medial hypothalamus a *satiety center.*

In addition to the hypothalamic centers, however, the conscious centers of the cerebral cortex and the amygdaloid nuclei also enter into regulation of food intake—not so much into the regulation of total quantity of intake, but into the specific choice of food, for it is in these areas that memories of previously eaten pleasant or unpleasant foods are stored, and these memories adjust the appetite for different foods accordingly.

LONG-TERM AND SHORT-TERM REGULATION OF FOOD INTAKE. There are two entirely different types of food intake regulation, called respectively "long-term regulation" and "short-term" regulation.

Long-term regulation means regulation of food intake in relation to the amount of nutritive stores in the body. For instance, a person who has been underfed for many weeks has intensified hunger until his normal nutritive stores have been replenished. Conversely, an animal that has been force-fed until it is greatly overweight has almost no hunger, a condition that can last for weeks until its normal body weight has returned. The precise mechanism by which the nutritive stores affect hunger is not known, though it is known that a decreased level of glucose in the body fluids, which usually goes along with decreased stores of other nutrients in the body, increases a person's hunger. Presumably, decreased quantities of amino acids and nonesterified fatty acids in the body fluids cause the same effect, in this way controlling the degree of hunger.

Short-term regulation means regulation of dietary intake in relation to the amount of food that can be assimilated in a given period of time. For instance, if a person overeats, he can so overload his gastrointestinal tract that he will become sick from this alone. Therefore, during the process of eating, two principal mechanisms prevent such overeating. These are (1) "metering" of food as it passes through the mouth and (2) reflexes caused by distention of the upper gastrointestinal tract. Metering of food means that sensory receptors in the mouth and pharynx detect the amount of chewing, salivation, swallowing, and tasting and thereby quantitate the amount of food that passes through the mouth. In some way that is not understood, this information passes to the hypothalamic feeding center to inhibit hunger for up to 30 minutes or an hour, but no longer. Likewise, as food fills the stomach and other regions of the upper gastrointestinal tract, visceral sensory impulses, caused mainly by distention of the gut, are transmitted to the feeding center and inhibit it. In this way, overfilling of the gastrointestinal tract is prevented until the food that is already there has had time to be digested.

OBESITY. Much obesity is caused simply by overeating as a result of poor eating habits. For instance, many people eat three meals a day simply because of habit rather than because of hunger at the time of the meal.

Many instances of obesity also result from *inherited* imbalance between the hunger and satiety centers of the hypothalamus. For instance, when an exceedingly obese person forces himself to diet until he reduces to many pounds below his obese weight, he develops voracious hunger and, if left to his own means, will regain weight almost exactly to his original obese level. Once he reaches this level, his hunger becomes essentially the same as that of a normal person. Thereafter, he eats merely enough to maintain his weight rather than to gain still more. This is analogous to setting the thermostat in a house to a high level. In other words, the "hungerstat" is set at a higher level in these persons than in others, causing excessive eating until the person becomes very obese.

STARVATION. During starvation, essentially all the stored carbohydrates in the body, which amounts to less than 300 grams of glycogen in the liver and muscles, becomes used up within the first 12 to 24 hours. Thereafter, the person exists on his stored fats and proteins. For the first 2 to 4 weeks, almost all of the energy used by the body is derived from the stored fats. But eventually even these are almost depleted, so that finally the proteins must also be used. Most tissues can give up as much as one half of their proteins before cellular death begins. Therefore, for at least another few days to a week, the body can derive its energy from proteins. But, finally, death ensues because proteins are the necessary chemical elements for performance of cellular functions. This usually occurs 4 to 7 weeks after starvation begins.

REFERENCES

Anand, B. K.: Nervous regulation of food intake. *Physiol. Rev.* 41:677, 1961.

Astwood, E. B.: The heritage of corpulence. *Endocrinology* 71:337, 1962.

Brobeck, J. R.: Regulation of feeding and drinking. *Handbook of Physiology*, Sec. I, Vol. II, p. 1197. Baltimore, The Williams & Wilkins Company, 1960.

Ciba Foundation Study Group No. 11: *The Mechanism of Action of Water-Soluble Vitamins.* Boston, Little, Brown & Company, 1962.

Comar, C. L., and Bronner, F. (eds.): *Mineral Metabolism.* 2 volumes. New York, Academic Press, 1963.

Cooper, L. F., Barber, E. M., Mitchell, H. S., Rynbergen, H. J., and Greene, J. C.: *Nutrition in Health and Disease.* 14th Ed. Philadelphia, J. B. Lippincott Company, 1963.

Davidson, S., Meiklejohn, A. P., and Passmore, R.: *Human Nutrition and Dietetics.* 2nd Ed. Baltimore, The Williams & Wilkins Company, 1963.

Fisher, A. E.: Chemical stimulation of the brain. *Sci. Amer.* 210:60, 1964.

Mottram, V. H.: *Human Nutrition.* 2nd Ed. Baltimore, The Williams & Wilkins Company, 1963.

Wohl, M. G., and Goodhart, R. S.: *Modern Nutrition in Health and Disease.* 3rd Ed. Philadelphia, Lea & Febiger, 1964.

CHAPTER 33

BODY HEAT AND TEMPERATURE REGULATION

HEAT PRODUCTION IN THE BODY

METABOLIC RATE. In the preceding chapter it was pointed out that essentially all of the energy released from foods eventually becomes heat. Therefore, the rate of heat production by the body, which is called the *metabolic rate,* is actually a measure of the rate at which energy is released from foods. The metabolic expenditure of energy is measured in Calories, which is the same term used to express the amount of energy in foods. When a normal individual is in a very quiet state, the metabolic rate may be as low as 60 to 70 Calories per hour. On the other hand, it may be as high as 1000 to 2000 Calories per hour for a few minutes at a time, or as high as 200 to 300 Calories per hour for many hours at a time.

Factors that Affect the Metabolic Rate

Any factor that increases the rate of energy release from foods also increases the metabolic rate. Some of the more important of these are the following:

EXERCISE. Exercise is perhaps the most powerful stimulus for increasing the meta-bolic rate. When muscles contract, a tremendous quantity of adenosine triphosphate is degraded to adenosine diphosphate, and this then enhances the rate of oxidation of the foodstuffs. During very strenuous exercise lasting only a moment or more the metabolic rate can actually increase to as much as 40 times normal.

SYMPATHETIC STIMULATION AND NOREPINEPHRINE. When the sympathetic nervous system is stimulated, norepineph-rine is released directly into the tissues by the sympathetic nerve endings. Also, large quantities of this hormone and of epinephrine are released into the blood by the adrenal medullae. These two hormones then exert a direct effect on all cells to increase their metabolic rates. The exact means by which this is accomplished is not known, but it does enhance the breakdown of glycogen into glucose and also increases the rates of some of the enzymatic reactions that promote oxidation of foods.

Strong sympathetic stimulation can increase the metabolic rate to as much as 160 per cent of normal, but the metabolic rate remains elevated for only a few minutes after the sympathetic stimulation ceases. By controlling the activity of the sympathetics, the central nervous system has a means for regulating the rates of activity of all the cells of the body, increasing their activity when this is required and decreasing their activity when the need no longer exists.

THYROID HORMONE. Thyroid hormone has an action on all cells of the body similar to that of norepinephrine, except that the action of thyroid hormone continues for as long as four to eight weeks after its release from the thyroid gland, rather than for only a few minutes. Thyroid hormone secreted in very large amounts can increase the metabolic rate to as much as 200 per cent of normal, and complete lack of secretion by the thyroid gland causes the metabolic rate to fall to as low as 50 per cent of normal. In other words, the overall span from total lack of thyroid hormone to great excess can increase the metabolic rate as much as 4-fold.

The precise mechanism by which thyroid hormone affects the cells is not known, but it is known to increase the quantities of most of the cellular enzymes, which perhaps explains its metabolic effects. The actions of thyroid hormone will be discussed at further length in Chapter 34.

Most of the other endocrine hormones, except norepinephrine and thyroxine, have only minor effects on the overall metabolic rate, though insulin from the pancreas, growth hormone from the anterior pituitary gland, testosterone from the testes, and adrenocortical hormones can increase the general rate of metabolism as much as 5 to 15 per cent.

BODY TEMPERATURE. The higher the temperature of a chemically reactive medium, the more rapid are the chemical reactions. This effect is observed also in the chemical reactions of all cells. Each degree Centigrade increase in temperature increases the rate of any chemical reaction, in the body or out of the body, approximately 10 per cent. With very high fever the metabolic rate may be as much as twice normal because of the fever itself.

SPECIFIC DYNAMIC ACTION OF FOODS. After a meal, the metabolic rate usually rises and remains elevated for the ensuing 2 to 10 hours. In general, a meal containing large quantities of fats and carbohydrates will increase the metabolic rate only 4 to 15 per cent, with the increase lasting 3 to 6 hours. In contrast, a meal containing large quantities of proteins can increase the metabolic rate as much as 30 to 60 per cent, and the increase may last as long as 10 to 12 hours.

The specific dynamic action of foods is believed to be caused at least partially by the increased metabolism required for digestion, absorption, and assimilation of the foods. But in addition to this effect, proteins probably increase the metabolic rate an extra amount because of direct stimulatory effects of some of the amino acids or of some other breakdown products of proteins.

Basal Metabolic Rate

Because of the many different factors that can affect the metabolic rate, it is extremely difficult to compare metabolic rates from one person to another. To have any valid comparison at all, a person's rate of energy utilization must be measured when he is in a so-called *basal* state. This means that (1) he is not exercising and has not been exercising for at least 30 minutes to an hour, (2) he is at complete mental rest so that his sympathetic nervous system is not overactive, (3) the temperature of the air is completely comfortable so that his body temperature is not too high nor too low and also so that his sympathetic nervous system is not unduly stimulated, (4) he has not eaten any food within the last 12 hours that could cause a specific dynamic action on the metabolic rate, and (5) his body temperature is normal to avoid the effect of fever on metabolism.

Under these *basal conditions* most of the factors that affect the metabolic rate are controlled, but this basal state does not remove the effect of the constantly secreted thyroid hormone. Therefore, the *basal metabolic rate* in essence is determined by two major factors: first, the inherent rates of chemical reactions of the cells, and, second, the amount of thyroid hormone acting on the cells. Because the inherent rate of cellular activity is relatively constant from one person to another, an abnormal basal metabolic rate is usually caused by abnormal secretory rate of thyroid hormone. For this reason basal metabolic rates are often measured to assess the degree of thyroid activity.

DIRECT METHOD FOR MEASURING BASAL METABOLIC RATE. Since the basal metabolic rate is actually a measure of the amount of heat produced by the body under basal conditions, it can be determined by measuring the heat given off from the body in a known period of time. To do this, human subjects are placed in a large chamber called a *human calorimeter.* This chamber is cooled by water flowing through a radiator system. The heat given off from the body is picked up by the cooling system and then measured by appropriate physical apparatus. This is called the *direct* method for measuring the basal metabolic rate because it measures the heat output directly. Obviously, it is very cumbersome, but it has been a very important tool in experimental studies.

INDIRECT METHOD FOR MEASURING BASAL METABOLIC RATE. An indirect method for measuring the basal metabolic rate is based on the amount of oxygen burned by the body in a given period of time. From this the rate of energy release can be calculated. The amount of energy released when 1 liter of oxygen burns with carbohydrates is 5.05 Calories. When 1 liter of oxygen burns with fats the amount released is 4.70 Calories. And when 1 liter of oxygen burns with proteins the amount is 4.60 Calories. It is obvious from these figures that every time 1 liter of oxygen is burned in the metabolic fires of the body, essentially the same amount of energy is released regardless of which one of the three different foods is being used for energy. Therefore, it is reasonable to use an approximate average of these values, 4.825 Calories, as the amount of energy released in the body every time 1 liter of oxygen is burned. When calculating the basal metabolic rate in this way, the value obtained is never more than 4 per cent in error, even though a great excess of one type of food or the other might be used for metabo-

lism momentarily, and 4 per cent is far less than the error of measurement anyway. Therefore, to determine the amount of energy being generated in the body, one needs only to determine the amount of oxygen being used. This is done with a respirometer.

The respirometer. Figure 320 shows a respirometer. The subject places a mouthpiece in his mouth and breathes in and out of an inverted can that rides up and down in a water bath. The respirometer contains oxygen that is breathed back and forth into the lungs. The oxygen is gradually absorbed into the blood, and in its place carbon dioxide is expired. Soda lime in the respirometer reacts chemically with the carbon dioxide to remove it from the respiratory gases. The net result, therefore, is a continual loss of oxygen from the respirometer; this loss causes the inverted can to sink deeper and deeper into the water. The recording pen also falls lower and lower, giving a record of the rate of oxygen utilization.

METHOD FOR EXPRESSING THE BASAL METABOLIC RATE. The basal metabolic rate is generally expressed in terms of Calories per square meter of body surface area per hour. The reason for expressing it in terms of body surface area is to allow comparisons between individuals of different sizes. Experimental studies have shown that the basal metabolic rate varies from one normal person to another approximately in proportion to the body surface area and not in proportion to the weight. Thus, someone who weighs 200 pounds does not have a basal

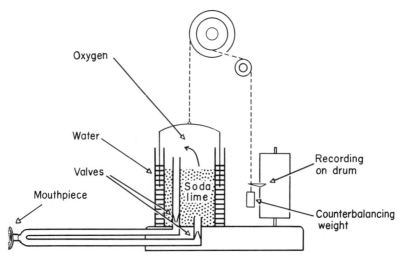

Figure 320.   A respirometer for measuring the rate of oxygen utilization.

metabolic rate twice that of someone weighing 100 pounds, but only about 30 per cent greater. His surface area also is about 30 per cent greater.

A representative calculation of basal metabolic rate might be the following: A person is found to use 1.8 liters of oxygen in six minutes. Therefore, in one hour he uses 18 liters of oxygen, and his total basal metabolic rate for this period would be 18 times 4.825, or 86.85 Calories per hour. To express this in Calories per square meter per hour one uses the chart in Figure 321A, which shows the square meters of surface area in relation to weight and height. If the person weighs 70 kg. and is 180 cm. tall, his surface area is 1.9 square meters. Therefore, his basal metabolic rate is 86.85 Calories per hour divided by 1.9, or 45.7 Calories per square meter per hour.

The basal metabolic rate is often also ex-pressed in per cent of normal. To do this one refers to the chart in Figure 321B, which shows the normal basal metabolic rate for males and females at different ages. If this person is a boy aged 18, his normal basal metabolic rate is 40 Calories per square meter per hour. However, his actual basal metabolic rate is 5.7 Calories greater than the normal value, or 14 per cent greater than normal. His basal metabolic rate, therefore, is stated to be +14. If his basal metabolic rate had been 25 per cent less than normal, it would have been expressed as −25, rather than plus.

When the thyroid gland is secreting an extreme quantity of thyroxine, the basal metabolic rate can at times go as high as +100. On the other hand, when the thyroid gland is secreting almost no thyroxine, the basal metabolic rate falls to as low as −40 to −50.

Figure 321. (A) Diagram for determining the body surface area when the weight and height are known. (From DuBois: *Basal Metabolism in Health and Disease.* Lea & Febiger.) (B) Normal basal metabolic rates for males and females at different ages.

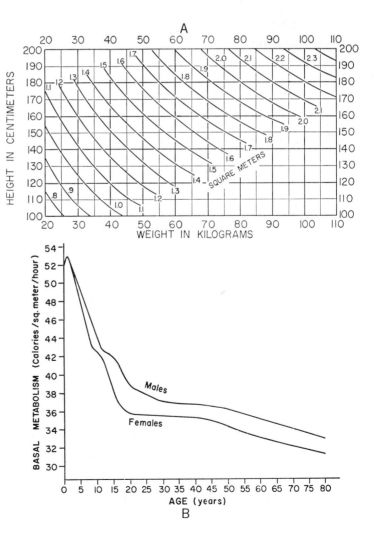

Major Loci of Heat Production in the Body

Though all the tissues of the body produce heat, those that have rapid chemical reactions produce the major amounts. In the resting state, the liver, heart, brain, and most of the endocrine glands produce large amounts of heat. This causes their temperatures to be a degree or so higher than that of most of the other tissues.

During rest the amount of heat produced by each skeletal muscle is not very great, but, because half of the entire mass of the body is composed of muscles, heat production of all the skeletal muscles together nevertheless accounts for about 40 per cent of all the body's heat production even at rest. A slight increase or decrease in the degree of muscular tone can have considerable effect on the amount of heat produced. During severe exercise the amount of heat produced by the muscles can rise to as great as 20 times that produced by all the remaining tissues together. For this reason, changes in the degree of muscular activity constitute one of the most important means by which the body regulates its temperature. This is discussed in detail later in this chapter.

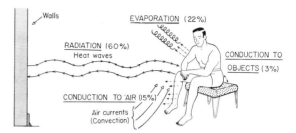

Figure 322. Methods by which heat is lost from the nude body. The percentage values are those for heat loss by each method when the person is in a room at 70° F.

radiates more heat to the walls than he receives from them.

The principle of radiation is used in radiant heating systems in commercial buildings and in many homes. The floors, ceilings, and walls of the rooms are heated to temperatures between 70 and 85° F. These are temperatures approximately comfortable for the human being, because they permit him to radiate heat at a rate that allows his body temperature to remain constant. A person is often comfortable in a radiantly heated room even though the air temperature may be in the 50's. This fact illustrates very forcefully how important radiation can be as a means of heat exchange.

HEAT LOSS FROM THE BODY

The heat produced in the body must be removed continually or otherwise the body temperature would continue rising indefinitely. The body loses heat to its surroundings in three ways: radiation, conduction, and evaporation.

Heat Loss by Radiation

About 60 per cent of the heat loss from a nude person sitting in a room at 70° F. is by radiation, as shown in Figure 322. Heat loss in this manner is based on the principle that objects near each other are always radiating heat toward one another. The human being radiates heat toward the walls, and the walls radiate heat toward him. However, since his body temperature is usually greater than that of the walls, he

Heat Loss by Conduction

Approximately 18 per cent of the heat lost by the nude person in Figure 322 leaves his body by conduction, 15 per cent by conduction to the air and 3 per cent to the floor and stool. These values are only average, for the colder the air and adjacent solid objects, the greater is the conduction of heat into them.

EFFECT OF CONVECTION CURRENTS ON LOSS OF HEAT BY CONDUCTION. Even though the temperature of the air remains constant at 70° F., the rate of conduction of heat into the air depends on how much the air is moving. If it is flowing past the body, every time the air adjacent to the body becomes warmed it is carried away to be replaced by cooler air. The more rapidly the air moves, the greater is the amount of heat conducted from the body. For this reason it is sometimes said that large quantities of heat are lost from the body by convection. This is not actually true; the heat is still being lost by conduction, though the convection currents carry the heated air away.

Heat Loss by Evaporation

A small amount of extracellular fluid continually diffuses through the skin and evaporates, and evaporation of each gram of water removes approximately one-half Calorie of heat from the body. Therefore, even normally, about 22 per cent of the heat formed in the body is removed by evaporation. Evaporation of only 150 ml. of water each hour would remove all the heat produced in the body under basal conditions. This shows how important evaporation can be as a cooling mechanism.

THE SWEAT MECHANISM. In addition to the continual diffusion of water through the skin, the sweat glands produce large quantities of sweat when the body is exposed to extreme heat. The sweat obviously increases the amount of heat that can be lost by evaporation. Under extreme conditions about one gallon of sweat can be secreted each hour, which can remove as much as 2000 Calories of heat from the body, or about 32 times the basal level of heat production.

NECESSITY FOR EVAPORATION IN TROPICAL CLIMATES. Evaporation as a means of heat loss is extremely important in tropical climates, for when the temperature of the air and of the surroundings rises above the temperature of the body, heat cannot be lost by either radiation or conduction. Instead, the body gains heat by these means. However, the evaporation of sweat can still keep the surface of the body at a temperature lower than that of the surroundings. Therefore, one's only means for maintaining normal body temperature in the tropics is by evaporation. Those few persons who are born without sweat glands must forever live in temperate or cold climates.

EFFECT OF CONVECTION ON EVAPORATION. Air currents have much the same effect on the removal of heat by evaporation as by conduction, for water evaporating from the skin quickly saturates the air immediately adjacent to the skin. If this air does not move away rapidly, the evaporation process will cease. However, if new air continually replaces the old, the air next to the skin never becomes saturated with moisture, and evaporation can continue unabated. This explains why in hot climates a fan is an important means for keeping cool; it also explains why one is usually much cooler outdoors under the shade of a tree, where the air is moving, than he is inside a house where the air is not moving, even though the air temperature may be the same.

Effect of Clothing on Heat Loss

Clothing is a barrier to the transfer of heat from the body to the surroundings. Heat radiated or conducted from the skin is absorbed by the inner surface of the clothing, but before it can be radiated or conducted to the surroundings it must pass through the mesh of the cloth. The inside of the clothing becomes warm in comparison with the outside; the inside warmth decreases the rate of conduction of heat from the body to the clothing and even radiates much of the heat back to the body.

The insulation properties of most clothing result mainly from air entrapped within the clothing mesh and not from the actual material comprising the clothing. A furred animal utilizes this principle in the winter time, for his hair grows long and stands on end, entrapping large amounts of air that become warm and act as a body insulator.

When clothing becomes wet, the spaces are filled with water, which conducts heat many hundred times as rapidly as does air. As a result, the clothing is no longer an adequate heat insulator; instead, heat is conducted almost as if the clothing did not exist. Therefore, wet clothing has almost no value in keeping the body warm in cold climates, and one of the most important lessons to be learned for arctic survival is always to keep the clothing dry whatever the circumstances.

EFFECT OF CLOTHING IN THE TROPICS. In the tropics, the effect of clothing on conduction and radiation usually is not important, because the temperature of the surroundings is close to that of the body. Instead, clothing must be designed for two other purposes: to protect one from direct exposure to the sun's heat rays and to allow maximum evaporation of sweat. Almost any clothing that blocks light rays also blocks the heat rays of the sun, but black clothing absorbs light rays and converts these into heat while white clothing reflects the rays. Also different types of clothing have vastly different effects on evaporation. Those types of clothing that rapidly absorb water, such as the cottons and the linens, usually do not depress the evaporative cooling of the body, because sweat is absorbed into the clothing and evaporated from its surface in the same manner as from the body. This is not true of wool and plastic materials, for they are not absorptive enough to allow satisfactory evaporative cooling.

THE BODY TEMPERATURE

The body temperature is determined by the balance between heat production and heat loss, as shown by Figure 323. If these two are exactly equal, the body temperature neither rises nor falls. When heat production is greater than heat loss, the body temperature rises; conversely, when heat loss is greater than heat production, the body temperature falls. Appropriate regulatory systems are always at work in the body to keep heat production and heat loss approximately equal, thereby maintaining a normal body temperature. These regulatory mechanisms are discussed as follows.

NORMAL VALUES OF BODY TEMPERATURE. The normal body temperature varies no more than a degree or so from one person to another. However, when a person is exposed to extremely cold or hot weather, his overall body temperature may vary as much as 1° F. Also, when he experiences extreme emotions that cause excessive stimulation of the sympathetic nervous system, the amount of heat production may become great enough

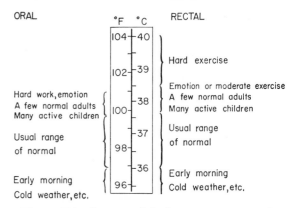

Figure 324. Ranges of body temperature under different normal conditions. (From DuBois: *Fever and the Regulation of Body Temperature*. Charles C Thomas.)

to raise the body temperature a degree or so. And, finally, extremely hard exercise can sometimes increase the body temperature as much as 5 to 6° F., though usually within 10 to 20 minutes after the exercise is over the temperature will have fallen back to that of the basal state.

Figure 324 depicts the normal temperature under different conditions as measured in the mouth and also as measured rectally. The average normal oral temperature is usually considered to be between 98.0 and 98.6° F., while the average normal rectal temperature is generally considered to be about 1° F. higher. The reason for the higher rectal temperature is that the mouth is continually cooled by the facial surfaces and by evaporation in the mouth and nose.

REGULATION OF BODY TEMPERATURE

Hypothalamic Regulation of Temperature

PREOPTIC TEMPERATURE SENSITIVE CENTERS. Nervous centers are capable of keeping the body temperature within the normal range essentially all the time. Located in the anteriormost portion of the hypothalamus, in the *preoptic area*, is a group of neurons that respond directly to temperature. When the temperature of the blood increases, the rates of discharge of these cells likewise increase. When the temperature decreases, the rates of discharge decrease.

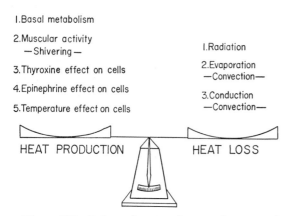

Figure 323. Balance between heat production and heat loss, illustrating that the body temperature remains constant as long as these two are equal.

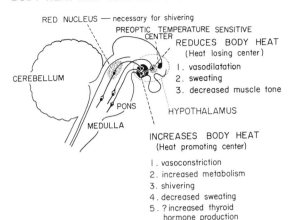

RED NUCLEUS — necessary for shivering

PREOPTIC TEMPERATURE SENSITIVE
CENTER
    REDUCES BODY HEAT
    (Heat losing center)

    1. vasodilatation
    2. sweating
    3. decreased muscle tone

CEREBELLUM

PONS

HYPOTHALAMUS

MEDULLA

INCREASES BODY HEAT
(Heat promoting center)

1. vasoconstriction
2. increased metabolism
3. shivering
4. decreased sweating
5. ? increased thyroid
   hormone production

Figure 325. The hypothalamic and brain stem mechanisms for regulating body temperature.

From this preoptic temperature sensitive area, signals radiate to various other portions of the hypothalamus to control either heat production or heat loss. In general, the hypothalamus can be divided into two major divisions: an anterior *heat losing center* which, when stimulated, reduces the body heat and a posterior *heat promoting center* which, when stimulated, increases the body heat. These areas are shown in Figure 325. The anterior center corresponds mainly with the parasympathetic centers, while the posterior center is principally sympathetic in nature.

HYPOTHALAMIC MECHANISMS FOR INCREASING THE BODY TEMPERATURE. Whenever blood colder than normal passes into the preoptic region of the hypothalamus, the posterior heat promoting center is strongly activated, and stimulation of this center automatically activates several different mechanisms to increase the body heat. These are the following:

Vasoconstriction. Stimulation of the heat promoting center acts through the sympathetics to cause the blood vessels of the skin to constrict very powerfully. This decreases the flow of warm blood from the internal structures to the skin, decreasing the transfer of heat to the body surface from those organs producing most of the heat. Very little heat diffuses directly from the internal structures to the body surface by means other than through the blood because of the fat beneath the skin which is a very adequate heat insulator. When vasoconstriction occurs, the temperature of the skin falls to approach the

temperature of the surroundings; this causes heat loss to be greatly diminished, allowing the internal body temperature to rise.

Increased metabolism. Stimulation of the sympathetic nerves releases norepinephrine throughout the body tissues and also causes both epinephrine and norepinephrine to be secreted into the blood by the adrenal medullae. These hormones in turn increase the rates of metabolism of all cells, thereby increasing heat production. This, too, tends to increase the body temperature.

Shivering. Stimulation of the heat promoting center increases the degree of wakefulness, and also causes transmission of large numbers of impulses into the bulboreticular formation and red nucleus of the hindbrain. Impulses passing through these regions increase the tone of all the muscles, which in turn increases the amount of heat produced by the muscles. When the tone becomes extremely great the stretch reflex begins to oscillate; that is, a slight movement stretches a muscle which elicits the stretch reflex and causes the muscle to contract. The contraction stretches the antagonistic muscle which also develops a stretch reflex. Its contraction then stretches the first muscle, and the cycle becomes repetitive so that continual shaking develops. All this muscular activity can increase the rate of heat production by several hundred per cent. Consequently, when the body is exposed to extreme cold, shivering is a very powerful force to maintain normal body temperature.

Pilo-erection. Pilo-erection means that hairs stand on end. This occurs when the sympathetic centers are stimulated, for the sympathetic nerves excite small pilo-erector muscles located at the bases of the hairs. In the human being this mechanism does not protect against heat loss because of the scarcity of hairs, but in lower animals pilo-erection entraps large quantities of air in the zone adjacent to the body and provides increased insulation against cold.

Increased thyroid hormone production. If the body is exposed to cold for several weeks, as at the beginning of winter, the thyroid gland begins to produce increased quantities of thyroid hormone. This is caused by formation of a neurosecretory hormone in the preoptic region of the hypothalamus (see Chapter 34) that passes in the blood to the anterior pituitary gland to increase its production of thyrotropic hormone that in turn excites the thyroid gland. The increase in thyroid hormone production over a period of weeks increases the rate of heat production 10 to 20 per cent and allows one to withstand the prolonged cold of wintertime.

HYPOTHALAMIC MECHANISMS FOR DECREASING THE BODY TEMPERATURE. When the temperature of the preoptic temperature sensitive region rises too high, the anterior hypothalamic heat losing center becomes stimulated, and because of reciprocal innervation between this center and the posterior hypothalamic heat promoting center, the latter is inhibited. As a result, all the mechanisms of the heat promoting center that tend to increase the body temperature become inoperative. For example, instead of vasoconstriction, the blood vessels to the skin dilate, allowing the skin to become very warm so that heat can be lost rapidly. Also, the increased metabolism caused by sympathetic stimulation ceases, and muscular tone greatly decreases. Also, the production of thyroid hormone diminishes gradually. The reversal of all of these effects allows the rate of heat loss to increase and the rate of heat production to decrease, thus making the body temperature fall.

In addition to reversing the heat promoting effects, stimulation of the heat losing center causes two effects of its own to decrease the temperature. These are sweating and panting.

Sweating. If the reversal of the heat promoting effects is not sufficient to bring the body temperature back to normal, the anterior hypothalamus initiates the sweating process. As much as a gallon of sweat can be poured onto the the surface of the skin in an hour, and under favorable conditions a large proportion of this will evaporate. By this means, a refrigeration mechanism is initiated to reduce the body temperature when it tends to rise too high.

Panting. In lower animals, though not in the human being, excessive stimulation of the hypothalamus by heat initiates a neurogenic mechanism in the pons of the hindbrain to cause panting. The animal breathes very rapidly but very shallowly, so that a tremendous quantity of air passes into and out of the respiratory passageways. Evaporation of water from the respiratory passageways is greatly increased, allowing a major amount of heat loss. Many lower animals—dogs, for instance—do not have well developed sweat mechanisms, and panting is often their only means of regulating body temperature in hot climates. Because of the shallowness of breathing during panting, most of the air entering the alveoli is dead space air, so that the alveolar ventilation of the animal remains essentially normal. This prevents overventilation despite the great turnover of air in the respiratory passageways.

SUMMARY OF HYPOTHALAMIC FUNCTIONS IN BODY TEMPERATURE REGULATION. The hypothalamus, when exposed to a temperature above normal, decreases heat production and increases heat loss. Conversely, when exposed to a temperature below normal, it increases heat production and decreases heat loss. The hypothalamus acts as a thermostat, maintaining the internal body temperature usually within one-half degree of the normal average value. The efficiency of the hypothalamus as a thermostat is indicated by the fact that the nude body can be exposed for many hours to dry air temperatures as low as 50° F. or as high as 170° F. without changing the internal body temperature more than 1 to 2 degrees.

Effect of Peripheral Reflexes on Body Heat

In addition to the hypothalamic mechanism for regulating body temperature, a few peripheral reflexes also aid in its regulation to a much lesser extent. Each area of the skin is supplied by spinal cord reflexes that dilate the blood vessels whenever the area is exposed to heat and constrict the vessels when the skin is exposed to cold. This explains why a hand placed in hot water soon becomes red, or a hand placed in cold water soon becomes blanched and white. These reflexes obviously aid in promoting heat loss from the skin in hot surroundings and in conserving heat in cold surroundings.

Despite this aid of the cord reflexes, a person whose spinal cord has been transected in the neck region, and who therefore has lost connections between the hypothalamus and the cord, is almost completely unable to regulate his body temperature. His hypothalamus is unable to transmit impulses into the sympathetics or into the fibers that control the sweat glands, and the bulboreticular formation and red nucleus cannot initiate the shivering mechanism. To keep his body temperature normal, the temperature of his surroundings must be maintained at a precise level and adjusted often to his body demands. Even a few minutes of exposure to the hot sun can increase his body temperature to almost lethal levels.

FEVER

Fever means a body temperature that is elevated beyond the normal range. It occurs in many disease states and, therefore, is of extreme importance in assessing the severity of a person's affliction. The most frequent cause of fever is severe bacterial or viral infection such as occurs in pneumonia, typhoid fever, tuberculosis, diphtheria, measles, yellow fever, mumps, poliomyelitis, and virus pneumonia. A less frequent cause of fever is the destruction of body tissues by some means other than infection. For instance, a person having a severe heart attack usually develops considerable fever, and a patient exposed to destruction of his tissues by x-ray or nuclear radiation may have fever for the ensuing few days.

Fever Caused by Resetting of the Hypothalamic Thermostat by Proteins

Fever is usually caused by abnormal proteins released into the body fluids during disease processes. These proteins have a direct effect on the hypothalamic thermostat to reset its normal operating range at a higher temperature level. For example, only one thousandth of a gram of protein derived from the bodies of typhoid bacilli can reset the level of the thermostat from normal up to as high as 110° F. When this happens, all the heat promoting mechanisms for increasing the body temperature become active and remain active until the body temperature reaches 110° F. Then the mechanisms for heat loss and heat production become equalized again, and the temperature continues to be regulated at this elevated level as long as the abnormal protein is present. In other words, the body temperature is still regulated

during fever, but it is regulated at a level considerably above normal because of resetting of the thermostat.

CHILLS AND SWEATING IN FEVERISH CONDITIONS. Figure 326 shows the effect of disease on the course of the body temperature. At the beginning of this chart, the oral temperature is 98.6° F., but after a few hours the disease suddenly sets the thermostat to a level of 103° F. During the ensuing hours, all the heat promoting mechanisms for increasing body temperature operate at full force. These include vasoconstriction, increased metabolism caused by norepinephrine secretion, and shivering. Even when the body temperature reaches 101° F., it still has not reached the setting of the thermostat, and the heat promoting phenomena continue to take place. Therefore, even though the temperature is high, the skin remains cold, and shivering occurs. This is called a *chill;* when someone is having a chill it is quite certain that his body temperature is actively rising.

After a number of hours the body temperature reaches the setting of the thermostat, and the chills disappear, but the temperature remains regulated at the high thermostatic setting until some factor breaks the disease process. In Figure 326 the disease process is corrected after another few hours, and the thermostat is suddenly set back to its normal value of 98.6° F. However, the body temperature is still high. This causes the heat losing mechanisms, including especially vasodilatation and sweating, to decrease body temperature. The person's skin suddenly becomes warm, and he begins to sweat, which effect is known as the *crisis.*

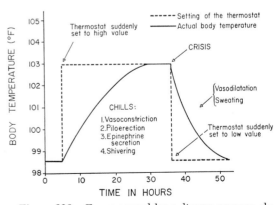

Figure 326. Fever caused by a disease process; development of chills when the temperature is rising and development of sweating when the temperature is falling.

When this happens one knows that the patient's temperature is beginning to fall. In the days before the miracle drugs, the sulfonamides and the antibiotics, many persons with bacterial diseases died, and the duty of the doctor was mainly to keep the sick person comfortable until he either died or a crisis appeared. If the crisis did appear, the doctor left the sickroom and announced to the family that all was now going to be well. The family in turn revered the doctor for his miraculous achievement. Unfortunately for the reputation of the present-day doctor, he too frequently cures the patient long before a crisis can develop.

POSSIBLE VALUE OF FEVER IN DISEASE. One wonders what the value of fever in disease might be. At present no clear-cut answer is available, but suggested values are: (1) Many bacterial and viral agents do not survive as well at high temperatures as at normal body temperatures. Therefore, elevation in temperature might well be a means for combating these infections. For example, high temperature has been shown to be especially lethal to gonococcal and syphilitic bacteria. (2) Because high rates of chemical reactions occur in the cells at high temperatures, it is possible that these increased rates allow the cells to repair the damage much more rapidly than they could otherwise.

Effect of Temperature Itself on the Body

When the body temperature rises above 106° to 110° F. it becomes very difficult, and often impossible, for the temperature regulatory mechanisms to return the body temperature to normal again. The main reason for this is that at these high temperatures the rates of cellular metabolism become so greatly increased because of the temperature itself that often no amount of regulation can overcome this very rapid rate of heat production. As a result, the body temperature rises still higher, the rate of heat production rises higher, and a vicious cycle develops, causing the body temperature to keep on rising.

Also, when the body temperature rises above 108° F. the metabolic rates of the cells become so great that the cells begin to "burn themselves out," and when the body temperature rises to 112 to 114° F. death almost always ensues simply because of the heat itself. The most damaging effect of a very high body temperature occurs in the neuronal cells of the brain, sometimes causing permanent destruction of these even though the person recovers.

When the body temperature falls far below normal, heat regulation again becomes impaired, but for opposite reasons. The low body temperature causes such slow rates of chemical reactions in the cells that no amount of regulation can bring the rate of heat production up high enough to return the body temperature to normal. The cold temperature decreases the rate of metabolism, and this allows the temperature to fall still lower, which decreases the rate of metabolism more, again creating a vicious cycle until the body temperature falls so low that the person dies. Usually death occurs when the body temperature reaches 70 to 75° F.

Body temperatures down to about 85° F. do not cause damage to the body, though the bodily functions become so greatly slowed that the person remains in a state of suspended animation until he is rewarmed.

REFERENCES

American Physiological Society: *Handbook of Physiology.* Sec. 4: Adaptation to Environment. Baltimore, The Williams & Wilkins Company, 1965.

Atkins, E.: Pathogenesis of fever. *Physiol. Rev.* 40:580, 1960.

Benzinger, T. H.: The human thermostat. *Sci. Amer.* 204(1):134, 1961.

Carlson, L. D.: Temperature. *Ann. Rev. Physiol.* 24:85, 1962.

Hammel, H. T.: Regulation of internal body temperature. *Ann. Rev. Physiol.* 30:641, 1968.

Hardy, J. D.: Physiology of temperature regulation. *Physiol. Rev.* 41:521, 1961.

Hardy, J. D. (ed.): *Temperature: Its Measurement and Control in Science and Industry.* Part 3: Biology Medicine Level. New York, Reinhold Publishing Corp., 1963.

Hemingway, A.: Shivering. *Physiol. Rev.* 43:397, 1963.

Leithead, C. S., and Lind, A. R.: *Heat Stress and Heat Disorders.* Philadelphia, F. A. Davis Company, 1964.

Pickering, G.: Fever and pyrogens. *Scient. Basis Med. Ann. Rev.* 97, 1961.

SECTION EIGHT

ENDOCRINOLOGY AND REPRODUCTION

CHAPTER 34

INTRODUCTION TO ENDOCRINOLOGY: THE HYPOPHYSEAL HORMONES AND THYROXINE

The Nature of Hormones and Their Function

A hormone is a chemical substance elaborated by one part of the body that controls or helps control some function elsewhere in the body. In general, hormones are divided into two types: first, the *local hormones*, which affect cells in the vicinity of the organ secreting the hormone, including such hormones as acetylcholine, histamine, and the gastrointestinal hormones, all of which have been discussed at different points in this text; and, second, the *general hormones*, which are emptied into the blood by endocrine glands and then flow throughout the entire circulation to affect cells and organs in far distant parts of the body. Some general hormones affect all cells almost equally; others affect specific cells far more than others. For example, growth hormone secreted by the pituitary gland and thyroxine secreted by the thyroid gland affect all cells of the body. On the other hand, the pituitary gland produces three gonadotropic hormones that affect the sex organs much more than other tissues, even though these hormones are secreted into the general circulation.

REGULATORY FUNCTIONS OF HORMONES. Some examples of the regulatory functions of the general hormones are: (1) The adenohypophyseal gland secretes six different hormones that regulate respectively the rate of growth of all tissues of the body, the rate of secretion of thyroid hormone by the thyroid gland, the rate of secretion of adrenocortical hormone by the adrenal gland, and the rates of secretion of several different sex hormones. (2) Thyroxine secreted by the thyroid gland controls the rate of metabolism of all cells. (3) Adrenal medullary hormones, which have been discussed in connection with the sympathetic nervous system, cause the same reactions throughout the body as stimulation of the general sympathetic nervous system. (4) Hormones secreted by the adrenal cortex regulate reabsorption of sodium by the kidneys, and also regulate some aspects of carbohydrate, fat, and protein metabolism. (5) Insulin secreted by the pancreas regulates the utilization of glucose throughout the body. (6) Hormones secreted by the testes control the sexual and reproductive functions of the male. (7) Hormones secreted by the ovaries control the sexual and reproductive functions of the female. (8) Hormones secreted by the parathyroid glands regulate the concentration of calcium in the body fluids.

In the present chapter and the following two chapters, the functions of the general hormones are considered in detail. The study of these and their functions is called *endocrinology,* and the glands that secrete the general hormones are called *endocrine glands.* The word *endocrine* means secretion to the interior of the body as opposed to secretion to the exterior as is true of the intestinal glands, the sweat glands, and others, all of which are called *exocrine glands.*

THE HYPOPHYSEAL HORMONES

The *hypophysis,* shown in Figure 327 and known also as the pituitary gland, is about the size of the tip of the little finger, and it lies in a small bony cavity beneath the base of the brain. It is divided into two completely separate parts, the *neurohypophysis,* known also as the posterior pituitary gland, which is connected by a stalk with the hypothalamus of the brain, and the *adenohypophysis,* called also the anterior pituitary gland, which lies anterior to the neurohypophysis and has functions not related to those of the neurohypophysis.

The Adenohypophyseal Hormones

The adenohypophysis secretes at least six different hormones, and various research workers have postulated as many as thirty more that might be produced by this gland. However, only the six have thus far proved to have major significance, for which reason it is believed that most of the other postulated hormones have really been combinations of these six, which are: *growth hormone, thyrotropin, corticotropin, follicle stimulating hormone, luteinizing hormone,* and *luteotropic hormone* (Fig. 328). The last three of these are called *gonadotropic hormones* because they regulate the functions of the sex glands.

The adenohypophysis is composed of three different types of cells: *chromophobe cells, basophilic cells,* and *acidophilic cells.* Most of the hormones are believed to be secreted by the basophils and acidophils, and the chromophobes seem to be only developmental stages of the other two types of cells.

GROWTH HORMONE. Growth hormone is secreted by the adenohypophysis in large quantities throughout life even though most growth in the body stops at adolescence. At this time the production of growth hormone diminishes but does not stop. The function of growth hormone during the

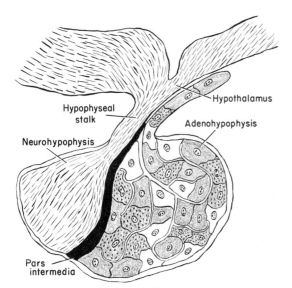

Figure 327. The hypophysis.

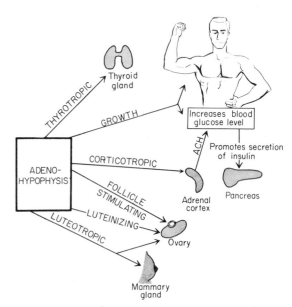

Figure 328. Adenohypophyseal hormones and their functions.

growing phase of an animal's life is to promote development and enlargement of all bodily tissues. It increases the sizes of all bodily cells and also their number. Consequently, each tissue and each organ become larger under the influence of growth hormone. The bones enlarge and lengthen, the skin thickens, and soft tissues, including the heart, liver, tongue, and all the other internal organs, increase in their size. In other words, growth hormone is exactly what its name implies; it causes the person to grow.

Growth hormone promotes growth by increasing the rates of synthesis of proteins and of other cellular elements, thereby increasing the sizes of the cellular structures. Also, the metabolic rates of the cells are increased about 15 per cent when normal amounts of growth hormone are secreted.

Function of growth hormone after adolescence. Growth hormone continues to function after adolescence as before, that is, to promote protein synthesis and formation of other cellular elements. However, by this age, most of the body's bones have grown as much as they can, so that continued growth is confined mainly to soft tissues such as the tongue, skin, liver, and so forth. In addition, the bones of the lower jaw and nose also continue to grow, increasing the prominence of these structures with age.

Basic cellular mechanism for growth hormone function. We still do not know the basic mechanism by which growth hormone causes increased synthesis of proteins and other cellular elements. However, two different theories are the following: First, it is postulated that growth hormone promotes transport of amino acids from the interstitial fluids into cells and that this excess of amino acids then promotes formation of protein enzymes and protein structures. There is much reason to believe this theory to be correct, for experiments have demonstrated that growth hormone does promote entrance of most amino acids into cells, though it does not have this effect on all amino acids.

The second theory is that growth hormone in some way activates specific genes of the cells, which in turn causes generalized growth of the cell. This theory also has much in its favor. For instance, growth hormone can cause growth of the long bones during childhood even in the face of a deficit of amino acids in the diet.

Control of growth hormone secretion.

The rate of growth hormone secretion increases and decreases markedly from day to day, which is quite contrary to the earlier belief that its secretion was constant throughout the years of growth. One situation that causes marked increase in growth hormone secretion is generalized nutritional deficiency. Therefore, growth hormone might be an important regulator of certain types of metabolic reactions, especially those that relate to protein formation. It is guessed that depressed protein formation might trigger growth hormone secretion, though such a supposition is still highly hypothetical.

Dwarfs. A person whose adenohypophysis fails to secrete growth hormone fails to grow. This causes the so-called "pituitary" type of dwarf (though most dwarfs have small stature because of their heredity rather than because of adenohypophyseal failure). The pituitary type of dwarf remains childlike in all physical respects. His organs as well as bones fail to grow, and his degree of sexual development remains that of a young child even after he has reached the age of an adult. He may never grow to more than twice the height of a newborn baby.

Giants. Secretion of excess growth hormone, if it occurs before adolescence, can cause a giant. After adolescence the growing parts of most bones have "fused" and are no longer capable of growing longer regardless of the amount of growth hormone available. Giantism usually results from a tumor of the acidophilic cells of the pituitary gland, this tumor secreting tremendous quantities of growth hormone.

Acromegaly. If an acidophilic tumor develops after adolescence, the excessive secretion of growth hormone cannot cause increased height; nevertheless, it can still cause increased size of the soft tissues. It can also cause thickening of the bones, and a few of the bones, the so-called "membranous bones," can even continue to grow longer. As a result, disproportionate growth occurs in different parts of the body, resulting in the condition called *acromegaly.* In acromegaly the lower jaw, which is formed of membranous bone, grows excessively long, causing the chin sometimes to protrude a half inch or more beyond the remainder of the face. The nose and lips also enlarge, and the internal organs such as the tongue, the liver, and the various glands enlarge. The bones of the feet and hands, unlike most bones of the body, continue to grow, so that the hands and feet become very large. In short, an acromegalic person is the same as a giant, except that failure of his already "fused" long bones to grow causes his height to be normal while other features of his body exhibit excessive growth.

THYROTROPIN. Thyrotropin, also known as *thyrotropic hormone,* controls the amount of hormone secreted by the thyroid gland. It increases the number of thyroid cells, the size of these cells, and also their rate of thyroxine production. When the adenohypophysis fails to secrete thyrotropin, the thyroid gland becomes so incapacitated that it secretes almost no measurable amount of hormone. In other words, the thyroid gland is almost completely controlled by thyrotropin. Further relationships of the thyroid gland to thyrotropin are discussed later in the chapter in connection with thyroxine.

CORTICOTROPIN. A third hormone secreted by the adenohypophysis, *corticotropin,* also known as *adrenocorticotropic hormone* or *ACTH,* controls secretion of adrenocortical hormones by the adrenal cortices in much the same manner as thyrotropin controls secretion by the thyroid gland. Corticotropin increases both the number of cells in the adrenal cortex and their degree of activity, resulting in an increased output of adrenocortical hormones. The relationship of corticotropin to the different adrenocortical hormones is discussed in the following chapter.

THE GONADOTROPIC HORMONES. The functions of the three gonadotropic hormones are summarized briefly in the following paragraphs, but their relationship to sexual function will be presented in Chapter 37.

Follicle stimulating hormone. In the female, follicle stimulating hormone initiates growth of *follicles* (fluid chambers) in the ovaries. In each of these a single ovum develops in preparation for fertilization. This hormone also helps to cause the ovaries to secrete *estrogens,* one of the female sex hormones. In the male, follicle stimulating hormone stimulates growth of the germinal epithelium in the testes, thus promoting the development of sperm that can then fertilize the female ovum.

Luteinizing hormone. In the female, luteinizing hormone joins with follicle stimu-lating hormone to cause estrogen secretion. It also causes the follicle to rupture, allowing the ovum to pass into the abdominal cavity and then through a fallopian tube, within which time fertilization may take place. And it causes the corpus luteum to secrete *progesterone.* In the male, luteinizing hormone causes the testes to secrete the male sex hormone *testosterone.*

Luteotropic hormone. In the female, luteotropic hormone, acting in concert with luteinizing hormone, causes the cells of the ruptured follicles to secrete progesterone as well as additional estrogens, which are necessary to prepare the uterine surface for implantation of a fertilized ovum. The function of luteotropic hormone in the male is yet unknown, though it probably has some important effect on either the production of male sex hormone or the production of sperm.

REGULATION OF ADENOHYPOPHYSEAL SECRETION—THE HYPOTHALAMIC-HYPO-PHYSEAL PORTAL SYSTEM. The adenohypophysis is a highly vascular organ which receives blood supply from two sources: (1) the usual arteriolar source and (2) the so-called *hypothalamic-hypophyseal portal system* which is illustrated in Figure 329. The hypothalamus, a part of the brain lying immediately beneath the thalamus, controls many of the automatic functions of the body. After the blood passes through the capillaries of the hypothalamus, particularly through the lower part called the *median eminence,* it leaves by way of small *hypothalamic-hypophyseal portal veins* which course down the anterior surface of the hypophyseal stalk to the adenohypophysis. At this point the veins dip into the gland where the blood flows through large numbers of *venous sinuses* that bathe the adenohypophyseal cells.

The hypothalamus probably secretes a *neurosecretory substance* for each of the major hormones formed in the adenohypophysis. It is these neurosecretory substances passing to the adenohypophysis by way of the portal system that control secretion by the adenohypophyseal cells. For instance, the neurosecretory substance *thyrotropin releasing factor* causes the adenohypophysis to secrete large quantities of thyrotropin. Likewise, the neurosecretory substance *corticotropin releasing factor* causes the adenohypophysis to secrete

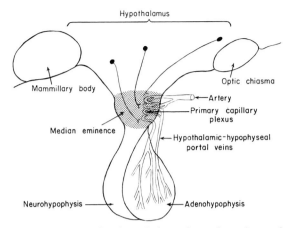

Figure 329. The hypothalamic-hypophyseal portal system.

large quantities of corticotropin. Similar neurosecretory substances control adenohypophyseal secretion of growth hormone and of each of the three gonadotropic hormones. Several of these neurosecretory substances are discussed in more detail in other parts of this text in relation to the functions of the thyroid gland, the adrenal cortex, and the sex glands.

Secretion of the neurosecretory substance that controls growth hormone release by the adenohypophysis, called *somatotropin releasing factor,* is controlled by a center in the lower midlateral portion of the hypothalamus close to the feeding and hunger areas of the hypothalamus. Deficiency of nutrients in the blood can increase the rate of secretion of somatotropin releasing factor which in turn increases the output of growth hormone.

Five out of six of the neurosecretory substances that control hormone release by the adenohypophysis have a positive effect to cause secretion of the respective adenohypophyseal hormones. However, one neurosecretory substance, *luteotropic inhibitory factor,* operates in a negative manner, inhibiting luteotropic hormone secretion rather than stimulating it. In the absence of luteotropic inhibiting factor, the adenohypophysis secretes luteotropic hormone continually, which is an effect that is different from that of the other five major adenohypophyseal hormones.

The Neurohypophyseal Hormones

The neurohypophysis does not secrete any hormones at all but simply stores two hormones called *antidiuretic hormone* and *oxytocic hormone* which are secreted by cells in the anterior hypothalamus and conducted through nerve fibers into the neurohypophysis. Transmission of nerve impulses from the anterior hypothalamus to the neurohypophysis causes the stored hormones to be released.

ANTIDIURETIC HORMONE AND CONTROL OF WATER REABSORPTION IN THE TUBULES OF THE KIDNEY. The importance of antidiuretic hormone to the function of the kidney was discussed in Chapter 18. Briefly, antidiuretic hormone prevents the body fluids from becoming too concentrated in the following way: When the concentration of electrolytes in the fluids begins to rise, the increasing osmotic pressure of these fluids is believed to cause special neurons in the anterior hypothalamus, called *osmoreceptors,* to shrink, which excites them and sends impulses to the neurohypophysis. The impulses cause antidiuretic hormone to be released from the neurohypophysis. This in turn passes by way of the blood to the collecting tubules of the kidney and causes increased quantities of water to be reabsorbed while allowing large quantities of electrolytes to pass on into the urine. The mechanism by which antidiuretic hormone increases the water reabsorption is to increase the pore size in the distal tubules and collecting ducts enough for water molecules to diffuse through, but not large enough for most other substances in the tubular fluid to pass through. Thus, the water of the body fluids is conserved, while the electrolytes gradually decrease. In this way, overconcentration of the extracellular fluids is corrected.

Pressor function of antidiuretic hormone. Antidiuretic hormone is also frequently called *vasopressin* because injection of very large quantities of purified hormone causes the arterial pressure to rise. However, only in rare instances is sufficient antidiuretic hormone *secreted* in the body to have significant effect on arterial pressure. Therefore, this pressor effect of antidiuretic hormone is more a pharmacological than a physiological effect.

OXYTOCIN. An "oxytocic" agent is a substance that causes the uterus to contract. This is one of the primary effects of *oxytocin,* also known as oxytocic hormone. This hormone is secreted in moderate quantities during the latter part of pregnancy and in especially large quantities at the time that the baby is born. It probably aids in the expulsion of the baby from the uterus. It is well known that a mother whose oxytocin secreting mechanism has been destroyed has considerable difficulty in delivering her baby. This will be discussed further in relation to childbirth in Chapter 38.

Effect of oxytocin on milk ejection. Oxytocin also has an important function in helping the mother to nurse the newborn infant as follows: Milk formed by the *glandular cells* of the breasts is secreted into the *alveoli* of the breasts where it is stored until the baby begins to nurse. For approximately the first minute after nursing begins the baby receives essentially no milk, but the suckling stimulus excites the mother's nipple and transmits impulses upward through the spinal cord, into the brain stem, and finally to the hypothalamus where they cause secretion of oxytocin. This hormone in turn flows by way of the blood to the breasts where it contracts many small *myoepithelial cells* surrounding the alveoli, thereby squeezing the milk into the ducts so that the baby can remove it by suckling.

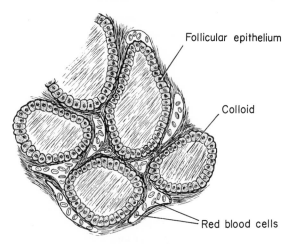

Figure 330. Cross-section of a portion of the thyroid gland, showing colloid material stored in the follicles.

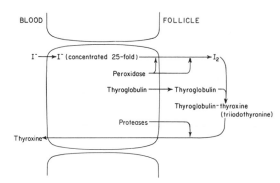

Figure 331. Mechanism for formation of thyroid hormone by the cells of the thyroid gland.

THYROXINE

FORMATION OF THYROXINE BY THE THYROID GLAND. Figure 330 illustrates the basic structure of the thyroid gland, showing it to be composed of large follicles. The cells lining the follicles secrete thyroid hormone to the interior of the follicles where it is temporarily stored. The mechanism of this secretion is illustrated in Figure 331, which shows that iodine is first absorbed into the follicular cell in the form of an iodide salt. The follicular cell converts the iodide to elemental iodine. At the same time the cell also secretes a protein called *thyroglobulin* into the follicle. The elemental iodine reacts with the thyroglobulin either before being released into the follicle or after release to convert much of the amino acid tyrosine in the molecule of the thyroglobulin into thyroxine. The chemical reaction for this process is the following:

Thyroxine formed in this manner in the follicle usually remains stored for several weeks as part of the thyroglobulin molecule before being released back through the follicular cells into the circulating blood. At the time of release, thyroxine breaks away from thyroglobulin and enters the blood in the form of thyroxine itself. However, in the blood it immediately combines with a plasma protein from which it is released over a period of several days to the tissue cells. Thus, this overall process assures a steady, slow flow of thyroxine into the tissues. Because of this slowness, some of the effects

$$2\left[HO-\!\!\left\langle\;\right\rangle\!\!-CH_2-CHNH_2-COOH \right]+4I\longrightarrow$$

Tyrosine

$$HO-\!\!\left\langle\;\right\rangle\!\!-O-\!\!\left\langle\;\right\rangle\!\!-CH_2-CHNH_2-COOH+Alanine$$

Thyroxine

of thyroxine are still apparent as long as 6 to 8 weeks after it is first formed in the thyroid gland.

All the preceding reactions related to thyroxine formation are under the control of another hormone, *thyrotropin*, secreted by the adenohypophysis; this will be discussed in more detail in following sections.

Mechanism of Action of Thyroxine

EFFECT OF THYROXINE ON CELLULAR METABOLISM AND ON THE CELLULAR ENZYMES. Thyroxine increases the rate of metabolism of all cells. Though the precise means by which this is accomplished are not known, studies have shown that tissues exposed to thyroxine develop greatly enhanced quantities of most of their enzymes. To date, at least 13 different cellular enzymes have been shown to be greatly increased under the influence of thyroxine, and because enzymes are the regulators of chemical reactions in the cells it is quite easy to understand how this could increase the metabolic rates of the cells. Because thyroxine affects all tissues of the body, it regulates the overall rate of activity of all functions of the body.

We still do not know the intracellular mechanism by which thyroxine increases the quantities of enzymes. However, since so many different enzymes are involved, and since the rate of enzyme formation is controlled primarily by genetic mechanisms of the cell, it is presumed that thyroxine activates specific genes to increase these enzymes.

EFFECTS OF THYROXINE ON SPECIFIC FUNCTIONS OF THE BODY. *Effect on total body metabolism.* Total lack of thyroxine production by the thyroid gland decreases the metabolic rate to about one-half normal. On the other hand, secretion of very large quantities of thyroxine can increase the rate of metabolism to as much as 2 times normal. Therefore, in all, the thyroid gland can change the rate of metabolism as much as four-fold. Measurements of the basal metabolic rate can be used to estimate the degree of activity of the thyroid gland.

Thyroxine causes the body to burn its available carbohydrates very rapidly, and then to make deep inroads on the stores of fats. A person with excess production of thyroxine usually loses weight, sometimes very rapidly. On the other hand, a person with less than normal production of thyroxine often develops rather extreme obesity.

Effect on the cardiovascular system. Thyroxine affects the cardiovascular system in two ways. First, because the metabolic rate rises, all the tissues of the body require increased quantities of nutrients, and cardiovascular reactions, particularly vasodilatation in all the tissues, cause the heart to pump greater quantities of blood than usual. Second, thyroxine has a direct effect on the heart, increasing its metabolism as well as its rate of beat and forcefulness of contraction. These effects also help to increase the cardiac output.

Effect on the nervous system. Thyroxine greatly increases the reactivity of the nervous system. The reflexes become very excitable with excess thyroxine, but very sluggish with diminished thyroxine. Thyroxine increases a person's degree of wakefulness, while lack of thyroxine sometimes makes him sleep as much as 12 to 15 hours per day.

A special effect of thyroxine on the nervous system is a continuous tremor of the muscles. The tremor is usually very fine but rapid, having a frequency of 10 to 20 times per second, which is considerably faster than the tremor caused by basal ganglia or cerebellar disease.

Effect on the gastrointestinal tract. Thyroxine increases the motility of the gastrointestinal tract and promotes copious flow of digestive juices. If these activities are enhanced sufficiently, diarrhea may develop. On the other hand, lack of thyroxine causes the opposite effects—sluggish motility and greatly diminished secretion—resulting in constipation. Excess production of thyroxine also causes a voracious appetite because of the rapid rate of metabolism. One eats a tremendous amount of food, digests it rapidly, absorbs large quantities of nutrients, but metabolizes these as rapidly as they become available.

Regulation of Thyroxine Production

Earlier in the chapter it was pointed out that thyroxine production is regulated almost entirely by *thyrotropin* from the adenohypophysis. In turn, the rate of secretion of thyrotropin is regulated by *thyrotropin releasing factor* secreted by the hypothalamus. Therefore, to describe the regulation of thyroxine production, we need only to discuss the factors that regulate secretion of thyrotropin releasing factor.

The primary stimulus controlling the rate of secretion of thyrotropin releasing factor is the level of metabolism in the body. If the metabolism falls to a low value, the rate of secretion of thyrotropin releasing factor increases automatically, in turn increasing the secretion of thyrotropin and consequently of thyroxine. The thyroxine then increases the metabolic rate of the body back toward normal. On the other hand, if the body's rate of metabolism rises above normal, the hypothalamus decreases its secretion of thyrotropin releasing factor, and an opposite sequence of events reduces the secretion of thyroxine, thereby reducing the metabolic rate back toward normal. Thus, as demonstrated in Figure 332, the thyrotropin-thyroxine mechanism ordinarily acts as a feedback system to regulate body metabolism at a normal mean value.

EFFECT OF COLD WEATHER ON THYROXINE PRODUCTION. When a person or an animal is exposed to very severe cold, the hypothalamus secretes increased quantities of thyrotropin releasing factor. Over a period of three to four weeks, the thyroid gland gradually enlarges, and the rate of thyroxine secretion increases. The basal metabolic rate can be increased by as much as 15 to 25 per cent by this mechanism,

which helps to warm the body and partially compensates for the cooling effect of the weather.

Abnormalities of Thyroid Secretion

HYPERTHYROIDISM. Failure of the adenohypophysis-thyroid regulatory system to function properly frequently leads to greatly increased production of thyroxine, sometimes to as much as 25 times normal. Almost always the increased production is caused by an initial increase in thyrotropin production by the adenohypophysis and not by an abnormality of the thyroid gland. But in a few persons a small *adenoma*, a tumor, develops in the thyroid gland and secretes thyroxine independently of control by the adenohypophysis. In either event the excess production of thyroxine causes *hyperthyroidism.*

In hyperthyroidism the basal metabolic rate rises very high—sometimes to as much as twice normal—the person loses weight, he develops diarrhea, he becomes highly nervous and tremulous, his heart rate is greatly increased, and the heart often beats so hard that he feels it palpitating in his chest. The state of hyperthyroidism is often so severe and so prolonged that it actually "burns out" the tissues, leading to degenerative processes in many parts of the body. One of the most common areas of degeneration is the muscle of the heart itself.

The usual methods for treating hyperthyroidism are (1) administration of a drug that suppresses thyroid function or destroys the thyroid gland, or (2) surgical removal of a major portion of the gland. Administration of *propylthiouracil* blocks the chemical reaction of tyrosine and iodine to form thyroxine, and hyperthyroidism can often be controlled with this drug. One of the more recent means for treating hyperthyroidism is the administration of radioactive iodine. Approximately two thirds of the ingested iodine is absorbed from the blood into the thyroid gland, so that a resulting intense radiation inside the gland destroys the overly active tissue, and normal thyroid function is reestablished.

Exophthalmos. Most patients with severe hyperthyroidism also develop protruding eyeballs, a condition called *exophthalmos*

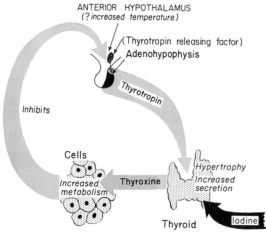

Figure 332. Interrelationships between the adenohypophysis and the thyroid gland for regulating the metabolic rate.

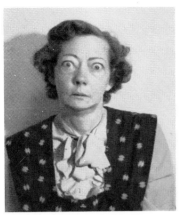

Figure 333. A hyperthyroid person with exophthalmos.

(Fig. 333). It is believed that thyroxine has nothing to do with the protrusion of the eyeballs but instead that this is caused by an abnormal adenohypophyseal hormone called *exophthalmos producing substance* that is secreted in large quantities as a side product of the excessive secretion of thyrotropin that occurs in hyperthyroidism. The cause of this association between exophthalmos producing substance and thyrotropin is not understood, though exophthalmos is indeed a very unfortunate and common side effect of hyperthyroidism. Exophthalmos producing substance causes edema, excessive growth, and degeneration of the tissues behind the eyeballs. Because exophthalmos results at least partially from increased quantity of tissue behind the eyes, elimination of the hyperthyroidism will not eliminate all the exophthalmos.

HYPOTHYROIDISM. Diminished production of thyroxine is called *hypothyroidism.* A person can live for many years with complete lack of thyroxine production, but the rate of metabolism in all his tissues is decreased to about one-half normal. He is extremely lethargic, sleeping sometimes as much as 12 to 15 hours a day. He usually is constipated, his mental reactions are sluggish, and he often becomes fat. In addition to increased deposition of fat throughout his body, a gelatinous mixture of protein and extracellular fluid is deposited in the spaces between the cells, giving an edematous appearance. For this reason the condition is often called *myxedema.*

GOITER. In hyperthyroidism the overactive thyroid gland usually enlarges two- to three-fold, and is then called a *goiter.* In hypothyroidism the gland frequently enlarges also, and here again the enlarged gland is called a goiter. Therefore, the term goiter is not synonymous with either hyper- or hypothyroidism.

Hypothyroidism is usually caused by some abnormality of the thyroid gland that makes it impossible for the gland, even when stimulated by thyrotropin, to secrete enough thyroxine. Yet the poorly secreting gland enlarges more and more in a futile attempt to produce adequate quantities of thyroxine, and large amounts of colloid substance containing almost no thyroxine are secreted into the follicles. For this reason this type of enlarged thyroid gland is called a *colloid goiter.*

Sometimes a colloid goiter becomes 15 times as large as the normal thyroid gland, weighing 500 or more grams and occupying a space in the neck or upper chest as large as one-half liter. Obviously, a gland this large can obstruct breathing and swallowing.

Endemic goiter. Persons residing in areas of the world where the food contains very little iodine cannot produce an adequate quantity of thyroxine. As a result, their metabolic rates decrease below normal, and this enhances the output of thyrotropin which in turn stimulates the thyroid gland in an attempt to produce increased quantities of thyroxine. Unfortunately, even this stimulus cannot enhance the output of thyroxine when iodine is lacking. But the adenohypophysis continues to produce large amounts of thyrotropin, so that the thyroid continues to enlarge, becoming progressively filled with colloid that contains almost no thyroxine. The enlarged gland is called an *endemic goiter* because everyone in the geographical region develops an enlarged gland. Endemic goiter was formerly widely prevalent in those regions of the world, such as the Great Lakes region of the United States and the Swiss Alps, where iodine is not present in the soil. More recently, however, a small amount of iodine has been added to most commercial table salts, so that now an inadequate intake of iodine is very rare.

REFERENCES

Farrell, G., Fabre, L. F., and Rauschkolb, E. W.: The neurohypophysis. *Ann. Rev. Physiol.* 30:557, 1968.

Fraser, T. R.: Human growth hormone. *Scient. Basis Med. Ann. Rev.* 36, 1963.

Guillemin, R.: The adenohypophysis and its hypothalamic control. *Ann. Rev. Physiol.* 29:313, 1967.

Heller, H., and Clark, R. B. (eds.): *Neurosecretion.* New York, Academic Press, 1962.

Kupperman, H. S.: *Human Endocrinology.* 3 volumes. Philadelphia, F. A. Davis Company, 1963.

McCann, S. McD., Dhariwal, A. P. S., and Porter, J. C.: Regulation of the adenohypophysis. *Ann. Rev. Physiol.* 30:589, 1968.

Pittman, J. A.: *Diagnosis and Treatment of Thyroid Diseases.* Philadelphia, F. A. Davis Company, 1963.

Trotter, W. R. (ed.): *The Thyroid Gland.* Vol. 2. London, Butterworth & Co., 1964.

von Euler, U. S., and Heller, H. (eds.): *Comparative Endocrinology.* 2 volumes. New York, Academic Press, 1963.

Werner, S. C., and Nauman, J. A.: The thyroid. *Ann. Rev. Physiol.* 30:213, 1968.

CHAPTER 35

ADRENOCORTICAL HORMONES AND INSULIN

ADRENOCORTICAL HORMONES

The adrenal glands, shown in cross-section in Figure 334, are elongated structures located immediately above each of the two kidneys. These glands produce hormones of two entirely different types. The *medulla* of each gland is part of the sympathetic nervous system, and it secretes *epinephrine* and *norepinephrine* in response to sympathetic stimuli; this was described in Chapter 28. The outer portion of the adrenal gland is the *cortex*, which secretes *adrenocortical hormones*. These hormones are all chemically similar, but they can be divided into three different categories on the basis of their functions: *mineralocorticoids, glucocorticoids,* and *androgens.* The mineralocorticoids control the electrolyte balances of sodium, chloride, and potassium; the glucocorticoids affect metabolism of protein, fat, and glucose; and the androgens cause masculinizing effects.

Mineralocorticoids

The function of mineralocorticoids is to regulate electrolytes, particularly sodium, in the extracellular fluids.

The adrenal cortex secretes at least three different hormones that can be classified as mineralocorticoids: aldosterone, corticosterone, and minute quantities of deoxy-

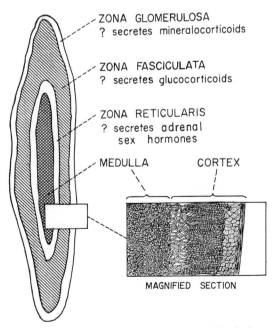

Figure 334. Cross-section of the adrenal gland, showing the three zones of the adrenal cortex that are believed to secrete the three types of adrenocortical hormones.

corticosterone. However, aldosterone accounts for at least 95 per cent of the total mineralocorticoid activity.

The function of mineralocorticoids in relation to electrolyte absorption from the kidney tubules has already been presented in Chapter 18; their overall function can be summarized briefly as follows:

EFFECT ON SODIUM. A direct effect of aldosterone is an increase in the rate of sodium reabsorption by the renal tubular epithelium. When large quantities of aldosterone are secreted, essentially all of the sodium entering the glomerular filtrate is reabsorbed into the blood, and almost no sodium passes into the urine. On the other hand, when minute amounts of aldosterone are secreted, much less sodium is reabsorbed, and as much as 20 to 30 grams of sodium may appear in the urine each day. Therefore, it is evident that aldosterone is essential to prevent rapid depletion of sodium in the body. Thus, aldosterone secretion represents a method for controlling the total quantity of sodium in the extracellular fluids.

Basic mechanism by which aldosterone increases sodium reabsorption. Aldosterone does not cause an immediate increase in sodium reabsorption. Instead, 30 minutes to an hour after aldosterone has been secreted, the various enzymes required for sodium reabsorption begin to increase in the tubular epithelial cells, and these reach a maximum level within about three hours. Therefore, it is believed that aldosterone activates the genetic structure of the cell to increase production of the specific enzymes required for sodium reabsorption. After disappearance of the aldosterone, these enzymes decrease back to a low level within another 2 to 4 hours.

EFFECT ON POTASSIUM. The enhanced sodium reabsorption from the tubules caused by aldosterone is accompanied by decreased reabsorption of potassium. This probably results from the following mechanism: When excessive amounts of sodium are reabsorbed under the influence of aldosterone, the positive charges of the sodium ions create electropositivity in the extracellular fluids surrounding the tubules. This electropositivity in turn opposes reabsorption of the positively charged potassium ions. As a result, an increased amount of potassium flows on through the tubules into the urine.

EFFECT ON CHLORIDE. Since sodium is by far the major alkaline ion of the extracellular fluid, enhanced reabsorption of sodium causes the extracellular fluid to become slightly alkaline. The alkaline reaction in turn increases the reabsorption of chloride by the kidney tubules, as explained in Chapter 18. Thus, another secondary effect of mineralocorticoid action is a decrease in the amount of chloride lost into the urine, and an increase in the concentration of chloride in the extracellular fluid. The net result is increased sodium chloride (salt) in the extracellular fluid and decreased potassium.

EFFECT ON WATER REABSORPTION IN THE KIDNEYS. The increased concentration of sodium chloride in the extracellular fluids has two effects on the absorption of water from the kidney tubules. First, absorption of sodium and chloride from the tubules diminishes the concentration of these substances in the tubular fluid and diminishes the tubular crystalloidal osmotic pressure. This allows increased absorption of water by *osmosis* into the peritubular fluids. Second, the increased concentration of sodium and chloride in the extracellular fluids increases the osmotic pressure of these fluids, which *stimulates the osmoreceptor antidiuretic mechanism* for enhancing the reabsorption of water by the tubules. This was explained in the previous chapter. Thus, by two means the quantity of water absorbed from the tubules is increased, and the total volume of extracellular fluid becomes greatly enhanced, sometimes as much as 100 per cent.

EFFECT ON FLUID VOLUMES AND CIRCULATORY DYNAMICS. The increased extracellular fluid enters both the interstitial spaces and the blood, *increasing the interstitial fluid volume and the blood volume.* The increased blood volume forces a greater than normal quantity of blood toward the heart, *increasing the cardiac output;* then, because of excess blood flow through the tissues, a local vasoconstrictive reaction occurs in the tissue vessels, thereby increasing the vascular resistance in the entire peripheral circulatory system. This combination of increased cardiac output and increased total peripheral resistance causes *elevated arterial pressure.* Therefore, because of the initial effect on sodium absorption in the kidney tubules, aldosterone can indirectly affect blood volume, cardiac output, and arterial pressure.

REGULATION OF ALDOSTERONE SECRETION. Any one of three different factors can cause the rate of aldosterone secretion to increase: low extracellular fluid sodium concentration, high extracellular fluid potassium concentration, and decreased cardiac output.

The mechanisms by which these stimuli cause the adrenal cortex to secrete increased aldosterone are not completely understood. However, at least three different mechanisms probably are important as follows:

(1) The adrenal cortex responds directly to concentration changes of sodium and potassium in the blood. When sodium concentration decreases or potassium concentration increases, aldosterone secretion increases.

(2) When the sodium concentration in the blood falls very low, and sometimes also when the arterial pressure falls low, the kidneys secrete the substance *renin*, which was discussed in Chapter 14 in relation to the development of hypertension. The renin in turn acts on a plasma protein to form the substance *angiotensin*. This substance then directly stimulates aldosterone secretion by the adrenal cortex.

(3) Decreased blood flow, and perhaps changes in sodium and potassium concentrations, can stimulate basal regions of the brain causing them to secrete hormones that in turn control adrenocortical secretion. For instance, decreased blood flow through the hypothalamus causes the hormone corticotropin (also called ACTH) to be released from the adenohypophysis. This, in turn, excites the adrenal cortex to secrete slightly increased quantities of aldosterone and greatly increased quantities of the glucocorticoids, as will be discussed in the following section. It has been claimed that similar stimuli can cause the mid-brain to secrete a substance called *glomerulotropin* which stimulates the *zona glomerulosa*, the outermost portion of the adrenal cortex, to secrete aldosterone.

Thus, we know several ways in which decreased sodium, increased potassium, or decreased circulatory function can increase aldosterone secretion. Yet we do not know the relative importance of each of these. Nevertheless, aldosterone in turn causes increased retention of sodium by the kidneys, increased loss of potassium, and increased retention of fluid. The fluid builds up the extracellular fluid volume and blood volume and thereby reestablishes normal circulatory function. Therefore, here again is an important feedback regulatory mechanism, regulating sodium and potassium ion concentrations in the body fluids and secondarily regulating function of the entire circulatory system.

Glucocorticoids

The precise functions of the glucocorticoids are not nearly as well understood as those of the mineralocorticoids. However, metabolic systems become greatly deranged when glucocorticoids are absent from the body, and one becomes unable to resist almost any traumatic or disease condition that tends to destroy tissues. Therefore, the most important function of glucocorticoids is to enhance resistance to physical "stress," though the means by which this is effected are yet very vague. Several different adrenocortical hormones exhibit glucocorticoid activity, but by far the most abundant one is *hydrocortisone*. Other much less important glucocorticoids are cortisone and corticosterone. The latter is a mineralocorticoid as well as a glucocorticoid.

EFFECT OF HYDROCORTISONE ON PROTEIN METABOLISM. Even though the name glucocorticoid implies that this type of adrenal hormone affects mainly glucose metabolism, it is becoming apparent that its action on protein metabolism is probably much more important than its action on glucose metabolism. Administration of hydrocortisone causes an increase in the amount of amino acids in the extracellular fluid and an increased rate of utilization of these acids for various purposes. The mechanism by which hydrocortisone increases the extracellular amino acids is not completely known, but experiments have shown that these hormones increase the rate of transfer of amino acids out of the cells into the blood. This makes the amino acids available for use wherever they are needed to provide energy, to repair damaged tissues, or even to synthesize necessary structures in new cells. This is a possible explanation for the ability of hydrocortisone and other glucocorticoids to enhance a person's resistance to factors that attempt to destroy his tissues.

In summary, even though the precise effects of glucocorticoids on protein metabolism are yet unknown, it seems that by increasing the transport of amino acids through cellular membranes these hormones increase the flux of amino acids throughout the body, allowing them to pass from one part of the body to another with increased ease, and especially allowing them to flow in great quantity to those areas where they are momentarily in special demand.

EFFECT OF HYDROCORTISONE ON FAT METABOLISM. Hydrocortisone mobilizes fat from the fat depots in much the same manner as it mobilizes amino acids from the cells. The net result is a decrease in the amount of fats in the storage areas and increased use of fats for energy and other purposes. It is believed, for example, that the mobilization of fats from the storage areas during periods of starvation is caused mainly by increased production of hydrocortisone.

During periods of rapid fat mobilization, the liver splits much of the fat into keto acids. If these are used immediately by the cells for energy, they cause no significant physiologic effect, but if they are not used immediately, their concentration in the extracellular fluid can become great enough to cause acidosis. This is occasionally one of the untoward effects of excessive hydrocortisone secretion.

EFFECT OF HYDROCORTISONE ON GLUCOSE METABOLISM. The earliest discovered effect of glucocorticoids was their ability to increase the concentration of glucose in the blood. Subsequent study indicates that this ability is actually an effect on protein and fat metabolism, for the increased glucose concentration is caused by increased conversion of proteins and, to a much less extent, fats into glucose, a process called *gluconeogenesis.* It is believed that when protein and fat are mobilized from their storage areas, their increased concentrations in the body fluids provide increased quantities of material for the liver to convert into glucose. Regardless of the exact mechanism by which gluconeogenesis occurs, this function of hydrocortisone is very important. It keeps the blood glucose concentration high even during starvation, and thereby provides adequate nutrition for the neurons that can use only glucose for energy.

REGULATION OF HYDROCORTISONE SECRETION. Figure 335 summarizes the regulation of adrenal secretion of hydrocortisone and other glucocorticoids. The mechanism can be explained as follows: The primary

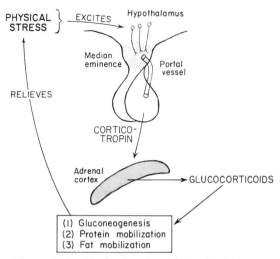

Figure 335. Mechanism by which physical stress induces adrenal secretion of hydrocortisone and other glucocorticoids.

stimulus that initiates glucocorticoid secretion is called "stress," which includes almost any physical damage to the body. For instance, a painful contusion of some part of the body, a broken bone, severe damage to large areas of cells by some diseased condition, or any other destruction of parts of the body sets off the sequence of events that leads to hydrocortisone secretion. The stress probably causes these reactions by initiating nerve impulses that are transmitted from the periphery into the hypothalamus. The hypothalamus then secretes the substance *corticotropin releasing factor* which passes by way of the hypothalamic-hypophyseal portal system into the adenohypophysis. Here, this factor causes the cells of the adenohypophysis to secrete *corticotropin* which flows in the blood to the adrenal cortex where it elicits hydrocortisone secretion and to a much less extent secretion of other glucocorticoids. The glucocorticoids then mobilize protein and fat from all over the body and also cause gluconeogenesis. The increased availability of amino acids, fats, and glucose helps in the repair of the physical damage, thus attenuating the initial stimulus that had set off the sequence of events leading to hydrocortisone secretion. Thus, this is also a feedback mechanism, which allows rapid mobilization of the necessary metabolic building blocks required for repair of tissues whenever physical damage occurs.

Androgens

Androgens are hormones that cause the development of male sex characteristics, as we shall see in Chapter 37. Though the male testes are the primary source of these hormones, the adrenal cortex also secretes minute quantities, quantities so small that in the normal person they have no significant effect. However, an androgen-producing tumor of the adrenocortical cells develops occasionally which secretes very large quantities of male sex hormones that cause serious masculinizing effects even in the female body. This is explained below.

Abnormalities of Adrenocortical Secretion

HYPOSECRETION OF ADRENOCORTICAL HORMONES — ADDISON'S DISEASE. The adrenal cortices are occasionally destroyed by disease, or sometimes they simply atrophy. Tuberculosis frequently destroys the adrenal cortices. Occasionally, excessive overstimulation of the adrenal gland by stress causes it to become, first, greatly enlarged, then hemorrhagic, and finally replaced almost completely by fibrous tissue. In each instance hyposecretion of adrenocortical hormone occurs, or complete lack of secretion, leading to a serious condition called *Addison's disease.*

Complete failure of the adrenal cortices usually causes death within three to five days unless appropriately treated. This early death is caused by the lack of aldosterone, because the kidney's sodium reabsorptive mechanism is highly dependent on this hormone. Without adequate sodium reabsorption, the level of sodium in the extracellular fluid and the volume of extracellular fluid decreases so greatly within only a few days that death ensues. This effect can be offset to some extent by having the afflicted person eat extra quantities of salt, and it can be overcome entirely by administering one of the mineralocorticoids; the one most usually used because of its easy synthesis is the compound *deoxycorticosterone.*

Even if the life of the person with Addison's disease is saved by administering deoxycorticosterone, he still remains unable to resist stress, and even a slight respiratory infection may prove lethal. He also has little energy. Usually, therefore, to provide completely adequate function of the body, treatment with hydrocortisone or some other glucocorticoid is necessary in addition to deoxycorticosterone.

HYPERSECRETION OF ADRENOCORTICAL HORMONES. Hypersecretion of the adrenal cortex can result either directly from a tumor in one part of an adrenal gland, or indirectly from increased production of corticotropin by the adenohypophysis, this in turn stimulating the adrenal cortices.

The effects of hyperadrenalism depend upon which part of the adrenal gland is secreting the excessive quantities of hormone. If it is the inner zone of the adrenal cortex, the *zona reticularis*, then excessive quantities of androgens are secreted, and the person, even a child or a female, develops masculine characteristics such as growth of hair on the face, deepening of the voice, sometimes baldness, changes in portions of the female sexual organs to resemble the sexual organs of the male, atrophy of the female breasts, and considerably enhanced muscular development.

If the middle zone of the cortex, the *zona fasciculata*, secretes excess hormones, there are symptoms of excess hydrocortisone secretion. These include excessive mobilization of proteins and fats from their storage areas. The mobilization of proteins causes weakness of the muscles and sometimes weakness of other protein structures such as the fibers that hold the tissues together beneath the skin. This allows laxity of the skin and predisposes to tears in the subcutaneous tissues that can be observed as long, linear, purplish *striae*. The excess mobilization of proteins and fats also increases gluconeogenesis, and raises the blood glucose concentration, sometimes enough to cause a very high blood sugar level, a condition known as *adrenal diabetes.*

If the increased secretion occurs in the outer layer of the adrenal cortex, the *zona glomerulosa*, the effects are those of excessive aldosterone secretion. The concentration of sodium chloride in the extracellular fluid increases, the concentration of potassium decreases, the blood volume increases, cardiac output increases, and the arterial pressure increases.

INSULIN

SECRETION OF INSULIN BY THE PANCREAS. The pancreas is a long gland that lies im-

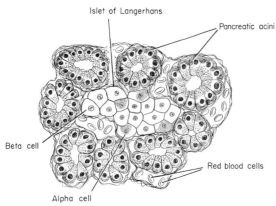

Figure 336.   Microscopic structure of the pancreas.

mediately beneath the stomach, a section of which is shown in Figure 336. It is composed of two different types of tissue. One type is the *acini*, which secrete digestive juices into the intestines. The other type is the *islets of Langerhans*, which secrete hormones directly into the body fluids; this secretion into the body fluids is an *endocrine* function of the pancreas, while the secretion of digestive juices is an *exocrine* function.

The islets of Langerhans are composed of two different types of cells, the *alpha* and *beta cells*. The beta cells secrete *insulin*, while the alpha cells secrete a hormone called *glucagon*, which is discussed later in the chapter.

Mechanism of Insulin Action

The basic mechanism of insulin action is to increase the rate of glucose transfer through the cellular membrane. The pores of the membrane are much too small for the glucose molecule to enter the cell by the process of diffusion. Instead, glucose must be carried through the membrane by an active transfer process such as that shown in Figure 337. In the absence of insulin, only a small amount of glucose can be transported into the cells, but in the presence of normal amounts of insulin the transfer is accelerated as much as 3- to 5-fold, and in the presence of large amounts of insulin the rapidity of glucose tranfer is increased as much as 15- to 25-fold. Therefore, insulin controls the rate of glucose metabolism in the body by controlling the entry of glucose into the cells.

EFFECT OF INSULIN ON BLOOD GLUCOSE. Because insulin accelerates the rate of glu-

cose transfer from the extracellular fluids to the interior of the cells, the concentration of glucose in the blood and extracellular fluids becomes diminished. Conversely, lack of insulin secretion causes glucose to dam up in the blood instead of entering the cells. Complete lack of insulin usually produces a rise in blood glucose concentration from a normal value of 90 mg. per 100 ml. up to about 350 mg. per 100 ml. On the other hand, a great excess of insulin can decrease the blood glucose to about 25 mg. per 100 ml., which is about one-fourth normal.

EFFECT OF INSULIN AND INSULIN LACK ON FAT   METABOLISM.   Insulin   has   almost equally profound effects on fat metabolism as on glucose metabolism. However, these effects probably occur secondarily to the carbohydrate effects in the following manner: Whenever large amounts of glucose are available, insulin causes some of this glucose to be transported into the fat cells. Products   of   glucose   metabolism,   especially *acetic   acid*   and   *alpha-glycerophosphate*, then promote fat storage. That is, acetic acid is polymerized into fatty acid which reacts with the glycerophosphate to form neutral fat, thereby promoting fat deposition. Conversely, in the absence of insulin, glucose does not enter the fat cells, which means that appropriate products are not available to cause fat storage. Instead, exactly opposite effects occur, namely, release of fatty acids into the blood.

Therefore, insulin has essentially the opposite effect on fat metabolism to its effect

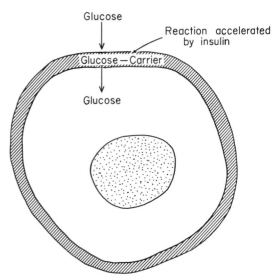

Figure 337.   Basic mechanism of insulin action.

on carbohydrate metabolism. That is, in the presence of insulin carbohydrates are utilized preferentially and excess carbohydrate is stored as fat, whereas in the absence of insulin fatty acids are mobilized and utilized in place of carbohydrates.

EFFECT OF INSULIN ON PROTEIN METABOLISM. Protein metabolism, like fat metabolism, probably is not affected directly by insulin, but indirectly it too is altered greatly by the action of insulin. When glucose cannot be utilized for energy, its *protein-sparing* effect is lost, and large quantities of protein are often utilized along with fats for energy in place of carbohydrates. As a consequence, fewer amino acids become available for synthesis of cellular and intercellular structures. For this reason, lack of insulin retards growth of an animal and also retards repair of damaged tissues. In fact, one of the serious consequences of insulin lack *(diabetes)* is diminished resistance to infections, trauma, and other types of physical stress.

Regulation of Insulin Secretion

When the blood glucose level rises, the pancreas begins secreting insulin within a few minutes. This is caused by a direct effect of glucose on the islet cells of the pancreas. That is, a high concentration of glucose automatically stimulates the beta cells to produce increased quantities of insulin. The insulin causes the excess glucose to be transported into the cells where it can be used for energy, stored as glycogen, or converted into fat.

GLUCOSE TOLERANCE TEST. A method often used for testing the ability of the pancreas to respond to glucose stimulation is the so-called glucose tolerance test, illustrated in Figure 338. In the normal person the blood glucose concentration at the beginning of the test is less than 100 mg. per cent. At this time 50 gm. of glucose is ingested, and during the ensuing half hour or more the blood glucose rises to a value of 130 to 160 mg. per cent. In less than one hour, however, large quantities of insulin will have been produced by the pancreas causing the blood glucose concentration to fall very rapidly, reaching normal once again about two hours after the ingestion of glucose. Then the concentration actually falls below normal, which is called the *hypoglycemic response*, because of continuing action of the insulin secreted during the hyperglycemic phase of the test.

A diabetic person whose pancreas cannot secrete insulin often exhibits a hyperglycemic response in which the blood glucose concentration rises to as high as 200 or more mg. per 100 ml., and the blood glucose level may require as long as five or more hours to return to normal. Also, because no insulin has been secreted, the glucose level never shows the hypoglycemic response.

Diabetes

Diabetes is the disease that results from failure of the pancreas to secrete insulin. Usually it is caused by degeneration of the beta cells of the islets of Langerhans, though

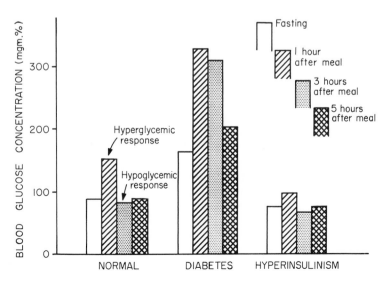

Figure 338. Effect on blood glucose concentration of administering 50 gm. of glucose to, first, a normal person, second, a diabetic person, and, third, a person with hyperinsulinism.

the cause of degeneration is usually not known. In lower animals certain chemical compounds, notably *alloxan*, can cause degeneration of these cells, and excess dietary carbohydrates in some instances can so overwork the secretory function of the beta cells that they actually "burn out." Therefore, it is possible that some persons develop diabetes secondary to exposure to abnormal chemicals, or secondary to eating excess quantities of carbohydrates. However, most diabetics are offspring of one or two parents who also had diabetes. Consequently, it is believed that the usual cause of diabetes is hereditary degeneration of the beta cells. This occurs in some persons at very early ages and in others at older ages.

EFFECTS OF DIABETES. The primary abnormality in diabetes is failure to utilize adequate quantities of glucose for energy. This causes the blood glucose level to rise to as high as three times normal. Large quantities of glucose are lost into the urine because the kidney tubules cannot reabsorb all that enters them in the glomerular filtrate each minute. The excess tubular glucose also creates a tremendous amount of crystalloidal osmotic pressure in the tubules that consequently diminishes the reabsorption of water. As a result, the diabetic person loses large quantities of water as well as glucose into the urine. In extreme cases the excess output of urine causes extracellular dehydration, which can in itself be very harmful.

Failure of the diabetic person to utilize glucose for energy deprives him of a major portion of the energy in his food. He loses weight and becomes weakened because of excess burning of his fat and protein stores. As a result of the nutrient deficiency in diabetes, the diabetic person usually becomes very hungry, so that he often eats voraciously even though the carbohydrate portion of his food contributes little to his nutrition.

Ketosis and diabetic coma. The extremely rapid metabolism of fats in diabetes sometimes increases the quantity of keto acids in the extracellular fluids to as high as 20 to 30 milliequivalents per liter, which is 25 to 50 times normal. On occasion this becomes sufficient to make the pH of the body fluids fall from its normal value of 7.4 to as low as 7.0, or rarely to as low as 6.9. This degree of acidosis is incompatible with life for more than a few hours. The person breathes extremely rapidly and deeply to blow off carbon dioxide, which helps to offset the metabolic acidosis, but despite

this the acidosis often becomes severe enough to cause coma, and unless the person is treated he usually dies in less than 24 hours. Treatment requires immediate administration of extreme quantities of insulin. Sometimes glucose is administered along with the insulin to help promote the shift from fat to glucose metabolism. Intravenous administration of alkaline solutions can be of great benefit for two reasons: First, alkaline solutions neutralize some of the acidosis, and, second, they correct the dehydration that often accompanies severe diabetic coma.

TREATMENT OF DIABETES WITH INSULIN. The diabetic person can usually be treated quite adequately by daily injections of insulin. Two principal types of insulin available for this purpose are *crystalline zinc insulin* and *protamine zinc insulin*. The duration of action of crystalline zinc insulin after injection is 4 to 6 hours, and that of protamine zinc insulin is 24 to 30 hours. Usually a person who has severe diabetes must take an injection of crystalline zinc insulin at meal times when his blood glucose concentration is likely to rise very high temporarily, and he must take an injection of protamine zinc insulin each morning to provide a steady rate of glucose inflow into his cells throughout the day.

ATHEROSCLEROSIS IN DIABETES. Prolonged diabetes usually leads to early development of atherosclerosis, and this subsequently causes heart disease, kidney damage, cerebral vascular accidents, or other circulatory disorders. The probable reason for the development of atherosclerosis is that, even with the best possible treatment of diabetes, glucose metabolism can never be maintained at a sufficiently high level to prevent some excess fat metabolism, and cholesterol deposition in the walls of the blood vessels is always an unfortunate accompaniment of rapid fat metabolism. Because of this fact the person who develops diabetes in childhood usually has a shortened life, regardless of how well he is treated.

Hyperinsulinism

Hyperinsulinism occasionally develops because of either too much insulin injected or too much insulin secreted by a pancreatic islet tumor. In either case the low blood glucose concentration causes overexcitability of the brain at first and then coma. The neurons require a constant supply of glucose because they cannot utilize fats or proteins for energy. Furthermore, the rate of glucose uptake by the neurons, unlike that of other cells, is dependent on the blood glucose concentration rather than on the amount of insulin available. Whenever excess insulin is available, the blood glucose becomes very low so that the neurons no longer receive the amounts of glucose needed to maintain their metabolism. This causes them first to become excessively excitable, and later depressed. In the excitement stage convulsions may occur, but in the depressed stage the person develops coma not unlike that of diabetes. Indeed, it is sometimes a problem to diagnose the cause of coma in a diabetic. It may result from too little insulin secretion, in which case it is diabetic coma, or from too much insulin, in which case the abnormality is hyperinsulinism.

Glucagon

The alpha cells of the islets of Langerhans secrete a hormone called *glucagon*. The action of glucagon is to cause *glycogenolysis* (breakdown of glycogen to glucose) in the liver cells and probably to a much less extent in other cells of the body where glycogen stores are much less available. This allows glucose to be dumped from the liver into the blood and extracellular fluids. The blood glucose concentration rises many milligrams per 100 ml. within a few minutes after injection of glucagon.

CONTROL OF GLUCAGON SECRETION, AND ITS FUNCTION IN THE BODY. When the blood glucose concentration falls below 60 to 80 mg. per 100 ml. of blood, the pancreas begins to pour glucagon into the blood; this effect probably results from direct stimulation of the alpha cells in the islets of Langerhans by the low glucose concentration. The glucagon in turn causes almost immediate release of glucose from the liver, thereby rapidly increasing blood glucose concentration back up toward the normal level of 90 to 100 mg. per 100 ml.

Thus, the glucagon mechanism acts as a limiting system to prevent too low blood glucose concentration. This keeps the glucose concentration high enough to prevent hypoglycemic convulsions or hypoglycemic coma, which were previously discussed.

REFERENCES

Bransome, E. D., Jr.: Adrenal Cortex. *Ann. Rev. Physiol.* 30:171, 1968.

Broom, W. A., and Wolff, F. W. (eds.): *The Mechanism of Action of Insulin.* Philadelphia, F. A. Davis Company, 1960.

Davis, J. O.: The control of aldosterone secretion. *Physiologist* 5:65, 1962.

Denton, D. A.: Evolutionary aspects of the emergence of aldosterone secretion and salt appetite. *Physiol. Rev.* 45(2):245, 1965.

Foa, P. P., and Galansino, G.: *Glucagon: Chemistry and Function in Health and Disease.* Springfield, Ill., Charles C Thomas, 1962.

Mills, I. C. H.: *Clinical Aspects of Adrenal Function.* Philadelphia, F. A. Davis Company, 1964.

Park, C. R., Reinwein, D., Henderson, M. J., Cadenas, E., and Morgan, H. E.: The action of insulin on the transport of glucose through the cell membrane. *Amer. J. Med.* 26:674, 1959.

Soffer, L. J., Dorfman, R. I., and Gabrilove, J. L.: *The Human Adrenal Gland.* Philadelphia, Lea & Febiger, 1961.

Symposium: The adrenal cortex. *Brit. Med. Bull.* 18(2), 1962.

Williams, R. H. (ed.): *Diabetes.* New York, Paul B. Hoeber, 1962.

CHAPTER 36

CALCIUM METABOLISM, BONE, PARATHYROID HORMONE, AND PHYSIOLOGY OF TEETH

CALCIUM METABOLISM

The adult human body contains about 1200 gm. of calcium, at least 99 per cent of which is deposited in the bones, but a very small and extremely important portion of which is in the blood plasma and interstitial fluid. Approximately one half of that in the plasma is ionized, and the remaining half is bound with the plasma proteins. It is the ionized calcium that diffuses into the interstitial fluid and enters into chemical reactions; it is this portion of the extracellular fluid calcium that is most important.

FUNCTIONS OF CALCIUM IONS. One of the principal functions of calcium ions occurs at the cellular membranes. In unicellular animals, calcium increases the thickness and strength of the membrane, and without calcium the membrane becomes very thin and friable (easily ruptured or crumbled). In the human being this basic effect of calcium on the membrane is not so obvious, but some of the secondary effects of its action on the cellular membrane are very important. For instance, lack of calcium causes the membranes of nerve fibers to become parti-

ally depolarized, and, therefore, to transmit repetitive and uncontrolled impulses. These may occur so rapidly that they actually cause spasm of the skeletal muscles, a condition called *tetany*. On the other hand, a great increase in the concentration of calcium ions depresses the neurons in the central nervous system. This presumably occurs because the membranes will not depolarize with normal ease.

A second effect of decreased calcium ion concentration is weakness of the heart muscle. Decreased calcium causes the duration of cardiac systole to decrease, and the heart dilates excessively during diastole. Usually, however, the person dies of tetany before his heart function is greatly impaired by calcium deficiency. An excess of calcium promotes overcontraction of the heart, causing the muscle to contract much too forcefully during systole and not to relax satisfactorily during diastole. Fortunately, this effect of excess calcium can be demonstrated only in experiments, for it never occurs to a severe extent even in disease conditions.

These effects of high and low calcium concentrations on the heart can be explained by the basic mechanism of muscle contraction which was discussed in Chapter 7. When the cardiac impulse passes over the cardiac muscle, a small amount of calcium ions is released from the sarcoplasmic reticulum, and the calcium ions set off the contractile process. When small amounts of calcium are available, the intensity of contraction is reduced, whereas excess calcium in the extracellular fluids causes overcontraction of the heart.

Another important function of calcium ions is the promotion of blood coagulation. It will be recalled from the discussion of blood coagulation in Chapter 10 that calcium enters into the chemical reaction between prothrombin, thromboplastin, and other factors to cause the formation of thrombin. The thrombin in turn catalyzes the formation of fibrin threads from fibrinogen, and thereby causes the blood to clot. Fortunately, the calcium ion concentration almost never falls low enough nor rises high enough to cause abnormalities of clotting.

Finally, an extremely important function of calcium ions is to react with phosphate ions to form bone salts, a function that is discussed in detail in subsequent parts of this chapter.

Reaction of Calcium Ions with Phosphate Ions

Calcium (Ca^{++}) and phosphate (HPO_4^{--}) ions react together to form *calcium phosphate* ($CaHPO_4$), a relatively insoluble compound. The mathematical product of the concentrations of calcium and phosphate ions in a solution can never be greater than a critical value called the *solubility product*, or they will precipitate in the form of calcium phosphate crystals. Therefore, the greater the concentration of calcium in the solution, the less can be the concentration of phosphate, and the greater the concentration of phosphate, the less can be the concentration of calcium.

One of the places where the precipitation of calcium and phosphate is especially important is in the gastrointestinal tract, for the simultaneous presence of both of these substances in the food often leads to precipitation of large quantities of calcium phosphate, which cannot be absorbed from the gut with ease.

Another instance in which calcium and phosphate are inextricably related to each other is in the deposition and reabsorption of bone, because the salts of bone are mainly calcium phosphate compounds. Every time calcium is deposited phosphate is deposited also, and every time bone is reabsorbed both calcium and phosphate are absorbed into the body fluids at the same time.

Function of Vitamin D in Calcium Metabolism

Vitamin D greatly accelerates the absorption of calcium from the gastrointestinal tract, as shown in Figure 339. This occurs even when calcium is in the intestine in the form of calcium phosphate, which is its usual form, such as in milk.

Even though the action of vitamin D seems to be principally on the absorption of calcium, when it is absorbed the phosphate with which it had been bound is almost

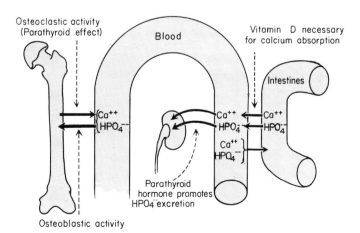

Figure 339. Absorption, utilization, and excretion of calcium and phosphate.

always left in the gut in a soluble form and is then easily absorbed. For this reason vitamin D secondarily enhances the absorption of phosphate in addition to calcium.

A vitamin D intake several hundred times that required for normal intestinal absorption of calcium causes bone absorption throughout the body, thereby releasing large quantities of both calcium and phosphate into the extracellular fluids. Vitamin D is actually any one of several similar steroid chemical compounds. Some of these have a more absorptive effect on bone than others, while still others affect mainly the absorption of calcium from the gut.

SOURCE OF VITAMIN D. Vitamin D is a fatty substance having a steroid structure, and it is usually formed by ultraviolet irradiation of various steroids in animal fat. For example, steroids in the skin, when irradiated, form vitamin D. Consequently, a person sufficiently exposed to sunlight needs no vitamin D in his diet. If he does not receive sufficient exposure to sunlight, then generally vitamin D must be provided in his food. Usually only foods of animal origin contain vitamin D. A cow exposed to sunlight forms vitamin D continually in the skin, and some of it is secreted into the milk. Also, most animals store vitamin D in great quantities in the liver; therefore, liver is usually an excellent source of this vitamin. An artificial source of vitamin D is irradiated milk, for milk contains steroids which can be converted into vitamin D by irradiation with ultraviolet light.

RICKETS. Rickets is a disease caused by prolonged calcium deficiency. Lack of dietary calcium for a short time will never cause rickets, for when the calcium ion concentration in the extracellular fluid falls below normal, calcium is automatically absorbed from the bones, thereby reestablishing an adequate calcium ion concentration. However, if insufficient calcium is absorbed from the gut for many months, then all or most of the calcium in the bones will have been absorbed, and little more is available. The calcium ion concentration of the extracellular fluids then falls to very low values. At this point the person develops rickets, the two major effects of which are (1) depletion of calcium salts from the bones and, therefore, severe weakening of the bones, and (2) tetany caused by the diminished extracellular fluid calcium ion concentration.

The most frequent cause of rickets is a deficiency of vitamin D rather than lack of calcium in the diet. The usual child living in temperate climates receives far too little sunlight during the winter months to provide the amount of vitamin D needed for absorption of calcium from the gut. Fortunately, the stores of vitamin D in the child's liver are usually sufficient to provide adequate calcium absorption for the early months of winter. Therefore, rickets usually develops in the early spring, before the child has been exposed to the sun again.

BONE AND ITS FORMATION

Figure 340 illustrates the general structure of a long bone, showing an articular surface at each end where it is jointed with the other bones, and a hollow shaft that is excellent for resisting mechanical stresses. To the right in the figure is shown a greatly magnified cross-section of bone; the white portion of the figure represents deposits of bone salts while the black areas represent spaces that contain blood vessels and tissue fluids. Bone, like other tissue, is continually supplied with an adequate flow of nutrients from the interstitial fluid.

CHEMICAL COMPOSITION OF BONE. Bone is composed of a strong protein *matrix* having a consistency almost like that of leather and of *salts* deposited in this matrix to make it hard and nonbendable. By far the major portion of the salts of bone have the following approximate chemical composition:

$$[Ca_3(PO_4)_2]_3 \cdot Ca(OH)_2$$

This is the chemical formula for *hydroxyapatite*, a hard marble-like compound. Small amounts of calcium carbonate ($CaCO_3$) are also present but probably are not of primary importance. The protein matrix prevents the bone from breaking when tension is applied, and the salts prevent the bone from crushing when pressure is applied. The matrix, therefore, is analogous to steel in reinforced concrete structures, and the salts are analogous to the concrete itself.

Deposition of Bone

Figure 341 shows, on the upper surface of a section of bone and also in some of the

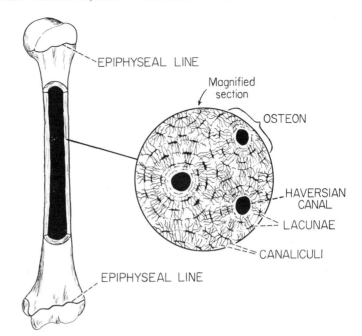

Figure 340.   Structure of bone.

cavities, a type of cell called *osteoblasts.* These are the cells that deposit bone. Bone deposition occurs in two stages as follows:

First, the osteoblasts secrete a protein substance that polymerizes to become very strong *collagen fibers.* These constitute by far the major portion of the matrix for the new bone.

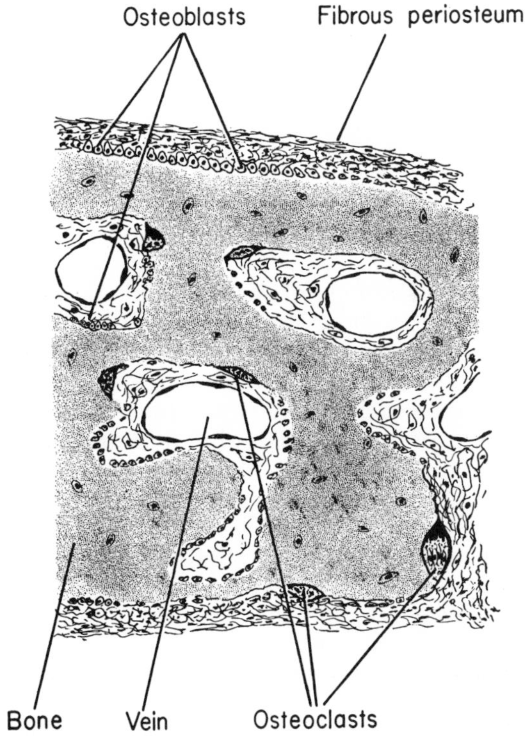

Second, after the protein matrix has been deposited, calcium salts precipitate in the matrix, making it the hard structure that we know to be bone. Deposition of the salts requires (1) combination of calcium and phosphate to form calcium phosphate, $CaHPO_4$, and (2) slow conversion of this compound into hydroxyapatite over a period of several more weeks. However, the concentrations of calcium and phosphate in the extracellular fluids are normally not sufficient to cause automatic precipitation of calcium phosphate crystals. It is believed that the newly formed collagen fibers of the bone matrix have a special affinity for calcium phosphate, causing calcium phosphate crystals to deposit even though the mathematical product of calcium ion and phosphate ion concentrations is not as great as the solubility product.

REGULATION OF BONE DEPOSITION.   The deposition of bone is regulated partially by the amount of strain applied to bone. That is, the greater the weight applied and the greater the bending of the bone the more active are the osteoblasts, the reason for which is yet unknown. As a result, a bone subjected to continuous and excessive loads usually grows thick and strong, while bones not used at all, such as the bones of a leg in a plaster cast, waste away.

Another important cause of bone production is a break in a bone. Injured osteoblasts

Figure 341.   Osteoblastic deposition of bone, and osteoclastic reabsorption of bone.

at the site of the break become extremely active and proliferate in all directions, secreting large quantities of protein matrix to cause deposition of new bone. As a result, the break is normally repaired within a few weeks.

Reabsorption of Bone

Figure 341 shows, in addition to osteoblasts, several giant cells, called *osteoclasts*, each of which contains many nuclei. These cells are present in almost all of the cavities of the bone, and have the ability to cause bone reabsorption. They do this probably by secreting enzymes that digest the protein matrix and also split the bone salts so that they will be absorbed into the surrounding fluid. Thus, as a result of osteoclastic activity, both calcium and phosphate are released into the extracellular fluid.

BALANCE BETWEEN OSTEOCLASTIC RE-ABSORPTION AND OSTEOBLASTIC DEPOSITION OF BONE. Osteoclastic reabsorption of bone occurs all the time, but this is offset by continued osteoblastic deposition of new bone. The strength of the bone depends on the relative rates of the two processes. If the rate of osteoblastic activity is greater than that of osteoclastic activity, the bone will be increasing in strength. This occurs in athletes and in others who subject their bones to excessive strain. On the other hand, osteoclastic activity is usually greater than osteoblastic activity when the bones are out of use, thereby causing the bones to become weakened.

An interesting effect of continual absorption and redeposition of bone is the tendency for crooked bones to become straight over a period of years. Compression of the inner curvature of a bent bone seems to promote rapid deposition of new bone, while stretching of the outer curvature seems to promote reabsorption. As a result, the inside of the curvature becomes filled with new bone while the outside is absorbed. In a child a broken bone may initially heal with many degrees of angulation, and yet will become essentially straight within a few years.

PARATHYROID HORMONE AND ITS REGULATION OF CALCIUM METABOLISM

SECRETION BY THE PARATHYROID GLANDS. Parathyroid hormone, also called

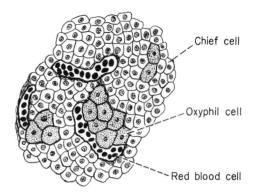

Figure 342. Histologic structure of a parathyroid gland.

parathormone, secreted by the parathyroid glands, causes release of calcium from bones, and is utilized by the body to regulate the concentration of ionic calcium in the extracellular fluids. Normally there are four different parathyroid glands lying respectively behind the four poles of the thyroid gland. Each of these glands weighs only 0.02 grams.

A histologic cross-section of a parathyroid gland is shown in Figure 342, illustrating two types of cells in the gland, *chief cells* and *oxyphil cells*. The parathyroid glands of some animals contain only chief glands, for which reason it is assumed that the chief cells are the ones that secrete parathyroid hormone. The oxyphil cells may be the same as the chief cells but in a different state of activity.

Effect of Parathyroid Hormone on Calcium and Phosphate Ion Concentrations in the Extracellular Fluid

Administration of parathyroid hormone causes bone to be absorbed. Simultaneously both calcium and phosphate ions are dumped into the extracellular fluid. Since calcium is slow to be excreted into the urine, its concentration in the extracellular fluid increases considerably. On the other hand, the rate of kidney excretion of phosphate is also enhanced by parathyroid hormone. This is often greater than the rate of phosphate absorption from the bone so that phosphate concentration in the extracellular fluid actually decreases.

BASIC MECHANISM OF BONE ABSORPTION CAUSED BY PARATHYROID HORMONE. Parathyroid hormone increases both the number and the sizes of the osteoclastic cells in bone.

Furthermore, after administration of parathyroid hormone, histologic study of the bone shows rapid invasion by the osteoclasts. When extreme quantities of parathyroid hormone are available, the osteoclasts become so plentiful and so large that they actually eat away enough bone to make the bones extremely weak.

Therefore, it is almost certain that parathyroid hormone removes calcium and phosphate from the bones simply by promoting osteoclastic activity.

Regulation of Parathyroid Secretion

Almost any factor that decreases the calcium ion concentration of the extracellular fluid causes the parathyroid gland to enlarge and to secrete increased quantities of parathyroid hormone. For example, a diet low in calcium increases the activity of the gland. Lack of vitamin D causes the same effect, so that the parathyroid glands frequently become four to five times normal size in persons with rickets.

In summary, parathyroid secretion responds directly to the concentration of calcium ion in the extracellular fluids. When the ionic level falls too low, the hormone is secreted, and calcium ions are absorbed into the blood from the bones, thereby replenishing the level of calcium ion. In this way, a relatively constant concentration of calcium ions is maintained in the extracellular fluids at all times.

THE VALUE OF PARATHYROID HORMONE TO THE HUMAN BEING. Some of the other functions of calcium ions besides that of bone formation are essential to life from minute to minute. For instance, if the calcium ion concentration falls more than 50 per cent below normal, the person develops immediate tetany and dies because of spasm of his respiratory muscles. On the other hand, if the calcium ion concentration becomes too great, he is likely to have rather severe mental or cardiac disturbances. Therefore, it is necessary that the calcium ion concentration remain almost exactly constant all the time. On the other hand, the bones have almost a thousand times the 1.5 gm. of calcium in all the extracellular fluids, so that the slight amount of calcium mobilized from the bones usually is not missed. The main value of parathyroid hormone, then, is to regulate calcium ion concentration in the extracellular fluid, so that the other functions of this ion besides that of bone deposition can continue normally all the time. The bone, on the other hand, can wait several months if necessary until adequate calcium is available to replenish its lost stores.

Abnormalities of Parathyroid Hormone Secretion

HYPERSECRETION OF PARATHYROID HORMONE. Occasionally a parathyroid tumor develops in one of the glands, and extreme secretion of parathyroid hormone causes tremendous overgrowth of the osteoclasts. Sometimes these cells grow so large and so numerous that they cause large honeycomb cavities in the bone, and even combine together to form large masses that resemble tumors. The result is often bones so weakened that even walking on a leg can cause it to break. Indeed, most hyperparathyroid persons first become aware of their disease through a broken bone.

Hypersecretion also increases the calcium ion concentration in the body fluids, but, if the hypersecretion is not too great, the level of calcium ion will not rise enough to cause untoward effects. Rarely, though, the level becomes great enough to cause precipitation of calcium phosphate in tissues other than the bone, such as the lungs, muscles, and heart, sometimes leading to death within a few days.

HYPOSECRETION OF PARATHYROID HORMONE. The most common cause of deficient parathyroid secretion is surgical removal of all the parathyroid glands, this usually occurring inadvertently when a surgeon is removing the thyroid gland, because of the close proximity of the parathyroid glands to this other gland.

Loss of parathyroid secretion allows the calcium ion concentration to fall so low that tetany occurs within about three days, and unless the person is treated he dies almost immediately. However, sufficient calcium to restore normal function can be mobilized from the bone within a few hours by administration of parathyroid hormone or certain of the vitamin D compounds.

A rare person has hereditary hyposecretion

of parathyroid hormone. Usually the degree of hyposecretion is not sufficient to cause tetany, but it may be sufficient to cause chronically depressed osteoclastic activity in the bones. This results in brittle bones, probably because without osteoclastic activity the constant absorption and reformation of new bone ceases, and the protein matrix becomes old and brittle and is not replaced as often as needed to maintain adequate bone strength.

CALCITONIN – A CALCIUM DEPRESSING HORMONE

A newly discovered hormone called *calcitonin* causes exactly the opposite effect on blood calcium ion concentration to that caused by parathyroid hormone, that is, decrease in calcium ion concentration rather than increase.

Calcitonin is secreted by the thyroid gland and to a less extent by the parathyroid glands. The thyroid calcitonin is sometimes called "thyrocalcitonin" to distinguish it from the parathyroid calcitonin.

The action of calcitonin is different from that of parathyroid hormone in another important respect: it begins to act almost immediately, whereas the effect of parathyroid hormone is hardly observable for several hours after its injection. It is believed that calcitonin causes some direct effect either on bones or on osteoblasts to cause rapid deposition of calcium phosphate salts in the bones.

Secretion of calcitonin is greatly enhanced when blood calcium concentration rises above normal. The calcitonin in turn causes some of the blood calcium to deposit in the bones, thereby returning the calcium ion concentration toward normal. Obviously, therefore, the calcitonin mechanism is a rapidly acting feedback mechanism that helps to stabilize calcium ion concentration in the extracellular fluids.

PHYSIOLOGY OF TEETH

The teeth cut, grind, and mix food. To perform these functions the jaws have ex-
tremely powerful muscles capable of providing an occlusive force of as much as 50 to 100 pounds between the front teeth, and as much as 150 to 200 pounds between the jaw teeth. Also, the upper and lower teeth are provided with projections and facets that interdigitate so that each set of teeth fits with the other. This fitting is called *occlusion*, and it allows even small particles of food to be caught and ground between the tooth surfaces.

Function of the Different Parts of Teeth

Figure 343 illustrates a *sagittal* section of a tooth, showing its major functional parts, *enamel*, *dentine*, *cementum*, and *pulp*. The tooth can also be divided into the *crown*, which is the portion that protrudes out of the gum into the mouth, and the *root*, which is the portion in the bony socket of the jaw. The collar between the crown and the root, where the tooth is surrounded by the gum, is called the *neck*.

DENTINE. The main body of the tooth is composed of dentine, which has a very strong bony structure. Dentine is made up principally of calcium salts of phosphate and carbonate embedded in a strong meshwork of *collagen fibers*. In other words, the

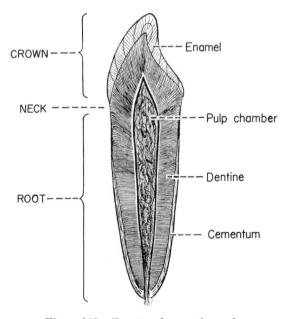

Figure 343. Functional parts of a tooth.

principal constituents of dentine are very much the same as those of bone. The major difference is its structure, for dentine does not contain any osteoblasts, osteoclasts, nor spaces for blood vessels or nerves. Instead, it is deposited and nourished by a layer of cells called *odontoblasts*, which line its inner surface along the wall of the pulp cavity.

The calcium salts in dentine make it extremely resistant to compressional forces, while the collagen fibers make it tough and resistant to tensional forces that might result when the teeth are struck by solid objects.

ENAMEL. The outer surface of the tooth is covered by a layer of enamel that is formed prior to eruption of the tooth by special epithelial cells called *ameloblasts*. Once the tooth has erupted no more enamel is formed. Enamel is composed of very small crystals of calcium phosphate and calcium carbonate embedded in a fine meshwork of *keratin* fibers. The smallness of the crystalline structure of the calcium compounds makes the enamel extremely hard, much harder than dentine. Also, the keratin meshwork makes enamel very resistant to acids, enzymes, and other corrosive agents, because keratin itself is one of the most insoluble and resistant proteins known.

CEMENTUM. Cementum is a bony substance secreted by cells that line the tooth socket. It forms a thin layer between the tooth and the inner surface of the socket, which itself is lined by an osteoblastic membrane called the periodontal membrane. Many collagen fibers pass directly from the bone of the jaw, through the periodontal membrane, and into the cementum. It is these collagen fibers and the cementum that hold the tooth in place. When the teeth are exposed to excessive strain the layer of cementum increases in thickness and strength. It also increases in thickness and strength with age, causing the teeth to become progressively more firmly seated in the jaws as one reaches adulthood and beyond.

PULP. The inside of each tooth is filled with pulp, which is composed of connective tissue with an abundant supply of nerves, blood vessels, and lymphatics. The cells lining the surface of the pulp cavity are the odontoblasts, which, during the formative years of the tooth, lay down the dentine but at the same time encroach more and more on the pulp cavity, making it smaller. In later life the dentine stops growing and the pulp cavity remains essentially constant in size. However, the odontoblasts are still viable and send projections into small *dentinal tubules* that penetrate all the way through the dentine; these are of importance for providing nutrition.

Dentition

Human beings and most other mammals develop two sets of teeth during a lifetime. The first teeth are called the *deciduous teeth* or *milk teeth,* and they number 20 in the human being. These erupt between the sixth month and second year of life and last until the sixth to the thirteenth year. After each deciduous tooth is lost, a permanent tooth replaces it, and an additional 8 to 12 molars appear posteriorly in the jaw, making the total number of permanent teeth 28 to 32, depending on whether the person finally grows his 4 *wisdom teeth*, which do not appear in everyone.

FORMATION OF TEETH. Figure 344 shows the formation and eruption of teeth. Figure 344A shows protrusion of the oral epithelium into the *dental lamina,* this to be followed by the development of a tooth-producing organ. The upper epithelial cells form ameloblasts which secrete the enamel on the outside of the tooth. The lower epithelial cells grow upward to form a pulp cavity and also to form the odontoblasts that secrete dentine. Thus, enamel is formed on the outside of the tooth, and dentine is formed on the inside, developing an early tooth as illustrated in Figure 344B.

Eruption of teeth. During early childhood, the teeth begin to protrude upward from the jawbone through the oral epithelium into the mouth, as in Figure 344C. The cause of eruption is unknown, though several theories have been offered. One of these assumes that an increase of the material inside the pulp cavity of the tooth causes much of it to be extruded downward through the root canal, and that this pushes the tooth upward. However, a more likely theory is that the bone surrounding the tooth hypertrophies progressively, and in so doing shoves the tooth forward toward the gum margin.

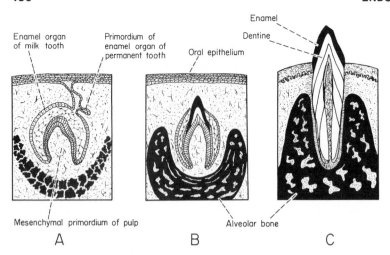

Figure 344. (A) Primordial tooth organ. (B) The developing tooth. (C) The erupting tooth. (Modified from Bloom and Fawcett: *A Textbook of Histology.* 8th Ed.)

DEVELOPMENT OF PERMANENT TEETH. During embryonic life a tooth-forming organ also develops in the dental lamina for each permanent tooth that will be needed after the deciduous teeth are gone. These tooth-producing organs slowly form the permanent teeth throughout the first 6 to 20 years of life. When each permanent tooth becomes fully formed, it, like the deciduous tooth, pushes upward through the bone of the jaw toward the gingival margin. In so doing it erodes the root of the deciduous tooth and eventually causes it to loosen and fall out. Soon thereafter the permanent tooth erupts to take the place of the original one.

Metabolic factors. The rate of development and the speed of eruption of teeth can be accelerated by both thyroid and growth hormones. Also, the deposition of salts in the early forming teeth is affected considerably by various factors of metabolism, such as the availability of calcium and phosphate in the diet, the amount of vitamin D present, and the rate of parathyroid hormone secretion. When all these factors are normal, the dentine and enamel will be healthy. When these factors are deficient, the ossification of the teeth also may be defective, so that the teeth will be abnormal throughout life.

Mineral Exchange in Teeth

The salts of teeth, like those of bone, are composed of calcium phosphate and calcium carbonate salts bound together in a single very hard crystalline substance. New calcium salts are constantly being deposited, while old salts are being reabsorbed from the teeth, as also occurs in bone. However, experiments indicate that this deposition and reabsorption occurs in the dentine and cementum, while almost none, if any at all, occurs in the enamel. The rate of absorption and deposition of minerals in the cementum is approximately equal to that in the surrounding bone of the jaw, while the rate of deposition and absorption of minerals in the dentine is approximately one-third that of bone. The characteristics of cementum, including the presence of osteoblasts and osteoclasts, are almost identical with those of usual bone, while dentine does not have these characteristics, as was explained above; this difference undoubtedly explains the different rates of mineral exchange.

The mechanism by which minerals are deposited in and reabsorbed from the dentine is unknown. It is possible that the small processes of the odontoblasts protruding into the tubules of the dentine are capable of absorbing salts and then providing new salts to take the place of the old. It is also possible, though this too has never been proved, that the odontoblasts provide continuous replacement of the collagen fibers of the dentine. This would be comparable to the rejuvenation of the bone matrix by osteoblasts, which is necessary for adequate maintenance of bone strength.

In summary, mineral exchange is known to occur in the dentine and cementum of teeth, though the mechanism of this exchange in dentine is unknown. On the other hand, enamel seems to maintain its original mineral complement throughout life, or until some abrasive or eroding action causes its destruction.

Dental Abnormalities

The two most common dental abnormalities are *caries* and *malocclusion*. Caries means erosions of the teeth, while malocclusion means failure of the facets of the upper and lower teeth to interdigitate properly.

MALOCCLUSION. Malocclusion is usually the result of a hereditary abnormality that causes the teeth of one jaw to grow in an abnormal direction. When malocclusion occurs, the teeth cannot perform adequately their normal grinding or cutting action. Occasionally malocclusion causes abnormal displacement of the lower jaw in relation to the upper jaw, causing such undesirable effects as pressure on the anterior portion of the ear or pain in the mandibular joint.

The orthodontist can often correct malocclusion by applying long-term gentle pressure against the teeth with appropriate braces. The gentle pressure causes absorption of alveolar jaw bone on the compressed side of the tooth, and deposition of new bone on the tensional side of the tooth. In this way the tooth gradually moves to a new position as directed by the applied pressure.

CARIES. Two major but differing theories have been proposed to explain the cause of caries. One of these postulates that acids formed in crevices of the teeth by acid-producing bacteria cause erosion and absorption of the protein matrix of the enamel and dentine. This theory assumes that the acids are formed by splitting carbohydrates into lactic acid. For this reason it has been taught that eating food high in carbohydrate content, and particularly eating large quantities of sweets between meals, can lead to excessive development of caries.

The second theory proposes that proteolytic (protein-digesting) enzymes secreted by bacteria in the crevices of the teeth, or in plaques that develop on the surface of unbrushed teeth, digest the keratin matrix of the enamel. Then the calcium salts, unprotected by their protein fibers, are slowly dissolved by the saliva.

Which of the foregoing theories correctly describes the primary process for development of caries is not known. Indeed, both processes might be operative simultaneously.

Some teeth are more resistant to caries than others. Teeth formed in children who drink water containing small amounts of *fluorine* develop enamel that is more resistant to caries than that of the teeth of children who drink water containing no fluorine. Fluorine does not make the enamel harder than usual, but is believed to inactivate proteolytic enzymes before they can digest the protein matrix of the enamel. Regardless of the precise means by which fluorine protects the teeth, it is known that small amounts of it deposited in enamel make teeth about twice as resistant to caries as are teeth without fluorine.

REFERENCES

Albright, F., and Reifenstein, E. C., Jr.: *The Parathyroid Glands and Metabolic Bone Disease.* Selected Studies. Baltimore, The Williams & Wilkins Company, 1948.

Arnaud, C. D., Jr., Tenenhouse, A. M., and Rasmussen, H.: Parathyroid hormone. *Ann. Rev. Physiol.* 29: 349, 1967.

Danowski, T. S.: *Clinical Endocrinology.* Vol. III: Calcium, Phosphorus, Parathyroids and Bone. Baltimore, The Williams & Wilkins Company, 1962.

Fourman, P.: *Calcium Metabolism and the Bone.* 2nd Ed. Philadelphia, F. A. Davis Company, 1965.

Greep, R. O., and Talmage, R. V.: *The Parathyroids.* Springfield, Ill., Charles C Thomas, 1961.

Hardwick, A. L.: *Advances in Fluorine Research and Dental Caries Prevention.* New York, The Macmillan Company, 1962.

McLean, F. C. (ed.): *Radioisotopes and Bone.* Philadelphia, F. A. Davis Company, 1963.

McLean, F. C., and Urist, M. R.: *Bone: An Introduction to the Physiology of Skeletal Tissue.* 2nd Ed. Chicago, University of Chicago Press, 1961.

Munson, P. L., Hirsch, F., and Tashjian, A. H., Jr.: Parathyroid gland. *Ann. Rev. Physiol.* 25:325, 1963.

Rasmussen, H.: Parathyroid hormone; nature and mechanism of action. *Amer. J. Med.* 30:112, 1961.

CHAPTER 37

SEXUAL FUNCTIONS OF THE MALE AND FEMALE, AND THE SEX HORMONES

The male and female play equal parts in initiating reproduction and in determining the hereditary characteristics of the baby—the male provides the *sperm* and the female the *ovum*. Combination of a single sperm with a single ovum forms a *fertilized ovum* that can grow into an *embryo*, then into a *fetus*, and eventually a newborn baby. The sexual functions of both the male and female responsible for initiating the process of reproduction are discussed in the present chapter, and reproduction itself is considered in the following chapter.

MALE SEXUAL FUNCTIONS

Figure 345 illustrates the male sexual organs, the most important of which are the testes, the prostate, and the penis. The functions of these are described in subsequent sections of the chapter.

The Sperm

CHARACTERISTICS OF SPERM. The *testes* of the male produce billions of *spermatozoa*,

also called *sperm*. Figure 346 shows the structure of a sperm, which itself is a single cell constructed principally of a *head* and a *tail*. The head is composed mainly of the nucleus of the cell, with only a very thin cytoplasmic and membrane layer around its surface. Most of the cytoplasm of the sperm is aggregated into the tail, and this is covered by a long extension of the cellular membrane.

The nucleus in the head of the sperm carries the genes of the male, while the tail provides motility. The tail moves back and forth (*flagellar movement*), propelling the sperm forward. Normal sperm move in a

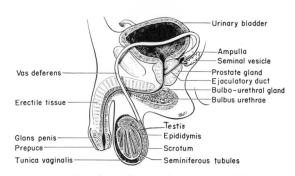

Figure 345. The male sexual organs. (Modified from Bloom and Fawcett: *A Textbook of Histology*. 8th Ed.)

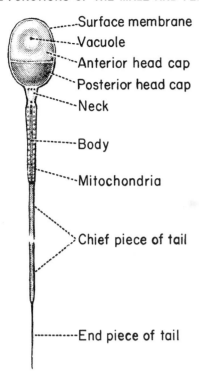

Figure 346. Structure of a sperm.

remaining cytoplasm at one end of the cell to form the tail.

Located among the germinal cells of the tubule are many large cells called *sustentacular cells.* These lie adjacent to the developing spermatocytes, and the spermatocytes remain attached to them until the head and tail of the sperm are formed. It is believed that the sustentacular cells secrete substances that are needed by the developing sperm, but their exact function is not known.

Division of chromosomes during sperm formation. When a secondary spermatocyte divides to form two spermatids, the cell division is not of the usual type, for reproduction of the chromosomes does not occur. Instead, each of the 23 pairs of chromosomes simply splits apart, allowing 23 unpaired chromosomes to pass into one of the two spermatids and the other 23 to pass into the other spermatid. In other words, only half of the genes enter each spermatid, and, similarly, only half of the genes from the parent cell are in each respective sperm. A similar division of genes occurs in the ovum. Thus,

straight line at a velocity of about 1 to 4 mm. per minute. This movement allows them to move up the female genital tract in quest of the ovum.

FORMATION OF SPERM — SPERMATOGENESIS. The testes are composed of many thousands of small tubules, called *seminiferous tubules,* which form the sperm. A cross-section of a typical seminiferous tubule is illustrated in Figure 347A, and the formation of sperm, which is called *spermatogenesis,* is shown in Figure 347B. The epithelioid cells lining the outer wall of the seminiferous tubules are called the *male germinal epithelium,* and it is from these that all sperm are formed.

The earliest form of cell in the development of spermatozoa is the *spermatogonium.* As this cell divides through several generations, the newly forming cells move toward the lumen of the tubule. The first stage in the development of the sperm is the *primary spermatocyte;* this divides again to form the *secondary spermatocyte,* which in turn divides to form two *spermatids.* The spermatid changes into a sperm by, first, losing some of its cytoplasm, second, reorganizing the chromatin material of its nucleus to form a compact head, and, third, collecting the

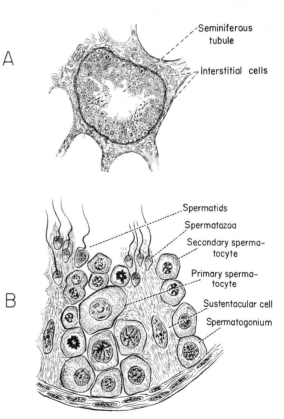

Figure 347. (A) Cross-section of a seminiferous tubule. (B) Development of spermatozoa by the germinal epithelium. (Modified from Arey: *Developmental Anatomy.* 6th Ed.)

each parent makes an equal contribution to hereditary characteristics of the child.

Sex determination by the sperm. Whether a baby will be male or female is determined by the sperm, for the following reasons. One pair of the chromosomes in all human cells is called the sex pair because these chromosomes determine whether the person will be male or female. One type of sex chromosome is known as a *Y* chromosome or a *male chromosome.* Another is known as an *X* chromosome or a *female chromosome.* A female has two X chromosomes. A male has one X chromosome and one Y chromosome. When the spermatocyte divides in the testis to form sperm, the sex chromosomes split; the Y chromosome passes into one sperm and the X chromosome passes into another. Since the unfertilized ovum contains only a single X chromosome, if a sperm containing an X chromosome combines with an ovum, the fertilized ovum then has a pair of X chromosomes. This combination leads to formation of a female. On the other hand, combination of a sperm containing a Y chromosome with an ovum creates an X-Y pattern in the fertilized ovum and causes the development of a male child.

STORAGE OF SPERM. Sperm are formed continually in the seminiferous tubules of the testes. After formation they are stored mainly in the lumen of these same tubules, but some of them pass along the male genital ducts, shown in Figure 345, to be stored in the *epididymis.* The sperm remain almost totally inactive during this period of storage because they release large quantities of carbon dioxide and other acidic end products that severely depress their motility. Nevertheless, they can live for as long as a month in the seminiferous tubules and epididymis, awaiting ejaculation.

The Male Sexual Act

In addition to the testes, the male sexual organs include, as shown in Figure 345: the *vas deferens,* which conducts the sperm from the testis to the urethra; the *seminal vesicles,* which secrete a mucoid material into the upper end of the vas deferens; the *prostate gland,* which secretes a milky fluid also into the upper end of the vas deferens; the *urethra,* which conducts the

sperm from the vas deferens to the exterior; and the *penis,* which provides the passageway into the female vagina. The penis in turn has two important parts for performance of the sexual act. These are the *glans,* which is the sensitive portion that elicits sexual excitement, and the *erectile tissue,* which surrounds the urethra and causes erection.

ERECTION AND LUBRICATION. Stimulation of the glans of the penis causes sensory impulses to pass into the sacral portion of the spinal cord, and, if the person simultaneously also has appropriate psychic stimulation to perform the sexual act, reflex impulses return through parasympathetic nerve fibers to the genital organs. These impulses dilate the arteries that supply the erectile tissue of the penis and probably also constrict the veins. As a result, a tremendous amount of blood enters the erectile tissue under high pressure and blows it up like a balloon. This causes the penis to become greatly enlarged, to become hard, and to extend forward.

In addition to causing erection, the parasympathetic impulses initiate mucus secretion by the *bulbo-urethral glands* located at the upper end of the urethra, and also by many small mucous glands along the course of the urethra. The mucus is expelled and helps to lubricate the movement of the penis and, therefore allows the vagina to massage the penis during movement rather than to abrade it. The massaging effect creates the necessary sexual stimulus to cause ejaculation; an abrasive effect causes pain, which inhibits sexual desire and blocks completion of the sexual act.

EJACULATION. When the degree of sexual stimulation has reached a critical level, neuronal centers in the tip of the spinal cord send impulses through the sympathetic nerves to the male genital organs to initiate rhythmic peristalsis in the genital ducts. The peristalsis begins in the testes themselves and then passes upward through the epididymis, the vas deferens, the seminal vesicles, the prostate gland, and the penis itself. This process, called *ejaculation,* moves sperm all the way from the testes out the tip of the penis into the vagina.

In addition to the sperm from the testes, mucus from the seminal vesicles and a milky serous fluid from the prostate are expelled during ejaculation. All these fluids plus the sperm are called *semen,* and the total quantity at each ejaculation is usually about 3 ml., each milliliter containing about 120 million sperm, a total of about one-third billion sperm.

The milky fluid from the prostate, which is highly alkaline, neutralizes the acidic fluid from the testes, in this way immediately stimulating the sperm to action, for sperm are immobile in acidic media but very active in alkaline media.

MALE STERILITY. Approximately one male out of every 25 to 30 is sterile. The most frequent cause of this is previous infection in the male genital ducts, though occasionally the seminiferous tubules of the testes may have been partially or totally destroyed by mumps infection, typhus infection, x-ray irradiation, or nuclear radiation. A few males have congenitally deficient testes that are incapable of producing normal sperm. The sperm may have two tails, two heads, or other less obvious abnormalities. These cannot fertilize an ovum, and when a large number of them occur in the ejaculate, it is usually an indication that all of the sperm are abnormal even though many appear to be completely normal.

Male sterility also can occur when the number of sperm in the ejaculate falls too low, even though all the sperm are normal. Usually a male is sterile when the number falls below about 150 million sperm in a single ejaculation. It is difficult to understand why sterility should exist in this instance, since only one sperm is required to fertilize the ovum. However, it is believed that the large number of sperm are necessary to provide enzymes or other substances that help the single fertilizing sperm to reach the ovum. It has been suggested, though not proved, that sperm secrete an enzyme, called *hyaluronidase,* which disperses the granulosal cells that normally cover the surface of the ovum. This action then allows the sperm to attack the ovum. Without a sufficient number of sperm to secrete hyaluronidase, the granulosal cells supposedly block the onslaught of all sperm.

HORMONAL REGULATION OF MALE SEXUAL FUNCTIONS

Role of the Adenohypophyseal Hormones

PUBERTY. The testes of the child remain dormant until they are stimulated at the age of 10 to 14 by gonadotropic hormones from the pituitary gland. At that age, for reasons not yet understood, the adenohypophysis begins to secrete both *follicle stimulating hormone* and *luteinizing hormone.* These cause an upsurge of testicular growth and function, causing the male sex life to begin. This stage of development is called *puberty.*

FOLLICLE STIMULATING HORMONE. Follicle stimulating hormone causes proliferation of the cells in the germinal epithelium, promoting the formation of sperm. The means by which follicle stimulating hormone promotes spermatogenesis are not known. Yet coincident with the proliferation of the germinal cells, estrogens—female sex hormones—are produced by the sustentacular cells. Therefore, it has been suggested that follicle stimulating hormone might act primarily on the sustentacular cells to cause estrogen production, and that it may be the estrogens diffusing into the adjacent germinal cells that make them proliferate. Indeed, one of the known functions of estrogens is to cause cellular proliferation; this is discussed in detail in relation to female sex functions later in the chapter.

LUTEINIZING HORMONE. Luteinizing hormone causes the production of testosterone by the *interstitial cells,* located in the testes as shown in Figure 348 in the interstices between the seminiferous tubules. Testosterone in turn is the major

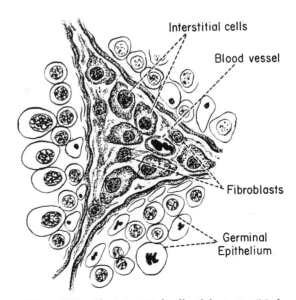

Figure 348. The interstitial cells of the testis. (Modified from Bloom and Fawcett: *A Textbook of Histology.* 8th Ed.)

hormone responsible for development of male sexual characteristics.

Functions of Testosterone

EFFECT ON SPERMATOGENESIS. Testosterone causes the testes to enlarge. Also, it must be present, along with follicle stimulating hormone, before spermatogenesis will occur. Unfortunately, though, the precise action of testosterone in spermatogenesis is unknown.

EFFECT ON MALE SEX CHARACTERISTICS. After a fetus begins developing inside its mother's uterus, its testes begin to secrete testosterone when it is only a few weeks old. This testosterone then helps the fetus to develop male sexual organs and male secondary characteristics. That is, it accelerates the formation of a penis, a scrotum, a prostate, the seminal vesicles, the vas deferens, and other male sexual organs. In addition, testosterone causes the testes to descend from the abdominal cavity into the scrotum; if the production of testosterone by the fetus is insufficient, the testes fail to descend, but instead remain in the abdominal cavity in the same manner that the ovaries remain in the abdominal cavity of the female.

Testosterone secretion by the fetal testes is caused by a hormone, called *chorionic gonadotropin*, that is formed in the placenta during pregnancy. Immediately after birth of the child, loss of connection with the placenta removes this stimulatory effect, so that the testes become dormant and the sexual characteristics remain at a standstill from birth until puberty. At puberty the reinstitution of testosterone secretion causes the male sex organs to begin growing again. The testes, scrotum, and penis then enlarge about ten-fold.

EFFECT ON SECONDARY SEX CHARACTERISTICS. In addition to the effects on the genital organs, testosterone exerts other general effects throughout the body to give the adult male his distinctive characteristics. It promotes *growth of hair* on his face, along the midline of his abdomen, on his pubis, and on his chest. On the other hand, it causes *baldness* in those male individuals who also have a hereditary predisposition to baldness. It increases the *growth of the larynx* so that the male, after puberty, develops a deeper pitch to his voice. It causes an *increase in the deposition of protein* in his muscles, bones, skin, and other parts of his body, so that the male adolescent becomes generally larger and more muscular than the female. Also, testosterone sometimes promotes abnormal secretion by the sebaceous glands of the skin, this leading to acne on the face of the post-puberal male.

FEMALE SEXUAL FUNCTIONS

Figure 349 shows the sexual organs of the female. The general function of these is the following: An *ovary* forms an *ovum*, which is then transported through a *fallopian tube* into the *uterus*. If a sperm is available to fertilize the ovum, the fertilized ovum develops in the uterus into a fetus and then into a child. The purpose of the present section is to explain the female sexual functions that are necessary to initiate the process of reproduction.

The Ovum

The ovaries, like the testes, have a germinal epithelium. However, this is not arranged in tubules as in the testes, but instead lies on the surface of each ovary. In the early stages of ovarian development some of the cells from this germinal epithelium migrate into the mass of the ovary

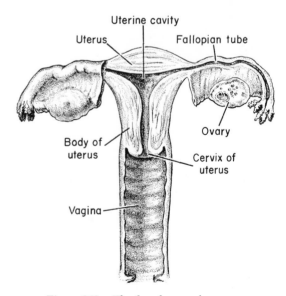

Figure 349.   The female sexual organs.

and after several divisions form large cells which become the *ova.* Approximately 400,000 ova in all are formed in the two ovaries, most of which are already present in the substance of the ovaries at birth and remain there in a dormant state until puberty. At this time, hormones from the pituitary gland begin promoting the release of ova one at a time, as explained below.

Each ovum in the ovary is surrounded by a layer of epithelioid cells called *granulosal cells,* and these along with the enclosed ovum are called a *primary follicle,* one of which is illustrated in Figure 350.

GROWTH OF THE FOLLICLE AND DISCHARGE OF THE OVUM. At puberty, when the ovaries begin to be stimulated by gonadotropic hormones from the adenohypophysis, a few of the primary follicles begin to enlarge. The granulosal cells proliferate, and adjacent stromal cells from the substance of the ovary also take on epithelioid characteristics and join the growing follicle. The follicular cells then secrete fluid, creating a cavity called an *antrum* in the follicle, as shown in Figure 350. The antrum grows until the follicle actually balloons outward on the surface of the ovary; several of these follicles are shown on the surface of the ovary to the right in Figure 349.

One of the follicles soon balloons outward more than the others and ruptures, and an ovum covered by a mass of granulosal cells is expelled into the abdominal cavity. As

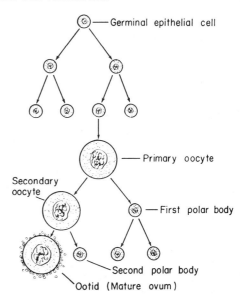

Figure 351. Formation of the mature ovum from the germinal epithelium. (Modified from Arey: *Developmental Anatomy.* 6th Ed.)

soon as this follicle ruptures, the remainder of the growing follicles suddenly begin to disappear without ever rupturing. Presumably the release of the follicular fluid from the ruptured follicle into the abdominal cavity and absorption of its hormones by the peritoneum in some way cause the involution of the other follicles. Regardless of the exact cause, only one ovum is usually expelled into the abdominal cavity each month, though occasionally a second or even more ova are expelled before the remaining follicles start to involute. This is the major cause of multiple births.

Division of chromosomes in the ovum. Figure 351 shows the development of the mature ovum from the germinal epithelium. After several divisions a germinal epithelial cell becomes a *primary oocyte,* which is the ovum of a primary follicle. Toward the end of growth of the follicle the nucleus of the primary oocyte divides, but without reproducing the chromosomes. Instead, each pair of chromosomes splits apart and 23 unpaired chromosomes remain in the ovum while the other 23 chromosomes are expelled in the so-called *first polar body.* The ovum now is called a *secondary oocyte,* and it has only half the genes of the mother. Then, approximately when the follicle ruptures, the secondary oocyte divides again; this time each of the unpaired chromosomes is reproduced, and

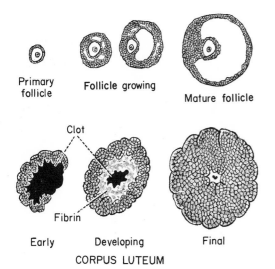

CORPUS LUTEUM

Figure 350. Growth of the follicle, and formation of the corpus luteum. (Modified from Arey: *Developmental Anatomy.* 6th Ed.)

half of them pass into the *second polar body* while half remain in the ovum. The ovum at this point is said to be mature, and it is now ready to be fertilized by a sperm.

TRANSPORT OF THE OVUM. Referring once again to Figure 349, it can be seen that the ends of the fallopian tubes lie in close proximity to the ovaries. Cilia on the epithelial surfaces of the fallopian tubes are always beating toward the uterus, causing fluids from the region of the ovary to be moving constantly out of the abdominal cavity into the tubes. The ovum, on being expelled into the abdominal cavity from the rupturing follicle, is transported into a fallopian tube along with this current of fluid flow. The epithelium of the fallopian tube, however, is pocked along its entire extent with large cavities that obstruct the movement of the ovum, so that usually three to seven days are required for its passage to the uterus.

The ovum must be fertilized within 8 to 24 hours after it is released from the ovary or it dies. Therefore, fertilization usually takes place in the upper portion of the fallopian tube, and the fertilized ovum begins to divide and to form the early stages of the embryo while it is still in the tube. It is believed that the fallopian tube epithelium secretes substances that provide some of the nutrition for the developing mass of cells. However, much if not most of the nutrition is supplied by the large amount of cytoplasm in the ovum itself.

FEMALE STERILITY. About one female out of 15 is sterile. As is true in the male, many of these instances of sterility are caused by previous infection. The infection sometimes blocks the fallopian tubes, or at other times involves the ovaries in a mass of scar tissue. The most common infection causing female sterility, as is true also in the male, is gonorrhea. Sterility is often caused also by failure of the ovaries to develop and expel ova into the abdominal cavity. Most frequently this results from insufficient stimulation by the adenohypophyseal hormones. At other times it is caused by excessively thick capsules on the surfaces of the ovaries, or occasionally by congenitally abnormal ovaries.

The Female Sexual Act

Part of the female's role in the sexual act is to help the male reach a sufficient degree of sexual stimulation for ejaculation to occur. She does this by providing appropriate conditions for maximum masculine stimulation, including especially female erection and the secretion of adequate lubricant.

FEMALE ERECTION AND SEXUAL LUBRICATION. Located around the opening of the vagina are masses of erectile tissue exactly like that in the male penis. Appropriate psychic excitation in the female, plus sexual stimulation of the female genital organs, causes parasympathetic impulses to pass from the caudal spinal cord to this erectile tissue, making it swell and provide a tight though distensible opening into the vaginal canal.

In addition to erection the parasympathetic impulses cause two small glands called *Bartholin's glands*, located immediately inside the vagina on either side, to secrete large quantities of mucus. It is this mucus from the female that is mainly responsible for lubrication of the sexual act, though mucus from the male helps to a less extent.

THE FEMALE CLIMAX. Most of the sexual stimulation in the female is provided by the massaging action of the penis against the *female clitoris*. This is the homologue in the female of the masculine penis, and is located at the anterior margin of the vulva immediately outside the vagina. The female, like the male, develops intense sexual stimulation, and when the degree of stimulation reaches sufficient intensity, the uterus and fallopian tubes begin rhythmic, upward-directed peristaltic contractions. This stage is called the *female climax*. The peristaltic contractions are believed to propel semen from the vagina into the fallopian tubes.

HORMONAL REGULATION OF FEMALE SEXUAL FUNCTIONS

Relation of the Adenohypophyseal Gonadotropic Hormones to the Ovarian Cycle

The adenohypophysis of the female child, like that of the male child, secretes essentially no gonadotropic hormones until the age of 10 to 14 years. However, at that time, for reasons not at all understood, the adenohypophysis begins to secrete three gonadotropic hormones. At first it secretes mainly *follicle stimulating hormone,* which initiates the beginning of sexual life in the growing female child, but later it secretes *luteinizing* and *luteotropic hormones,* which help to control the monthly female cycle.

FOLLICLE STIMULATING HORMONE. Follicle stimulating hormone causes a few of the primary follicles of the ovary to begin growing each month, promoting very rapid proliferation of the epithelioid cells surrounding the ovum. These cells then begin to secrete estrogens, one of the two major female sex hormones. Thus, the two functions of follicle stimulating hormone are to cause proliferation of the ovarian follicular cells and to cause secretory activity by these cells which causes the follicular cavities to develop and grow. As soon as the follicles grow to about one-half their maximum size, the adenohypophysis begins to secrete *luteinizing hormone* instead of follicle stimulating hormone.

LUTEINIZING HORMONE. This hormone increases still more the rate of secretion by the follicular cells, and soon makes one follicle grow so large that it ruptures, expelling its ovum into the abdominal cavity. Simultaneously, luteinizing hormone causes the follicular cells to grow in size and to develop a fatty, yellow appearance. These cells are then known as *lutein cells,* and the entire mass of lutein cells is called a *corpus luteum,* which means a "yellow body." This change of follicular cells into lutein cells is illustrated in the lower part of Figure 350.

LUTEOTROPIC HORMONE. The adenohypophysis also secretes a third hormone called *luteotropic hormone,* though little is known about the quantities secreted and the time in the cycle that it is secreted. This hormone further enhances the development of the corpus luteum and, in concert with luteinizing hormone, stimulates it to secrete large quantities of both estrogens and progesterone, the two female sex hormones. The corpus luteum persists for about two weeks, at which time it degenerates. Then, the adenohypophysis begins secreting large quantities of follicle stimulating hormone once again, which initiates a new cycle of growing follicles.

Ovarian Hormones

The two ovarian hormones, the *estrogens* and *progesterone,* are responsible for sexual development of the female and also for the female monthly sexual changes. These hormones, like the adrenocortical hormones and the male hormone testosterone, are both steroid compounds, and they are formed mainly from the fatty substance cholesterol. Estrogens are actually several different hormones called *estradiol, estriol,* and *estrone,* but they have identical functions and almost but not exactly identical chemical structures. For this reason they are considered together as if they were a single hormone.

FUNCTIONS OF ESTROGENS. Estrogens cause many types of cells in certain parts of the body to *proliferate*—that is, to increase in number. For example, they cause the smooth muscle cells of the uterus to proliferate, making the female uterus, after puberty, enlarge to about two to three times that of a child. Also, estrogens cause enlargement of the vagina, development of the *labia* surrounding the vagina, growth of hair on the pubis, broadening of the hips, conversion of the pelvic outlet into an ovoid shape rather than the funnel shape of the male, growth of the breasts, proliferation of the glandular elements of the breasts, and finally deposition of fatty tissues in characteristic female areas such as on the thighs and hips. In summary, essentially all of the characteristics that distinguish the female from the male are caused by estrogens, and the basic reason for the development of these characteristics is the ability of estrogens to promote proliferation of the respective cellular elements in certain regions of the body.

Estrogens also increase the growth rate of all bones immediately after puberty, but they cause the growing portions of the bone to "burn out" within a few years, so that growth then stops. As a result, the female grows very rapidly for the first few years after puberty and then ceases growing entirely. On the other hand, the male child continues to grow even beyond this time and grows to a taller height than the female—not because of more rapid growth but because of more prolonged growth.

Estrogens also have very important effects on the internal lining of the uterus, the *endometrium,* which is discussed below in relation to the menstrual cycle.

FUNCTIONS OF PROGESTERONE. Progesterone has little to do with the development of the female sexual characteristics; instead, it is concerned principally with preparing the uterus for acceptance of a fertilized ovum and preparing the breasts for secretion of milk. In general, progesterone increases the degree of secretory activity of the glands of the breasts and also of the cells lining the uterine wall. This latter effect is also discussed below in relation to the menstrual cycle. Finally, progesterone inhibits the contractions of the uterus, and prevents the uterus from expelling a fertilized ovum that is implanting or a fetus that is developing.

Regulation of the Female Sexual Cycle

OSCILLATION BETWEEN THE ADENOHYPOPHYSEAL AND OVARIAN HORMONES. The monthly female sexual cycle is caused by alternating secretion of follicle stimulating hormone by the adenohypophysis and estrogens by the ovary. The cycle of events that causes this alternation is the following: During the first part of the monthly cycle, the adenohypophysis secretes follicle stimulating hormone. This causes the ovaries to secrete estrogens. Then the estrogens exert an inhibitory effect on the adenohypophysis to make it stop secreting follicle stimulating hormone. Lack of this hormone eventually stops the ovaries from secreting estrogens. Lack of estrogens removes the inhibition of the adenohypophysis and it begins secreting follicle stimulating hormone again, thus initiating a new cycle.

Before puberty the adenohypophysis is unable to secrete any gonadotropic hormones. Consequently, until that time the alternation between follicle stimulating hormone and estrogens cannot take place. As soon as a sufficient quantity of follicle stimulating hormone begins to be secreted, the sexual life of the female begins.

After about 30 years of sexual life, the ovaries "burn out" because all the primary follicles finally have grown into mature follicles and ruptured or involuted. The sexual cycles cease because the ovaries no longer have any follicular cells to secrete estrogens, even though the adenohypophysis continues to secrete large quantities of follicle stimulating hormone for the remainder of the woman's life. This cessation of the monthly sexual cycles occurs in the average woman at an age of about 45, and this stage of her life is called the *menopause*.

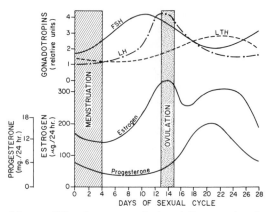

Figure 352. Secretion of the different adenohypophyseal and ovarian hormones during the normal monthly sexual cycle.

RELATION OF LUTEINIZING HORMONE, LUTEOTROPIC HORMONE, AND PROGESTERONE TO THE OSCILLATING CYCLE. It is probable that the primary controlling mechanism of the female sexual cycle is the simple oscillation between follicle stimulating hormone and estrogens, for the cycle continues even when the other hormones are secreted in very small quantity during anovulatory cycles (cycles in which ovulation fails to occur). However, if ovulation occurs, the other three hormones are secreted in a cyclic manner each month just as is true for the estrogens and follicle stimulatory hormone. Figure 352 shows the secretion of the various adenohypophyseal and ovarian hormones at different stages during the monthly cycle.

THE ENDOMETRIAL CYCLE. It is evident from Figure 352 that during the first half of the monthly cycle the only *ovarian* hormone secreted in large quantity is the estrogens. Then during the latter half both estrogens and progesterone are secreted.

Estrogens cause the endometrium—that is, the lining of the inside of the uterus—to grow in thickness. Both the epithelial cells on the surface and the deeper layers of the endometrium proliferate approximately three-fold. Also, the glands of the endometrium increase greatly in depth and in tortuosity. These changes constitute the *proliferative phase* of endometrial development and are illustrated in the first portion of Figure 353, occurring during the first 10 to 14 days of the monthly cycle.

At approximately the fourteenth day of the cycle the corpus luteum begins secreting progesterone, which then has the following

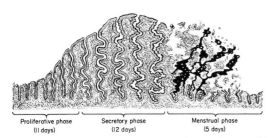

Proliferative phase   Secretory phase   Menstrual phase
    (11 days)   (12 days)   (5 days)

Figure 353. Endometrial changes during the monthly sexual cycle, showing especially the process of menstruation.

effects on the endometrium: the endometrial glands begin secreting a nutrient fluid that can be used by a fertilized ovum before it implants; large quantities of fatty substances and glycogen deposit in the deeper cells of the endometrium; and blood flow to the endometrium increases. In other words, the major function of progesterone is to make available an adequate supply of nutrients for an embryo should it begin to develop in the uterus.

Menstruation. If an ovum becomes fertilized and implanted in the uterus, special hormones are then released by the developing ovum itself to stimulate the corpus luteum. These cause it to continue to produce estrogens and progesterone, and pregnancy begins. However, if toward the end of the monthly cycle implantation of an ovum has not occurred, the corpus luteum dies, and the production of both estrogens and progesterone either ceases or almost ceases. The sudden lack of these two hormones causes the blood vessels of the endometrium to become spastic so that blood flow to the endometrium ceases. As a result, the endometrial tissues die and slough into the uterine cavity, a process called *menstruation.* Blood oozes from the denuded endometrial wall, causing a blood loss of about 50 ml. during the several days of menstruation. The necrotic endometrial tissue, plus the blood and much serous exudate from the denuded uterine surface, is gradually expelled for about three to five days.

The deepest pits of the endometrial glands remain intact during menstruation, despite the sloughing of all the outer layers of the endometrium. During the first few days after menstruation, new epithelium grows from the edges of the glands to cover the inner surface of the uterus. Then, under the influence of renewed estrogen production by the ovaries, the endometrial cycle begins again. These sequential changes during the entire cycle are shown in Figure 353.

THE PERIOD OF FERTILITY DURING THE SEXUAL CYCLE. An ovum can be fertilized by a sperm for a period of 8 to 24 hours after ovulation. Also, sperm can live in the female genital tract usually for 24 to 72 hours. Therefore, for successful fertilization, sexual exposure must occur either shortly before ovulation so that sperm are already available when ovulation occurs, or within a few hours after ovulation.

Ovulation usually occurs on the fourteenth day of the normal 28-day sexual cycle. However, many women, instead of having normal 28-day cycles, have cycles as short as 21 days or as long as 40 days, and in some women the periodicity of the cycles is very irregular. Yet despite the length of the cycle, ovulation occurs almost exactly 14 days *before* menstruation begins. Therefore, in those women who have regular cycles, regardless of how long they may be, the day of ovulation can be calculated to be 14 days before the predicted day of menstruation. And because of the viability of sperm and the ovum, the period of fertility is usually between 17 days and 12 days *before* menstruation begins.

REFERENCES

Albert, A. (ed.): *Human Pituitary Gonadotropins.* Springfield, Ill., Charles C Thomas, 1961.

Beech, F. A.: Cerebral and hormonal control of reflexive mechanisms involved in copulatory behavior. *Physiol. Rev.* 47(2):289, 1967.

Bishop, D. W.: Sperm motility. *Physiol. Rev.* 42:1, 1962.

Carey, H. M.: *Modern Trends in Human Reproductive Physiology.* London, Butterworth & Company, 1963.

Donovan, B. T.: The timing of puberty. *Scient. Basis Med. Ann. Rev.* 53, 1963.

Drill, V. A.: *Oral Contraceptives.* New York, McGraw-Hill Book Company, 1968.

Everett, J. W.: Central neural control of reproductive functions of the adenohypophysis. *Physiol. Rev.* 44:373, 1964.

Kleegman, S. J., and Kaufman, S. A.: *Sterility.* Philadelphia, F. A. Davis Company, 1965.

Kristen, B. E.: Effects of gonadotrophins on secretion of steroids by the testis and ovary. *Physiol. Rev.* 44:609, 1964.

Lloyd, C. W.: *Human Reproduction and Sexual Behavior,* Philadelphia, Lea & Febiger, 1964.

Pincus, G.: Reproduction. *Ann. Rev. Physiol.* 24:57, 1962.

CHAPTER 38

REPRODUCTION AND FETAL PHYSIOLOGY

FERTILIZATION OF THE OVUM AND EARLY GROWTH

ENTRY OF THE SPERM INTO THE OVUM. After the ovum is expelled from the ovary, it remains viable for 8 to 24 hours. During this time it usually moves approximately one quarter of the distance down the fallopian tube toward the uterus. Obviously, then, fertilization must occur either in the abdominal cavity before the ovum enters the tube, or in the upper portion of one of the tubes. Sperm travel at a velocity of only 1 to 4 mm. per minute, and the total length of a fallopian tube is approximately 15 cm. Therefore, the minimum time usually required for a sperm to reach the ovum is approximately 40 minutes.

The ovum, when released from the ovary, carries on its surface a large number of granulosal cells which collectively are called the *corona radiata*, shown in Figure 354. It is claimed, though not proved, that the corona radiata is dispersed by an enzyme called *hyaluronidase* secreted by the sperm. This enzyme, which was discussed in the previous chapter, supposedly digests the protein linkages that hold the cells together, and allows the cells to break away from the ovum. In other words, the sperm supposedly

removes this barrier so that it can attack and enter the ovum.

After a single sperm has entered the ovum, the membrane of the ovum suddenly becomes impermeable to entry of additional sperm. Sperm that do reach the ovum are inactivated as they attempt to enter. This effect prevents more than one set of male chromosomes from combining with the chromosomes of the ovum.

Shortly after a sperm has entered the ovum the head of the sperm begins to swell, forming a *male pronucleus*, and the original

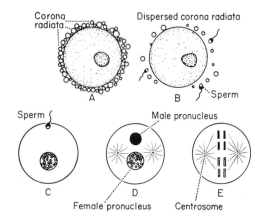

Figure 354. Fertilization of the ovum by the sperm, and beginning cleavage. (Modified from Arey: *Developmental Anatomy.* 6th Ed.)

448

nucleus of the ovum is called the *female pronucleus*. It will be recalled from the previous chapter that each of these pronuclei has only 23 *unpaired* chromosomes. However, these two pronuclei soon combine so that the fertilized ovum now has 46 *paired* chromosomes in a single nucleus, as do all other cells of the human body. A few hours later the chromosomes reproduce and split apart, initiating division of the ovum into two daughter cells.

EARLY DIVISION OF THE OVUM. The first cleavage of the fertilized cell occurs approximately 30 hours after the sperm has entered the ovum, and the next few generations occur every 10 to 15 hours. By the time the ovum reaches the uterus, the total number of cells in the cellular mass is usually about 16 to 32. By this time the cells have actually begun to *differentiate*, which means that some of them have developed different characteristics from the others.

IMPLANTATION OF THE OVUM. Figure 355 shows the *blastocyst* stage of a developing human ovum 1½ weeks old. It illustrates the method by which the mass of cells attaches itself to the inner wall of the uterus. At this stage the cells on the outside of the blastocyst, called *trophoblasts*, secrete large quantities of proteolytic enzymes that digest the endometrium, and then the trophoblasts phagocytize the digested products. Thus, they eat their way into the wall of the endometrium. The trophoblasts grow and divide very rapidly at the same time, and soon they and adjacent cells begin forming the placenta and fetal membranes while the embryo develops on the inside of the blastocyst.

NUTRITION OF THE FETUS IN THE UTERUS

The Trophoblastic Phase of Nutrition

During the first few weeks after implantation of the ovum, the placenta and its blood supply are not adequately formed to supply the fetus with nutrients. During this phase nutrition is provided only by trophoblastic digestion and phagocytosis of the endometrium. It will be recalled that prior to implantation of the ovum, the endometrium stores large quantities of proteins, lipid materials, and glycogen. Also, small quantities of iron and vitamins are stored awaiting phagocytosis by the developing ovum. Even so, the embryo can obtain nutrition in this manner for only the first few weeks of its development. By that time the placenta will have developed to a stage such that it can begin supplying nutrition. Figure 356 shows that the trophoblastic period of phagocytic nutrition lasts until approximately the twelfth week of pregnancy.

Placental Nutrition of the Fetus

Figure 357 shows the *placenta* and the *fetal membranes*. The fetal membranes at-

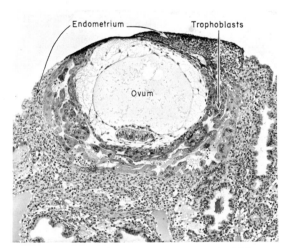

Figure 355. Implantation of a 1½ week old human ovum in the endometrium. (Courtesy of Dr. Arthur Hertig.)

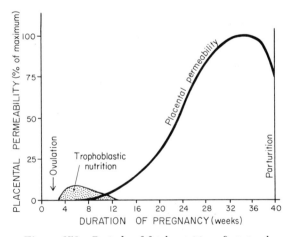

Figure 356. Periods of fetal nutrition: first, trophoblastic phagocytosis of the endometrium lasting for the first few weeks of pregnancy and, second, diffusion through the placenta during the remainder of pregnancy.

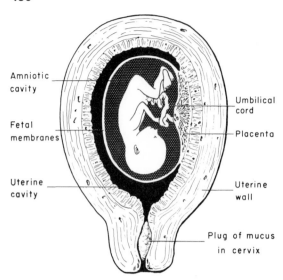

Figure 357. The fetus, the fetal membranes, and the placenta in the pregnant uterus.

villi into the fetal blood almost entirely by the process of diffusion. The gaseous pressure of oxygen in the blood of the maternal sinuses is normally 60 to 80 mm. Hg, whereas the gaseous pressure of oxygen in the fetal capillaries is only 20 to 30 mm. Hg. Because of this pressure difference, oxygen simply diffuses through the tissues of the villus from the mother's blood to the baby's blood. Likewise, glucose, some amino acids, some fats, many of the vitamins, and most of the minerals are present in greater concentration in the mother's blood than in the baby's blood because the baby utilizes these substances almost as rapidly as they enter his blood. As a consequence, all of them diffuse inward and thereby supply the fetus with nutrition.

Figure 356 shows the progressive increase in placental permeability as pregnancy proceeds. The total amount of nutrients

tach to the entire inner surface of the uterus, forming a cavity called the *amniotic cavity.* The placenta covers about one-fourth the surface of the uterus, and the fetus is connected with it by means of the *umbilical cord.* The fetus floats freely in *amniotic fluid* that fills the amniotic cavity. The function of the umbilical cord is to transport blood from the baby to the placenta and then back again to the baby. In this way the baby's blood picks up nutrition from the placenta and transports it into the developing child.

Figure 358 illustrates the gross organization of the placenta and, at the bottom, a cross-section of a *placental villus.* The mother's blood flows into large placental chambers called *placental sinuses;* the blood passes through these so rapidly that it remains fresh essentially all the time. The fetal portion of the placenta is composed of many small cauliflower-like projections of tissue extending into the placental sinuses. Each of these in turn is covered with tremendous numbers of small villi containing blood capillaries. The baby's blood flows into the capillaries of the villi where it receives nutrients from the mother's blood, and then it passes back through the umbilical veins to the fetus. The cross-section of a villus in Figure 358 shows the close proximity of the mother's blood in the placental sinuses to the baby's blood in the fetal capillaries.

DIFFUSION OF NUTRIENTS THROUGH THE VILLI. Nutrients pass through the placental

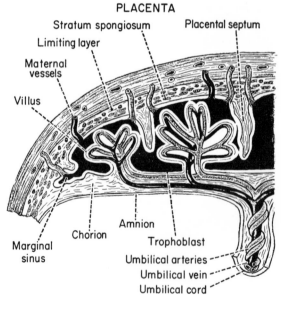

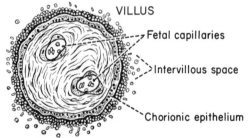

Figure 358. Gross anatomical structure of the placenta and microscopic cross-section of a placental villus. (Modified from Goss: *Gray's Anatomy of the Human Body.* Lea & Febiger; and from Arey: *Developmental Anatomy.* 6th Ed.)

that can be transported through the placenta reaches a maximum 32 to 36 weeks after pregnancy begins. At this point the placental tissues begin to grow old and degenerate. As a result, even though the fetus still requires progressively increasing quantities of nutrients, the ability of these nutrients to enter the baby becomes lessened. Fortunately birth of the baby occurs at this time, and the baby assumes independent existence.

Excretion through the placenta by diffusion. In addition to the diffusion of nutrients into the baby, excretory products diffuse through the placenta from the fetal blood to the maternal blood. For instance, the baby's metabolism continually forms large quantities of carbon dioxide, urea, uric acid, creatinine, phosphates, sulfates, and other normal excretory products. The concentration of each of these substances rises in the baby's blood until it is greater than the concentration in the mother's blood. Then, because of this reverse concentration gradient, these substances diffuse backward through the villi, and the mother excretes them through her kidneys, lungs, and other excretory organs.

ACTIVE TRANSPORT THROUGH THE VILLI. The chorionic epithelium that covers the surface of the placental villi is formed from the trophoblastic cells of the early fetus. These cells continue their phagocytic activity throughout the life of the placenta, so that some nutrients are actively transported through the villi all through the period of gestation. In the first few weeks of placental development this aids greatly in increasing the amount of amino acids, fatty substances, and some minerals that can be supplied to the baby, but after 12 to 20 weeks the amount of nutrients supplied in this manner becomes far less than that supplied by the simple process of diffusion.

THE HORMONES OF PREGNANCY

Secretion of Estrogens during Pregnancy

During the normal menstrual cycle moderate quantities of estrogens are secreted first by the cells lining the developing ovarian follicles and then by the corpus luteum that forms from the ruptured follicle. If the ovum becomes fertilized and implants in the uterus, the corpus luteum, instead of degenerating as it normally does at the end of the normal monthly cycle, grows even larger and produces increasing amounts of estrogens for an additional four to five months. At this time, the placental tissues begin to form extreme quantities of estrogens, the amount progressively increasing to a maximum immediately before birth, as shown in Figure 359. The peak production reaches about 50 times as much as the production during a normal monthly sexual cycle.

Figure 359. Secretion of hormones during the course of pregnancy.

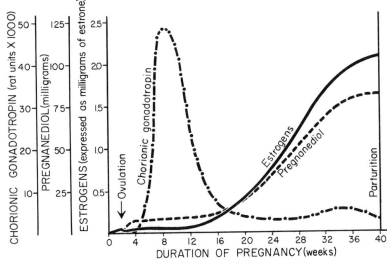

FUNCTIONS OF ESTROGENS DURING PREG-NANCY. Estrogens cause rapid proliferation of the uterine musculature; greatly increased growth of the vascular system supplying the uterus; enlargement of the external sex organs and of the vaginal opening, providing an appropriately enlarged passageway for birth of the baby; and probably also some of the relaxation of the pelvic ligaments that allows the pelvic opening to stretch as the baby is born.

In addition to their effects on the reproductive organs, estrogens also cause the breasts to grow rapidly. The ducts, especially, enlarge, and the glandular cells proliferate. Finally, estrogens cause a pound or more of extra fat to deposit in the breasts.

Another effect of estrogens not yet completely understood is their effect on the fetus itself. It is believed that they cause much of the rapid proliferation of fetal cells and also aid in the differentiation of some of these cells into special organs. In particular they are believed to control development of some of the female sex characteristics. If the production of estrogens ever fails during pregnancy, the fetus dies almost immediately, and is expelled from the uterus.

Secretion of Progesterone during Pregnancy

Figure 359 also shows the amounts of progesterone secretion during pregnancy (expressed in terms of *pregnanediol*, which is the form in which progesterone is excreted in the urine). Progesterone, like estrogens, is secreted in moderate quantities by the corpus luteum during the early part of pregnancy. At approximately the sixteenth week of gestation the placenta begins producing very large quantities of progesterone as well as estrogens, reaching a maximum production toward the end of pregnancy of as much as 10 times that produced by the corpus luteum.

FUNCTIONS OF PROGESTERONE DURING PREGNANCY. The first function of progesterone during pregnancy is to make increased quantities of nutrients available to the early endometrium for use by the developing ovum. It does this by causing the endometrial cells to store glycogen, fat, and amino acids. In addition, progesterone has a strong inhibitory effect on the uterine musculature, causing it to remain relaxed throughout pregnancy. It is believed that it is this effect that allows pregnancy to continue until the fetus is large enough to be born and live an independent existence. Despite these important functions of progesterone, some of the lower animals even among the mammals do not secrete this hormone.

Progesterone complements the effects of estrogens on the breasts. It causes the glandular elements to enlarge and to develop a secretory epithelium, and it promotes deposition of nutrients in the glandular cells, so that when milk production is required the appropriate materials will be available.

Function of Chorionic Gonadotropin during Pregnancy

If the corpus luteum is removed from the ovary at any time during the first three to four months of pregnancy, the loss of estrogen and progesterone secretion causes the fetus to stop developing and to be expelled within a few days. For this reason it is important that the corpus luteum remain active at least during the first third and preferably during the first half of pregnancy. Beyond that time removal of the corpus luteum does not affect pregnancy because the placenta by then is secreting many times as much estrogens and progesterone as the corpus luteum.

It will be recalled from the previous chapter that the corpus luteum normally degenerates and is absorbed at the end of each monthly sexual cycle. To keep the corpus luteum intact when the ovum implants, a special hormone, *chorionic gonadotropin*, is secreted by the developing fetal tissues—by the trophoblasts. This hormone has almost exactly the same properties as luteinizing and luteotropic hormones combined. It not only keeps the corpus luteum from involuting, but actually stimulates it so greatly that it enlarges several fold during the first 2 to 4 months of pregnancy.

Chorionic gonadotropin begins to be formed from the day that the trophoblasts implant in the uterine endometrium. Its concentration is highest approximately at the eighth week of pregnancy, as shown in Figure 359. Thus, the concentration is greatest at the time of pregnancy when it is essential to prevent involution of the corpus luteum. In the middle and later parts of pregnancy the secretion of gonadotropin falls to very low values. Its only known function at this time of pregnancy is to stimulate production of testosterone by the testes of the male fetus, which was discussed in the previous chapter.

FETAL PHYSIOLOGY

In general, the physiology of the growing fetus is not far different from that of the normal child. All the organs of the body assume their final anatomic form, with minor exceptions, by four to five months after the beginning of pregnancy, and most of them can function almost completely normally by six months after the beginning of pregnancy. For example, long before birth the kidneys of the fetus become active, the gastrointestinal tract imbibes and absorbs fluid from the amniotic cavity, breathing is attempted even though it cannot be accomplished because the fetus is immersed in amniotic fluid, the heart pumps blood quite normally from approximately the third month on, and the metabolic systems operate very much the same as they do after birth.

Growth of the Fetus

Figure 360 depicts the increase in length and weight of the fetus during the 40 weeks of pregnancy. The length of the fetus increases almost directly in proportion to its age, while the weight increases in proportion to the third power of the age. The weight is

almost infinitesimal until the twelfth to sixteenth week, but at that time it begins to increase extremely rapidly. Two months before birth the fetus usually weighs about one half its birth weight, and one month prior to birth it weighs approximately three-fourths its birth weight. In other words, the major increase in weight occurs during the last two to three months. This is an especially important factor when one is considering the appropriate nutrition for the mother during pregnancy, because essentially all of the nutrients required by the baby are needed during the last three months. However, for several months prior to that time, growth of the uterus, placenta, and fetal membranes requires additional nutrients.

SPECIAL NUTRIENTS REQUIRED BY THE FETUS. The fetus requires especially large quantities of iron, calcium, phosphorus, amino acids, and vitamins. Iron begins to be used to form red blood cells within the first weeks of development of the fetus. Much of this early iron enters the fetus by active absorption from the endometrium by the trophoblasts. Throughout the remainder of pregnancy, still larger quantities of iron diffuse through the placental membrane to be used by the liver, spleen, and bone marrow for production of the fetus' blood.

Calcium is needed to ossify the bones. During the first two thirds of pregnancy the fetal bones contain mainly organic matrix and almost no calcium salts. During the last three months of fetal development, ossification occurs very rapidly, approximately one-half to two-thirds occurring in the final month. It is at this time, therefore, that the mother needs an especially abundant amount of milk or other foods containing calcium and phosphate.

The large quantities of amino acids and vitamins required by the fetus provide the necessary building materials for growth of the fetal tissues. During the last three months of pregnancy, a mother can become depleted of protein and vitamins if they are not in the diet in adequate amounts.

The Fetal Nervous System

The major anatomic parts of the nervous system are formed during the first few months of fetal growth, but complete func-

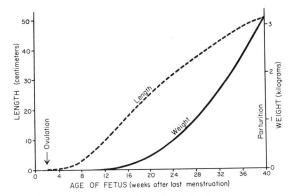

Figure 360.   Growth of the fetus.

tion of this system is not reached even by the time the child is born. A premature baby always exhibits signs of poor nervous system function. As an example, one of the most difficult problems in treating a premature baby is to keep his body temperature normal, because the hypothalamic centers of a 6- or 7-month fetus usally have not developed sufficiently to regulate temperature. The child must be placed in an incubator for several weeks or sometimes a month or more until his nervous system has developed to a greater degree.

By the time of birth most, if not all, of the peripheral nerve fibers are completely developed, but in the central nervous system the deposition of myelin around many of the large nerve fibers is far from complete. Though myelin is not always necessary for nerve fibers to function, it has been inferred from this lack of complete myelination that certain portions of the central nervous system probably are far from functional at the time of birth. On the other hand, all the neurons that will ever be formed by the child are believed to be present by birth. This means that every time a neuron is destroyed thereafter, no new neuron takes its place.

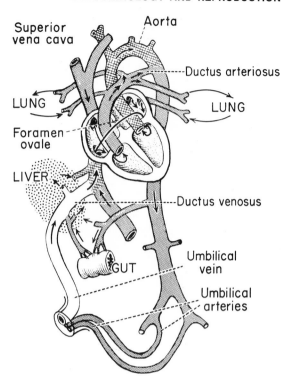

Figure 361. The fetal circulation, showing especially the ductus arteriosus, ductus venosus, and foramen ovale. (Modified from Arey: *Developmental Anatomy.* 6th Ed.)

Changes in the Circulation at Birth

Figure 361 illustrates the peculiarities of the circulatory system in the fetus. Special *umbilical vessels* allow blood to flow into and out of the placenta. The blood returning from the placenta through the umbilical vein enters the fetal circulatory system through a special vessel called the *ductus venosus,* which passes directly through the liver. Also, blood from the portal veins of the fetus flows through the ductus venosus, by-passing the liver entirely and emptying directly into the general venous system.

On reaching the heart, fetal blood by-passes the non-aerated lungs by two routes. First, it can flow from the right atrium through an opening called the *foramen ovale* directly into the left atrium, thus by-passing both the right ventricle and the lungs. Second, the remainder of the blood enters the right ventricle and is pumped into the pulmonary artery. However, most of this blood, instead of going through the lungs, passes through a vessel called the *ductus arteriosus* directly into the aorta.

Thus, by utilizing the ductus venosus, the foramen ovale, and the ductus arteriosus, blood flow through the liver and lungs occurs only to a minor extent in the fetus. This is a means for conserving the energy of the fetal heart, because the functions normally required of the fetal liver are performed instead by the mother's liver, and the fetal lungs are completely nonfunctional until birth.

After birth all three of these openings soon close. The ductus venosus and the ductus arteriosus become gradually occluded during the first few days or weeks of life. Presumably the ductus venosus becomes occluded because it no longer carries the tremendous blood flow from the umbilical vein, and the sluggish blood flow from the portal system is not sufficient to keep it open. The ductus arteriosus probably closes because the functioning lungs increase the oxygen in the blood which in turn causes increased contraction of the ductus muscle.

The foramen ovale is covered by a thin valve on the left atrial side. As long as the pressure in the right atrium is greater than that in the left atrium, which is the case before birth of the fetus, blood flows from right to left, by-passing the right ventricle and lungs. However, when the lungs expand, the right heart can then pump blood so easily through the lungs that the right atrial pressure falls to about 3 mm. Hg less than the left atrial pressure. This backward pressure differential thereafter keeps the valve of the foramen ovale closed. In about two thirds of all persons the valve gradually adheres to the opening so that it becomes permanently closed; in about one third it does not, but, regardless of which is true, blood ceases to flow through it after birth.

Thus, because of a series of changes in the dynamics of the circulation, blood begins to flow through both the liver and lungs immediately after birth, even though in the fetus the heart had conserved its energy by by-passing these two organs.

PHYSIOLOGY OF THE MOTHER DURING PREGNANCY

The mother's physiology is changed during pregnancy in several ways. First, accessory changes occur in her reproductive organs and breasts to provide for development of the fetus and to provide nutrition for the newborn child. Second, all her metabolic functions are increased to supply sufficient nutrition to the growing fetus. Third, tremendous production of certain hormones by the placenta during pregnancy causes many side effects not directly associated with reproduction.

CHANGES IN WEIGHT. The pregnant mother gains an average of about 20 pounds during pregnancy. In general, this gain is accounted for in the following manner: fetus, 7 pounds; uterus, 2 pounds; placenta and membranes, 2.5 pounds; breasts, 2 pounds; and the remainder is fat and increased quantities of extracellular fluid and blood. The amount of increase in these last two varies tremendously from one mother to another, depending upon her eating habits and upon the amounts of hormones secreted during pregnancy.

CHANGES IN METABOLISM. The mother's metabolic rate in general increases approximately in proportion to her increase in weight, plus perhaps an additional 5 to 10 per cent. Much of this increase is occasioned simply by the increased amount of energy required for the mother to carry the growing load. However, rapid growth of the fetus also demands increased activity of most of the mother's functions, such as rapid intermediary metabolism in her liver, rapid pumping of blood by her heart, increased respiration, and increased digestion and assimilation of food.

CHANGES IN THE BODY FLUIDS AND CIRCULATION. The female sex hormones and the extra adrenocortical hormones produced during pregnancy cause the mother usually to gain about 5 to 7 pounds of fluid, or, in other words, about 3 liters. About 0.5 liter of this is in the plasma, and another 0.5 liter is red blood cells, making a total gain in blood volume of about 1 liter. About one third of the extra blood is needed to fill the sinuses of the placenta, but the other two thirds of a liter collects in the circulation, causing blood to flow toward the heart with far greater ease than usual. As a result, the cardiac output becomes roughly 25 per cent more than normal.

During birth of the baby, the mother loses an average of 200 to 300 ml. of blood as the placenta separates from the uterus. This ordinarily causes no physiological inconvenience because of the extra blood that has been stored during pregnancy.

BIRTH OF THE BABY (PARTURITION)

DURATION OF PREGNANCY AND ONSET OF PARTURITION. The duration of pregnancy, between the time of the last menstrual period until birth of the baby, is normally 40 weeks, though occasionally surviving babies are born as early as 28 weeks or as late as 46 weeks. Approximately 90 per cent of all babies are born within 10 days before or after the 40-week interval.

The reason for the relatively constant duration of pregnancy has never been completely understood. Presumably, growth of the baby and of the placenta to a certain size and state of maturity initiates birth. The probable factors that cause the onset of parturition are the following:

When the baby becomes large, pressure of its body against the uterus stretches the uterine musculature, which in turn initiates uterine contractions. In addition to this, movements of the baby in the uterus—such as the feet and hands striking the uterine wall—also initiate contractions. Obviously, the larger the baby becomes, the more likely are these initiated contractions to become strong enough to cause birth.

Perhaps more important than the mechanical factors, however, are hormonal factors. The concentration of progesterone secreted by the placenta begins to decrease a few weeks prior to birth, and, since progesterone normally inhibits the uterus, this change allows an increase in uterine contractions. On the other hand, the concentration of active estrogens increases shortly before birth, and this increases the activity of the uterus, in contrast to the inhibition caused by progesterone. These two changes in hormonal secretion by the placenta probably account for much of the progressively increasing contractility of the uterus shortly before birth.

A third factor possibly helping to initiate parturition is the secretion of oxytocin by the hypothalamic-neurohypophysis system. This hormone causes extreme contractility of the uterine musculature, and absence of its secretion in animals usually makes parturition difficult.

MECHANISM OF PARTURITION. Uterine musculature, like almost all smooth muscle, undergoes rhythmic contractions all the time. However, because of the influence of progesterone, these contractions are so weak during the early months of pregnancy that they can hardly be noted. During the last three months of pregnancy, they increase steadily and reach extreme intensity a few hours before birth. These very strong contractions wedge the head against the cervix, stretch the cervical ring and vaginal canal, and expel the baby. This period, from the onset of the strong contractions until the baby is born, is called the period of *labor*, and the actual expulsion of the child is called *parturition*.

Approximately 19 times out of 20 the portion of the baby that pushes against the cervix is the head, and this acts as a wedge to open the cervical and vaginal canals. In most of the remaining cases the buttocks are the presenting portion of the baby, though occasionally a leg, a shoulder, or even the side of the baby may be against the cervical canal. In all of these instances the cervix cannot be wedged open nearly as effectively as by the head, and birth of the baby is considerably more difficult. When the head comes first, the remainder of the body slips through the vaginal canal within a few seconds after the head is born; when the head is not the presenting part, the body portion of the baby may be born relatively easily, but the head, the largest part of the child, then has difficulty in passing through the canal. One of the problems of bottom-first birth is often a period of interrupted umbilical blood supply to the baby because the cord can become compressed against the wall of the delivery canal for a minute or more by the large head.

THEORETICAL CAUSE OF THE SUDDEN ONSET OF LABOR. The reason the rhythmic uterine contractions suddenly become intense enough to cause labor is yet undetermined, but the following explanation has been offered as a possiblity.

Irritation of the *cervix* is known to cause a muscular reaction over the entire uterus, making it become very excitable and causing its contractions to increase. It is presumed that when the natural contractions of the uterus reach a certain level of intensity, which occurs at the end of approximately nine months of pregnancy, they push the baby's head against the cervix, which stretches it and thereby makes the entire uterus contract more forcefully. This contraction in turn pushes the baby's head into

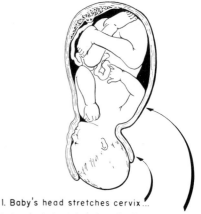

1. Baby's head stretches cervix...
2. Cervical stretch excites fundic contraction...
3. Fundic contraction pushes baby down and stretches cervix some more...
4. Cycle repeats over and over again...

Figure 362. A postulated mechanism for initiation of labor.

the cervix still more and initiates still further intensity of uterine contraction. Thus, a vicious cycle develops, as illustrated in Figure 362, and the uterine contractions become stronger and stronger until finally the baby is expelled.

The increased uterine excitability caused by stretching or irritating the cervix is believed to result from two mechanisms. First, cervical stimulation is believed to transmit impulses upward through the uterine musculature itself, thus resulting in uterine contraction. Second, in lower animals, and therefore presumably in man, cervical irritation transmits nerve impulses to the hypothalamus causing increased secretion of oxytocin, which in turn increases the contractility of the uterus, as was explained in Chapter 34.

PRODUCTION OF MILK (LACTATION)

Hormonal Control of Lactation

The onset of secretion of the estrogens and progesterone after puberty in a girl begins to prepare the breasts for lactation. The breasts enlarge and the glandular elements begin to develop. However, the degree of glandular development is relatively slight compared with that achieved during pregnancy. The tremendous quantities of estrogens secreted by the placenta during pregnancy cause rapid development of the glands in the breasts, and the large quantities of progesterone change the glandular cells into actual secreting cells. By the time the baby is born, the breasts will have reached a degree of development capable of producing milk. Yet these same hormones, despite their developmental effects on the breasts, inhibit the actual formation of milk until after the baby is born.

Figure 363 shows the changes in hormone production at the time of parturition. Sudden expulsion of the placenta from the uterus removes the source of estrogens and progesterone, so that the availability of these two hormones falls immediately almost to zero. Throughout pregnancy the production of gonadotropic hormones by the adenohypophysis has been inhibited by the placental hormones, but now that they are gone the adenohypophysis begins to secrete large quantities of *lactogenic hormone*, which is the same as *luteotropic hormone* and which is also known as *prolactin*. This hormone causes the breasts to produce milk.

As long as the breasts are emptied of milk as rapidly as it is formed, the adenohypophysis will continue to produce lactogenic hormone, but when the milk is not emptied from the breasts, the production of lactogenic hormone also ceases. Though the reason for this is not known, it has been suggested that suckling sensations from the baby cause nerve impulses to pass from the nipples of the breasts, up the spinal cord, and then to the hypothalamus, which in turn maintains lactogenic hormone formation.

As long as the adenohypophysis continues

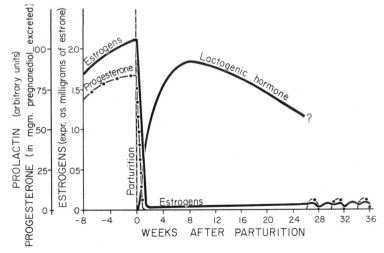

Figure 363. Changes in hormonal secretion that lead to the production of milk after parturition.

to produce lactogenic hormone it usually cannot produce sufficient follicle stimulating and luteinizing hormones to cause ovulation by the ovaries. Therefore, if the mother lactates, the normal female sexual cycle usually remains dormant for many months after parturition. But, if she does not lactate, her next ovulation appears about four weeks after birth and her next menstrual period about six weeks after birth.

MILK EJECTION. When a baby sucks on the nipples, it usually obtains no milk for approximately the first 45 seconds to a minute. Then suddenly milk appears in the ducts of both breasts even though suckling occurs on only one breast, indicating that some general phenomenon has occurred to cause milk to flow toward the nipples. Experiments have shown that suckling causes sensory impulses to pass first into the spinal cord, then upward through the brain stem, and finally into the hypothalamus to stimulate the production of oxytocin. This then circulates through the blood to the breasts, where it causes the myoepithelial cells surrounding the alveoli to contract and express the milk collected in the alveoli into the ducts leading to the nipples.

This milk ejection mechanism can be adversely affected tremendously by psychic factors. For example, a mother's fear that she might not be able to nurse her baby can actually keep her from doing so. Disturbances caused by other children in the family or by overly concerned relatives may lead to difficulty in milk ejection and cause the breasts to fail to empty. Failure to empty in turn allows the adenohypophysis to cease its production of lactogenic hormone, and the breasts stop secreting milk.

Composition of Milk

Milk contains the usual substances needed for energy and growth by the baby. These include two types of protein, *casein* and *lactalbumin;* an easily digested sugar, *lactose,* composed of one molecule of glucose and one of galactose; and large quantities of fatty substances such as *butter fats, cholesterol,* and *phospholipids.*

In addition, milk contains small quantities of vitamins and large quantities of calcium phosphate. Yet it has a conspicuous lack of iron. This lack usually is not detrimental to

TABLE 14. *Composition of Milk in Per Cent*

	HUMAN	COW
Water	88.5	87.0
Fat	3.3	3.5
Sugar	6.8	4.8
Casein	0.9	2.7
Lactalbumin and other protein	0.4	0.7
Electrolytes	0.2	0.7

the baby in early life, because a sufficient quantity of this mineral is stored in the fetal liver prior to birth to continue the formation of hemoglobin for about two months. Beyond this time, however, iron must be in the baby's diet, or he will develop progressive anemia.

Table 14 shows the relative compositions of human and cow's milk. When cow's milk is substituted for human milk, the baby's metabolic systems must adjust to a large increase in electrolytes—to the phosphates in particular. Also, it is desirable to fortify cow's milk with extra quantities of easily digested sugar such as pure glucose (dextrose).

Effect of Lactation on the Mother

Production of milk by the mother in some ways is as great a drain on her metabolic systems as pregnancy itself. Particularly does she lose tremendous amounts of stored proteins and fats during lactation. Also, if large quantities of calcium phosphate are not in her diet, her parathyroid glands become greatly overactive. This causes reabsorption of her bones, releasing the necessary calcium and phosphate for milk formation. As a result, her bones are likely to become weakened. Fortunately, though, the amount of calcium phosphate stored in the bones of the average mother is tremendous in comparison with the amount that the baby will need during the first few months of life. Therefore, only when the mother already has some degree of decalcification will this loss cause her any difficulty.

Frequently mothers develop extensive dental caries during pregnancy, which has often been ascribed to the loss of calcium phosphate from the teeth. Experiments, however, have shown that no significant amount of calcium phosphate leaves the teeth, but that the caries are probably caused by enhanced bacterial growth in the mouth during pregnancy.

REFERENCES

Austin, C. R.: *The Mammalian Egg*. Philadelphia, F. A. Davis Company, 1961.

Blechschmidt, E.: *The Stages of Human Development Before Birth*. Philadelphia, W. B. Saunders Company, 1961.

Harrison, R. G.: *A Textbook of Human Embryology*. 2nd Ed. Philadelphia, F. A. Davis Company, 1964.

Kon, S. K., and Cowie, A. T. (eds.): *Milk: The Mammary Gland and Its Secretion*. 2 volumes. New York, Academic Press, 1961.

Metcalfe, J., Bartels, H., and Moll, W.: Gas exchange in the pregnant uterus. *Physiol. Rev.* 47(4):782, 1967.

Nalbandov, A. V., and Cook, B.: Reproduction. *Ann. Rev. Physiol.* 30:245, 1968.

Parkes, A. S.: *Marshall's Physiology of Reproduction*. 3rd Ed. 3 volumes. Boston, Little Brown & Company, 1956–1965.

Schaffer, A. J.: *Diseases of the Newborn*. 2nd Ed. Philadelphia, W. B. Saunders Company, 1965.

Short, R. V.: Reproduction. *Ann. Rev. Physiol.* 29:373, 1967.

Smith, C. A.: The first breath. *Sci. Amer.* 209(4):27, 1963.

INDEX